AF294260

Maria G. Carvalho · Fred Lockwood
Jean Taine (Eds.)

Heat Transfer in Radiating and Combusting Systems

Proceedings of EUROTHERM Seminar No. 17,
8-10 October 1990, Cascais, Portugal

With 372 Figures

Springer-Verlag
Berlin Heidelberg New York
London Paris Tokyo
Hong Kong Barcelona Budapest

Prof. Maria da Graça Carvalho

Technical University of Lisbon
Instituto Superior Técnico
Secção de Termodinâmica Aplicada
Av. Rovisco Pais
1096 Lisboa Codex
Portugal

Prof. F. C. Lockwood

Imperial College of Science, Technology and Medicine
Department of Mechanical Engineering
Exhibition Road
London SW7 2BX
U.K.

Prof. J. Taine

Laboratoire E.M.2.C. UPR 288
École Centrale de Paris
92295 Chatenay Malabry Cedex
France

ISBN-13:978-3-642-84639-7   e-ISBN-13:978-3-642-84637-3
DOI: 10.1007/978-3-642-84637-3

Library of Congress Cataloging-in-Publication Data
Eurotherm Seminar (17th : 1990 : Cascais, Portugal)
Heat transfer in radiating and combusting systems
proceedings of Eurotherm Seminar no. 17, 8-10 October 1990, Cascais, Portugal /
M. G. Carvalho, F. Lockwood, J. Taine (eds.).
(Eurotherm seminars : v. 17)
Includes bibliographical references.
ISBN-13:978-3-642-84639-7

1. Heat--Transmission--Congresses. 2. Combustion chambers--Congresses.
I. Carvalho, M. G. (Maria da Graca) II. Lockwood, F. (Fred) III. Taine, J. (Jean) IV. Title. V. Series.
TJ260. E9   1990   91-38270
621.402'2--dc20

This work is subject to copyright. All rights are reserved, whether the whole or part of the material is concerned, specifically the rights of translation, reprinting, reuse of illustrations, recitation, broadcasting, reproduction on microfilm or in any other way, and storage in data banks. Duplication of this publication or parts thereof is permitted only under the provision of the German Copyright Law of September 9, 1965, in its current version, and permission for use must always be obtained from Springer-Verlag. Violations are liable for prosecution under the German Copyright Law.

© Springer-Verlag Berlin, Heidelberg 1991
Softcover reprint of the hardcover 1st edition 1991

The use of general descriptive names, registered names, trademarks, etc. in this publication does not imply, even in the absence of a specific statement, that such names are exempt from the relevant protective laws and regulations and therefore free for general use.

Typesetting: Camera ready by editors

61/3020-5 4 3 2 1 0 Printed on acid-free paper.

Preface

This volume contains the selected papers presented at the EUROTHERM SEMINAR No. 17 — Heat Transfer in Radiating and Combusting Systems held at Cascais from October 8th-10th, 1990.

The EUROTHERM COMMITTEE was created by representatives of the member countries of the European Communities for the organization and coordination of European Scientific events in the field of thermal sciences and their applications.

The book is focused on the integration of the heat transfer and combustion. These two subjects have traditionally been considered separate disciplines. In reality, the two are closely interwoven. The central purpose of the book is to generate an effective cross fertilisation of the two at both the fundamental and applied levels. The book reports on: mathematical simulations of heat transfer in reacting systems, new measurements of and measurement techniques for the radiation properties of the intervening medium, and data and theoretical analyses which clarify the physical nature of the complex interactions between the radiation/convection heat transfer processes and the combustion and turbulence of real reacting flows.

The book contains the following Chapters:

Chapter 1:   Gas Radiation
Chapter 2:   Heat Transfer in Flames
Chapter 3:   Particle Phase — Gas Phase Radiative and Convective Interaction
Chapter 4:   Measurements of Radiative and Convective Heat Transfer
Chapter 5:   Heat Transfer in Combustion Equipment
Chapter 6:   Heat Transfer in Fires

The papers were selected following review of extended abstracts by the members of the Scientific Committee. The Editors are grateful to the members of this Committee listed on the following page, to the authors and to all participants who helped to make the Seminar a success. We wish to acknowledge the financial support of the organizations that made the Seminar possible. The assistance of Ms. Salvina Ribeiro, Mr. Tiago Farias, Mr. Pedro Saraiva, Mr. Chen Xiquing and Mr. Jorge Coelho in the typing, proof-reading and art graphics of this volume is most appreciated.

Lisbon, July 1991   *The Editors*

Scientific Committee:

Prof. P. Anglesio, Politecnico di Torino, Torino, Italy
Dr. Claude Bertrand, Lafarge Coppee Recherche, Viviers-sur-Rhone, France
Prof. R. Borghi, Faculté des Sciences et Techniques de Rouen, France
Dr. Geoff Cox, Fire Research Station, Borehamwood, United Kingdom
Dr. R.M. Davies, British Gas Corp.-Midlands Res. Station, United Kingdom
Dr. Michel Desaulty, SNECMA, France
Prof. J.J. Delgado Domingos, Instituto Superior Técnico, Lisboa, Portugal
Prof. D.F.G. Durão, Instituto Superior Técnico, Lisboa, Portugal
Prof. G. Gouesbet, INSA de Rouen, France
Prof. A. Linan, Universidad Politecnica de Madrid, Spain
Prof. N.C. Markatos, National Technical University of Athens, Greece
Prof. Hector L.J. Meunier, Faculté Polytechnique de Mons, Belgium
Prof. J.K. Nieuwenhuizen, Eidhoven University of Technology, The Netherlands
Prof. H.S. Pfeifer, German French Research Inst., France
Dr. P.A. Pilavachi, DG XII, C.E.C., Brussels, Belgium
Prof. P. Roberts, Inst. Flamme Res. Foundation, The Netherlands

Sponsoring Organizations:

- Banco Comercial Português
- Centro de Termodinâmica Aplicada e Mecânica dos Fluidos da Universidade Técnica de Lisboa
- Direcção Geral de Energia
- Direcção Geral do Ensino Superior
- ERCOFTAC Lisbon Pilot Centre
- Instituto da Energia - INTERG
- Instituto Nacional de Investigação Científica
- Instituto de Promoção Turística
- Instituto Superior Técnico
- Junta Nacional de Investigação Científica e Tecnológica
- Junta de Turismo da Costa do Estoril
- Ministério da Indústria e Energia
- Portuguese Section of Combustion Institute
- TAP-Air Portugal

CONTENTS

CHAPTER 1
GAS RADIATION

GAS RADIATION SPECTRAL CORRELATED APPROACHES FOR INDUSTRIAL APPLICATIONS

A. Soufiani
Laboratoire d'Energétique Moléculaire et Macroscopique, Combustion;
du CNRS et de l'ECP, Ecole Centrale Paris
Grande Voie des Vignes, 92295 Chatenay-Malabry, France

ABSTRACT

The errors introduced by spectrally non correlated methods in the prediction of thermal radiation from real gases may be very important and non controlled. Two kinds of correlated narrow-band models are presented here. First, statistical narrow-band (SNB) model principles are recalled and the derivation of the required parameters is discussed. Some applications of SNB models to onedimensional and multidimensional radiative transfer are shown. An other approach, called the correlated-K distribution (C-K) method, which consists in reordering the absorption coefficient and integrating radiative quantities over the cumulative distribution function, is extended here with the fictitious gas mixture (FG) idea, to account for strong temperature gradients. This method is particularly suitable for gas-particle absorbing, emitting and scattering media where SNB models do not apply. It is shown to be also superior to SNB models with the Curtis-Godson approximation in the case of high pressure gradients.

1. GENERAL CONSIDERATIONS

In most of the engineering applications involving infrared radiative transfer through participating media, the considered medium is a gas-particle mixture. Gases such as H_2O, CO_2 and CO absorb and emit radiation, while particles may also scatter if their sizes are not too small in comparison with the electromagnetic wavelength [1]. Gas absorption spectra are characterized by a high dynamic fine structure resulting from bound vibrational and/or rotational molecular transitions. On the other hand, particle absorption and scattering coefficients present generally a smooth dependence on wavelength, due to their size distribution. The spectral correlation problem appears then to be crucial, especially for gas dominated radiative transfer.

Two different properties of gas absorption spectra are to be considered. The first one is the low resolution spectral dependence on wavelength, resulting from the location and the structure of different absorption bands, and the second one is the high resolution structure related to different rotational quantum numbers involved in the transitions (*see* Figure 1). We will see in the following that some radiative models may account for one of these properties but not for the other. The most accurate approach for gas radiative transfer calculations is the line by line (LBL) approach which consists in considering, at a given wavenumber ν, the

4

contribution of each particular line, centered at v_j, of each absorbing species i, to the spectral absorption coefficient K_v:

$$K_v = \sum_{\text{species } i} \sum_{\text{line } j} x_i p \, S_j(T) F(v - v_j)$$

(1)

where x_i is the molar fraction of species i, p the total pressure, $S_j(T)$ the line intensity at temperature T and F the line normalized profile [2,3]. This approach is easy to implement if all the spectroscopic parameters (line positions, intensities, widths and shapes) are known. These parameters are collected in data banks [4,5] for the easily observable lines of the most useful molecules. Nevertheless, further developments are needed for high temperature and pressure data ([6,7], for example).

It is seen from Figure 1 that the spectral resolution required for a line by line calculation is about $\delta v = 10^{-2}$ cm^{-1}. The line density is typically $1/\delta$ = few lines per cm^{-1} and the line overlapping parameter is about $\beta = 2\pi \bar{\gamma}/\delta = 0.5$, where $\bar{\gamma}$ designates the mean line half-width at half-maximum and δ the mean line spacing. In the particular case $\beta \gg 1$, corresponding to high pressure for example, the high resolution spectral structure disappears partially, and the spectral treatment of radiative transfer is less crucial. The LBL approach remains a powerful tool for some specific applications like infrared teledetection in a narrow spectral range [8], or as a reference model for checking the validity of approximate models. But the use of LBL approach in radiative transfer calculations is generally prohibited because of the required resolution δv and the integration over all the spectrum.

The desirable spectral resolution Δv in engineering calculations is the spectral range over which the blackbody intensity I_v^b may be considered as constant (generally from 10 to 100 cm^{-1}). This range contains many absorption lines and a mean absorption coefficient

$$K_{\Delta v} = \frac{1}{\Delta v} \int_{\Delta v} K_v \, dv$$

(2)

is not meaningful since attenuation of radiative intensity $\bar{I}_v$, averaged over Δv, does not obey the Beer-Lambert's law (exponential attenuation with the pathlength). This is the reason why a great research effort has been devoted, during the last forty years, to the development of spectrally correlated radiative models.

The use of total emissivity charts, collected by Hottel and Sarofim [9], has been useful in estimating radiative fluxes from hot gases considered as isothermal

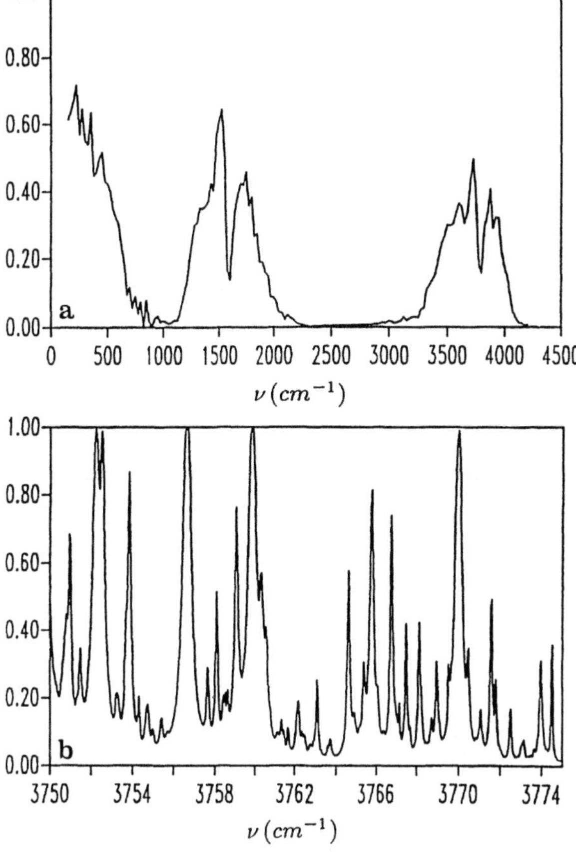

Fig. 1: H_2O - N_2 absorption spectra at T=1000K and p=1 atm. with x_{H_2O}=0.1 and L=100 cm. (a) low resolution Δv=25 cm^{-1}, (b) high resolution δv=0.01 cm^{-1}.

media. But this gray gas isothermal assumption cannot be used to accurately predict, for example, wall flux distribution in real systems or to account for the transmission of radiation emitted by a hot region and partially absorbed by a cold region. In fact, this assumption ignores both low resolution and high resolution spectral correlations.

The first and simplest idea dealing with gas nongrayness consists in considering the medium as a weighted sum of gray gases [9]. The total emissivity for a pressure pathlength product pL is written in a series form:

$$\varepsilon_g(T) = \frac{\pi \int_0^\infty I_v^b(T)\,(1 - \exp(-K_v L))\,dv}{\sigma T^4} = \sum_{i=1}^{i=n} a_i(T)\,(1 - \exp(-K_i pL)) \tag{3}$$

where K_i is the absorption coefficient, per unit pressure, of the fictitious gas i and a_i the temperature dependent weighting coefficient. If ε_g tends to 1 as pL tends to infinity, we have:

$$\sum_{i=1}^{i=n} a_i = 1 \tag{4}$$

but as a result of the windows in the spectrum (*see* Figure 1), the ε_g limit may be less than one, and a clear gas is generally assumed among the gray gases (there is an index k, $1 \le k \le n$ for which $K_k=0$). Four terms in eq. (3) are generally sufficient for an accurate fit of ε_g in a wide range of the product pL. Parameter couples (a_i,K_i) have been calculated by Taylor and Foster [10, 11] for CO_2 — H_2O — soot mixtures from total emissivities deduced themselves from statistical narrow-band models and the NASA parameters [12]. Some different parameters for the same mixtures may be found in Refs. [13, 14]. The "one clear plus gray gases" approach accounts for low resolution spectral correlations in emissivity calculations but not in the precise determination of absorptivities. On the other hand, it does not account for the fine structure of the spectrum. The use of this approach in transfer problems can lead to considerable errors, especially for steep temperature gradients [15].

A second partially correlated model, widely used in the literature is the exponential wide-band model developed by Edwards and Menard [16]. Starting from some physical arguments related to the variations of line intensities inside a band and using a statistical narrow-band model (*see* section 2), the total band absorption:

$$A = \int_{band} (1 - \exp(-K_\nu L))\, d\nu \tag{5}$$

is expressed as a function of three parameters: the band width ω, the optical depth at band heat τ_H and the line width to spacing ratio parameter η. The function $A(\omega,\tau_H,\eta)$ was first given by a four region approximate expression [16], and other closed-form relations were developed later (*see* [17] for example). The parameters (ω,τ_H,η) which give the best fit to experimental data, and their dependence on temperature and pressure may be found in Ref. [18] for the most important bands of H_2O, CO_2, CO, NO, SO_2, and CH_4.

The exponential wide-band model presents a great advantage on gray gas or weighted sum of gray gases models since the high resolution spectral correlations are accounted for, via the statistical narrow-band model. It has been successfully used in many applications. Nevertheless, three difficulties, inherent to this model, are to be noticed: (i) A wide-band model cannot be applied in teledetection problems which require narrow spectral ranges. (ii) For very wide bands, such as the H_2O band at 6.3 μm, the low resolution correlation between blackbody intensity and absorption spectra is not accounted for. (iii) The implementation of this model in radiative transfer calculations is not easy in the case of partially or totally reflecting walls, as it was recognized by Edwards [18].

In the following sections, we will assume that high resolution spectra are available from a line by line procedure, and will consider some correlated approaches, constructed from these spectra, in order to enable the use of narrow band calculations. In section 2, the principles of statistical narrow-band models are recalled both for isothermal and nonisothermal media, and some validity tests for radiative properties are given. Two different practical ways for the implementation of statistical narrow-band models in radiative transfer calculations are presented in section 3. Finally, an original approach, using both the absorption coefficient distribution function and the idea of fictitious gases, is presented. This approach, compatible with particle scattering, is tested in the case of the 2.7 μm water vapor band.

2.　STATISTICAL NARROW-BAND (SNB) MODELS

SNB models lead generally to analytical expressions of the transmissivity $\bar{\tau}_v$ averaged over a spectral range Δv containing many absorption lines. Assuming that the number N of lines inside Δv is sufficiently high, statistical assumptions are made concerning line locations, shapes and intensities. The result is a simple expression of $\bar{\tau}_v$ as a function of few parameters characterizing the high resolution spectrum.

2.1. ISOTHERMAL AND HOMOGENEOUS CASE

The analytical developments and full demonstrations are out of the purpose of this paper. They can be found in reference books or papers (e.g. [12,19]). The assumptions made and the useful formula obtained with different models are just recalled here.

The first classification is related to line center positions inside Δv. The Elsasser model assumes that the lines are regularly spaced and have the same intensity S. For Lorentzian lines with the same half-width γ, the result is:

$$\bar{\tau}_v = \frac{1}{2\pi} \int_{-\pi}^{+\pi} \exp\left[-\frac{\beta u \sinh \beta}{\cosh \beta - \cos z} \right] dz$$

$$(6)$$

8

with:

$$\beta = \frac{2\pi\gamma}{\delta}, \ u = \frac{S x_i \, pL}{2\pi\gamma}, \text{ and } z = \frac{2\pi\nu}{\delta} \tag{7}$$

The integral in expression (6) may be approximated by simple functions of the parameters β and u (Godson's approximation for example [19]). The Elsasser model is a reasonable approximation for diatomic molecules at low temperature. Its counterpart has been developed by Golden for Doppler [20] and Voigt [21] line profiles, but there is no simple analytical expression of $\bar{\tau}_\nu$ in these cases.

As temperature increases, the regular spectral structure disappears partially or totally, as a result of the increasing influence of absorption bands starting from higher vibrational levels. Another extreme assumption consists then in considering random locations of line centers inside $\Delta\nu$. If, in addition, we assume that the lines have the same shape, it is shown that $\bar{\tau}_\nu$ is only a function of the mean equivalent width to spacing ratio [19,22]:

$$\bar{\tau}_\nu = \exp\left[-\frac{\overline{W}(L)}{\delta}\right] \tag{8}$$

with, for a single line producing the absorption coefficient K_ν

$$W(L) = \int_{-\infty}^{+\infty} \left(1 - \exp(-K_\nu L)\right) d\nu \tag{9}$$

is the equivalent line width, i.e., the width of the saturated rectangular line producing the same total absorption. Equation (8) is the general expression for random narrow-band models, applied to isothermal and homogeneous paths. The computation of $\bar{\tau}_\nu$ from this equation does not require line positions but still requires the information about each line intensity. If we assume now a continuous line intensity distribution law P(S), the mean equivalent width becomes:

$$\overline{W}(L) = \int_0^\infty W(L,S) P(S) \, dS \tag{10}$$

The most representative distribution functions used in the literature, and the resulting expressions of $\overline{W}(L)/\delta$ in the case of Lorentzian lines are summarized in Table 1. The mean parameters for $\bar{\tau}_\nu$ calculations are:

- The mean intensity to spacing ratio:

$$\bar{k} = \frac{\overline{S}}{\delta} = \frac{\Sigma_{j=1}^{N} S_j}{\Delta\nu} \tag{11}$$

- The mean half-width:

$$\bar{\gamma} = \frac{1}{N} \sum_{j=1}^{N} \gamma_j \tag{12}$$

- The weighted mean line spacing:

$$\bar{\delta} = \frac{\bar{k}\,\bar{\gamma}}{\left((\Delta v)^{-1} \sum_{j=1}^{N} \sqrt{S_j \gamma_j}\right)^2} \tag{13}$$

where N is the number of lines inside Δv. The parameter $\bar{\delta}$ reduces to δ if all the lines are assumed to have the same intensity and the same width. In fact, the parameters $\bar{k}$, $\bar{\gamma}$ and $\bar{\delta}$ appear only through the ratio $\beta = 2\pi\bar{\gamma}/\bar{\delta}$, and, generally, only the parameters $\bar{k}$ and $\bar{\delta}$ are stored, assuming a constant value of $\bar{\gamma}$ over all the spectrum. It is interesting to notice that the three expressions of $-\text{Log}(\bar{\tau}_v) = \overline{W}(L)/\delta$ have the same asymptotic developments:

- weak absorption:

$$-\text{Log}(\bar{\tau}_v) = xpL\bar{k} \quad \text{for} \quad \frac{xpL\bar{k}\,\bar{\delta}}{\bar{\gamma}} \ll 1 \tag{14}$$

- strong absorption:

$$-\text{Log}(\bar{\tau}_v) = \sqrt{xpL\bar{k}\,\frac{4\bar{\gamma}}{\bar{\delta}}} \quad \text{for} \quad \frac{xpL\bar{k}\,\bar{\delta}}{\bar{\gamma}} \gg 1 \tag{15}$$

Disagreements between different models may occur only for intermediate cases.

The validity of SNB models has been studied in [26] and it was found that the random, exponential-tailed-inverse model gives the best agreement with line by line calculations for H_2O and CO_2, even at low temperature. Figure 2 presents some results for the 3775 cm^{-1} band of H_2O. It shows that the Elsasser model overestimates absorption since line overlapping, that lower absorption, is not correctly accounted for.

It is worth noticing that some SNB models have also been developed with Doppler and Voight (mixed Doppler-Lorentz) profiles [12, 27, 28] for high temperature and low pressure applications.

2.2. NONISOTHERMAL MEDIA

Radiative transfer equation for an infinite nonisothermal and nonhomogeneous medium, averaged over Δv, may be written:

$$\bar{I}_v (s) = \int_{-\infty}^{s} I_v^b (s') \frac{\partial \bar{\tau}_v}{\partial s'} (s',s) \, ds' \tag{16}$$

where s is the abscissa along the optical path and I_v^b the black-body intensity, assumed to be constant over Δv. Equation (16) leads in a discretized form to:

$$\bar{I}_v (s) = \sum_{-\infty}^{s} I_v^b \left(s' + \frac{\Delta s'}{2}\right)\left(\bar{\tau}_v (s' + \Delta s',s) - \bar{\tau}_v (s',s)\right) \tag{17}$$

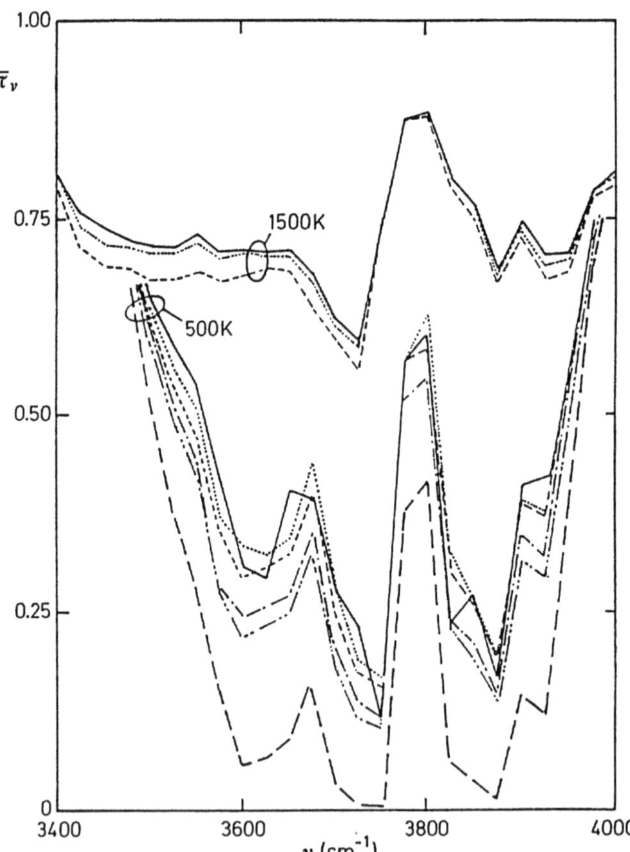

Fig. 2: pure H$_2$O transmissivities for L=9 cm. ———— line by line, — — Elsasser model, ... general random model, random statistical model with the distribution: — ·· — ·· uniform, — · — exponential, ------ exponential-tailed-inverse (from [26]).

It is seen from Eqs. (16,17) that a given radiative transfer problem can be solved if the transmissivity $\bar{\tau}_\nu (s_1, s_2)$ is known for any couple of points (s_1, s_2).

Curtis — Godson approximation

The first attempt to express the transmissivity of a nonisothermal column by using statistical narrow-band models was the Curtis-Godson (C.G.) approach. It requires that $\bar{\tau}_\nu$ be formally expressed as a function of two parameters p_1 and p_2 representing integrals of gas properties between s_1 and s_2:

$$\bar{\tau}_\nu (s_1, s_2) = f(p_1, p_2) \tag{18}$$

Table 1: Some random narrow-band models for Lorentz profile (to be used with equation (8)). (a) LR is the Ladenburg-Reiche function which can be approximated by [25]

$$LR(z) \cong z\left(1 + \left(\frac{\pi z}{2}\right)^{1.25}\right)^{-0.4}$$

Model	Uniform	Exponential	Exponential-tailed-inverse
$P(S)$	1 if $S = \bar{S}$, 0 if $S \neq \bar{S}$	$\sigma^{-1}\exp\left(-\dfrac{S}{\sigma}\right)$	$\dfrac{1}{S \operatorname{Log} R}\left[\exp\left(-\dfrac{S}{S_M}\right) - \exp\left(-\dfrac{RS}{S_M}\right)\right]$
$\dfrac{\overline{W}(L)}{\delta}$	$\dfrac{2\pi\bar{\gamma}}{\delta} LR\left(\dfrac{xpL\delta\bar{k}}{2\pi\bar{\gamma}}\right)^{(a)}$	$\dfrac{xpL\bar{k}}{\sqrt{1 + xpL\bar{k}\delta/4\gamma}}$	$\dfrac{2\bar{\gamma}}{\delta}\left(\sqrt{1 + \dfrac{xpL\bar{k}\delta}{\gamma}} - 1\right)$
Ref.	[23]	[19]	[24]

The function $f(p_1, p_2)$ must reduce to the expression of the classical narrow-band model in the limit case of an isothermal path, and must give accurate asymptotic approximations for weak and strong absorption. The last requirement, associated to Eqs. (14, 15) shows that radiative properties are to be averaged in the way:

$$\int_{s_1}^{s_2} xp\bar{k}(s)ds \quad \text{and} \quad \int_{s_1}^{s_2} xp\bar{k}(s)\bar{\beta}(s)ds$$

Then, the Curtis-Godson approximation uses the same expression for $\bar{\tau}_\nu(s_1, s_2)$ than that for the isothermal band model, with the average parameters

$$v = \int_{s_1}^{s_2} xpds \quad \text{(amount of absorber)} \tag{19}$$

$$\bar{k}_{\text{C.G.}} = \frac{1}{v} \int_{s_1}^{s_2} xp\bar{k}(s)ds \tag{20}$$

$$\bar{\beta}_{\text{C.G.}} = \frac{1}{v\bar{k}_{\text{C.G.}}} \int_{s_1}^{s_2} xp\bar{k}(s)\bar{\beta}(s)ds \tag{21}$$

This leads, in the case of the exponential-tailed-inverse distribution model for example, to:

$$\bar{\tau}_v(s_1, s_2) = \exp\left[-\frac{\bar{\beta}_{\text{C.G.}}}{\pi}\left(\sqrt{1 + \frac{2\pi v\bar{k}_{\text{C.G.}}}{\bar{\beta}_{\text{C.G.}}}} - 1 \right) \right] \tag{22}$$

Lindquist — Simmons approximation

The C.G. approximation is a direct approximation of $\bar{\tau}_v(s_1,s_2)$ or $\bar{W}(s_1,s_2)/\delta = -\text{Log}[\bar{\tau}_v(s_1,s_2)]$. But it appears from eq. (16) or (17) that only transmissivity differences (or derivative) are required to solve the transfer equation. A direct approximation to the derivative $[\partial\bar{\tau}_v/\partial s_1](s_1,s_2)$ is then expected to give a more accurate approximation for radiation intensity. If we notice that the number of lines N is constant over the optical path, differenciation of eq. (8) leads to:

$$\frac{\partial\bar{\tau}_v}{\partial s_1}(s_1, s_2) = -\bar{\tau}_v(s_1, s_2)\,\frac{1}{\delta}\,\frac{\partial\bar{W}}{\partial s_1}(s_1, s_2) \tag{23}$$

Lindquist and Simmons [29] give an approximation to $\partial\bar{W}/\partial s$ for a random array of Lorentz lines with equal strengths and Young [30, 31] extended this approach to the case of the exponential-tailed-inverse intensity distribution. The result obtained in this case is:

$$\frac{1}{\delta}\frac{\partial\bar{W}}{\partial s}(0,s) = x(s)p(s)\bar{k}(s)\,y\left[\frac{\pi\bar{k}_{\text{L.S.}}(s)v(s)}{\bar{\beta}_{\text{L.S.}}(s)}, \frac{\bar{\gamma}(s)}{\bar{\gamma}_{\text{L.S.}}(s)} \right] \tag{24}$$

with the new path averaged parameters:

$$\bar{k}_{\text{L.S.}}(s) = \frac{1}{v(s)} \int_0^s x(s')p(s')\bar{k}(s')ds' \tag{25}$$

$$\bar{\delta}_{\text{L.S.}}(s) = \frac{1}{v(s)\,\bar{k}_{\text{L.S.}}(s)} \int_0^s x(s')p(s')\bar{k}(s')\bar{\delta}(s')ds' \tag{26}$$

$$\bar{\gamma}_{L.S.}(s) = \frac{1}{v(s)\,\bar{k}_{L.S.}(s)\,\bar{\delta}_{L.S.}(s)} \int_0^s x(s')p(s')\bar{k}(s')\bar{\delta}(s')\bar{\gamma}(s')ds' \qquad (27)$$

$$\bar{\beta}_{L.S.}(s) = \frac{2\pi\bar{\gamma}_{L.S.}(s)}{\bar{\delta}_{L.S.}(s)} \qquad (28)$$

and the function y defined by:

$$y(a,b) = \frac{2b(1+a) + (1+b^2)\sqrt{1+2a}}{\sqrt{1+2a}\left(b + \sqrt{1+2a}\right)^2} \qquad (29)$$

Transmissivity calculations with the Lindquist-Simmons (L.S.) approximation require more computer time than Curtis-Godson approximation since an integration of eq. (23) is needed. But it was shown by Young [32] that the L.S. approximation corrects some C.G. approximation failures, especially in the case of very strong variations of line widths inside the optical path (max-min ratio up to 10). Comparisons between C.G., L.S. and other nonisothermal approximations may be found in [26, 32]. Figure 3 (from [26]) shows the compared spectra in the case of H_2O-CO_2-air two adjacent cells at different temperatures. C.G. and L.S. approximations lead in this case to practically the same transmission spectrum, which is in good agreement with line by line calculations.

2.3. BAND MODEL PARAMETERS

The parameters required for statistical narrow-band model calculations are $\bar{k}$ and $\bar{\beta}$, or $\bar{k}$ and $\bar{\delta}$ if we assume a constant value of $\bar{\gamma}$ over the spectrum. These parameters are functions of the spectral range Δv and temperature.

The experimental method for their determination consists in measuring low resolution transmissivities of gaseous mixtures in isothermal cells. Measurements in the weak absorption limit lead to the parameter $\bar{k}$ and those in the strong absorption limit give $\sqrt{\bar{k}\bar{\beta}}$ as it is seen from eqs. (14) and (15). The parameters may also be determined from a least square fit of the curve of growth $-\text{Log}\bar{\tau}_v = f(\bar{k}\,\bar{\beta})$. The experimental method has been used for example by Ferriso et al. [33] and Ludwig [34] for H_2O in the temperature range [300, 3000K]. Their results have been used for H_2O high temperature parameters listed in Ref. [12]. CO_2 parameters in the 4.3 μm region have measured by Kunitomo and Osumi [35] for temperatures up to 1200 K, and recently, Phillips [36] has measured H_2O parameters in the spectral range [3000, 4500 cm^{-1}] and the temperature range [300, 1000K], with a Fourier transform interferometer.

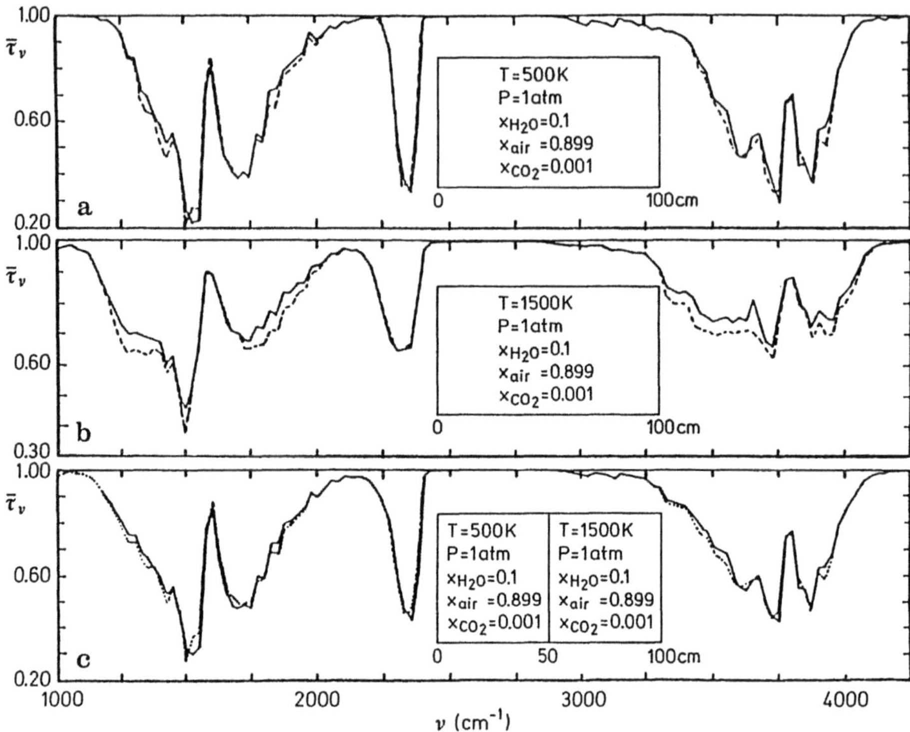

Fig. 3: H_2O-CO_2-air mixture transmissivities for isothermal paths ((a) and (b)) and nonisothermal path (c). ——— line by line, ------ statistical model with the exponential-tailed-inverse distribution, ... LS and CG approximations (from [26]).

Band model parameters can also be deduced theoretically from spectral data banks like HITRAN [4] or GEISA [5], directly by using the definitions given in (11-13). However, these banks have been created for atmospheric applications and contain then only few lines starting from high rotational and vibrational levels. They must be completed for high temperature applications. This procedure has been used to compute the NASA parameters [12] for CO_2 and diatomic molecules such as CO, NO and OH. Bernstein *et al.* [37] give some theoretical improvements for CO_2 computations in the 4.3 µm region and in the temperature range [200, 3000K], while Young [38] constructed H_2O parameters in the [2500, 4500 cm^{-1}] spectral range by combining the high temperature NASA parameters with low temperature parameters derived from the AFGL tape. Other band model parameters have been constructed by Hartmann *et al.* [39], and Soufiani *et al.* [26] for CO_2 and H_2O in the spectral range [150, 8000 cm^{-1}]. These parameters were derived from high temperature line by line calculations [39, 40]. They have been recently completed in the temperature range [300, 2400K].

A direct derivation of band model parameters from data banks (containing S_j, v_j, γ_j) must be handled carefully since it depends on the number N of the considered lines, and then, on the higher vibrational level accounted for in the calculations. Bernstein [41] has shown that a maximum vibrational energy cutoff, up to 15,000 cm^{-1} is required for the convergence of the parameter $\bar{k}$.

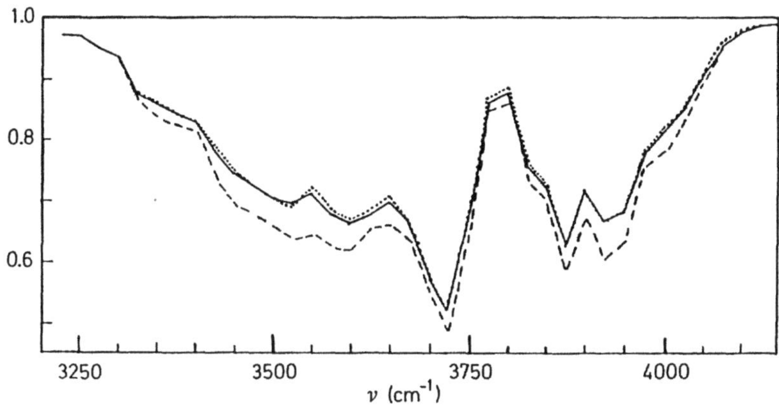

Fig. 4: H_2O transmissivities in the conditions T=1200K, p=756 mm Hg, L=7.75 cm, $x_{H_2O}=1$. ——— Line by line, and statistical narrow-band model with the exponential-tailed-inverse distribution. ----- parameters from the definitions (11-13), (.....) parameters fitted from line by line transmissivities.

An other alternative procedure consists in line by line computing the transmissivity, averaged over Δv, for various optical depths, and extracting the parameters from a least square fit of transmissivity curves with the chosen narrow-band formula. Figure 4 shows a comparison between H_2O absorption spectra computed with parameters from eqs. (11-13) and the fitted parameters. The difference between these spectra may be attributed to a nonconvergence of the parameter $\bar{k}$ in the first procedure (too small cutoff criterion).

3. APPLICATION OF SNB MODELS TO RADIATIVE TRANSFER

Statistical narrow-band models have been often used in radiative transfer problems. Some applications may be found in [42, 43] in the field of flame radiation, in [44, 45] for the study of turbulence-radiation interaction, or in [46, 47] for radiative transfer in the atmosphere. The two quantities generally required in these applications are the wall radiative flux q_w and the volumetric radiative source term in the energy conservation equation P. They can be deduced from the radiative intensity field by integration over solid angle and frequency:

$$q_w = \int_0^\infty \int_{4\pi sr} I_\nu(s,\vec{u})\, \vec{u}.\vec{n}\, d\Omega d\nu \tag{30}$$

$$P = \int_0^\infty \int_{4\pi sr} \frac{\partial I_\nu}{\partial s}(s,\vec{u})\, d\Omega d\nu \tag{31}$$

where $d\Omega$ is an elementary solid angle around the propagation direction $\vec{u}$ and $\vec{n}$ is the normal to the wall. The directional integration in eqs. (30, 31) requires the resolution of the radiative transfer equation for each direction $\vec{u}$. Unfortunately, the approximate methods, developed in order to simplify the directional variations of I_ν (such as spherical harmonics and P_N approximations [48]), cannot be applied with SNB models, since these methods are based on the radiative transfer equation written in terms of the absorption coefficient K_ν. Indeed, as it was noticed before, the intensity averaged over $\Delta\nu$ is not attenuated following the Beer-Lambert's law (*see* eq. (8) and Table 1). The radiative transfer equation is generally solved by using direct numerical methods such as the discrete ordinates method [49]. We present in this section some numerical methods for correlated radiative transfer calculations in planar media and then in multidimensional media. In these procedures, if the radiative model is not mentioned, we use the SNB model with the exponential-tailed-inverse intensity distribution and the Curtis-Godson approximation.

In the case of a bounded medium, the radiative transfer equation, averaged over $\Delta\nu$, gives:

$$\bar{I}_\nu(s) = \overline{I_\nu(0)\tau_\nu(0,s)} + \int_0^s \overline{I_\nu^b}(s')\, \frac{\partial \bar{\tau}_\nu}{\partial s'}(s',s)\, ds' \tag{32}$$

where $s=0$ designates a wall point. Fully correlated treatment of the wall reflected intensity, included in the first term of the right side of eq. (32), is extremely complicated. The non correlated reflection approximation

$$\overline{I_\nu(0)\tau_\nu(0,s)} \approx \bar{I}_\nu(0)\bar{\tau}_\nu(0,s) \tag{33}$$

is generally used. This approximation holds for weakly reflecting walls or in the case of hot walls and cold medium. In other cases, it may introduce appreciable errors, especially for optically thin media and low values of the overlapping parameter β. But, to our knowledge, there is no simple and practicable alternative to this approximation.

Replacement of (33) in (32) and differenciation with respect to the abcissa s leads to:

$$\frac{\partial \bar{I}_v}{\partial s}(s) = \bar{I}_v(0)\frac{\partial \bar{\tau}_v}{\partial s}(0,s)^{(a)} + I_v^b(s)\frac{\partial \bar{\tau}_v}{\partial s'}(s'=s,s)^{(b)}$$

$$+ \int_0^s I_v^b(s')\frac{\partial^2 \bar{\tau}_v}{\partial s'\partial s}(s',s)\,ds'^{(c)} \tag{34}$$

where (a) represents the local absorption of the leaving wall intensity, (b) is the local emission, and (c) is the intensity emitted by the gas between 0 and s and absorbed at s. The radiative source term $P_{\Delta v}$, integrated over Δv is then given by

$$P_{\Delta v} = \Delta v \int_{4\pi sr} \frac{\partial \bar{I}_v}{\partial s}\,d\Omega \tag{35}$$

3.1. ONEDIMENSIONAL CASE

We consider now the particular case of a onedimensional gas mixture bounded by two infinite parallel walls at temperatures T_1 and T_N. The gaseous medium is divided into N-2 isothermal layers as shown in Figure 5. Centered discretization of eq. (34) and integration over the propagation directions θ (or $\mu = \cos\theta$) gives for the layer i:

$$d_i \; P_{\Delta v,i} = \sum_{j=1}^N A_{ij} B_j \tag{36}$$

where d_i is the thickness of layer i, B_j is defined by:

$$B_j = \pi \, I_v^b(T_j)\,\Delta v, \qquad 2 \leq j \leq N\text{-}1 \tag{37}$$

and A_{ij}, for non adjacent layers, is related to the emission by j, transmission, and then absorption by i. For $j \geq i + 2$ for example:

$$A_{ij} = 2\int_0^1 \left[\bar{\tau}_{vi+1,j-1}(\mu) - \bar{\tau}_{vi,j-1}(\mu) - \bar{\tau}_{vi+1,j}(\mu) + \bar{\tau}_{vi,j}(\mu)\right]\mu\,d\mu \tag{38}$$

where $\bar{\tau}_{vm,n}$ is the transmissivity of the nonisothermal column of length $\sum_{l=m}^n d_l/\mu$.

The expressions of A_{ij} for $|i\text{-}j|=1$ (adjacent layers), or i=j (emission from the layer i), as well as the wall radiosities B_1 and B_N may be found in Ref. [26]. Instead of discretizing eq. (34), the radiative source term may also be derived from a direct radiative balance for the considered layer. The flux $q_{\Delta vj \to i}$ emitted by the layer j and directly absorbed by the layer i is given by:

$$q_{\Delta vj \to i} = 2\pi \, I_v^b(T_j) \int_0^1 \left(\int_{\Delta v} \varepsilon_{vj} \, \tau_{vj-1} \dots \tau_{vi+1} \varepsilon_{vi} \, dv \right) \mu \, d\mu \tag{39}$$

The integration over Δv is solved by replacing the emissivities ε_{vj} and ε_{vi} by $(1-\tau_{vj})$ and $(1-\tau_{vi})$, respectively, executing the product, and then integrating. This yields:

$$q_{\Delta vj \to i} = A_{ij} \, B_j \tag{40}$$

Flux and source term calculations reduce, in the case of a planar medium, to the computation of the integrals:

$$\int_0^1 \bar{\tau}_{vm,n}(\mu) \, \mu \, d\mu$$

which replace the classical integroexponential function E_3.

The procedure described above has been applied to combined conductive and radiative transfer in a planar medium [26]. Figure 6 shows the temperature profiles and the conductive fluxes obtained with various radiative models for a CO_2-air mixture for which only the 4.3 µm band was considered. The best agreement with LBL calculations is obtained for the SNB model with the exponential-tailed-inverse distribution and the Curtis-Godson approximation. Non correlated models lead to wall conductive flux discrepancies as large as 65% in the case of the wide box model (K_v averaged over the hole band). The same radiative transfer procedure has been used to study forced laminar convection-radiation combined transfer in channel flows [50], and mixed laminar convection-radiation transfer on a vertical plate [51]. Disagreements between the results of SNB and exponential wide-band models up to 30% were observed [50].

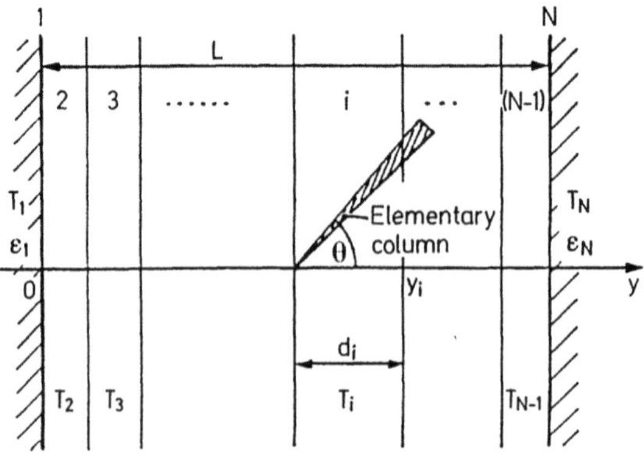

Fig. 5:   Spatial discretization and coordinates in 1D

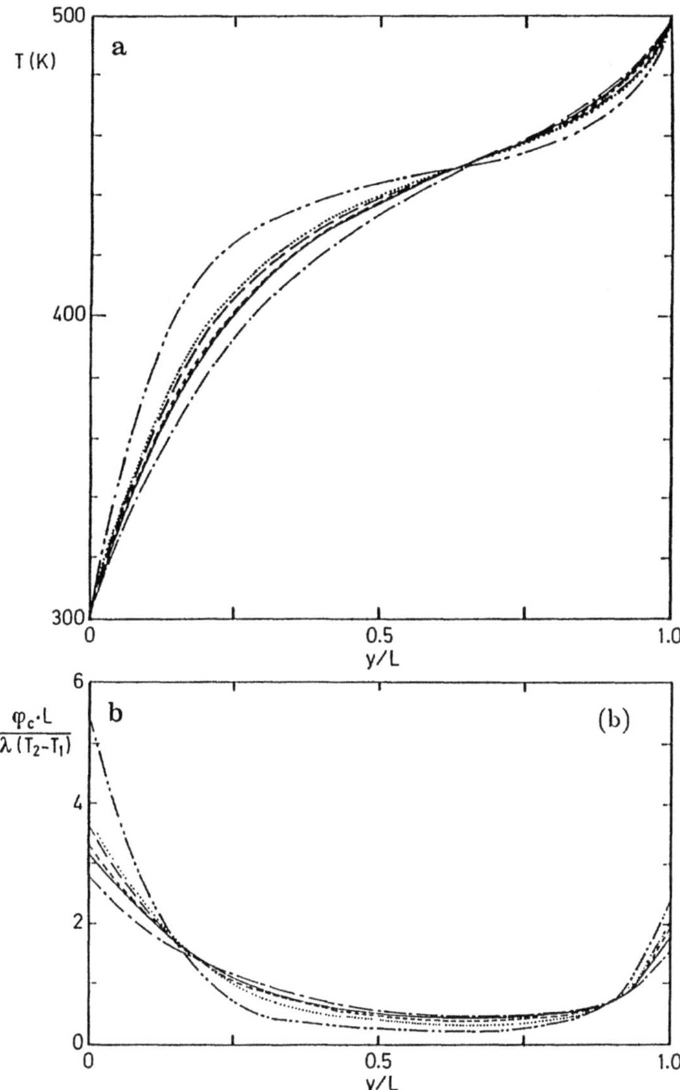

Fig. 6: Temperature profiles (a) and normalized conductive fluxes (b) for a onedimensional radiating-conducting CO_2-N_2 mixture (p=1 atm, x_{CO2}=0.001, T_1=300K, T_N=500K, L=100 cm). —— line by line, — — — Elsasser model, random statistical models with the distributions ------- exponential-tailed-inverse, — · — · —· exponential and Edward's parameters. .... narrow box model, — ·· — wide box model (from [26]).

3.2. MULTIDIMENSIONAL PROBLEMS

We suppose that the gaseous medium is discretized into a cartesian grid and look forward a correlated numerical procedure to compute the radiative field on the same grid. This procedure is then compatible with finite difference schemes for flow or other computations.

The 1D procedure described above cannot be easily generalized to multidimensional problems since the line of sight Os will generally not intercept the grid points of the discretized medium. An exact correlated approach requires, for each point M and each direction $\vec{u}$, the reconstitution of the optical path from the wall point W intercepting the line $(M,\vec{u})$ and the intensity computation from eqs. (32, 33). This method is numerically very heavy and impracticable in complex geometries. A "marching procedure", in which the intensity at M is computed from the intensity at the neighbour points will be prefered. But with this procedure, the spectral correlations are necessarily lost since the information about radiation "history" is lost. An approximate method consists then in carrying out non correlated calculations and correcting the intensity with a correlation coefficient, deduced from the exact calculations in some specific directions. This method has been applied for a general axisymmetric system [52], but it can be generalized easily to any geometry.

Figure 7 shows that, for an axisymmetric system, any discrete propagation direction may be reconstructed in one of the discrete planes parallel to the system axis. Then, intensity calculations may be carried out in these discrete planes. The non correlated intensity $\overset{*}{I}_\nu$ is given by:

$$\overset{*}{I}_\nu (M,\vec{u}) = \overset{*}{I}_\nu (M',\vec{u})\; \bar{\tau}_\nu (M',M) + (1- \bar{\tau}_\nu (M',M))\; I_\nu^b (T_{M'M}) \tag{41}$$

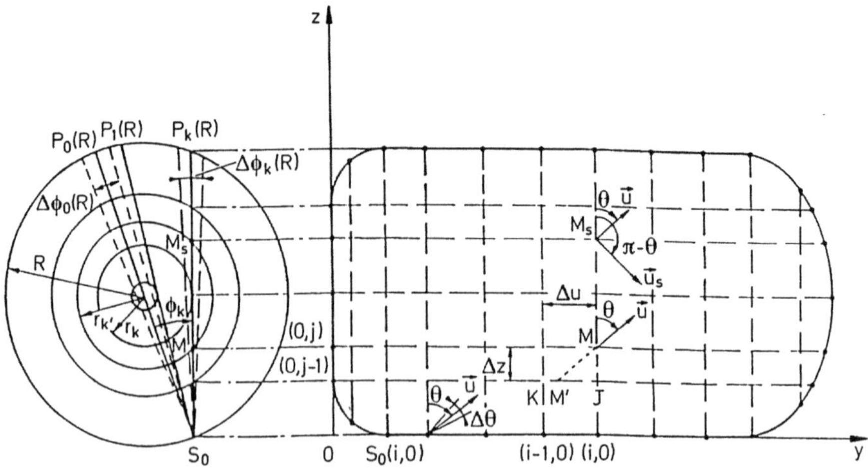

Fig. 7: Axisymmetric geometry and discretization into planes parallel to the system axis (from [52]).

where $I_\nu^*(M',\vec{u})$ can be interpolated from $I_\nu^*(K,\vec{u})$ and $I_\nu^*(J,\vec{u})$ (*see* Figure 7). The corrective coefficient is defined by the ratio between correlated and non correlated intensities:

$$C(M,\vec{u}) = \bar{I}_\nu(M,\vec{u})/\bar{I}_\nu^*(M,\vec{u}) \tag{42}$$

In the approximate correlated method developed in [52], $C(M,\vec{u})$ is calculated exactly for the directions $\vec{e}_y$, $-\vec{e}_y$ and $\vec{e}_z$, and approximated for a general direction $\vec{u}$ by:

$$C(M,\vec{u}) \cdot \left[\left(C(M,\vec{e}_z)\cos\theta\right)^2 + \left(\left(C(M,\vec{e}_y)\sin\theta\right)^2\right)\right]^{\frac{1}{2}} \quad \text{for } \theta \geq 0$$

$$C(M,\vec{u}) \cdot \left[\left(C(M,\vec{e}_z)\cos\theta\right)^2 + \left(\left(C(M,-\vec{e}_y)\sin\theta\right)^2\right)\right]^{\frac{1}{2}} \quad \text{for } \theta \leq 0 \tag{43}$$

This approximation has been first tested in a planar medium. Figure 8 shows a comparison between the exact correlation coefficient and the approximate coefficient calculated from the exact one at $\theta=0$ and $\theta=70$ (instead of 90). A good agreement is obtained, except for θ close to $\pi/2$, where correlation phenomena disappear since there is no temperature gradient in y direction and then no radiative transfer. The full details of the implementation of the approximate correlated method in the case of axisymmetric systems are given in [52] where the method is applied to finite cylindrical absorbing and emitting CO_2—H_2O—

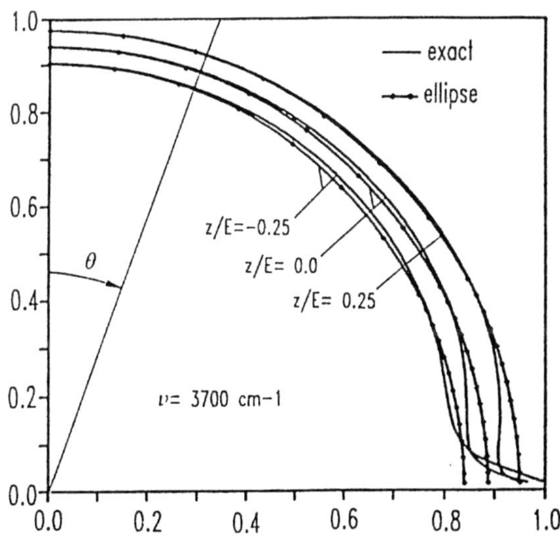

Fig. 8: Exact and approximate correlation coefficient in the case of a planar medium (from [52]).

CO—N_2—O_2—particle mixtures. Figure 9 shows some results obtained for homogeneous H_2O—N_2 mixtures with a temperature profile.

$$T(r,y) = 800 + 1200 \ (1\text{-}r/y)y/L$$

Strong discrepancies on the lateral wall flux are shown when the non correlated approach is used.

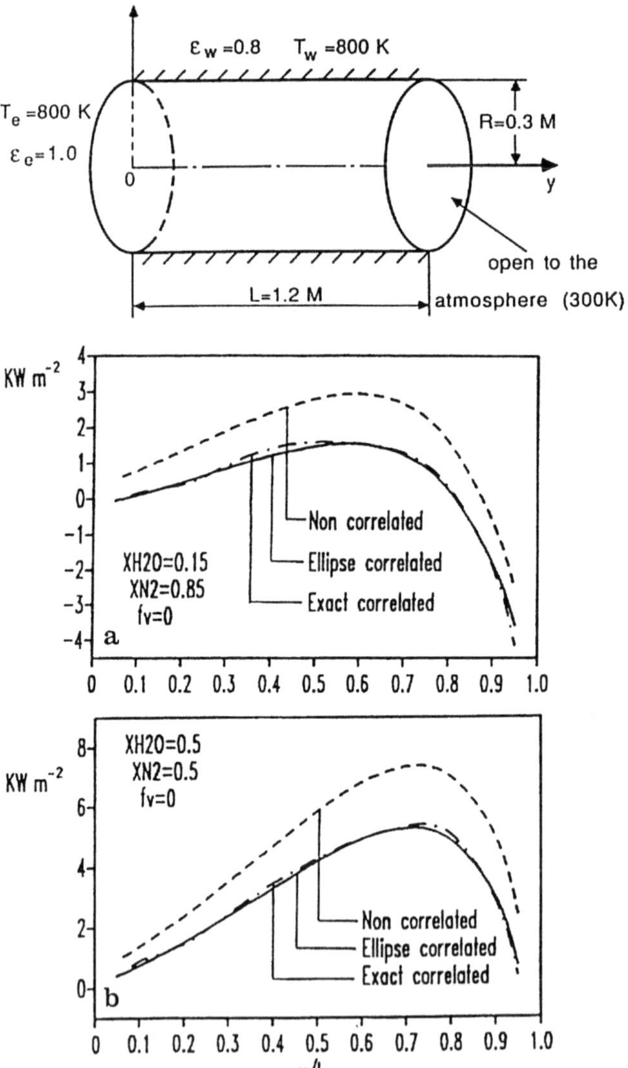

Fig. 9:   Lateral radiative wall fluxes computed for the cylindrical geometry (shown above), containing H_2O—N_2 mixtures (from [52]).

4. C-K AND FICTITIOUS GAS METHODS

The failures of statistical narrow band models are of two kinds: (i) their application in the case of strong pressure gradients, or generally of line width γ (typically $\Delta\gamma/\gamma > 10$), leads to appreciable errors when the Curtis-Godson approximation is used (*see* Ref. [32] and Figs. 14 and 15). Nevertheless, it is worth noticing that temperature gradients alone in the range [300, 3000K] cannot produce such an effect. (ii) In the case of combined radiative transfer by absorption-emission and scattering, the transfer equation is:

$$\frac{\partial I_\nu}{\partial s}(\vec{u}) = K_\nu \left[I_\nu^b - I_\nu(\vec{u}) \right] - \sigma_\nu I_\nu(\vec{u}) + \frac{\sigma_\nu}{4\pi} \int_{4\pi sr} P_\nu(\vec{u'}, \vec{u}) \, I_\nu(\vec{u'}) \, d\Omega' \tag{44}$$

where σ_ν is the spectral scattering coefficient and P_ν the phase function. In the previous section, we have seen how to deal with the first two terms on the right side of eq. (44) with SNB models, but the extension of these models to account for the last scattering term is very complicated. In fact, radiation from each direction $\vec{u'}$ has a different "history", and only a time consuming Monte Carlo method can account for scattering in a correlated manner. Nevertheless, an attempt to extend SNB model to scattering is presented in [53] for planar media and regular models. But the generalization of this approach is not easy.

The C-K fictitious gas mixture method is a different approach, developed in order to deal with these failures. We present first the correlated-K distribution method (C-K) and then its extension with the idea of fictitious gases.

4.1. CORRELATED-K DISTRIBUTION METHOD

Isothermal and homogeneous gases

This method starts from the observation that for a given spectral range $\Delta\nu$, what is significant for the averaged transmissivity or any other radiative quantity, is not so the positions of the absorption lines inside $\Delta\nu$, but the contribution of each fraction of $\Delta\nu$ to the absorption coefficient K_ν. The order of K_ν inside $\Delta\nu$ may be ignored. If we define $f(K) \, dK$ as the fraction of $\Delta\nu$ where the absorption coefficient K_ν lies between K and $K + dK$, the averaged transmissivity of a column of length L is given by:

$$\bar{\tau}_\nu(L) = \int_0^\infty f(K) \exp(-KL) \, dK \tag{45}$$

which is an alternative expression to the exact definition of $\bar{\tau}_\nu$. The average formula (45) applies in fact for any radiation quantity which depends on the absorption coefficient K [54]:

$$\overline{G} = \frac{1}{\Delta\nu} \int_{\Delta\nu} G(K_\nu)\, d\nu = \int_0^\infty f(K)G(K)\, dK \tag{46}$$

This is the K distribution method. The determination of the distribution function f(K) can be derived from its mathematical definition. If we divide the high resolution spectrum into N domains where K_ν is a monotonic function of ν, f(K) is given by (*see* Figure 10):

$$f(K) = \sum_{i=1}^{i=N} \frac{1}{\Delta\nu} \left| \frac{d\nu}{dK_\nu} \right|_i \left[h(K-K_{\min\,i}) - h(K-K_{\max\,i}) \right] \tag{47}$$

where $K_{\min\,i}$ and $K_{\max\,i}$ are the minimum and the maximum values of K_ν in the ith domain, and h the Heaviside function. Equation (47) can be used analytically or numerically to compute f(K) from a line by line calculation. But the K distribution method can also be used if the assumptions of statistical narrow-band models are made. Equation (45) shows that $\bar{\tau}_\nu(L)$ is the Laplace transform of f(K), and then f(K) may be calculated analytically by using inverse Laplace transformation of the band model transmissivity formula. For the exponential-tailed-inverse distribution, this leads to:

$$f(K) = \frac{1}{2\pi K} \sqrt{2\overline{\beta}\,\frac{\overline{xpk}}{K}} \, \exp\left[\frac{\overline{\beta}}{2\pi}\left(2 - \frac{K}{\overline{xpk}} - \frac{\overline{xpk}}{K} \right) \right] \tag{48}$$

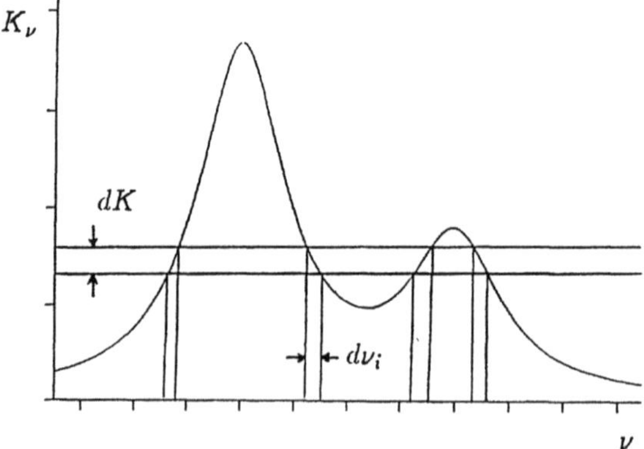

Fig. 10: Definition of the K distribution function. f(K) dK is the sum of the narrow ranges $d\nu_i$.

Equivalent expressions for regular models and random models with some other intensity distributions are given in [55].

Nonisothermal media

The K distribution method, as described above cannot be directly applied to nonisothermal media since the information about line positions is scrambled and spectral correlations are lost. The exact average transmissivity of a column of length L is given by:

$$\bar{\tau}_v = \frac{1}{\Delta v} \int_{\Delta v} \exp\left[-\int_0^L K_v(s)\,ds\right] dv \qquad (49)$$

and it is not easy to express, in the general case, the distribution function of $1/L \int_0^L K_v(s)\,ds$ versus the local distribution functions. The correlated K distribution method (C-K) uses the cumulative distribution function:

$$g(K) = \int_0^K f(K')dK' \qquad (50)$$

instead of the distribution function $f(K)$. $g(K)$ is a monotonic function of K from $[K_{min}, K_{max}]$ to $[0,1]$ and the relation between g and K may be inverted for any point s. The C-K method consists in approximating the general expression (49) by:

$$\bar{\tau}_v \approx \int_0^1 \exp\left[-\int_0^L K(g,s)\,ds\right] dg \qquad (51)$$

It is seen from eq. (50) that the inverse function $K(g)$ is nothing else than the reordered absorption coefficient versus the wavenumber scaled by $dg=dv/\Delta v$. Then, this function can be easily computed from a line by line computation, just by reordering increasingly the absorption coefficient.

Goody *et al.* [55] show that the approximation (51) tends asymptotically to the exact expression (49) for the weak absorption limit, since the exponential may be linearized. This approximation holds also for scaling conditions, where the variations of K_v with temperature, pressure and wavenumber may be factorized in the form:

$$K_v(T,p) = \Phi(T,p)\,\Psi(v) \qquad (52)$$

It is obvious, in these conditions, that reordering of K_v is made in the same manner for any couple (T,p).

The C-K distribution method is an accurate approximation for small temperature gradients and leads to good results for atmospheric applications [55] where the scaling conditions are quasi satisfied. Nevertheless, an obvious case where the approximation (51) may be not valid is schematically shown in Figure 11. For strong temperature gradients, the scaling conditions do not hold because of the coexistence of absorption lines starting from very different energy levels. An extension of the C-K distribution method is required.

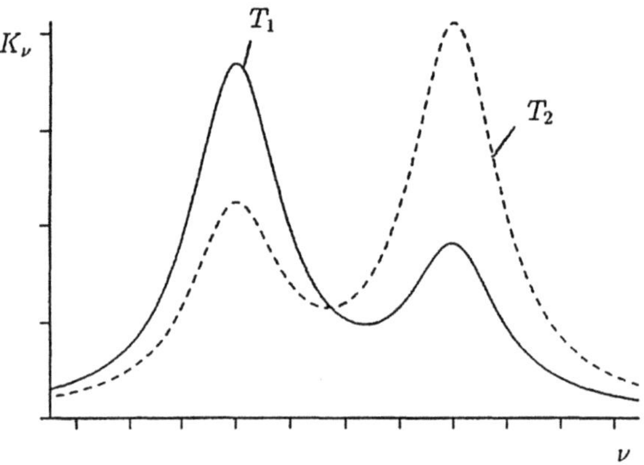

Fig. 11:   Schematic evolution with temperature of two absorption lines characterized by very different energies of the lower level of the transition.

4.2. C-K FICTITIOUS GAS METHOD

At a given wavenumber, the most significant parameter for line intensity evolution with temperature is the energy E" of the lower level of the transition. Indeed, S(T) is directly proportional to the population of this level, and then to exp(-E"/kT) at local thermodynamic equilibrium. A simple way to obtain similar variations with temperature for the absorption coefficient resulting from different lines consists then in dividing the gas into several fictitious gases [56]. Each fictitious gas is characterized by the lines corresponding to E" in a given range ΔE". The transmissivity of the real gas is obtained by the product of fictitious gas transmissivities, as it is generally done for real gas mixtures.

Figure 12 shows the reordered absorption coefficient resulting from 5 low energy classes for H_2O in the spectral range $\Delta\nu$=25 cm^{-1} around 3500 cm^{-1}, at 300K and 1500K. The considered energy classes are [0, 1000 cm^{-1}] for the first class, [1000, 2000 cm^{-1}] for the 2nd, [2000, 3000 cm^{-1}] for the 3rd, [3000, 4000 cm^{-1}] for the 4th and E" > 4000 cm^{-1} for the 5th class. It appears from this figure that, while K_ν is mainly due to the first class at 300K, the major contributions at 1500K are those of the 5th and the 2nd classes. These different evolutions show that the C-K distribution method does not correctly account for the spectral correlations.

In the C-K fictitious gas method (C-K F.G.), the line by line procedure is used to provide N high resolution spectra corresponding to the N energy classes. Each spectrum is then reordered increasingly. This leads to the cumulative distribution function $g_i(K)$ for each class i. The mean transmissivity is then given by:

$$\bar{\tau}_\nu = \prod_{i=1}^{N} \int_0^1 \exp\left[-\int_0^L K(g_i, s)\, ds\right] dg_i \tag{53}$$

The product in this expression introduces a new source of errors if the high resolution spectra of the fictitious gases are statistically correlated. This will be surely the case if the intervals $\Delta E''$ are very small. But for $\Delta E''$ sufficiently high (N sufficiently low), and triatomic molecules such as H_2O and CO_2, the spectral range $\Delta\nu$ contains generally many lines belonging to absorption starting from different vibrational levels.

The statistical correlation between lines of different fictitious gases is then expected to produce small effects. The number N of energy classes must be chosen in order to improve the C-K distribution method but must be sufficiently low to avoid statistical correlation phenomena.

Implementation of the method and discussion

The C-K and C-K F.G. methods would be of few interest if the cumulative distribution $g(K)$ is to be deduced in each application from line by line calculations. It is highly desirable to get a set of parameters as it is generally done for SNB models.

The integrals of the form $\int_0^1 F(g)\, dg$ may be approximated with a Gauss quadrature:

$$\int_0^1 F(g)\, dg = \sum_{m=1}^{M} a_m\, F(g_m) \tag{54}$$

to any degree of accuracy. The coefficients a_m and g_m are then constant integration parameters depending on the method of quadrature and the nature of the quantity F to be calculated. In fact, the quantity F depends on g only through the value of K at g. For a mixture of fictitious gases, the real parameters of the method are then $K_i(g_m)$, i.e. the values of the absorption coefficient for each fictitious gas i and each discrete scaled wavenumber g_m. For a given temperature and pressure conditions, and each wavenumber range, there are N x M model parameters.

Goody *et al.* [55] studied the accuracy of the C-K method as a function of the number of quadrature points M. They found that M=10 leads generally to acceptable errors (±1%) in atmospheric applications. We use here the Gauss-Lobatto quadrature with 7 integration points including the fixed points $g_1=0$ and $g_5=0.9$, to study the validity of the C-K F.G. method, applied to H_2O near 2.7 μm. The other quadrature points are shown on Figure 12. A similar quadrature used in [56] gave excellent results for CO_2 transmissivities.

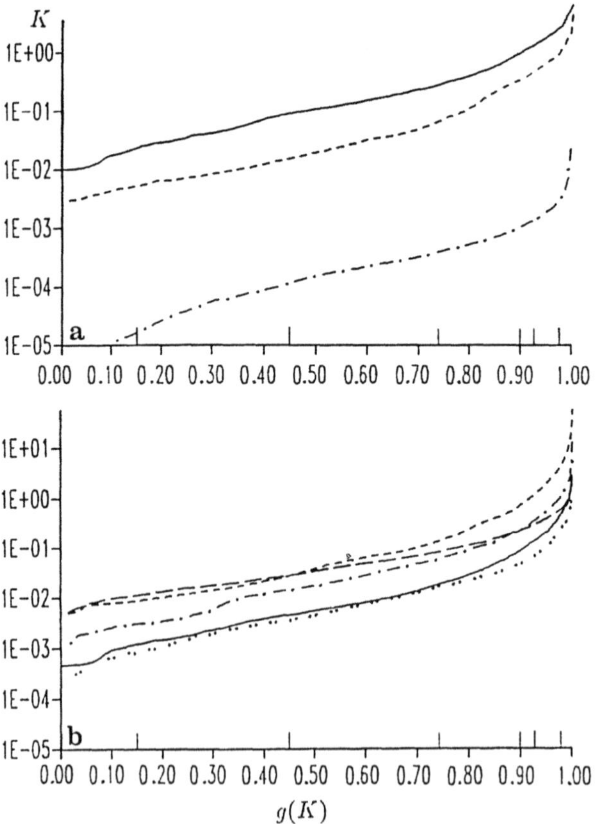

Fig. 12: Contribution of each lower energy class to the reordered absorption coefficient for H_2O at 300K (a) and 1500K (b). The quadrature integration points are also shown on the g abscissa. —— class 1, ------ class 2, ———· class 3, ... class 4 and — — — class 5. Classes 4 and 5 are not represented at 300K since their contribution is very small.

The application of C-K F.G. method to isothermal and homogeneous mixtures containing H_2O yields excellent agreements with line by line calculations, only some results corresponding to nonisothermal and/or non constant pressure paths are presented here, with the 5 energy classes listed above for the C-K F.G. method. The transmission spectra shown in Figs. 13, 14 and 15 correspond to two adjacent isothermal and homogeneous cells containing H_2O — N_2 mixtures with x_{H2O}=0.03, at different temperatures and pressures. Figure 13 shows that the fictitious gas method corrects the defects of the C-K method when applied to steep temperature gradients. The errors introduced by statistical correlations between different fictitious gas lines are shown to be much more smaller than those inherent to the C-K method. The SNB results, not shown in Figure 13, are in good agreement with line by line calculations. Figs. 14 and 15 correspond to strong pressure gradients and combined strong pressure and temperature gradients (encountered for example in shocks), respectively. SNB model with the Curtis-Godson approximation overestimates absorption in this case, while the C-K F.G. results are in good agreement with line by line calculations.

At this stage, the practical implementation of the C-K F.G. method is still difficult since the parameters $K_i(g_m)$ depend on temperature and pressure. These dependences have been fitted with analytical functions by Levi Di Leon and Taine [56] for CO_2 in the 4.3 µm region, and by Riviere [57] for H_2O near 2.7 µm. For CO_2, the analytical fit function was deduced from the expression of the

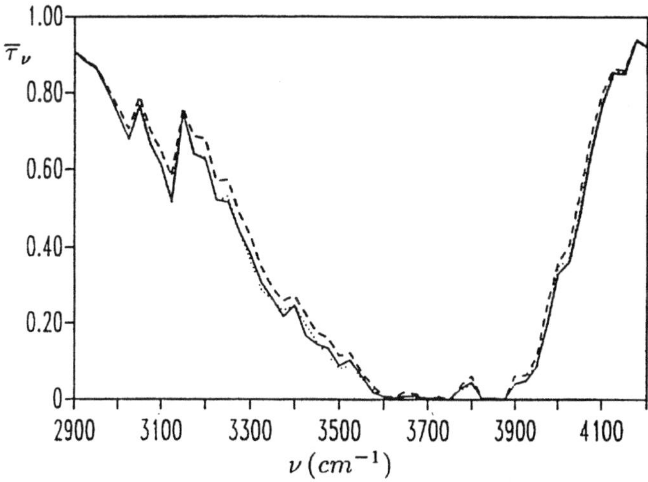

Fig. 13: H_2O — N_2 nonisothermal spectra computed with, — line by line, ------ C-K method, and ...... C-K F.G. method. T_1=300 K, T_2=1500 K, p_1=1 atm, p_2=1.34 atm, L_1=33 m, L_2=33 m.

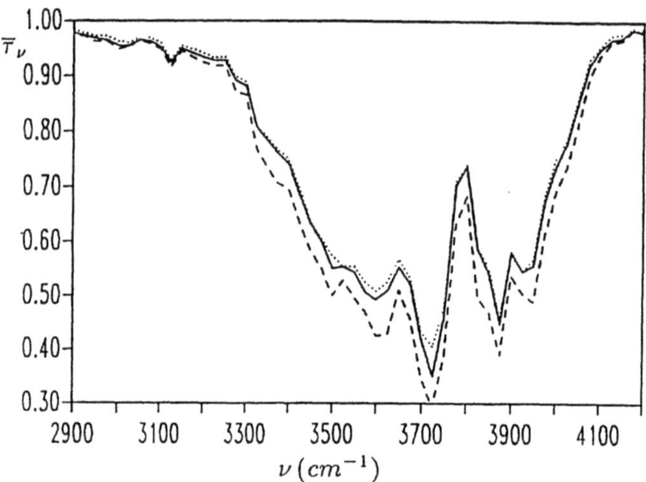

Fig. 14: H_2O — N_2 spectra computed with ____ line by line, ... C-K F.G. method, and ------ the statistical narrow band model with the exponential-tailed-inverse distribution and the Curtis-Godson approximation. T_1=1000K, T_2=1000K, p_1=0.25 atm, p_2=4.93 atm, L_1=33 m, L_2=0.33 m.

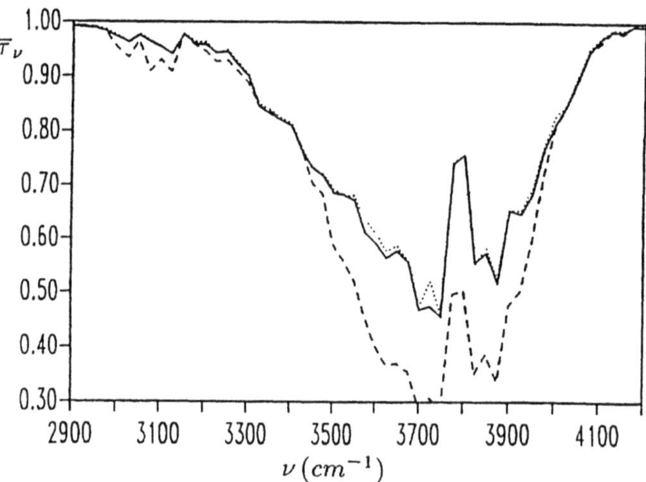

Fig. 15: same comparisons as in Figure 14 for steep pressure and temperature gradients. T_1=300K, T_2=1500K, p_1=0.1 atm, p_2=6.69 atm, L_1=33 m, L_2=0.33.

absorption coefficient due to a single line. The result was written in the form:

$$K_i(g_m) = \frac{\exp\left[1 - \exp\left(-\frac{h\nu_0}{kT}\right)\right] K_i^*(g_m)}{T^2\left[1 - \exp\left(-\frac{T_1}{T}\right)\right]\left[1 - \exp\left(-\frac{T_2}{T}\right)\right]\left[1 - \exp\left(-\frac{T_3}{T}\right)\right]} \tag{55}$$

corresponding to a simplified development of the partition function of CO_2. In (55), ν_0 may be approximated by the center of the spectral range $\Delta\nu$, T_1=1996.0K, T_2=960.0K, and T_3=3416.2K. The dependance of the remaining coefficient $K^*_i(g_m)$ was then approximated by:

$$K_i^*(g_m) = \exp\left(a_1 + a_2 \, \mathrm{Log}\, G + \frac{a_3}{T} + \frac{a_4}{G} + \frac{a_5}{TG} + \frac{a_6}{G^2} + \frac{a_7}{TG^2}\right) \qquad (56)$$

with

$$G = \frac{p}{p_s}\left(\frac{T_s}{T}\right)^\beta \qquad (57)$$

where β is the mean temperature dependence coefficient of Lorentz line widths, and p_s and T_s are standard pressure and temperature. The coefficients $(a_j)_{j=1,7}$, specified for each $\Delta\nu$, m and i, are now the model parameters, independent of temperature and pressure. This method yielded excellent results for CO_2 near 4.3 μm [56].

SOME CONCLUDING REMARKS

Statistical narrow-band models are powerful tool for most of the engineering applications involving radiative transfer. Their use leads to a very important reduction of computer times when compared to line by line calculations (the ratio is about $\Delta\nu/\delta\nu \simeq 2500$), and to very important prediction improvements when compared to non correlated radiative models. Line by line calculations remain the reference approach for validity studies and for the derivation of band model parameters, when the latter are not available from experiments.

The correlated-K method with the fictitious gas mixture idea (C-K F.G.) is an intermediate alternative between LBL and SNB models. We have shown that this approach is preferable to SNB models in the case of some specific applications including (i) scattering and (ii) relative variations of the line widths in the studied medium as high as ten or more (such variations may result only from strong pressure gradients since a temperature ratio of 10 leads generally to a line width ratio about 5).

However, the C-K F.G. method needs further developments and systematic derivation of the parameters for the most often encountered molecules such as H_2O and CO_2. On the other hand, the number of energy classes and the spectral range $\Delta\nu$ must be optimized in order to provide the minimum number of parameters and faster computations without any loss of accuracy.

Finally, in the SNB models, the range Δv is limited by blackbody intensity variations and by the fact that lines inside Δv must be statistically homogeneous. The last limitation is avoided in the C-K F.G. method [58].

REFERENCES

1. Bohren C.F., Huffman D.R.: Absorption and scattering of light by small particles. New York, Wiley (1983).

2. Herzberg G.: Molecular spectra and molecular structure. 1 Spectra of diatomic molecules. Van Nostrand, Princeton (1950). 2 Infrared and Raman spectra of polyatomic molecules. Van Nostrand, Princeton (1945).

3. Penner S.S.: Quantitative molecular spectroscopy and gas emissivities. Pergamon, London (1959).

4. Rothman L.S., Gamache R.R., Barbe A., Goldman A., Gillis G.R., Brown L.R., Toth R.A., Flaud J.M., Camy-Peyret C.: Appl. Optics 22 (1983) 2247.

5. Chedin A., Husson N., Scott N., Jobard I., Cohen-Hallalet I., Berroir A.: La banque de données GEISA, Description et logiciel d'utilisation. Lab. Météorologie Dynamique du CNRS, note interne LMD 108, palaiseau (1980).

6. Hartmann J.M., Rosenmann L., Perrin M.Y., Taine J.: Appl. Optics. 27 (1988) 3063.

7. Rosenmann L., Hartmann J.M., Perrin M.Y., Taine, J.: Appl. Optics. 27 (1988) 3902.

8. Papineau N.: La recherche Aerospatiale 3 (1987) 31.

9. Hottel H.C., Sarofim A.F.: Radiative transfer. McGraw-Hill, New York (1967).

10. Taylor P.B., Foster P.J.: Int. J. Heat Mass Transfer 17 (1974) 1591.

11. Taylor P.B., Foster P.J.: Int. J. Heat Mass Transfer 18 (1975) 1331.

12. Ludwig C.B., Malkmus W., Reardon J.E., Thompson A.L.: Handbook of infrared radiation from combustion gases. NASA SP-3080, Scientific and Technical Information Office, Washington D.C. (1973).

13. Smith T.F., Shen Z.F.: 20th Nat. Heat Transfer Conf., ASME paper n 81-HT-55 (1981).

14. Felske J.D.: Int. J. Heat Mass Transfer 25 (1982) 1849.

15. Grosshandler W.L.: Int. J. Heat Mass Transfer 23 (1982) 1447.

16. Edwards D.K., Menard W.A.: Appl. Optics 3 (1964) 621.

17. Felske J.D., Tien C.L.: J. Quant. Spectrosc. Radiat. Transfer 14 (1974) 35.

18. Edwards D.K.: Molecular gas band radiation. in Adv. in Heat Transfer. Academic Press, New York 12 (1976) 115.

19. Goody R.M.: Atmospheric radiation. Clarendon, Oxford (1964).

20. Golden S.A.: J. Quant. Spectrosc. Radiat. Transfer 8 (1967) 887.

21. Golden S.A.: J. Quant. Spectrosc. Radiat. Transfer 9 (1969) 1067.

22. Plass G.N.: J. Opt. Soc. Am. 48 (1958) 690.

23. Tien C.L.: Thermal radiation properties of gases. in Adv. in Heat Transfer. Academic Press, New York 5 (1968) 254.

24. Malkmus W.: J. Opt. Soc. Am. 57 (1967) 323.

25. Goldman A.: J. Quant. Spectrosc. Radiat. Transfer 8 (1968) 829.

26. Soufiani A., Hartmann J.M., Taine J.: J. Quant. Spectrosc. Radiat. Transfer 33 (1985) 243.

27. Malkmus W.: J. Opt. Soc. Am. 58 (1968) 1214.

28. Roney P.L.: J. Quant. Spectrosc. Radiat. Transfer 42 (1989) 169.

29. Lindquist G.H., Simmons F.S.: J. Quant. Spectrosc. Radiat. Transfer 12 (1972) 807.

30. Young, J.: J. Quant. Spectrosc. Radiat. Transfer 15 (1975) 483.

31. Young J.: J. Quant. Spectrosc. Radiat. Transfer 15 (1975) 1137.

32. Young J.: J. Quant. Spectrosc. Radiat. Transfer 18 (1977) 1.

33. Ferriso C.C., Ludwig C.B., Thomson A.L.: J. Quant. Spectrosc. Radiat. Transfer 6 (1966) 241.

34. Ludwig C.B.: Appl. Optics 10 (1971) 1057.

35. Kunitomo T., Osumi M.: J. Quant. Spectrosc. Radiat. Transfer 15 (1975) 345.

36. Phillips W.J.: J. Quant. Spectrosc. Radiat. Transfer 43 (1990) 13.

37. Bernstein L.S., Robertson D.C., Conant J.A.: J. Quant. Spectrosc. Radiat. Transfer 23 (1980) 169.

38. Young J.: J. Quant. Spectrosc. Radiat. Transfer 18 (1977) 29.

39. Hartmann J.M., Levi Di Leon R., Taine J.: J. Quant. Spectrosc. Radiat. Transfer 32 (1984) 119.

40. Taine J.: J. Quant. Spectrosc. Radiat. Transfer 30 (1983) 371.

41. Bernstein L.S.: J. Quant. Spectrosc. Radiat. Transfer 23 (1980) 157.

42. Gore J.P., Jeng S.M., Faeth G.M.: J. Heat Transfer 109 (1987) 165.

43. Gore J.P., Faeth G.M.: J. Heat Transfer 110 (1988) 173.

44. Soufiani A., Mignon P., Taine J.: 5th AIAA/ASME Thermophysics and Heat Transfer Conference, Seattle, HTD 137 (1990) 141-148.

45. Soufiani A.: 5th AIAA/ASME Thermophysics and Heat Transfer Conference, Seattle HTD 138 (1990) 45-51.

46. Coantic M., Simonin O.: J. Atmospheric Sciences 41 (1984) 2629.

47. Tsay S.C., Stamnes K., Jayaweera K.: J. Quant. Spectrosc. Radiat. Transfer 43 (1990) 133.

48. Özisik M.N.: Radiative transfer and interaction with conduction and convection. Wiley, New York (1973).

49. Kim T.K., Menart J.A., Lee H.S.: 5th AIAA/ASME Thermophysics and Heat Transfer Conference, Seattle HTD 137 (1990) 149-156.

50. Soufiani A., Taine J.: Int. J. Heat Mass Transfer 30 (1987) 437.

51. Zhang L., Soufiani A., Petit J.P., Taine J.: Int. J. Heat Mass Transfer 33 (1990) 319.

52. Zhang L., Soufiani A., Taine J.: Int. J. Heat Mass Transfer 31 (1988) 2261.

53. Malkmus W.: J. Quant. Spectrosc. Radiat. Transfer 40 (1988) 201.

54. Goody R.M., Yung Y.L.: Atmospheric radiation, Theoretical basis. Second Edition, Oxford University Press, Oxford (1989).

55. Goody R., West R., Chen L., Crisp D.: J. Quant. Spectrosc. Radiat. Transfer 42 (1989) 539.

56. Levi Di Leon R., Taine J.: Revue Phys. Appl. 21 (1986) 825.

57. Riviere P.: Etude d'une methode de modelisation du rayonnement des gaz par des gaz fictifs. Laboratoire EM2C, Internal report (1990).

58. Wang W.C., Shi G.Y.: J. Quant. Spectrosc. Radiat. Transfer 39 (1988) 387.

SPECTRAL GAS EFFECTS IN GAS-FIRED FURNACES

J.A. Wieringa, J.J.Ph. Elich, C.J. Hoogendoorn
Delft University of Technology, Delft, The Netherlands

ABSTRACT

In recent years calculations showed that the emissivity of the furnace refractory can have an influence on heat transfer, because of banded properties of gas radiation. Some aspects of this phenomenon have been studied using a spectral well-stirred furnace model of a combustion compartment. The influence of spectral lines in the emission bands has been examined. These lines have been found to increase spectral effects considerably. Further, a discrepancy that exists between grey and spectral calculations has been explained in terms of the definition of gas emissivity. The spectral properties of the refractory appear to raise the heat flux to the load when compared with a grey refractory.

LIST OF SYMBOLS

d	distance between two surfaces	m
E	hemispherical black-body emissive power	W/m^2
e_3	exponential integral (eq. 2)	
k	absorption coefficient	m^{-1}
ℓ	length	m
L_m	mean beam length	m
n	integer number	
q	net heat flux per m^2	W/m^2
r	dimensionless ratio of black body emissive powers	
t	integration parameter in eq. (2)	
W	equivalent line width	cm^{-1}
x	variable in eq. (2)	
δ	mean distance between spectral lines	cm^{-1}
Δ	amplitude of simulated, block-like spectral lines	
Δ^*	value of Δ in agreement with flux calculations according to eq. (7)	
ε	emissivity	
λ	wavelength	m
ν	wavenumber	cm^{-1}
ρ	reflectivity	
τ	transmissivity	

subscripts:

ν	wavenumber
1, 2	surface numbers
g	gas
gl	glass
i	internal Planck mean
r	refractory
t	Planck mean

superscripts

'   mean value as defined in eqs. (12) and (13)

     overlined quantities denote averaged values.

1. INTRODUCTION

In gas-fired high-temperature furnaces like glass melting furnaces, heat transfer from the flames to the load takes place mainly by radiation in the infrared. Convection and conduction contribute to less than 5% of the heat transfer. More knowledge of the radiative processes should therefore be used to optimize furnace design and the conditions under which the furnace is operated, with beneficial effects for the environment.

This study is concerned with effects on heat transfer that arise from the banded nature of gas radiation. In non-luminous gas flames, emission of radiation can to a large extent be ascribed to carbon dioxide (CO_2) and water vapour (H_2O). These gases emit and absorb in several wavelength bands in the infrared. In fact the bands consist of thousands of spectral lines, originating from the energy levels of the gas molecules. Between the emission bands, the gas is nearly transparent.

Due to the banded structure of gas radiation, the emissivity of the furnace refractory may have a noticeable influence on the heat flux to the load. Here we think of a geometry in which the flames are confined between a load surface and a refractory. When the emissivity of the refractory is low, radiative heat emitted by the flames will mostly be reflected at its surface. The wavelength distribution of the reflected flux is unchanged, so that much of the heat is absorbed in the flames again. A black refractory (emissivity equal to 1), however, absorbs all radiation and re-emits the energy apart from a small loss. Now this flux is distributed spectrally as a Planck curve (maximum around

1.6 μm) and therefore more of the energy can reach the load surface through the transparent regions in the spectrum. A dark furnace roof is therefore favourable. When the gas is grey, the effect described here will not take place. Then the refractory temperature will rise a few Kelvin if its emissivity is raised, but the increase of the losses through the refractory will be so small that the influence on heat transfer is negligible.

The effect of the refractory emissivity is shown in Figure 1. Raising the refractory emissivity (ε_r) increases the part of the flux that is absorbed by the glass via the roof. This causes the gas temperature to be lowered and consequently also the direct heat flux from the flames to the load will drop. The net result is a small increase of the total heat flux to the load. Also the fuel saving at constant heat flux can be calculated, but in this article we will regard only the first.

In recent years, several publications appeared on this subject. Docherty and Tucker [1,2], Elliston *et al* [3] and Alexander *et al* [4] studied the effects of spectral radiation in steel furnaces. Their theoretical predictions of fuel savings were up to 5 or 10%, if the refractory emissivity would be increased to 1. An experimental validation in a recuperative furnace [3], however, did not confirm the expectations. More recently a study has been performed by Wieringa *et al* [5,6]. It has been concluded that in regenerative glass furnaces the possible fuel saving is in the order of 2%, if the refractory emissivity can be raised to (almost) 1. Because the temperature of the combustion gases is lowered when the heat transfer is improved, the improvement will partly be compensated by the regenerator. The higher the regenerator efficiency, the lower the fuel saving. Furthermore, savings were found to be higher in smaller furnaces.

In this study, we are not so much concerned about the precise value of the increase of the furnace efficiency, but more about some theoretical problems. It can be noticed in Figure 1 that there is a discrepancy between the heat transfer of a grey gas and that of a non-grey gas, when the grey-gas emissivity is equal to the total (or Planck mean) emissivity of the non-grey gas. An explanation of this will be given in section 6.

Another problem has to do with the influence of spectral lines. As mentioned before, the effect of the refractory emissivity is caused by the non-greyness of the gas. As a first step, the gas spectrum can be represented by a number of block-like bands. Each of these bands, however, is not grey in itself but consists of many spectral lines, that will increase the influence of ε_r on heat transfer. A

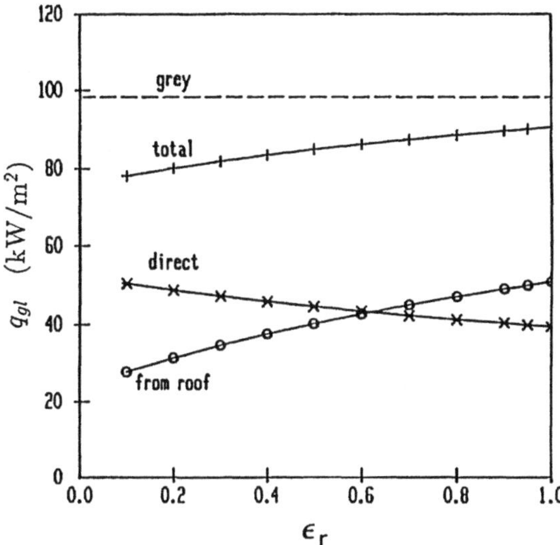

Fig. 1:   The net heat flux to the load (glass) q_{gl} as a function of the total emissivity of the refractory ϵ_r. Indicated are: the direct flux from flame to the glass, the flux that is either reflected or absorbed and re-emitted by the roof, and the total of these two. Further the flux from a grey gas with same (total) emissivity is indicated. Calculations were performed using the 15 bands spectrum.

quantitative estimation of this effect is given in section 5. Also some attention will be paid to the effect of the non-greyness of the refractory on heat transfer predictions.

2.   THE COMPUTATIONAL MODEL OF THE COMBUSTION CHAMBER

The glass furnace combustion chamber and the computational model have been discussed earlier [5,6]. For clarity, however, they will be described here shortly.

The combustion chamber (Figure 2) is bounded by the flat surface or the load (glass in our case), two walls at front and end, a curved roof and two planes of symmetry. The average height is 2.1 m, the length is 7.3 m and the width is 1.7 m. The total heat input is 3.8 MW, yielding a flame temperature of about 2000 K. The load temperature is taken 1773 K and is independent of the heat flux in our calculations. Heat losses through roof and walls are 5 kW/m^2. The refractory temperature is 1800 to 1850 K. At that temperature the refractory emissivity of a glass furnace was found to be 0.4 [5]. The load emissivity has been taken 0.8.

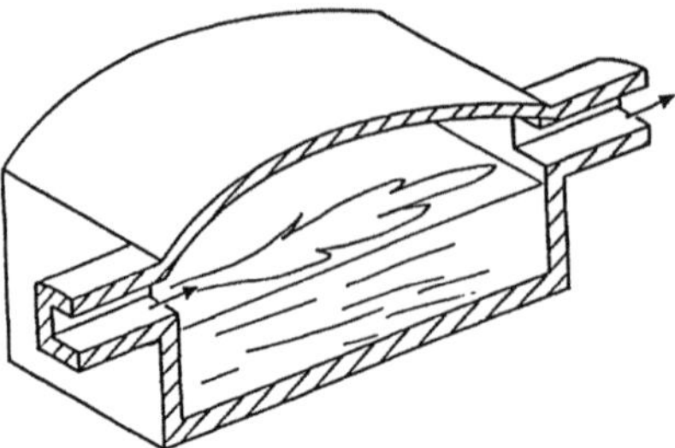

Fig. 2: Sketch of glass furnace section, of which the combustion compartment has been modelled.

The radiative transfer in this system has been modelled using a well-stirred furnace model, which assumes the gas volume and surfaces to have uniform properties. Despite its simplicity, it was found to give good predictions of the heat transfer, compared with a more complete simulation including flow and combustion modelling of the flame [7]. The spectral model can be regarded as a set of Oppenheim networks [8], where each network is used to calculate radiative transport in one spectral region. The networks are coupled by demanding that the temperatures of roof, load and gas are equal in all spectral bands. The total fluxes are obtained adding the spectral fluxes. Alternate solving of the networks and a heat balance of the combustion chamber yields values of the heat flux to the load, gas temperature, roof temperature, etc.

3. SPECTRA OF GAS AND REFRACTORY

In the introduction it was mentioned that the influence of the refractory emissivity ε_r is an effect of the variation of the gas emissivity with wavelength. Optically dense emission bands and transparent "windows" will lead to large effects of ε_r. In order to give an accurate estimation of the influence of ε_r the gas emissivity spectrum should be known well. Also the influence of spectral lines in the emission bands should be accounted for.

Basically two different gas spectra (Figure 3) have been used in the computations. The first has been taken from the study by Alexander *et al* [4] on spectral effects in steel furnaces. The total emissivity of this spectrum has been computed according to data by Taylor and Foster [9], whereas the relative contribution of each band has been found from Edwards [10]. The spectrum consists of 15 bands of which 7 are clear and the total emissivity at 2000 K is 0.14.

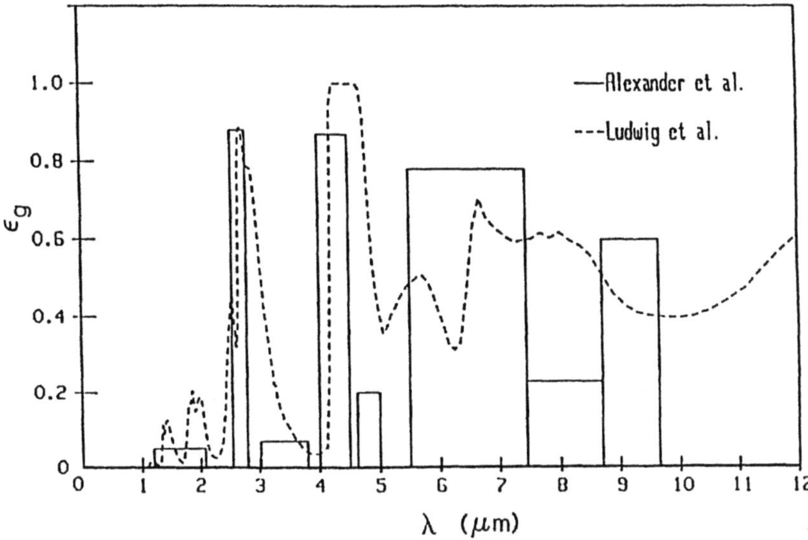

Fig. 3:   The two gas emissivity spectra, one with 15 bands as calculated by Alexander *et al* [4], and one consisting of 372 spectral regions, calculated with the narrow band model of Goody [11] and experimental data from Ludwig *et al* [14].

The other spectrum has been computed using the statistical narrow band model by Goody [11]. Assuming a random distribution of spectral lines, this model relates the transmission $\bar\tau_\nu$ in a wavenumber region around ν to the properties of a single line. For the line strength distribution function an exponential-tailed inverse distribution [12] has been taken. For Lorentz broadened lines, Soufiani *et al* [13] showed that this leads to very accurate results. We also took Doppler broadening into account. Ludwig *et al* [14] provide the necessary experimental data in wavenumber intervals of 25 cm^{-1}. Using these data, the spectrum which is also shown in Figure 3 has been computed. The spectrum consists of 372 spectral ranges, and has been calculated for a gas temperature of 2000 K, a mean beam length of 2.83 m and partial pressures of CO_2 and H_2O of 0.08 and 0.15 atm. At 2000 K the total emissivity of the spectrum is 0.22.

Differences between the two spectra may be explained partly from different temperatures, gas compositions, and geometries that were used in the computations of the spectra. More important, however, may be the fact that the 15 bands spectrum has been calculated from wide band data, whereas narrow band data were used for the other spectrum. It should be noticed that the largest discrepancies were found at high wavelengths, where the emissive power of the gas is small.

42

The influence of spectral lines has been examined by superposing a block-like line structure onto the narrow band spectrum. Each 25 cm^{-1} interval has been split up into three smaller bands. The emissivity in the middle interval has been increased by amplitude Δ, the other two have been lowered by $\Delta/2$, as in Figure 4. In some ranges, Δ is limited by the fact that $\varepsilon_{g,v}$ has to be between 0 and 1. This operation has a completely negligible influence on the total emissivity of the gas. Results will be shown in the next section. A similar calculational experiment has been carried out with more of these sub-intervals per 25 cm^{-1}, but it was found that the number of these sub-intervals in a specified wavenumber range is irrelevant. Then the amount of energy that is transferred in the refractory from intervals with high gas emissivity to intervals with low gas emissivity is not changed.

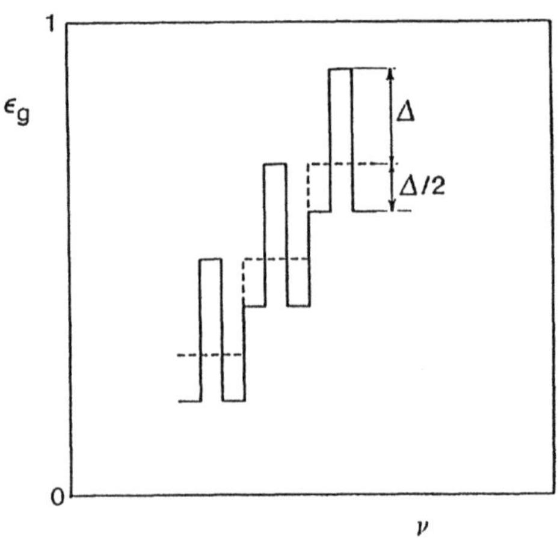

Fig. 4:   Construction of the spectral block-like lines (solid line) on the original gas emissivity spectrum (dashed line).

In many situations the refractory surfaces in furnaces are considered grey. In reality, of course, this idealized behaviour does not exist. Measurements of the spectral emissivity at room temperature of a typical sample of a glass furnace refractory have been carried out and have been described more extensively in a previous paper [5]. The spectrum, shown in Figure 5, shows a behaviour that has also been found in other materials like concrete: at high wavelengths ($\lambda=1/\nu$) the emissivity is high, but at lower wavelengths $\varepsilon_{r,\lambda}$ drops to much lower values. At high temperatures the Planck curve shifts towards these wavelengths, so that the total emissivity ε_r, is lowered. At a temperature of around 1850 K, which is found in glass furnaces, the measured spectrum yielded an ε_r-value of 0.4.

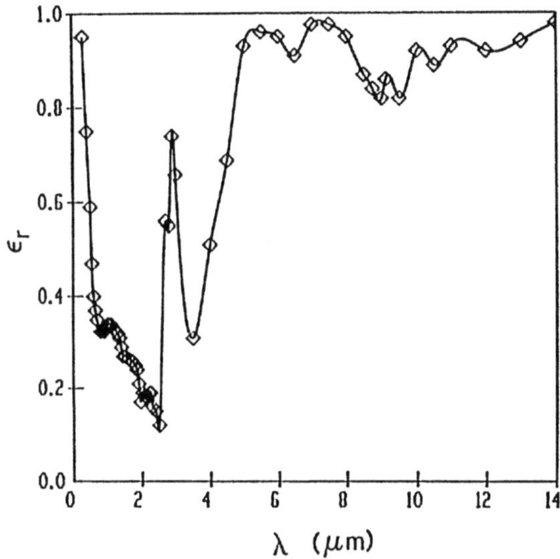

Fig. 5: The measured emissivity of the refractory at room temperature, as a function of wavelength.

4. SOME RESULTS

The first results, shown in Figure 1, have been obtained using the 15 bands spectrum from Alexander *et al* [4]. The gas spectrum using the data by Ludwig et al. [14] yields a similar pattern, but the difference in heat transfer between low and high refractory emissivity is smaller, mainly because the perfect "windows" of the 15 bands spectrum are less clear in the narrow band spectrum. When ε_r is increased from 0.4 to 0.95, the 15 bands spectrum gives a flux increase of 7.9%, whereas the narrow bands spectrum yields only 3.6%. This confirms our expectation that the gas emissivity spectrum is a very important parameter in the prediction of the effect that the refractory emissivity has on heat transfer.

The refractory spectrum of Figure 5 has been applied in the heat transfer calculations. It has been found that, because of the interaction between the gas and refractory spectra, the heat flux to the load comes out somewhat higher than when a grey refractory with ε_r=0.4 is taken. We find a heat flux of 98.2 instead of 97.1. This implies that the flux increase that can be achieved by increasing ε_r is somewhat lower than with a grey refractory.

The simulation of spectral lines has been found to have a considerable effect on the dependence of heat transfer on ε_r. This is shown in Figure 6, where the heat

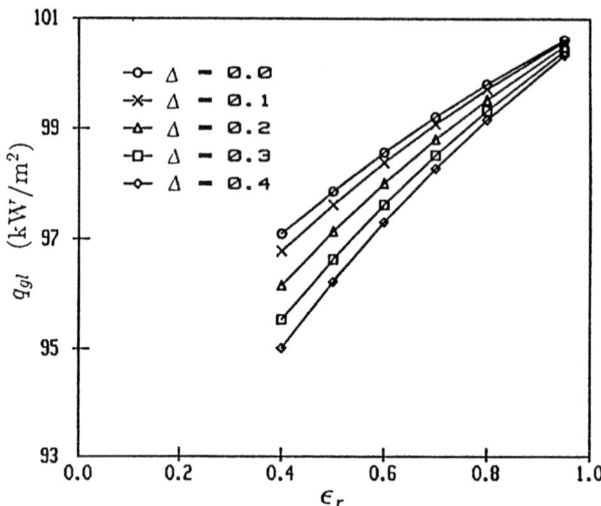

Fig. 6.   The effect of refractory emissivity on net heat transfer to the glass using the narrow band spectrum, for several values of amplitude Δ.

flux is shown versus ε_r for various values of Δ. When ε_r is low, the spectral lines hamper the heat transfer via the roof. Increasing ε_r diminishes this effect, but even when ε_r is equal to 1 the heat transfer still depends on Δ because some reflection of radiation at the glass surface ($\varepsilon_{glass}=0.8$) causes different heat losses through the roof. When the glass is black, all lines converge into one point for $\varepsilon_r=1$.

5.   THE VALUE OF Δ

The value of "line" amplitude Δ to be used in heat transfer calculations, has been derived from computations in a simplified geometry: an infinite plane slab of homogeneous gas, confined between two grey plates of uniform, known temperature, separated at distance d. We found that the effect of this plane geometry on heat transfer is very small: the simpler geometry leads to a heat flux to surface 2 (the load) which is about 2% lower than in our furnace geometry. Therefore we assume that the obtained value of Δ is also valid for the more complex geometry of the combustion chamber. In order to find Δ, we have compared the computations with the 25 cm^{-1} intervals split up into three smaller ranges, with an expression that accounts for the non-greyness in each spectral range.

In the plane geometry, the net *monochromatic* radiative heat transfer to surface 2 (the load) can be written as:

$$q_{2,\nu} = [\varepsilon_2(1 - 2\varepsilon_1 e_3(k_\nu d) - 4\rho_1\, e_3^2(k_\nu d))\,(E_g - E_2)$$

$$+ 2\varepsilon_1\varepsilon_2 e_3^2(k_\nu d)(E_1 - E_2)]/(1 - 4\rho_1\rho_2 e_3^2(k_\nu d)), \tag{1}$$

where ε_1 and ε_2 are emissivities of both surfaces, ρ_1 and ρ_2 are the reflectivities ($\rho = 1 - \varepsilon$), k_ν is the spectral absorption coefficient of the gas, and E_g, E_1 and E_2 are emissive powers of gas and surfaces. Further, e_3 is an exponential integral:

$$e_3(x) = \int_1^\infty \frac{e^{-xt}}{t^3}\, dt. \tag{2}$$

The first term in the numerator represents the gas radiation; the second is the flux from surface 1, attenuated by the gas absorption. Following the reasoning by Hottel and Sarofim [15] about the mean beam length, $2e_3(k_\nu d)$ can be approximated by $\exp(-k_\nu L_m) = \tau_\nu$, the gas transmissivity, with mean beam length $L_m = 1.76d$.

Then we have:

$$q_{2,\nu} = \frac{\varepsilon_2\left(1 - \varepsilon_1\tau_\nu - \rho_1\tau_\nu^2\right)(E_g - E_2) + \tau_\nu\,\varepsilon_1\varepsilon_2(E_1 - E_2)}{1 - \rho_1\rho_2\tau_\nu^2}. \tag{3}$$

The denominator of eq. (3) can be written as a series expansion, where each power of τ_ν represents a number of reflections at a surface. In order to obtain an average value of the heat flux in a spectral range around wavenumber ν containing a large number of spectral lines, we follow the calculation procedure of Goody [11] (pp. 152-154).

We have:

$$\tau_\nu^n = \tau_\nu(nL_m) = \exp(-k_\nu\, nL_m) = \exp(-nL_m \sum_i k_{\nu i}) \tag{4}$$

where each $k_{\nu i}$ is due to a single spectral line. For constant line intensity a statistical analysis yields a mean value of the transmissivity:

$$\bar{\tau}_\nu(nL_m) = \exp\left(-\frac{1}{\delta}\int_{-\infty}^\infty \{1 - \exp(-k_\nu\, nL_m)\}\, d\nu\right) \tag{5a}$$

46

or

$$\bar{\tau}_v(nL_m) = \exp(-W(nL_m)/\delta). \tag{5b}$$

Here δ is the mean line distance and W is the equivalent width of one line. In eq. (5a) integration is performed over the absorption profile of one spectral line. More generally, for a distribution of line strengths we get:

$$\bar{\tau}_v(nL_m) = \exp(-\overline{W}(nL_m)/\delta). \tag{6}$$

We see that $\bar{\tau}_v(nL_m)$ is not equal to $(\bar{\tau}_v(L_m))^n$. Therefore the series expansion of eq. (3) becomes, dropping the subscript v for simplicity:

$$
\begin{aligned}
\bar{q}_2 = (E_g - E_2) \Bigg\{ &\varepsilon_2 \Big[1 + \rho_1\rho_2\bar{\tau}(2L_m) + (\rho_1\rho_2)^2\bar{\tau}(4L_m) + \ldots \Big] \\
&- \varepsilon_1\,\varepsilon_2 \Big[\bar{\tau}(L_m) + \rho_1\rho_2\bar{\tau}(3L_m) + (\rho_1\rho_2)^2\bar{\tau}(5L_m) + \ldots \Big] \\
&- \rho_1\,\varepsilon_2 \Big[\bar{\tau}(2L_m) + \rho_1\rho_2\bar{\tau}(4L_m) + (\rho_1\rho_2)^2\bar{\tau}(6L_m) + \ldots \Big] \Bigg\} \\
&+ (E_1 - E_2)\,\varepsilon_1\,\varepsilon_2 \\
&\qquad \cdot \Big\{ \bar{\tau}(L_m) + \rho_1\rho_2\bar{\tau}(3L_m) + (\rho_1\rho_2)^2\bar{\tau}(5L_m) + \ldots \Big\}
\end{aligned}
\tag{7}
$$

This expression takes the effects of spectral lines with reflections between grey surfaces into account. For each number of mean beam lengths in eq. (7) a narrow band spectrum has been calculated using the data from Ludwig [14]. In practice 4 or 5 terms in each series were enough to obtain a reasonable accuracy.

Taking, for instance, $\rho_1{=}1$ and $\rho_2{=}0$ in eq. (7), only

$$\bar{q}_2 = (1 - \bar{\tau}(2L_m))\,(E_g - E_2) \tag{8}$$

is left, which is just what would be expected for a column of gas with double mean beam length.

A comparison between expression (7) and the model with each 25 cm^{-1} interval split up (with values of Δ ranging from 0 to 0.4) is shown in Figure 7. For these curves the fixed temperatures T_1, T_2 and T_g were 1850, 1773 and 2000 K respectively, and ε_2 was 0.8. Let Δ^* be the value of Δ for which the heat fluxes to surface 2 from both representations are equal. Then the comparison yields values of Δ^* between 0.30 and 0.38 for $L_m{=}2.83$ m and $\varepsilon_2 \geq 0.5$, lightly depending on values of ε_1 and ε_2. If either ε_1 or ε_2 is 1, we always find 0.38. Δ^* depends on the amount of radiation that is reflected at least once between the surfaces. If

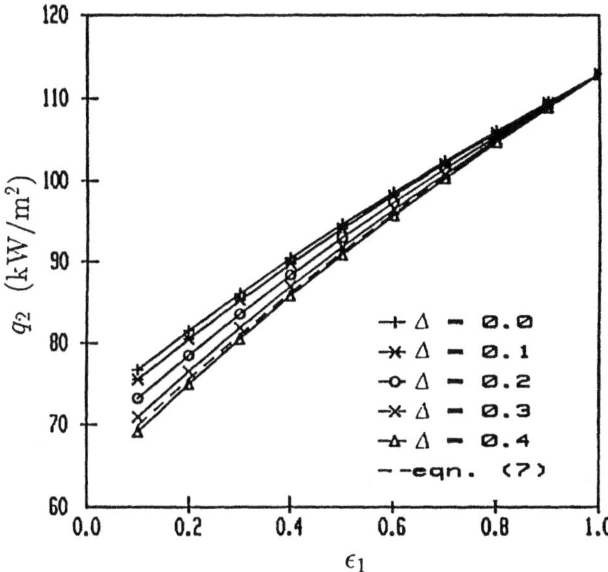

Fig. 7: The effect of the refractory emissivity ε_1 on net heat transfer to the load using the narrow band spectrum, in the plane geometry and keeping temperatures fixed, for several values of Δ. Use of eq. (7) yielded the dashed line. Here load emissivity ε_2 was 0.8.

$\varepsilon_2=1$, gas radiation can be reflected only once, and then Δ^* is independent of the quantity $r=\dfrac{E_g-E_2}{E_1-E_2}$. If both ε_1 and ε_2 are unequal to 1, the parameter r has some influence, but only lightly. For the temperatures mentioned r is 3.3, but when for instance r is -1.3, Δ^* is still around 0.3. There is also a small dependence on the mean beam length L_m. Dividing L_m by 2 reduces Δ^* from 0.38 to 0.36.

It is also possible to consider the block-line structure as a two-grey-gas representation of the emissivity in each 25 cm^{-1} interval. Putting:

$$\bar{\varepsilon}_{g,v}(\ell) = a_1\left(1 - e^{-k_{1,v}\ell}\right) + a_2\left(1 - e^{-k_{2,v}\ell}\right) \tag{9}$$

where ℓ is measure of length, we have:

$$a_1 = \frac{1}{3} \qquad\qquad a_2 = \frac{2}{3}$$

$$k_{1,v} = -\frac{\ln(1 - \bar{\varepsilon}_{o,v} - \Delta_v^*)}{L_m}$$

$$k_{2,v} = -\frac{\ln(1 - \bar{\varepsilon}_{o,v} + \Delta_v^*/2)}{L_m} \tag{10}$$

where $\bar{\varepsilon}_{0,\nu}$ is the gas emissivity before the line structure has been imposed. It is easy to see that for $\ell = L_m$, we find $\bar{\varepsilon}_{g,\nu} = \bar{\varepsilon}_{0,\nu}$. By equating eq. (9) for $\ell = 2L_m$ with a narrow band spectrum for double mean beam length, the spectral value Δ_ν^* has been found. The spectrum of Δ_ν^* is shown in Figure 8, together with the emissivity spectrum. The total value (i.e. integrated over the spectrum) for Δ^* of 0.38 that we found before, appears high but it must be reminded that the construction of the block-lines already accounted for the fact that the amplitude was limited when $\bar{\varepsilon}_{0,\nu}$ was near 0 or 1. When Δ is high, this limitation is important in large parts of the spectrum.

Having found these values for Δ^*, we conclude that in the furnace combustion chamber the heat flux enhancement that arises when ε_r is raised, is amplified by spectral lines. Figure 6 shows that for Δ is almost 0.4, the flux enhancement is about 1.5 times as large as when the spectral intervals are treated as grey ($\Delta=0$).

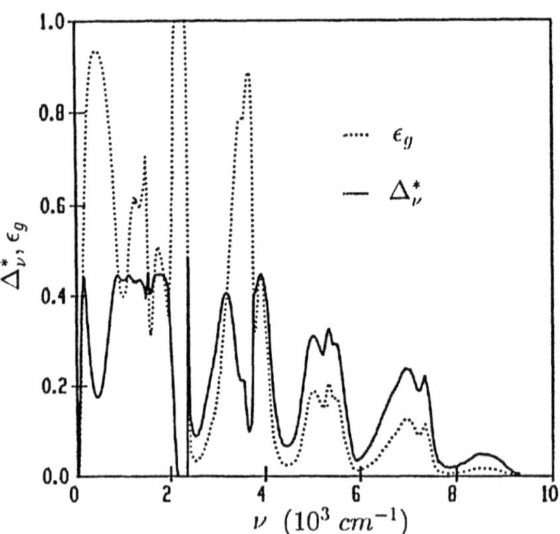

Fig. 8: Spectra of gas emissivity (narrow band spectrum) and calculated values of Δ_ν^*.

6. THE DISCREPANCY BETWEEN GREY AND SPECTRAL CALCULATIONS

As mentioned before in the introduction, the heat transfer that is found in calculations accounting for spectral gas emission is always lower than when the gas is treated as grey. Here the emissivity in the grey case is taken equal to the total emissivity of the spectral gas. Even when both the refractory and the glass surface are taken black, this discrepancy that has been found before by Alexander *et al* [4] does not disappear (Figure 1). In this case the difference is 8% of the spectral heat flux using the 15 bands spectrum, and 6% with the narrow band spectrum. (The influence of spectral lines does disappear when glass and refractory are black, but that is because the wavenumber width of these lines is much smaller than of the emission bands. In one 25 cm^{-1} interval the Planck function can be approximated by a linear function with ν. Then the change in heat transfer by increasing ε_g in one sub-interval is compensated by lowering ε_g in the neighbouring sub-intervals. For complete bands, this is not the case.)

For simplicity, the case with $\varepsilon_{gl} = \varepsilon_r = 1$ will be considered. Then eq. (7) reduces to:

$$\bar{q}_{gl,\nu} = \bar{\varepsilon}_{g,\nu} (E_g - E_{gl}) + \bar{\tau}_\nu (E_r - E_{gl}). \tag{11}$$

So the heat transfer to the glass can be characterized by two quantities:

$$\varepsilon'_{g\text{-}gl} = \frac{\int_0^\infty (E_{g,\nu} - E_{gl,\nu})\bar{\varepsilon}_{g,\nu}\, d\nu}{\int_0^\infty (E_{g,\nu} - E_{gl,\nu})\, d\nu}, \tag{12}$$

$$\varepsilon'_{r\text{-}gl} = \frac{\int_0^\infty (E_{r,\nu} - E_{gl,\nu})(1 - \bar{\tau}_\nu)\, d\nu}{\int_0^\infty (E_{r,\nu} - E_{gl,\nu})\, d\nu}. \tag{13}$$

The so defined emissivities appear to be much lower than the usual total or Planck mean gas emissivity:

$$\varepsilon_t = \frac{\int_0^\infty E_{g,\nu}\bar{\varepsilon}_{g,\nu}\, d\nu}{\int_0^\infty E_{g,\nu}\, d\nu}. \tag{14}$$

For instance, $\varepsilon'_{g\text{-}gl}$ was found to be 0.098 at the appropriate gas and glass temperatures, and using the 15 bands spectrum, whereas the Planck mean yielded 0.136. $\varepsilon'_{r\text{-}gl}$, which could better be regarded as an absorptivity, is 0.11. Similarly, such quantities could be defined to describe the flux to the refractory.

The effective emissivities defined here yield the correct heat flux in the case of a black confinement. Although the Planck mean emissivity is a useful measure for radiative properties of gases, it is not a good quantity to characterize the heat transfer from strongly non-grey flames to enclosures, as soon as the temperatures of the gas and its surroundings become of comparable magnitude. When the temperatures approach each other, ε' becomes equal to the internal Planck mean emissivity, which is defined [16] as:

$$\varepsilon_i = \frac{\int_0^\infty \varepsilon_{g,\nu} \frac{\partial E_{g,\nu}}{\partial T_g} \, d\nu}{\int_0^\infty \frac{\partial E_{g,\nu}}{\partial T_g} \, d\nu} .$$

(15)

If, finally, glass and refractory are not black, reflections at these surfaces will complicate the situation. Although the effective or internal emissivity will still give a better heat transfer prediction, it is not easily possible to account for all interactions by a grey gas approximation.

One might think that the observed difference between calculations with spectral and grey refractory, mentioned in section 4, could have an explanation similar to the one given here for the spectral gas. However, this appears not to be the case. Taking gas and glass grey, a difference in the heat flux between grey and spectral refractory has only been found when the refractory temperature is fixed. In the present heat transfer computations the *heat loss* through the refractory is (virtually) constant and consequentially there is no effect on the heat balance of the combustion chamber and on the heat transfer to the load. Therefore, the effect on heat transfer of the spectral emissivity of the refractory mentioned earlier must be explained from the interaction between that spectrum and the spectral emission and absorption of the gas.

7.   CONCLUSIONS

In this article some aspects of spectral radiation effects have been discussed. Using a spectral well-stirred furnace model, it has been found that, because of the banded nature of the gas emissivity, the refractory emissivity influences the heat transfer to the load. The heat flux enhancement or fuel saving that can be obtained by raising the refractory emissivity is determined by a number of parameters, of which the gas spectrum is a very important one. This has been shown by using two different gas spectra. A gas spectrum with some transparent wavelength regions and well-defined bands will lead to larger effects of the refractory emissivity ε_r than a gas spectrum with less perfectly transparent regions.

In order to examine the influence of spectral lines, a block-like wave form has been superposed onto the gas emissivity spectrum. Increasing the amplitude Δ

these "lines" showed that the spectral lines will substantially enhance the importance of the refractory emissivity. Comparing the heat fluxes using the block-like lines and from an expression that takes the influence of spectral lines in each spectral interval into account, a value of Δ of 0.30 to 0.38 has been found. Over a reasonable range of surface emissivities, mean beam lengths and temperatures, the value of Δ was found to change only slightly, so that the superposed structure appears to be an effective way to describe spectral line effects. The lines were found to enhance the influence of ε_r on the heat transfer to the load by a factor 1.5.

The discrepancy that has been found between grey and spectral heat transfer calculations has been explained from the definition of gas emissivity. The Planck mean emissivity, which is normally used to characterize a gas, has been found not to describe the heat transfer correctly when the gas has banded radiative properties and temperatures of confinement and gas are of the same order of magnitude. Using an effective emissivity, defined in this article, or the internal Planck mean emissivity, the differences between grey and spectral calculations can be eliminated (when the surfaces are black) or reduced considerably.

Finally, it has been shown that the spectral properties of the refractory also influence the heat transfer. A measured refractory emissivity spectrum yielded a slightly higher heat flux than from a grey roof. Increasing ε_r in a real furnace will therefore have a slightly smaller effect than in an idealized one with a grey roof.

ACKNOWLEDGEMENT

This study was performed under contract with the Netherlands Agency for Energy and Environment (NOVEM).

REFERENCES

1. Docherty P., Tucker R.J.: The influence of wall emissivity on the thermal performance of furnaces in Proc. Int. Gas Research Conf., ed. T.L. Kramer, Toronto (1986).

2. Docherty P., Tucker R.J.: The influence of wall emissivity on furnace performance. J. of the Inst. of Energy (1986) 35-37.

3. Elliston D.G., Gray W.A., Hibberd D.F., Ho T-Y., Williams A.: The effect of surface emissivity on furnace performance. J. of the Inst. of Energy (1987) 155-167.

4. Alexander I., Gray W.A., Hampartsoumian E., Taylor J.M.: Surface emissivities of furnace linings and their effect on heat transfer in an enclosure in Proc. 1st European Conf. on Industrial Furnaces and Boilers, Lisbon (1988).

5. Wieringa J.A., Elich, J.J.Ph., Hoogendoorn C.J., The spectral emissivity of glass furnace roofs and its effect on heat transfer in Proc. Conf. on Ceramics in Energy Applications, Sheffield (1990).

6. Wieringa J.A., Elich, J.J.Ph., Hoogendoorn C.J.: Spectral effects of radiative heat-transfer in high-temperature furnaces burning natural gas. J. of the Inst. of Energy (sept. 1990) 101-108.

7. Post L.: Modelling of flow and combustion in a glass melting furnace, Ph.D. thesis, Delft University of Technology (1988).

8. Oppenheim A.K.: Radiation analysis by the network method. Trans. ASME 78 (1956) 725-735.

9. Taylor P.B., Foster P.J.: The total emissivities of luminous and non-luminous flames. Int. J. Heat Mass Transfer 17 (1974) 1591-1605.

10. Edwards D.K.: Absorption by infrared bands of carbon dioxide gas at elevated pressures and temperatures. J. Opt. Soc. Am. 50 (1960) 617-626.

11. Goody R.M.: Atmospheric Radiation I, theoretical basis. Clarendon Press, Oxford (1964).

12. Malkmus W.: Random Lorentz band model with exponential-tailed S^{-1} line-intensity distribution function. J. Opt. Soc. Am. 57(3) (1967) 323-329.

13. Soufiani A., Hartmann J., Taine J.: Validity of band-model calculations for CO_2 and H_2O applied to radiative properties and conductive-radiative transfer. J. Quant. Spectrosc. Radiat. Transfer 33(3) (1986) 243-57.

14. Ludwig C.B., Malkmus W., Reardon J.E., Thomson J.A.L.: Handbook of infrared radiation from combustion gases. NASA SP-3080, Washington D.C. (1973).

15. Hottel H.C., Sarofim A.F.: Radiative Transfer. McGraw-Hill, New York (1967).

16. Edwards D.K.: Molecular gas band radiation in Advances in Heat Transfer 12. Edited by Irvine T.F., Hartnett J.P.. Ac. Press N.Y. (1976) 115-193.

EXPERIMENTAL AND THEORETICAL STUDIES ON THE RADIATIVE PROPERTIES OF CO_2 AT HIGH TEMPERATURE: APPLICATION TO DIAGNOSTIC

Grisch F., Coppalle, A., Vervish, P.
UA 230 CNRS Faculté des Sciences
76134 Mt. St. Aignan
France

ABSTRACT

The 4.3 µm molecular band of CO_2 was studied. Measurements were carried out on a high temperature flame (T > 1800 K). Statistical model and line by line calculations were used to predict the CO_2 emissivities. These theoretical values will be shown to agree well with the experimental ones. A new method is proposed in order to make temperature and concentration measurements from emission spectra alone.

INTRODUCTION

During hydrocarbon fuel combustion, large amounts of CO_2 are produced. So this molecule can be detected by optical means. If pure oxygen is used as the oxidant, the gas temperature may be very high. The main purpose of this work is to show the possibility of temperature and concentration measurements in very hot gases (T > 2000 K) using the radiation emitted by the CO_2 present in the jet.

The most important molecular band of CO_2 is located at 4.3 µm. Two spectral regions are of particular interest: the hot band region which lies between 4.4 and 4.8 µm, and the band head at 4.18 µm. The first zone is used to measure the temperature of the gases by the standard emission-absorption method. It will be shown that a comparison between the experimental spectrum and the calculated emissivities, using the Bernstein statistical model, give a good estimation of the CO_2 concentration in the jet.

Up to now, few measurements have been made in the second zone, the band head. This spectral region is interesting because it is not disturbed by the absorption of cold layers. These, for example, may exist between the jet and the detector. It is not possible to apply the statistical model in this region because the line overlap is too high. A line by line method is used and the comparison with experiments shows a good agreement. This has provided us with the means to develop a new method for temperature and CO_2 concentration measurements. It will, therefore, be shown that this spectral region is interesting for diagnostic predictions.

The measurements are performed at different temperatures between 1800 and 2800 K. This range is obtained in a laminar boundary layer developed along a vertical plate which is placed above a burner fed by pure oxygen and natural gas.

Finally, numerical studies of the parabolic laminar flow are made. The mass, momentum and energy conservation equations are resolved using standard method. The calculated temperatures will be presented and compared with the experimental values in the boundary layer.

EXPERIMENTAL METHODS AND MEANS

INFRARED SPECTROSCOPY

Both emission and absorption measurements were carried out and to do that a spectroscopic device was built. This is shown in Figure 1. The spectroscopic measures are standard and are briefly described. Mirrors and fluor lenses are used to carry out the optical configurations, the source S with the center of the burner for the absorption measurements and this center with the entrance slit of the monochromator for the emission measurements. During the emission measurements the source is blocked. The air is pumped out of the boxes in which the optical system is located in order to avoid an absorption of the signal by the air. A black body furnace is put instead of the burner and is used to calibrate the signal in correct units. The temperature is given by:

$$\frac{1}{T} = \frac{1}{1.439\,\omega} \log\left\{ 1 + \frac{\varepsilon_\omega L^0_{\omega,Tb}}{L^0_\omega} \left(\exp\left[\frac{1.439\,\omega}{Tb}\right] - 1 \right) \right\}$$

where the subscript refers to the black-body furnace, ω is the wavenumber, L_ω the emitted signal and ε_ω the spectral emissivity determined from an absorption measurement. In order to obtain the experimental accuracy, several measurements were made at the same place in the flame. The dispersion of the results is about 50 K, this value is assumed to be the experimental error.

BURNER AND PLATE

The burner is similar to that used in previous studies [1,2], it is designed to produce a high temperature methane-oxygen diffusion flame. As was deduced from these studies, the burner system produces an exhaust jet of uniform temperature and uniform mole fractions of the major species (H_2O, CO_2, CO). It is used for gas mixtures near the stochiometric proportion (3.5 m^3/h and 11 m^3/h for CH_4 and O_2 flow-rates respectively), and runs in a vertical position.

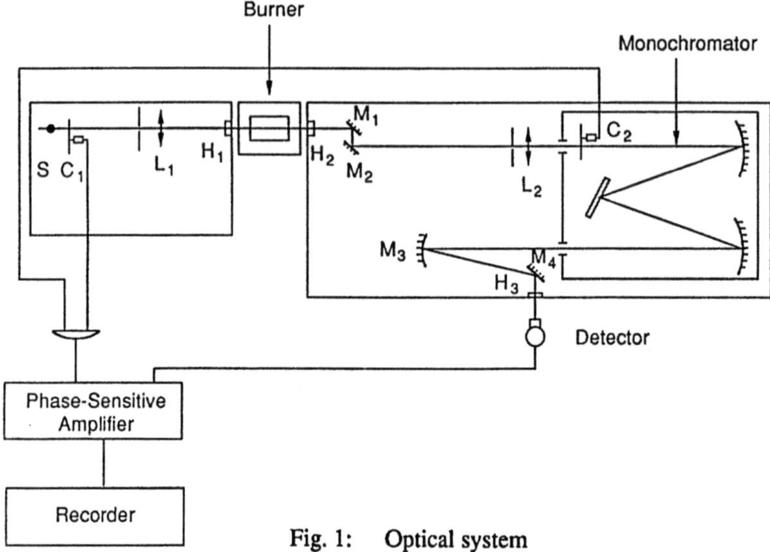

Fig. 1:   Optical system

The plate is made of two parallel stainless steel pieces. A water cooling operates inside. The temperature plate is constant and controlled by a chromel-alumel thermocouple. The temperature value is equal to 374 K, thus avoiding water condensation on the plate.

The reference axes used in this work are shown in Figure 2, and the X axis has been taken in the flow direction. The plate's dimensions are 160 mm and 120 mm in the X and Z directions respectively. Those of the burner exit are 60 and 100 mm.

After several centimeters the jet becomes unstable. This is mainly due to the interaction of the flame with the air drag. Consequently, only regions near the burner exit were analysed to avoid this effect.

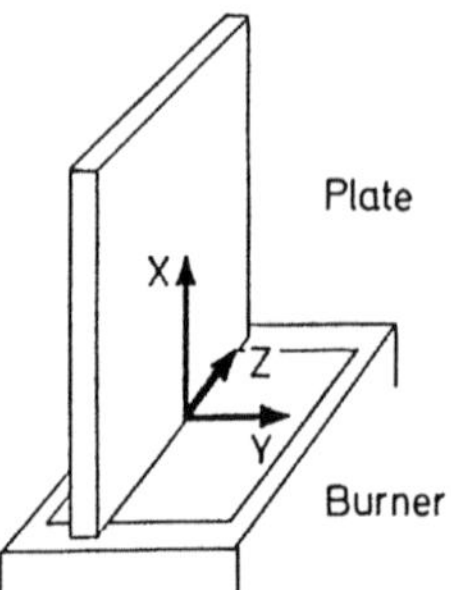

Fig. 2:   Reference axes used

EMISSIVITY RESULTS

Using our optical device the 4.3 μm CO_2 band was analysed at different positions in the boundary layer. In Figure 3 we can see the spectral emissivity values at 50 mm above the burner exit and for two different points in the flame, corresponding to near and far from the plate.

For wavenumbers less than 2300 cm^{-1}, the emissivity is mainly due to the hot bands of CO_2, which are active at high temperature. It was in this spectral range that the temperatures were determined. The unresolved lines located at 2100 cm^{-1} correspond to the emissions of the v2 and v3-v1 H_2O bands.

In the 2380-2400 cm^{-1} region, we can see two sharp peaks which are the band head of the $00^01 \rightarrow 00^00$ and $01^11 \rightarrow 01^10$ transitions (according to Herzberg [3] notation). The rotational transitions, lying in this spectral range, correspond to high rotational energy levels and therefore are interesting because their emission spectrum is not disturbed by the absorption of cold layers. The same remark could also have been made about the first spectral region described just above.

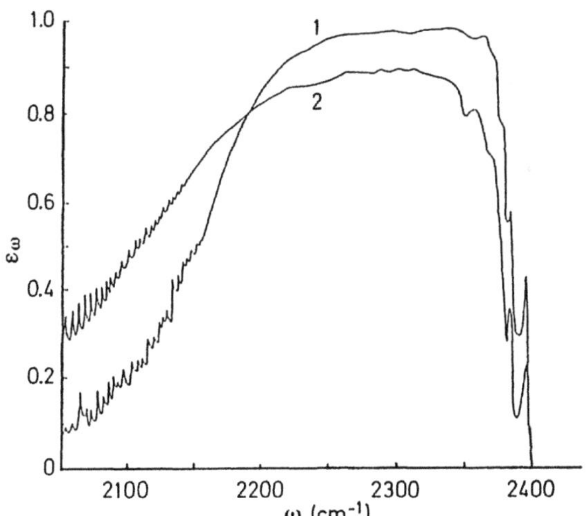

Fig. 3:   Spectral emissivity of CO2: X=50 mm; 1 Y=1 mm; 2 Y=10 mm.

MEASUREMENTS IN THE BOUNDARY LAYER

TEMPERATURE RESULTS

With our spectroscopic method, we were able to determine the temperatures in the boundary layer. This was done at two heights above the burner exit, X = 24 mm and 50 mm. The results are shown in Figure 4, where the experimental error is also reported. As we can see, the boundary layer is thin. The thickness is about 5 and 10 mm for the two points. The temperature value far from the plate is equal to 2850 K. Other measurements, not reported here but carried out at Y=15, 20 and 25 mm, gave the same result. These confirm our first assumption about the flame homogeneity outside the boundary layer.

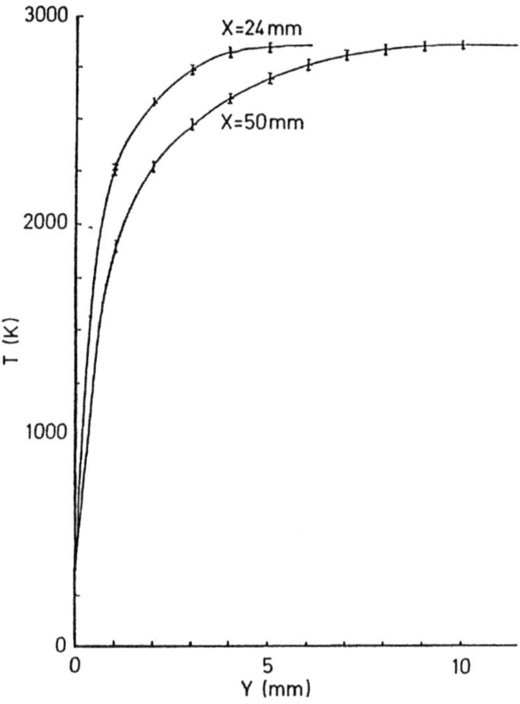

Fig. 4: Temperature profile in the boundary layer

CO_2 CONCENTRATION

In order to obtain the CO_2 concentrations in the jet, the experimental spectrum is compared to a theoretical one. The concentration value is that which gives the smallest discrepancy between the experimental and theoretical spectra.

We have used the statistical model to predict the CO_2 emissivity. This theory, which is a particular way to calculate the emissivity of gases, was developed a few years ago. Its description is out of the scope of this paper, and complete information is available in the literature [4-10]. The most important point to remember, is that the calculated emissivity is the average of the true spectrum over a narrow spectral width. We have chosen the Bernstein method [11] which is the most recent work carried out on CO_2. The temperature is needed in order to perform the emissivity calculations. We take the experimental one.

Examples of comparison between measurements and theoretical results are shown in Figure 5, at 50 mm above the burner. Only the values at two points, near and far from the plate, are reported. For wavenumbers less than 2300 cm^{-1}, the agreement is good. The best fit is given by a CO_2 concentration equal to 0.18 (in mole fraction), and at the two positions. The concentration does not seem to vary in the boundary layer. Other measurements, performed between these two positions but not reported here, give values which confirm this assumption. As we can see in Figure 5, two theoretical spectra are shown at the same point, Y=1 mm near the plate; these results give us the sensitivity of the method, and the error which is about 10%.

For wavenumbers greater than 2300 cm^{-1}, the observed discrepancy is easily explained. At the band head (2380-2400 cm^{-1}), the Bernstein method is not valid because the space between each line is very small, and the lines overlap. The 2300-2380 cm^{-1} interval is the region where the absorption of the cold layers surrounding the jet is high, and the experimental values are not much reliable.

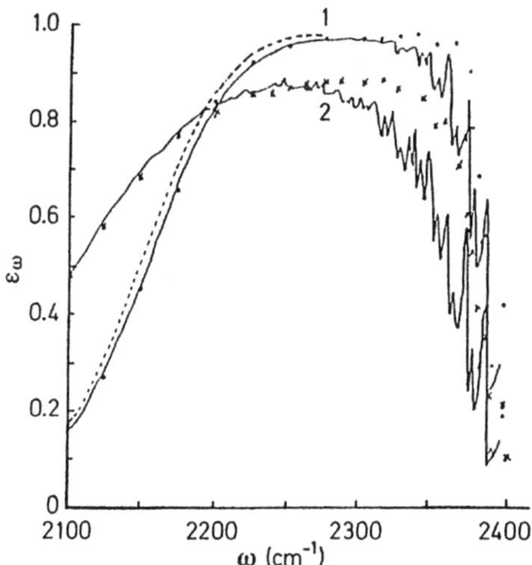

Fig. 5: CO_2 experimental emissivities and results of the statistical model X=50 mm.

1	Y=1 mm T = 1880 K		2	Y=10 mm T=2860 K
*	experimental		*	experimental
______	statistical model XCO$_2$=0.18		------	statistical model XCO$_2$=0.18
------	statistical model XCO$_2$=0.20			

In order to obtain the H_2O concentration in the layer, the same method was also applied. The experimental values of H_2O emissivity and the statistical model results may be seen in the publication by Grisch [19]. The conclusion of these studies on H_2O is the same as that which was given for CO_2; that is to say, the H_2O concentration does not vary in the layer and is equal to 0.50.

BAND HEAD

As was noted above, our experimental device enables us to measure the emissivities at the two band heads, lying in the 2380-2400 cm^{-1} spectral region. They correspond to the returning portion of the R branch of the $00^01 \rightarrow 00^00$ and $01^11 \rightarrow 01^10$ vibrational transitions. In this part of the spectrum, the statistical model cannot be applied and we use the line by line method. This has been developed recently at high temperatures [12,13], and it gives the monochromatic emissivity value (while the statistical model gives a mean value over a narrow spectral width).

A lot of spectroscopic data is needed to perform the emissivity calculations. In the case of the CO_2 band heads, the energy levels and line strengths are provided by Rothmann et al [14,15], the line widths are assumed to be independent of the rotational quantum number and are taken from the works of Burch et al [16], Taine [12] and Papineau [13]. A line shape is supposed to be Lorentzian. More details on the numerical procedures are given by Grisch [17]. This method and the measured temperatures and CO_2 concentrations therefore make it possible to calculate the monochromatic emissivity of CO_2. Because of the response of the monochromator, the theoretical spectrum has to be convoluted with a slit function of 2 cm^{-1} spectral width. After this, the measured and calculated spectra can be compared.

Figures 6 and 7 show examples carried out at 50 mm above the burner and at different points in this section. For wavenumbers greater than 2385 cm^{-1}, the theoretical values agree well with the experimental ones, and we can assume that our numerical procedure is valid and available at high temperature (> 1800 K). For wavenumbers less than 2385, the discrepancy is easily explained by the atmospheric absorption which has not been taken into account.

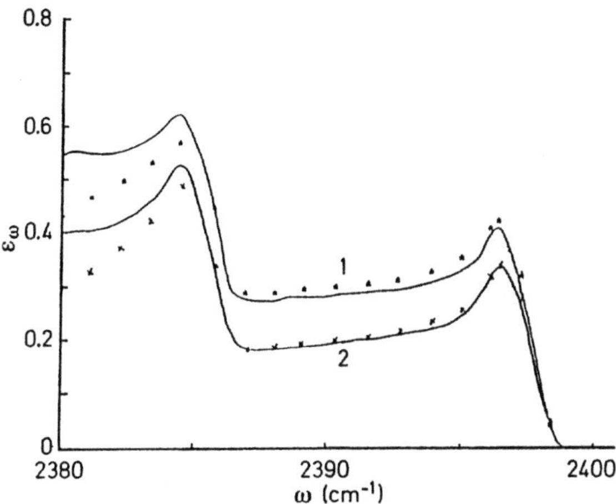

Fig. 6: Band head measurements and results of the line by line method spectral resolution 2 cm^{-1}*.8 cm^{-1} X=50 mm.

1	T = 1880 K XCO_2=0.18		**2**	T = 2260 K XCO_2=0.18
*	experimental		*	experimental
------	line by line model		------	line by line model

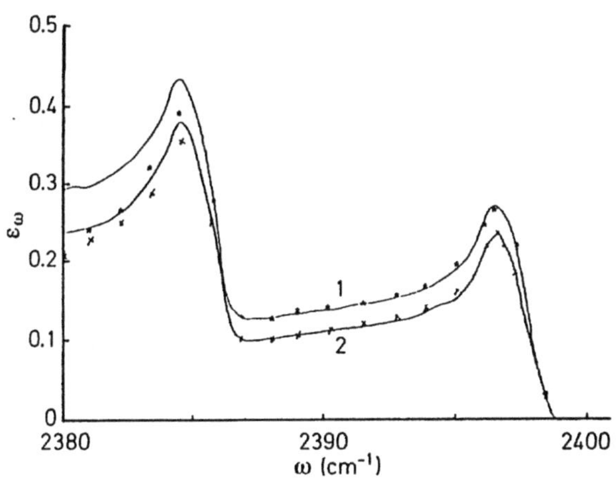

Fig. 7: Band head measurements and results of the line by line method spectral resolution 2 cm^{-1}*.8 cm^{-1} X=50 mm.

1	T = 2600 K XCO_2=0.18		**2**	T = 2810 K XCO_2=0.18
*	experimental		*	experimental
------	line by line model		------	line by line model

NEW EXPERIMENTAL METHOD FOR TEMPERATURE AND CO_2 MEASUREMENTS

We seek a method which gives the temperature and the concentration from emission measurements alone. In our previous experimental method (section 2), we need both emission and absorption measurements. The latter are not easy to perform. A source must be placed behind the jet under study. This is not possible in all situations, for instance for a moving object and when its trajectory is located far from the experimental facilities. Emission measurements, however, are easily achieved in all situations.

As we have shown in the section above, the line by line calculations are in agreement with the experimental results at the band head. They have been used to develop the new method. In Figures 6 and 7, the sharp peak located at 2398 cm^{-1} is produced by the effect of the accumulation of a lot of lines, and it becomes sensitive at high temperature. The height of this emissivity peak varies with the temperature value. Unfortunately we cannot use it directly, for this height depends also on the CO_2 concentration value. Therefore, another quantity must be known and has to vary with the concentration. If we look at Figures 6 and 7, we can see that the 2386-2400 cm^{-1} spectral region only corresponds to the 00^01 $\rightarrow 00^00$ vibrational transition, and the surface under the emissivity curves is proportional to the CO_2 concentration. The height to surface ratio should be sensitive to the temperature, but independent of the concentration. To look at the reality of this proposition, the ratio was determined at different temperatures and for CO_2 concentrations between 5% and 25%, this range corresponding to realistic values encountered in flames. The results are reported in Figure 8 and we can see that the ratio is sensitive to the temperature but less to the concentration value, and the uncertainty will be about 150 K in the 2000-2500 K range and 80 K in the 2500-3000 K range.

We can test this method from our measurements in the flame. The ratio is determined from the experimental emissivities and then the temperature is obtained from the curve reported in Figure 8. The results can be compared to those obtained by emission absorption. This was done in some cases (those of Figures 6 and 7). In the following table we can see the comparison.

emis./abs. (K)	band head (K)
2260	2220-2360
2600	2590-2670
2810	2800-2860

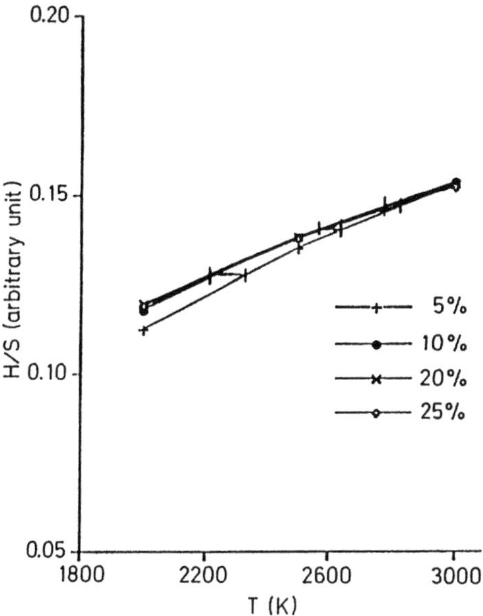

Fig. 8:   H/S band head ratio optical length L=10 cm, spectral resolution 2 cm^{-1}*0.8 cm-1

We must remember that the absorption emission measurements are carried out with an accuracy which is about 50 K, so the two methods agree well.

Finally, the CO_2 concentration is easily determined. Indeed, for each desired temperature, we plot the curve of the band head surface calculated as a function of the concentration. Then the CO_2 concentration in the flame is provided from the previous curve, knowing the experimental surface value. Figure 9 shows this

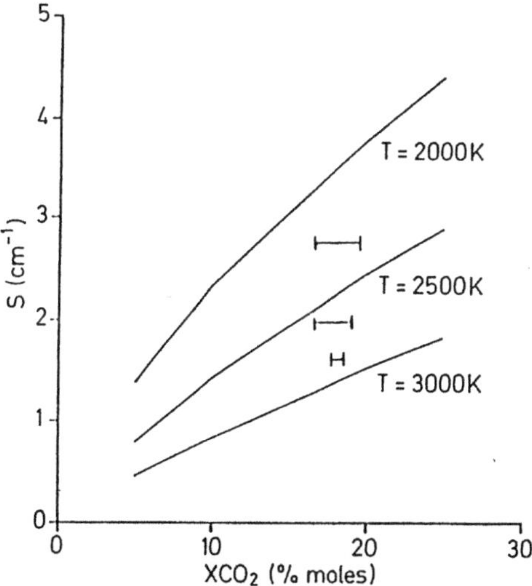

Fig. 9:   Emissivity surface S, optical length L=10 cm, spectral resolution 2 cm^{-1}*0.8 cm^{-1}.

method applied to different temperatures mentioned earlier. With the method used in the section 3.2, we have measured a constant value equal to 0.18 (in mole fraction). The values which are given by the curve in Figure 9 are very close.

It must be noticed that the accuracy of this new method can be improved by determining a new temperature value from the curve shown in Figure 8, but now with a known concentration.

TEMPERATURE ANALYSIS IN THE BOUNDARY LAYER

INTRODUCTION

If we look again at Figure 4, we feel obliged to ask ourselves about the nature of the boundary layer. We assume that it is a laminar flow. First, the visual observation of the flame shows an excellent stability, which excludes low frequency fluctuations in the flow. Second, the Reynolds number is easy to calculate. According to the method described below, it is about 5000. This value corresponds to a laminar plane jet. Consequently, the temperature evolution

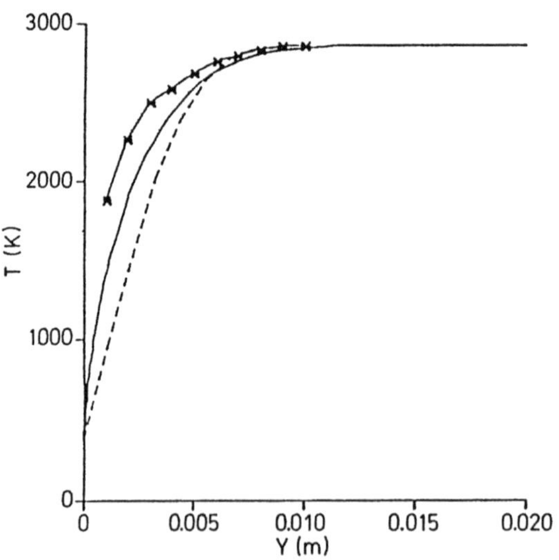

Fig. 10:   Temperature profiles X= 50 mm

—×—   experimental
------   Blasius
          values of the references [18-20] taken at T=1600 K for the different parameters
_____   variable thermophysical properties: C_p, v, λ

should be of a Blasius kind. The comparison between this analytical profile and the experimental values is shown in Figure 10. A discrepancy exists between the two evolutions.

The theoretical predictions are improved if the variations of the thermophysical properties of the gases with the temperature are taken into account.

THERMOPHYSICAL PROPERTIES OF HOT GASES

The different parameters, which are needed, are the heat capacity Cp, the heat conductibility λ and the viscosity v, and for each compounds of the mixture. We use the data of Kee *et al* [18] for the Cp values, those of Svelha [19] for λ and v. As was noticed in section 3, the CO_2 and H_2O concentrations do not vary in the layer. The constant values are XCO_2=.18 and XH_2O=.50 in mole fraction. The other gas concentrations (CO, O_2, H_2) are given by the work of Coppalle et al [1], in which a similar burner was used under the same conditions. Finally, the mixture properties are determined by using the relations recommended by Amder [20].

NUMERICAL SCHEME

In the case of varying properties, numerical predictions are recommended, and are easier to perform than analytical resolutions, which are only solved in special cases.

The flow near the plate is two dimensional, steady-state and laminar. With the usual boundary layer assumptions, the mass, momentum and energy equations are written as follows:

$$\frac{\partial u}{\partial x} + \frac{\partial v}{\partial y} = 0$$

$$u \frac{\partial u}{\partial x} + v \frac{\partial v}{\partial y} = v \frac{\partial^2 u}{\partial y^2}$$

$$u \frac{\partial T}{\partial x} + v \frac{\partial T}{\partial y} = a \frac{\partial^2 T}{\partial y^2}$$

with $\quad u = v = 0 \quad y = 0$

$\qquad u \rightarrow 7.6\,\text{m/s} \qquad T \rightarrow 2850K \quad y \rightarrow \infty$

We use the procedure of Patankar [21]. As the above equations are parabolic, a marching solution procedure in the X direction is applied. The space is divided into a grid with constant mesh for X and a geometrical progression for Y. The numerical results are shown in Figure 10 where the experimental and Blasius values have already been reported. The agreement is better than in the section above, but a discrepancy remains between the experimental and numerical profiles.

This could be explained by the wall reactions which provide a heat release in the boundary layer. This phenomenon is not included in our energy equation and is very large in flames. In fact, at high temperature, a lot of radicals are produced (OH,H,O) and their recombinations on the plate are exothermic reactions. The corresponding term must therefore be included in the energy equation.

CONCLUSION

The experimental and theoretical studies on the CO_2 emission at high temperature have given the following results

- the emissivity values provided by the statistical model are good on the whole band of CO_2 except in the band head region;

- in this spectral range, the line by line method is reliable;

- these line by line calculations and the measurement of the CO_2 emission spectra make it possible to determine the temperature and the concentrations in hot jet.

These experimental methods have enable us to measure the temperature profile in a boundary layer, developed on a water cooling stainless steel plate. The numerical predictions have shown that the chemical reactions are of importance and have to be included in the energy equation.

REFERENCES

1. Coppalle A., Vervisch P.: J.Q.S.R.T. 33(5)(1983) 465-473.

2. Coppalle A., Vervisch P.: J.Q.S.R.T. 35(2) (1985) 121-125.

3. Herzberg G.: Infrared and raman spectra. Van Nostrand (1945).

4. Ludwig C.B., Malkmus W., Reardon J.E., Thomson J.A.L.: Handbook of infrared radiation from combustion gases. NASA SP3080 (1973).

5. Elsasser W.M.: Phys. Rev 54 (1938) 126.

6. Plass G.N.: J.O.S.A. 48 (1958) 690.

7. Goody R.M.: Atmospheric radiation. Clarence Press Oxford (1964).

8. Malkmus W.: J.O.S.A. 57 (1967) 323.

9. Malkmus W., Thomson A.: J.Q.R.S.T. 2 (1962) 17

10. Soufiani A., Hartman J.M., Taine J.: J.Q.R.S.T., 3(3) (1985) 243.

11. Bernstein L.S.: J.Q.R.S.T. 23 (1980) 157.

12. Taine J.: J.Q.R.S.T. 30 (1983) 371.

13. Papineau N.: Thesis, Paris 4, (1985).

14. Rothman L.S., Young L.D.G.: 25 (1981) 505.

15. Rothman L.S.: Applied Optics 25 (1986) 1795.

16. Burch D.E., Gryvnac D.A., Patty R.R., Bartky C.E.: J.O.S.A., 59 (1969) 267.

17. Grisch F.: Thesis, Rouen (1988).

18. Kee R.J., Rupley F.M., Miller J.A.: Sandia report, sand87-8215 (1987).

19. Svella R.A.: Washington, D.C.. Nasa TR R 132 (1962).

20. Amder I., Mason E.A.: The Physics of Fluid 1 (5) (1958) 370.

21. Patankar S.V.: Numerical Heat Transfer and Fluid Flow. McGraw-Hill, (1980).

EXPERIMENTAL STUDIES OF A METHANE/AIR COUNTERFLOW FLAME DOPED WITH NO

Didier Schreiber, Magnus Lenner, Ove Lindgren, Jim Olsson
Department of Physical Chemistry
Chalmers University of Technology
S-412 96 Göteborg
Sweden

ABSTRACT

The results of experimental studies of the behaviour of a CH_4 (in N_2)/air counterflow diffusion flame with additions of nitric oxide are reported. Optical spectrometric emission techniques were employed to measure emission in UV-vis from flame radicals and in IR from stable molecules. Emission intensities for $NH(A^3\Pi)$ increased almost linearly with added amounts of nitric oxide, even at very low (about 200 ppm) values.

INTRODUCTION

The structure of laminar flames gives fundamental information, pertinent to many domains within combustion science as a whole. For example, the most advanced theories for the modelling of combustion employ laminar flames as a basic element.

Previous investigations of hydrocarbon-oxygen/NO flames by UV-vis and/or IR techniques [1-4], have concentrated on premixed flames, dealing with the mechanisms which govern the origin and fate of important flame radicals. In particular the presence of the excited N-containing species NH* and CN* which are prominent emitters in NO doped flames, has been studied. In that context the relationship of NH* reactions with the prompt-NO mechanism has been penetrated. In the present work, similar problems were addressed, with respect to a more complex flame type, the counterflow diffusion flame.

In this study, radicals (OH, CH, CN and NH) in electron excited states, were measured with rapid-scanning fiber-optic UV-vis spectrometry [5]. An FTIR spectrometer, equipped for emission measurements, was used to register vibrationally excited IR-emitting stable molecules (CO_2 and CH_4). Species can be determined by emission and/or transmission methods, as described by Solomon *et al* [6-8].

The main goal of the present work was to apply UV-vis and IR emission techniques to a methane/air counterflow diffusion flame doped with nitric oxide, concentrating on obtaining emission intensity profiles for various flame species including OH*, CH*, CN*, NH*, CO_2 and CH_4. The profiles provide important information to aid understanding of the fundamental mechanism underlying the combustion mechanism for methane-NO/air. Therefore the results potentially may lead to the improvement of many practical systems, such as detection of pollutants.

EXPERIMENTAL METHODS

Counterflow Burner. The burner, which represents a modified version of that employed by Smooke *et al* [9], is depicted in Figure 1. Fuel premixed with nitrogen streams upwards centrally through the bottom of the burner while air enters downwards through the top. Upon ignition a diffusion flame, positioned between the two burner halves, is formed. To shield the flame from turbulence and separate it from the oxygen in the surrounding atmosphere, a shroud of nitrogen gas streaming downwards through peripheral holes in the upper burner half, completely surrounds the flame. Accurate vertical motion of the whole burner, which stands on a lifting device, is achieved by a stepper motor run via a Keithley 500A Measurement and Control System (Keithley Instruments, Cleveland, OH, USA), interfaced with a personal computer.

The flames investigated consisted of a 2.0 lpm CH_4/N_2 (72.4%/27.6%) from the fuel side and 3.4 Ipm O_2/N_2 (15.5%/84.5%) from the oxidant side, and with

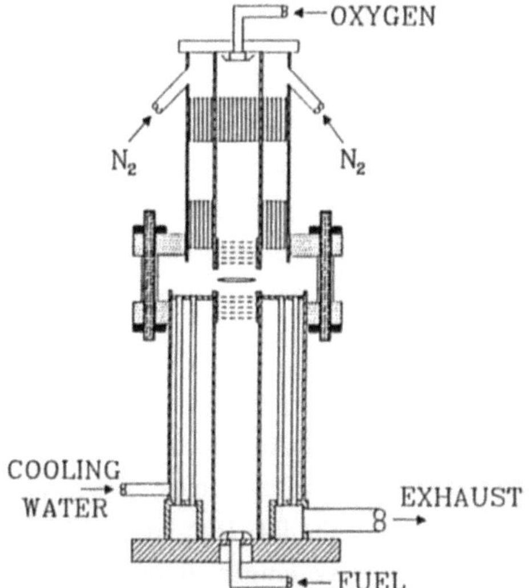

Fig. 1: Counterflow Burner

addition of varying amounts (0-1 lpm) of NO at the fuel side. Profiles of temperatures and emission intensities versus height for flame molecules and radicals, were obtained using two different spectroscopic techniques.

Temperature Analysis. Temperature analysis using a Pt-Pt10%Rh(type-S) thermocouple, was performed with the thermocouple junction situated in the center of the flame. The temperatures, uncorrected for radiation losses, were registered by a CRL 252 digital display precalibrated for type-S thermocouples with internal cold junction compensation (Pentronics, Gunnebobruk, Sweden).

Rapid-Scanning UV-vis Spectrometry. For measurements of flame emission from electronically excited radicals in the UV and visible spectral regions, a rapid-scanning spectrometer based on fiber-optic probes (OPSIS AB, Lund, Sweden) [5], was used. The light is imaged into the spectrometer via a fiber-optic entrance and dispersed by a holographic grating (3600 rulings/mm). An approx. 10 nm wide, preset spectral region is projected on the 25 mm wide exit slit, where it is scanned over 1000 channels by 20 peripheral slits on a wheel rotating 5 times/sec., registered by a side-on photomultiplier detector and a subsequent A/D converter. A built-in personal computer performs signal-averaging of the resulting 10^5 data points/sec., presents the result on the screen and stores spectra on a fixed disk.

The instrument performs 100 scans/sec. and signal averaging of 5×10^3, 20×10^3 and 45×10^3 scans was tried to achieve a satisfactory noise level. The signal/noise ratio is inversely proportional to the square root of n (number of scans), so that eg. 45×10^3 scans will reduce noise by a factor of 3, compared to 5×10^3 scans, and thus yield a corresponding greater accuracy and reproducibility.

Emission intensities, vs. wavelength for the methane-NO/air diffusion flame specified above, from CH- and OH-, CN- and NH- radicals were registered using a fused silica optical fiber probe (dia. 1 mm). Measurement series were performed at various heights in the flame, by in each case averaging of 20×10^3 scans. The spectra were evaluated by subtraction of baseline value from peak intensity. The net intensities for each height thus obtained, were then plotted as a function of the amount of NO addition.

FTIR Emission Measurements. A Mattson Polaris FTIR spectrometer (Mattson Inc., Madison, WI, USA) was used for registration of IR emission. An infrared spectrometer using an interferometer has several advantages over dispersive instruments, including better frequency precision and the fact that each point of the interferogram contains information from all wavelength elements. Our instrument was adapted for emission experiments by removal of the spherical mirror next to the regular IR source, and instead allowing radiation emitted by the flame to enter the instrument through an adjacent emission port fashioned with a 10 mm i.d. dia. brass tube. The end of the tube, which was fitted with a 1.0 mm high horizontal slit, was positioned at the periphery of the flame, and single beam flame emission spectra at 2 cm^{-1} resolution were obtained by co-adding and signal averaging of, in each case, 32 scans, using a liquid nitrogen cooled narrow-band MCT detector. By moving the burner vertically, with the probe fixed, it was possible to obtain profiles of emission intensity for CO_2 at 2250 cm^{-1} and for CH_4 at 3000 cm^{-1}, when integrated peak areas were plotted against height in the flame.

RESULTS AND DISCUSSION

A methane-air/NO counterflow diffusion flame was investigated by UV-vis and IR spectroscopy.

Temperature profiles as function of NO addition, for a number of heights, were constructed by vertical relative movement of the thermocouple in the flame. The results are given in Figs. 2 and 3.

By moving the burner vertically, with the probe fixed, it was possible to obtain profiles of emission intensity for CO_2 at 2250 cm^{-1} and for CH_4 at 3000 cm^{-1}, when integrated peak areas were plotted against height in the flame. An infrared spectrum over the mid-IR frequency range, showing the peaks mentioned above as well as other CO, CO_2 and H_2O bands is shown in Figure 4. Additions of NO

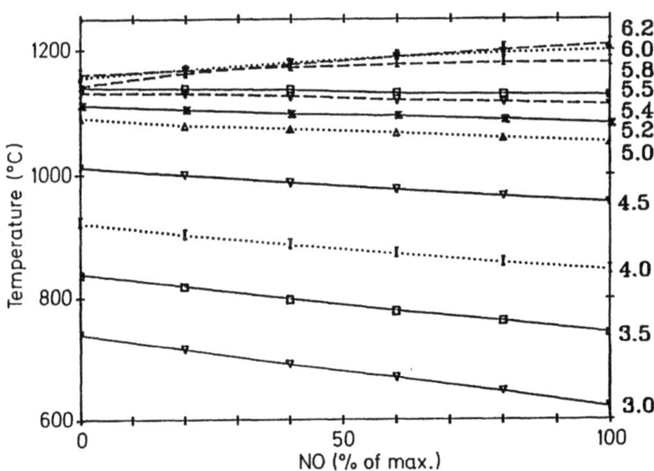

Fig. 2: Temperature as function of NO addition. Profiles for 3.0 mm to 6.2 mm height.

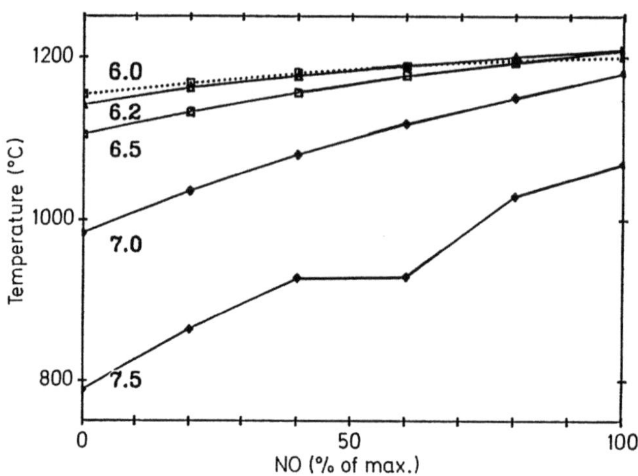

Fig. 3: Temperature as function of NO addition. Profiles for 6.0 mm to 7.5 mm height.

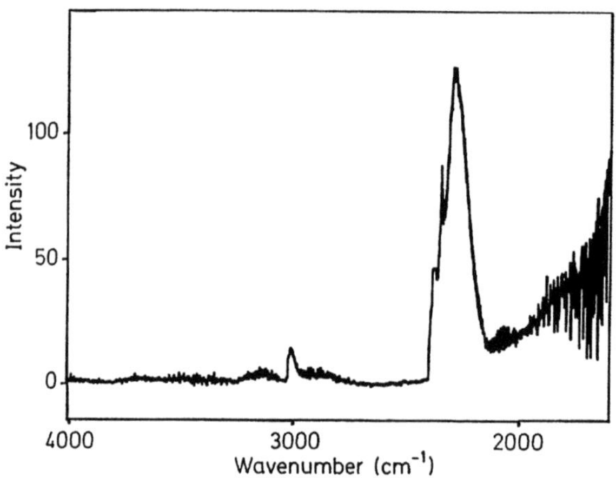

Fig. 4: Infrared emission spectrum

did not influence the appearance of the IR spectra measured. The resulting profiles are shown in Figure 5 and Figure 6 respectively.

The electronic transitions of the four radicals, for which emission intensities were calculated are given below:

CH: $Q(0,0)$ band $A^2\Delta$ - $X^2\Pi$ transition at 431.4 nm
OH: $Q_2(0,0)$ band $A^2\Sigma^+$ - $X^2\Pi$ transition at 309.0 nm
CN: $S(0,0)$ band $B^2\Sigma^+$ - $X^2\Sigma^+$ transition at 388,3 nm
NH: $Q(0,0)$ band $A^3\Pi$ - $X^3\Sigma^-$ transition at 336.4 nm

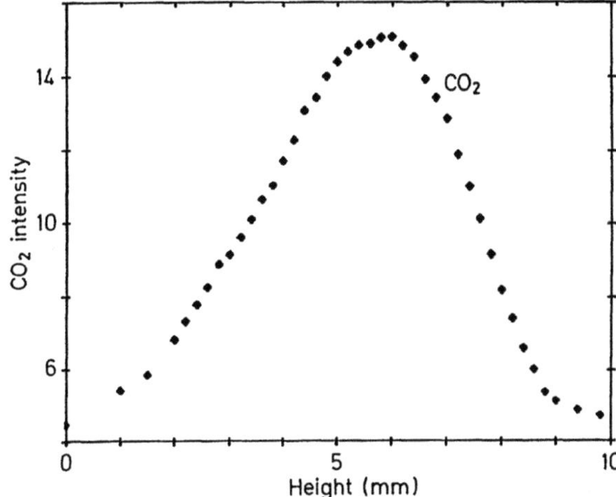

Fig. 5: Integrated CO_2 emission intensity as function of height.

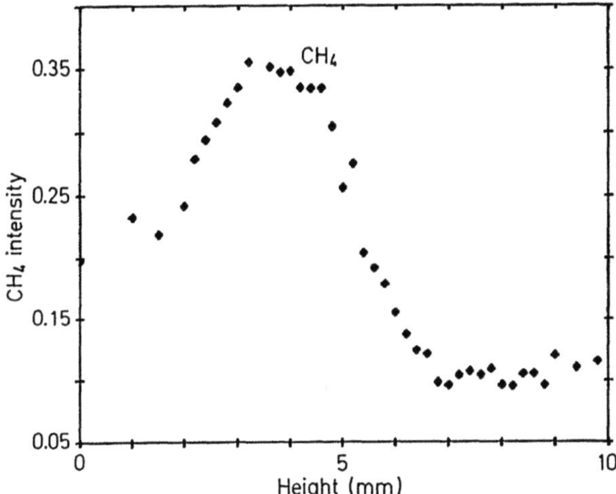

Fig. 6: Integrated CH_4 emission intensity as function of height.

Figure 7 shows an OH spectrum over the spectral range 304 nm-314 nm. Figs. 8 through 11 show emission intensity profiles for the measured radicals at heights between 4 and 9 mm, with NO amounts from 0% to 100% (100% = 1 lpm) added. More detailed mappings of NH* emission intensities for very low NO addition (0%-2%, 0%-10%) for the visible flame region between 5 mm to 7 mm and 4 mm to 7.5 mm respectively, are shown in Figs. 12 and 13. The NH* signal strength shows a quite linear dependence with respect to the amount of NO addition in these low NO ranges. The OH*, CH* and CN* intensities were quite insensitive to such small additions of NO.

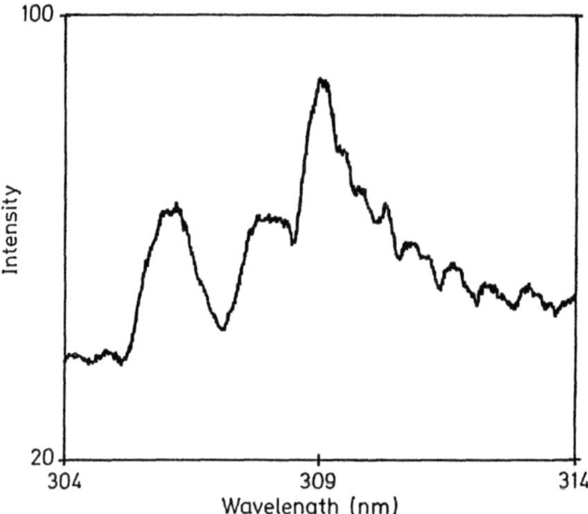

Fig. 7: Ultraviolet emission spectrum of the OH* radical.

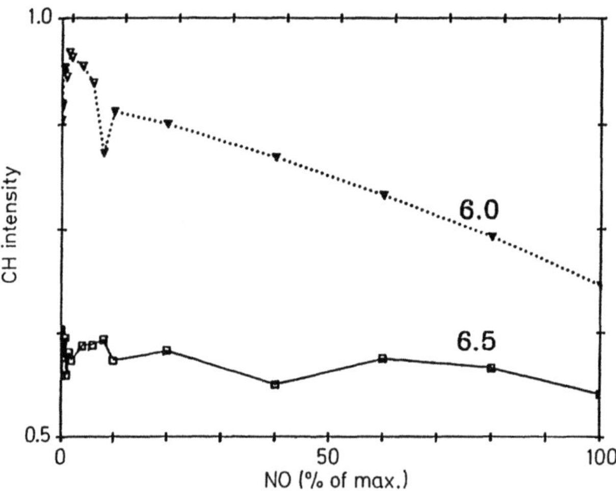

Fig. 8: CH* emission intensities as function of 0-100% of max. NO addition.

Upon addition of nitric oxide on the fuel side NH* and CN* were formed [1,2]:

$$CH + NO \rightarrow NH^* + CO$$
$$C_2 + NO \rightarrow CN^* + CO$$

The first reaction removes CH radicals and the second reaction competes with: $C_2 + OH \rightarrow CH + CO$ [3], by with CH radicals are produced.

From the measured intensity profiles in Figs. 8 through 12, where emission intensities were plotted as function of amount of NO addition, it is evident that

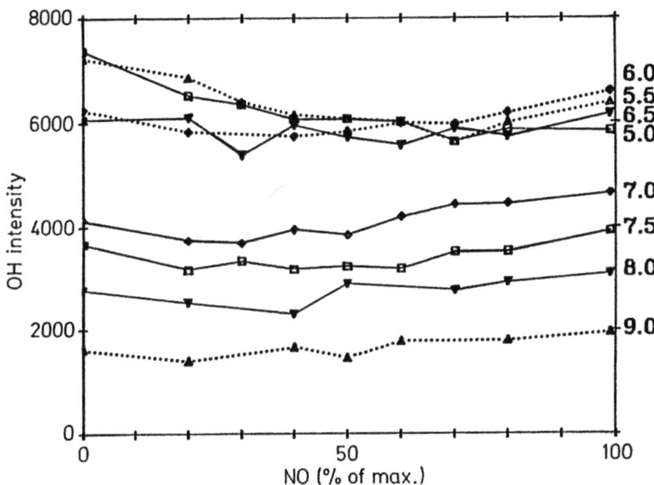

Fig. 9: OH* emission intensities as function of 0-100% of max. NO addition.

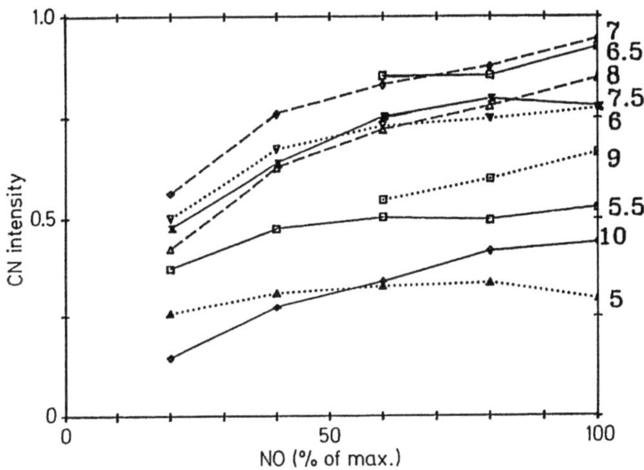

Fig. 10: CN* emission intensities as function of 0-100% of max. NO addition.

NH* and CN* increase which NO increase, while CH* decreases. OH* emission intensities are approximately constant through the flame. From Figure 12 it can be concluded that NH* is sensitive down to very small (200 ppm) additions of NO, the NH* emission intensity rising linearly for 0.001-0.020 lpm of NO added with the fuel.

The measured profiles, although related to concentrations of the respective flame radicals, cannot be put to straightforward quantitative interpretation. As pointed out by Gaydon [10] the emission intensity is a complex function of concentration,

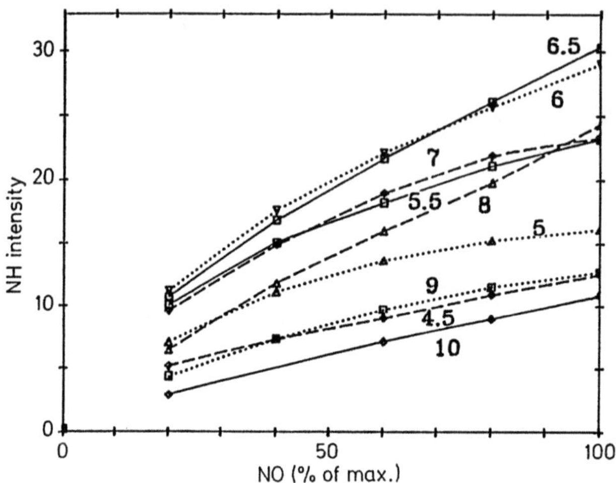

Fig. 11: NH* emission intensities as function of 0-100% of max. NO addition.

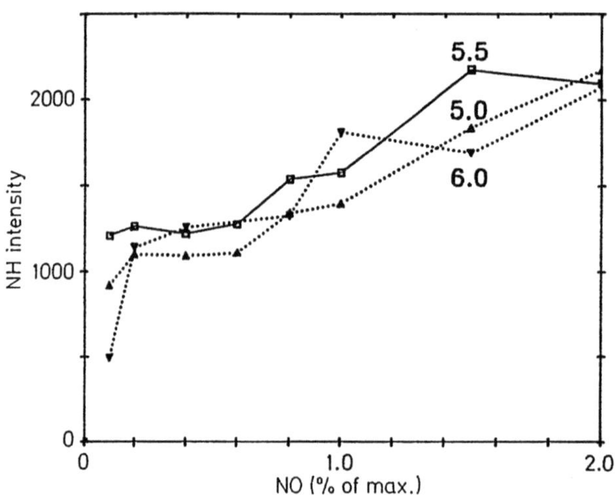

Fig. 12: NH* emission intensities as function of 0-2% of max. NO addition.

depending on eg. the effective excitation temperature and the presence of self-absorption.

The sensitivity of the present technique for detecting radical emission can be improved by at least a factor of 10, by use of a filter spectrometer to measure a band from the radical spectrum, and employing a cooled photomultiplier for effective suppression of noise.

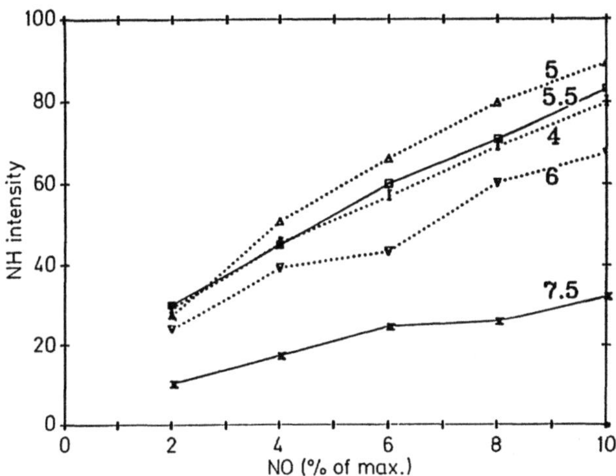

Fig. 13: NH* emission intensities as function of 0-10% of max. NO addition.

The IR data prove the feasibility of detecting IR radiation emitted from CO_2 and fuel molecules using FTIR. Since the excitation in this case is of thermal origin, the profiles seen in Figs. 5 and 6 can be interpreted as reflective of the relative concentrations. It can be concluded from the IR spectrum demonstrated in Figure 4 however, that the CO_2 signal suffers somewhat from self-absorption, two absorption shoulders around 2350 cm^{-1} being clearly visible on the high frequency side of the emission peak.

CONCLUSIONS

The IR data prove the feasibility of detecting IR radiation emitted from CO_2 and fuel molecules using FTIR. Emission intensities for $NH(A^3\Pi)$ increased almost linearly with added amounts of nitric oxide, even at very low (about 200 ppm) values.

ACKNOWLEDGEMENTS

This work was supported by the Swedish Energy Administration and the Swedish Natural Research Council.

We would also like to thank Prof. K. Seshadri at UC San Diego who constructed the counterflow burner.

The participation in the project of Didier Schreiber was made possible by a grant from Centre National d'Etudes Spatiales, France.

A generous grant from the Erna and Victor Hasselblad Foundation, which facilitated aquisition of our Mattson Polaris FTIR spectrometer is gratefully acknowledged.

REFERENCES

1. Matsui Y., Nomaguchi T.: Spectroscopic Study of Prompt Nitrogen Oxide Formation Mechanism in Hydrocarbon-Air Flames. Combust. Flame 32 (1978) 205-214.

2. Cummings G.A. McD., Hutton E.: Ionization in Ethylene-Air-Nitric Oxide Flames. Symp. (Int.) on Combustion 11 (1966) 335-341.

3. Quang L.N., Vanpee M.: A Spectroscopic Investigation of the Premixed Acetylene-Nitric Oxide Flame. Symposium (Int.) on Combustion 19 (1982) 293-301.

4. Darian S.T., Vanpee M.: A Spectroscopic Study of the Premixed Acetylene-Nitrous Oxide Flame. Combust. Flame 70 (1987) 65-77.

5. Olsson J.O., Lenner M., Wallin S., Uneus, L.: Rapid-Scanning Spectrometer Based on Fiber-Optics Applied to Flames. Rev. Sci. Instrum. 60 (1989) 2601-2605.

6. Solomon P.R., Carangelo R.M., Best P.E., Markham J.R., Hamblen D.G.: The Spectral Emittance of Pulverized Coal and Char. Symp. (Int.) on Combustion 21 (1986) 437-446.

7. Best P.E., Carangelo R.M., Markham J.R., Solomon P.R.: Extension of Emission/Transmission Technique to Particulate Samples Using FT-IR. Combust. Flame 66 (1986) 47-66.

8. Solomon P.R., Best P.E., Carangelo R.M., Markham J.R., PO-Liang Chen, Santoro R.J., Semerjian H.G.: FT-IR Emission/Transmission Spectroscopy for In Situ Combustion Diagnostics. Symp. (Int.) on Combustion 21 (1986) 1763-1771.

9. Smooke M.D., Puri I.K., Seshadri K.: A Comparison Between Numerical Calculations and Experimental Measurements of the Structure of a Counterflow Diffusion Flame Burning Diluted Methane in Diluted Air. Symp. (Int.) on Combustion 21 (1988) 1783-1792.

10 Gaydon A.G.: The Spectroscopy of Flames. London, Chapman and Hall, (1974).

CHAPTER 2:
HEAT TRANSFER IN FLAMES

THE CALCULATION OF LOCAL FLUCTUATIONS IN NON PREMIXED TURBULENT FLAMES

R. Borghi[1], L. Vervisch[2] and D. Garréton[3]

[1] CORIA Faculté des Sciences de Rouen, URA 230/CNRS
[2] LMFN INSA Rouen, URA 230/CNRS and EDF DER LNH
[3] Direction des Etudes et Recherches - Electricité de France.
   LNH - 6, Quai Watier - 78400 CHATOU, France.

ABSTRACT

Turbulent diffusion flames take place in many combustion equipments: furnaces, combustion chambers of jet engines. Very noticeable improvements have been obtained recently concerning the prediction methods.

This paper presents first a synthetic review of these prediction methods.

More details and results are given secondly for a particular approximate method, which provides an interesting compromise for engineering predictions.

In the end of this paper, it is shown how this approximate method can be extended to account, to some extend, for the radiation phenomena including the influence of the fluctuations of temperature and concentration.

1. INTRODUCTION

Turbulent diffusion flames take place in many combustion equipments: furnaces, combustion chambers of jet engines. The design of these practical devices, requires an increasing use of computing codes, devoted to the prediction of the properties of such turbulent flames. However, that prediction is not easy. The interaction between turbulence and chemical reaction plays a very crucial role in this case. The chemistry itself is very complex, and not yet well understood in flames, but even if it would be completely known, the prediction will not be possible, because the temperature as well that the concentrations of chemical species do fluctuate, due to turbulence, in a way that cannot be followed by the numerical codes. The calculations have to take into account the fluctuations of velocity, temperature and concentration, even if only mean values have to be predicted, as a consequence of the high non-linearity of chemistry. Very noticeable improvements have been obtained recently concerning the prediction methods. While in the past only very fast chemistry could be handled, the methods concerning finite rate chemistry are now in encouraging agreement with the experiments.

This paper presents first a synthetic review of these prediction methods. More details and results are given secondly for a particular approximate method, which provides an interesting compromise for engineering predictions. It allows, with a moderate amount of computer time, to get approximate predictions for non infinitely fast chemistry, in particular for the chemistry associated with pollutants formation in flames. The existing methods, up to now, are dealing only with adiabatic flames, neglecting in particular the radiative losses. In the end of this paper, it is shown how this approximate method can be extended to account, to some extend, for the radiation phenomena including the influence of the fluctuations of temperature and concentration.

2. REVIEW OF EXISTING METHODS

2.1. BASIC EQUATIONS AND PROBLEMS

The 3-D calculations of the mean flow field, mean temperature and mean species concentration field require the solution of the coupled set of partial differential equations that comprises:

- The continuity equation,
- The mean momentum equation,
- The mean energy equation,
- The mean species mass fraction equations,
- The equations for the turbulence model.

Very often, diffusion flames in furnaces are fed with a liquid fuel, and then involve some droplets. But these droplets are also often not explicitly considered in the reacting medium; if they are not too big, a jet with droplets can be conveniently represented by a gaseous jet (see the review of Faeth [1].) So we do not consider explicitly the presence of droplets in this paper.

The mean equations are classically written in terms of Favre, or density weighted averages; i.e. each variable g (except ρ and p) is written as $g = \tilde{g} + g'$, where $\tilde{g} = \dfrac{\overline{\rho g}}{\overline{\rho}}$.

It is not our intention to write and discuss here all equations of this type. We just focus our discussion on the mean species mass fraction:

$$\frac{\partial}{\partial t}\,\overline{\rho}\,\widetilde{Y}_i + \frac{\partial}{\partial x_1}\left(\overline{\rho}\,\widetilde{U}_1\,\widetilde{Y}_i\right) = \frac{\partial}{\partial x_1}\left(\overline{\rho}\,D_i\frac{\partial \widetilde{Y}_i}{\partial x_1} - \overline{\rho}\,\widetilde{U'_1 Y'_i}\right) + \overline{\rho}\,\widetilde{\omega}_i \tag{1}$$

In this equation, several characteristics numbers can be pointed out:

- The Reynolds number, ratio of the advective time scale to the molecular diffusion one,
- The Peclet number, ratio of the turbulent diffusion time scale to the molecular one $\left(\dfrac{k^{1/2}\,l_t}{D_i}\right)$,
- The Damköhler number, ratio of the fluid dynamic time scale τ_{ex} to the chemical reaction one τ_c.

For large Peclet numbers, the turbulent fluxes need to be modelled: this well-known problem may be treated, for instance, with the definition of a turbulent diffusion coefficient, which has to be computed as a function of the local turbulence characteristics. Another modelling problem lies in the mean reaction rate ω_i, this rate is not simply given by the chemistry, but must take into account the fluctuations that are present in the turbulent flame. For very large Damköhler numbers, the assumption of chemical equilibrium allows to determine the mean composition and temperature (2.2.). The crucial problem occurs when the Damköhler number is of order of unity. Indeed, the exact definition of ω_i involves the joint probability density function of the fluctuations of all the chemical species involved in the production or consumption of the species i, and the temperature:

$$\widetilde{\omega}_i = \int_{\psi_1}\cdots\int_{\psi_N}\;\dot{\omega}_i\left(\psi_1,\,\ldots,\,\psi_N\right)\widetilde{P}\left(\psi_1,\,\ldots,\,\psi_N\right)d\psi_1\,\ldots\,d\psi_N \tag{2}$$

Where the ψ_i are random variables associated with the species mass fraction Y_i and temperature.

Then, some knowledge, at least approximate, of $\widetilde{P}(\psi_1,\ldots,\psi_N)$ is needed to allow the computation of $\widetilde{\omega}_i$ (2.3.4.); or $\widetilde{P}$ has to be directly determinated (2.3.3.).

The numerical integration of the set of partial differential equations can be done only after the solution of the three modelling problems: the modelling of the diffusion fluxes (for species, for enthalpy and momentum), the modelling of the mean reaction rate and finally the modelling of the turbulence itself, i.e. for instance, its kinetic energy k and its dissipation rate ε, that is needed for the two first models.

We will focus our discussion, in the following, on the problem of the mean reaction rates (eq. (2)), needed in eq. (1). Concerning the problems related to the diffusion fluxes and to the turbulence model, the reader may refer to a recent review [2].

2.2. THE MODELS FOR VERY FAST CHEMISTRY

The case of very fast chemistry, when any characteristic time of the chemical processes is much smaller than any characteristic time of the other physical processes, is the simplest one, although it appears, at first sight, as a limiting case to be treated carefully. This can be emphasized very clearly if we consider the simple case of a single reaction, as:

$$A + B \rightarrow C$$

with reaction rate (massic)

$$\dot{\omega}_A = \dot{\omega}_B = -K(T)\, Y_A Y_B \tag{3}$$

and with assuming that the Lewis numbers of all species are equal to one.

In this case, when the chemistry is very fast, K is infinite (whatever can be the temperature) and the reaction rates $\dot{\omega}_A$ and $\dot{\omega}_B$ remain finite only if $Y_A Y_B = 0$, and are then under an undeterminate form $\infty \times 0$! This is also the case for $\dot{\omega}_A = \dot{\omega}_B$, which are involved in the equation (2) for $\tilde{Y}_A$ and $\tilde{Y}_B$.

But with the linear combination $\Phi = Y_A - Y_B$, this new variable satisfies an equation without any chemical term:

$$\frac{\partial}{\partial t}\, \rho\Phi + \frac{\partial}{\partial x_1}\left(\rho\, U_i \Phi\right) = \frac{\partial}{\partial x_1}\left(\rho D\, \frac{\partial \Phi}{\partial x_1}\right) \tag{4}$$

Then Φ is a purely convective-diffusive quantity, and may, in principle, be predicted by a suitable model fo turbulence and turbulent diffusion, irrespective to the chemical rate (3).

In addition, one can remark that, because $Y_A Y_B = 0$, and that in consequence Y_A and Y_B cannot be non zero at the same time at the same position, Φ represents in fact either Y_A (when $\Phi > 0$), or Y_B (when $\Phi < 0$).

The direct consequence of that is that $\tilde{Y}_A$ or $\tilde{Y}_B$ can be computed without the solution of the corresponding eq. (2). It suffices to write:

$$\begin{cases} \tilde{Y}_A = \int_{\psi > 0} \psi\, \tilde{P}(\psi)\, d\psi \\ \tilde{Y}_B = \int_{\psi < 0} \psi\, \tilde{P}(\psi)\, d\psi \end{cases} \tag{5}$$

The crucial quantity here is the pdf $\widetilde{P}(\psi)$ of the random variable ψ associated to the scalar Φ. This property has been discovered first by Toor [3], and Lin and O'Brien [4]. This property does not hold only in that simple case. It holds also with a complex chemistry, if the flame is assumed to be adiabatic (so without any radiation losses) and at low Mach number. In this case, eq. (5) becomes:

$$\widetilde{Y}_i = \int_0^1 Y_i^{eq}(\Psi) \, \widetilde{P}(\Psi) \, d\Psi \tag{6}$$

where $Y_i^{eq}(\Psi)$ is the value of Y_i at the chemical adiabatic equilibrium of a mixture composed of Φ grammes of the fuel and $(1 - \Phi)$ grammes of the oxidant jet. Recently this method was extended also to the case of non adiabatic flow field by Mechitoua and Viollet [5]. More details about this approach can be found in the review of Bilger [6]. The problem is then solved if one is able to compute $\widetilde{P}(\psi)$. This can be done in two different ways. The simplest one uses a "presumed shape" for $\widetilde{P}(\psi)$. Many forms have been used:
- Sinusoidal, Spalding [7],
- Clipped gaussian function, Lockwood [8],
- Triangular, Rhodes [9],
- ...

A convenient method uses an n-parameter function and relates the first n moments to these parameters. One of the most widely used shape is the so called β function distribution.

$$\widetilde{P}(\Psi) = \frac{\Psi^{\alpha-1}(1-\Psi)^{\beta-1}}{\int_0^1 \varphi^{\alpha-1}(1-\varphi)^{\beta-1} \, d\varphi} \tag{7}$$

where:

$$\alpha = \widetilde{\Phi}\left(\frac{\widetilde{\Phi}(1-\widetilde{\Phi})}{\widetilde{\Phi'\Phi'}} - 1\right) \tag{8}$$

$$\beta = \left((1-\widetilde{\Phi})\right)\frac{\alpha}{\widetilde{\Phi}} \tag{9}$$

Then it suffices to know $\widetilde{\Phi}$ and $\widetilde{\Phi'}^2$ by partial differential equations as the one of $\widetilde{Y}_i$, in order to be able to compute an approximation of $\widetilde{P}(\psi)$. The balance equation for $\widetilde{\Phi}'^2$ has been studied and found as:

$$\frac{\partial}{\partial t}\overline{\rho}\,\widetilde{\Phi'}^2 + \frac{\partial}{\partial x_i}\left(\overline{\rho}\,\widetilde{U}_i\widetilde{\Phi'}^2\right) = \frac{\partial}{\partial x_i}\left(-\overline{\rho}\,\widetilde{U'_i\Phi'}^2\right) - 2\overline{\rho}\,\widetilde{U'_i\Phi'}\frac{\partial\widetilde{\Phi}}{\partial x_i} - 2\overline{\rho}\varepsilon_\Phi \tag{10}$$

This equation has to be closed, and a very classical closure assumption states that:

$$\bullet \ \overline{U'_i \Phi'} = - D_t \frac{\partial \widetilde{\Phi}}{\partial x_i}$$

$$\bullet \ \overline{U'_i \Phi'}^2 = - D_t \frac{\partial \widetilde{\Phi'}^2}{\partial x_i}$$

$$\bullet \ \varepsilon_\Phi = C_D \frac{\widetilde{\Phi'}^2}{\tau_t}$$

where C_D is a constant, D_t and τ_t are respectively the turbulent diffusivity and the turbulent integral time scale. A more precise method to get $\widetilde{P}(\psi)$ consists in using a balance equation for $\widetilde{P}(\psi)$ itself. This balance equation has to be modelled, with some physical or mathematical assumptions. This question will be detailed in the next section. Janika and Kollmann [30], to our knowledge, have been the first one to compute $\widetilde{P}(\psi)$ with a balance equation.

When the velocity of the jet flames is not too high, the chemistry may effectively be assumed as very fast, and the method works quite well. Experimental results assessing the occurrence of relations Y_i^{eq}, or T^{eq} exist, see for instance [10] (Figure 1). Example of computations can be found in ref [5] - [8]. This method is often called the method of the conserved scalar, because Φ is a scalar conserved by the chemical process.

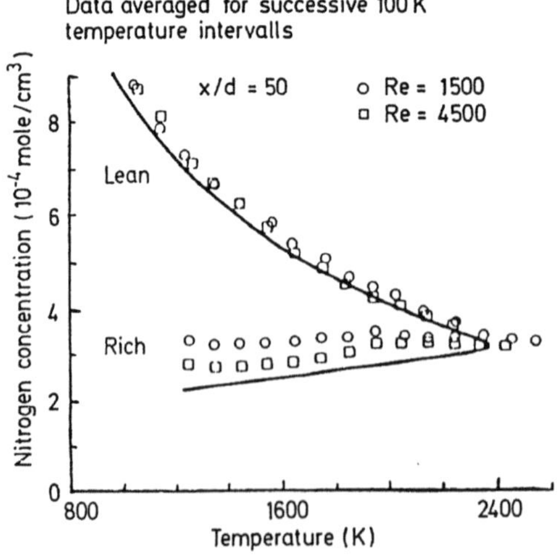

Fig. 1:   Instantaneous measurements in a H_2 - AIR diffusion flames. (The equilibrium is well verified on the lean side)

Because this method requires the computation of $\tilde{P}(\psi)$, which is not very simple, even with the presumed shape, a different model, first proposed by Magnussen [11], is also often used. This model assumes a single reaction and uses an equation like (2) for $\tilde{Y}_A$ and $\tilde{Y}_B$, and represents $\tilde{\omega}_A$ in a very simple manner.

$$\tilde{\omega}_A = -\frac{1}{\tau_t} \mathrm{Min}\!\left[\tilde{Y}_A, \tilde{Y}_B\right] \tag{11}$$

Indeed this formula does not possess a well definite proof. It states only that the mean reaction rate is not controlled by the chemistry (that does not play any role here, as in the conserved scalar method), but is controlled by the turbulence time τ_t and becomes zero when either the oxidant or the mean fuel disappears. This formula provides always realistic results, although they are only qualitative.

2.3. THE MODELS FOR FINITE RATE CHEMISTRY

2.3.1. Preliminaries

The assumption of very fast chemistry (or large Damköhler numbers) holds with large size flames of not too high velocity. But if the velocity of the jet is too high, and its reference length scale too small, the time scale associated with the convection phenomena can be too low and the large Damköhler assumption fails. This occurs when the flames are close to the limits of their stabilization domain. An another interesting case in which low Damköhler numbers effects occur is the prediction of pollutant formation like NO, which is clearly generally far from equilibrium, or like CO or unburned hydrocarbons, due very often to low temperature zones close to the walls. So the extension of prediction methods towards non infinitely fast chemistry is of very practical interest.

We shall review here three types of methods. The first one is called "flamelets models", because it is based on the assumption that the turbulent flame brush is composed of many flamelets, wrinkled and stretched by the turbulence. The second one is based on the computation of the joint probability function: $\tilde{P}(\psi_1,...\psi_N)$ which appears in eq. (2), and uses a modelled balance equation for this purpose. The third one, which deals with lagrangian models, is a simplification of the pdf approach.

2.3.2. The "Flamelets models"

The idea to represent the turbulent reacting field as a collection of "reacting laminae", or "flamelets", has been proposed as early as 1970 by Toor [12] and

Gibson and Libby [13]. It has become very popular, more recently after the review of Peters [14]. The method is based on the assumption that these flamelets are reaction-diffusion-layers, in quasi-steady state, which are continuously displaced and stretched within the turbulent medium. The layers are assumed to be thinner than all the turbulent scales, so that their internal structure can be studied with all chemical details, apart from the turbulent flow. A "library" of flamelets can be built and used for the turbulent flame computations.

At this stage of the approach, two types of strategies have been proposed. The one of Marble and Broadwell [15] consits in computing the mean reaction rate with:

$$\tilde{\dot{\omega}}_A = V_{D_i} \Sigma \tag{12}$$

V_{D_i} is a flux of reactant by unit of surface that is related only to the flamelet structure, and Σ is the so called "mean flamelet surface area by unit of volume". Σ is supposed to follow its own balance equation, wich has been proposed for the first time in [15] and intends to incorporate all the important physical phenomena: stretching by the turbulence, consumption by the molecular diffusivity and by the reaction.

The second type of approach uses again the conserved scalar, and has been proposed by Liew and Bray [16], and Peters [14]. Here any mean value is directly computed by:

$$\tilde{Y}_i = \int_\Psi Y_i^{Fl}(\Psi)\, \tilde{P}(\Psi)\, d\Psi \tag{13}$$

where $Y_i^{Fl}(\Psi)$ is feeded by the library of flamelets. In fact, eq. (13) constitutes only the first version of the model. In order to take into account the stretching of the flamelets by the turbulence (and not only their displacement, represented by $\tilde{P}(\psi)$), a two dimensional pdf $\tilde{P}(\psi,\gamma)$ is needed, where γ is the turbulent stretching rate; then eq. (14) replaces eq. (13):

$$\tilde{Y}_i = \int_\Psi \int_\gamma Y_i^{Fl}(\Psi,\gamma)\, \tilde{P}(\Psi,\gamma)\, d\Psi d\gamma \tag{14}$$

This new joint pdf is not easy to calculate; Peters proposes to assume statistical independence between Φ and γ, with a lognormal distribution for $\tilde{P}(\gamma)$. These aspects have to be studied further.

A good example of the possibilities of this method is given by the results of Warnartz and Rogg [17]. Figures 2 and 3, extracted from this work, show recent results concerning mean profiles compared with the experiments of ref [18].

These "flamelet models", whatever may be their exact form, allow to take into account the complex chemical mechanism that is actually occurring; they are based on physically clear assumptions. However, in their present form, they neglect important phenomena. The picture of the turbulent reactive flow on which they are based is not complete: the curvature of laminae is clearly neglected, the re-ignition of extinguished flamelets, as well as their transient evolutions between the several states of strain that are imposed by the turbulence, are not considered; in addition, interactions of adjacent laminae, which are likely to occur, in particular within coherent vortices present in

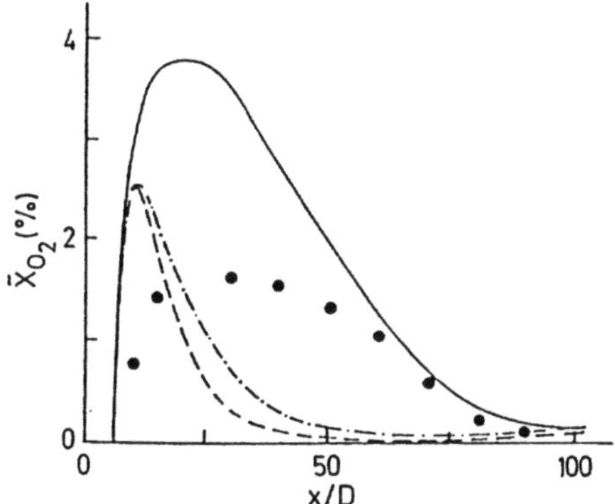

Fig. 2:   Axial profiles of oxygen

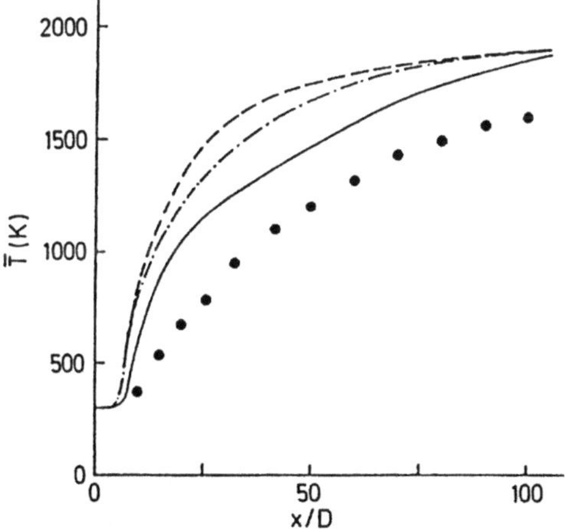

Fig. 3:   Axial profiles of temperature

many mixing layers, are also neglected. These approaches are then restricted to the cases where such phenomena are negligible, that is essentially for fast enough reactions, wrinkled flames and not too large turbulence Reynolds numbers.

Indeed, during their attempt to apply these ideas to a chemically reacting shear layer (with negligible heat release), Broadwell and Dimotakis [19] have found that considering only "flamelets" in order to represent the mixing with reactions was not sufficient to explain the experimental results. They have proposed that two parallel processes have to be taken into account in the model, one being the mixing and reaction occurring in molecularly mixed homogeneous fluid pockets. They have found that the percentage of each process depends on the Peclet number, and this holds even if the chemistry is very fast. This theory has not been put into a form such that the mean reaction rate can be found, but has been written into an integral form.

2.3.3. The "Pdf methods"

Pdf (probability density function) methods for predicting turbulent flows with chemical reactions have been developed and applied successfully. First, Lundgren [20] devised a method to derive a hierarchy of transport equations for a N points probability density function in a turbulent incompressible flow. Dopazo and O'Brien [21] and then Pope [22] extended this method to reacting flows. Chen and Kollman [23] have predicted turbulent diffusion flames with this approach. In USSR also, as early as 1970, Frost has proposed a model that involves an equation for the pdf of one inert species [24]. The statistics of a scalar field can be described by a single point pdf $\widetilde{P}(\psi_1,...\psi_N;x,t)$ corresponding to the thermochemical variables $Y_1,..., Y_{N-1},T$. The knowledge of this function allows the calculations of all moments:

$$\widetilde{Y}_i = \int_{\Psi_i}...\int_{\Psi_N} \Psi_i \, \widetilde{P}(\Psi_1,...,\Psi_N) \, d\Psi_1...d\Psi_N \tag{15}$$

and

$$\widetilde{Y'^n_i} = \int_{\Psi_i}...\int_{\Psi_N} \left(\Psi_i - \widetilde{Y}_i\right)^n \widetilde{P}(\Psi_1,...,\Psi_N) \, d\Psi_1...d\Psi_N \tag{16}$$

The single point pdf P satisfies the following transport equation (see [2], [36]):

$$\bar{\rho}\frac{\partial \widetilde{P}}{\partial t} + \bar{\rho}\,\widetilde{U}_1\frac{\partial \widetilde{P}}{\partial x_1} + \bar{\rho}\,\overline{U'_1\frac{\partial P}{\partial x_1}} =$$

$$-\sum_{i=N}^{N}\frac{\partial}{\partial \Psi_i}\left(\rho\,\dot{\omega}_i(\Psi_1,...,\Psi_N)\widetilde{P}\right) - \sum_{i=1}^{N}\sum_{j=1}^{N}\frac{\partial}{\partial \Psi_i}\,\frac{\partial}{\partial \Psi_j}\left(\overline{\rho\,D_i\frac{dY_i}{dx_1}\frac{dY_j}{dx_1}}P\right) \tag{17}$$

The most noticeable feature of equation (17) is that the chemical source term:

$$CST = \sum_{i=1}^{N} \frac{\partial}{\partial \Psi_i} \left(\rho \, \dot{\omega}_i(\Psi_1,...,\Psi_N) \tilde{P} \right)$$

(18)

which represents a flux of $\tilde{P}$ in the probability space due to the reaction, does not contain any unknown correlations; so there are no problem of modelling as it was the case in the conservation equation of the mean species $\tilde{Y}i$.

Modelling the small scale mixing term:

$$SSMT = \sum_{i=1}^{N} \sum_{j=1}^{N} \frac{\partial}{\partial \Psi_i} \frac{\partial}{\partial \Psi_j} \left(\rho \, D_i \frac{dY_i}{dx_l} \frac{dY_j}{dx_l} P \right)$$

(19)

is the crucial issue in the pdf approach and several mixing models have been developed.

- The Dopazo and O'Brien model [21] is the first proposed. They have shown that the mixing term can be written in term of two point pdf

$$SSMT = - \sum_{i=1}^{N} \frac{\partial}{\partial x'_i} \lim_{x' \to x} \frac{\partial}{\partial x'_1} \int_{\Psi_1} ... \int_{\Psi_N}$$
$$\rho D_i \frac{\partial \Psi'_i}{\partial x_l} \tilde{P}\left(\Psi_1,...,\Psi_N, \Psi'_1,...,\Psi'_N \right) d\Psi'_1 ... d\Psi'_N$$

(20)

Introducing the conditional pdf:

$$\tilde{P}_c \left(\Psi'_1,...,\Psi'_N / \Psi'_1,...,\Psi'_N \right) = \frac{\tilde{P}\left(\Psi_1,...,\Psi_N, \Psi'_1,...,\Psi'_N \right)}{\tilde{P}\left(\Psi_1,...,\Psi_N \right)}$$

(21)

and the assumption that the conditional expectation:

$$E\left(\Psi'_i / \Psi_i \right) = \int \left(\Psi'_i - \tilde{Y}_i \right) P_c\left(\Psi'_i / \Psi_i \right) d\Psi'_i$$

(22)

is linearly related to $(\Psi_i - \tilde{Y}_i)$ (exact for some pdf shapes, including the gaussian one, see Borghi [2]), they suggested the following model:

$$SSMT = +\frac{1}{2} \sum_{i=1}^{N} C_{Y_i} \frac{\partial}{\partial \Psi_i} \left(\frac{\Psi_i - \tilde{Y}_i}{\tau_t} \tilde{P} \right)$$

(23)

where C_{Y_i} is an appropriate constant; $\tau_t = k/\varepsilon$ is the integral time of turbulence. This model is attractive in its simplicity, and when applied by Borghi and Bonniot [25] has given very realistic results, but there is no relaxation to a gaussian distribution in decaying turbulence. This incorrect behaviour might be expected since the model contains no information about the shape of the distribution, only the mean value $\tilde{Y}_i$ appears.

- When the number of species taken into account increases, traditional finite difference schemes are not feasible for solving the pdf equation. Monte Carlo simulation can be used to obtain a statistical solution for the pdf equation. In this simulation (proposed by Pope [26]) the pdf is represented by an ensemble of particles per grid cell and the evolution of these particles is obtained by simulating the effect of various terms in (17):
- Transport in physical space is simulated by exchanging particles between neighbouring nodes,
- Reaction and scalar dissipation are represented by moving particles in the probability space.

This numerical technique is well suited for the use of stochastic models for the mixing term. Like, for instance, the Curl's coalescence redispersion model and the Modified Curl Model.

- Curl [27] expressed a mixing model in the pdf form while Spielman and Levenspiel [28] devised the equivalent stochastic model. If N_p is the number of pair particles randomly selected for the mixing $(\phi^a,\phi^b),...,(\phi^y,\phi^z)$, the mixing is performed by changing the value ϕ of to:

$$\overset{a}{\phi}(t + \Delta t) = \overset{b}{\phi}(t + \Delta t) = \frac{1}{2}\left(\overset{a}{\phi} + \overset{b}{\phi}\right) \tag{24}$$

This simulation leaves the mean value unchanged while the variance decreases. In term of pdf, the mixing term becomes

$$SSMT = \frac{1}{\tau_{ex}}\left| \int_{\Psi'} \int_{\Psi''} \tilde{P}(\Psi'') \tilde{P}(\Psi') \delta\left(\Psi - \frac{1}{2}(\Psi' + \Psi'')\right) d\Psi' d\Psi'' - \tilde{P}(\Psi) \right| \tag{25}$$

It is known that the behaviour of this model is incorrect in the short time limit and, like the Dopazo O'Brien model, leads to unbounded flatness and superskewness values in decaying turbulence as time goes to infinity. To overcome this problem Dopazo [29], Janika, Kolbe and Kollman [30] have suggested a modification to Curl's model and Pope [31] has suggested to bias the sampling probability of particles according to the time (age) elapsed since their last mixing events. In term of pdf, this can expressed by:

$$SSMT = \frac{1}{\tau_{ex}}\left[\int_{\Psi'} \int_{\Psi''} X(\Psi') X(\Psi'') T(\Psi,\Psi',\Psi'') d\Psi' d\Psi'' - \tilde{P}(\Psi) \right] \tag{26}$$

where

$$X(\Psi) = \int_0^\infty Z(\eta)\, Q(\Psi,\eta)\, d\eta \qquad (27)$$

and $Q(\Psi,\eta)$ is the joint pdf of Ψ and η (age of the element), the transition probability is

$$T(\Psi,\Psi',\Psi'') = \int_0^1 F(\theta)\, \delta\left(\Psi - 1(1-\theta)\,\Psi' - \frac{1}{2}\theta(\Psi' + \Psi'')\right) d\theta \qquad (28)$$

The role of the parameter θ is to control the extend of the mixing. $F(\theta)$ is the pdf of θ. With a continuous pdf $F(\theta)$, the model itself produces continuous pdf. Pope proposed an appropriate $Z(\eta)$ function. This model was used with success by Pope [31], Kollman and Chen [23].

- The change in properties of the particles representing the pdf due to chemical reaction is the effect of the chemical reaction rate. Unfortunately its direct calculation would requires an enormous computer time. A way for solving the problem is to set up a "look-up" table with an appropriate reduced scheme [32].
- Finally, the turbulent diffusion in physical space due to the velocity fluctuations has to be modeled. This has be done with the use of a diffusion

coefficient: $\overline{\rho\, U'_\iota \dfrac{\partial P}{\partial x_\iota}} = \dfrac{\partial}{\partial x_\iota}\left(\dfrac{\mu_t}{\sigma_t}\dfrac{\partial \widetilde{P}}{\partial x_\iota}\right)$; or a second order closure model:

$$\overline{\rho\, U'_\iota \frac{\partial P}{\partial x_\iota}} = Cs\overline{\rho}\,\frac{K}{\epsilon}\,\frac{\partial}{\partial x_\iota}\left(\overline{U'_\iota U'_m}\frac{\partial \widetilde{P}}{\partial x_m}\right).$$

Pope [33] avoids this new closure problem by using the joint pdf that includes the velocity components as random variables:$P(\Omega_1,\Omega_2,\Omega_3,\Psi_1,...,\Psi_N)$. The equation for this new pdf raises additional closure problem, particularly due to the pressure influence, but it allows to handle the problem of the turbulence model at the same time.

The pdf approach has a great potential, this method gives all the informations about the statistical description of the reacting flow field: all the moments of all the thermochemical variables are known. But the Monte Carlo simulations are first order in space and time and so are not easy to use in industrial devices, because they are yet too time consuming. As an example of the possibilities of the pdf approach, one can see the recent results of Chen and Kollmann [34], compared with the experiments of [44].

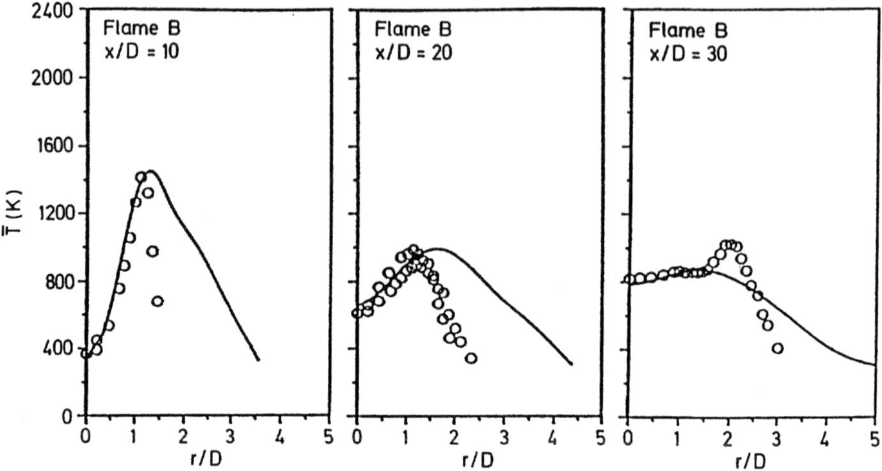

Fig. 4:   Mean temperature profiles predicted for the flame B of [44]

2.3.4. The "Lagrangian models"

The Lagrangian closures for the diffusion term have been first proposed in the field of chemical engineering as the Curl model. The simplest example is the IEM model of Villermaux et al [35]; it represents the small scale mixing term as a mixing with a mean fluid particle, introducing a mixing exchange coefficient τ_{ex}, in the case of well stirred reactors.

As the turbulent Lagrangian models propose to determine the evolution of a fluid particle in the local discretized probability space, these closures are now used for the trajectory of a fluid particle:

$$\frac{dY_i}{dt} = \frac{\tilde{Y}_i - Y_i}{\tau_{ex}} + \dot{\omega}_i \tag{29}$$

One can deduce from the pdf equation, without too much difficulties (see for example [36]), that this model is exactly the first one of Dopazo and O'Brien (that they call LMSE). The model (29) has been explicitly used by Frost [24], to find his pdf equation.

Other Lagrangian closures, the ESCIMO model of Spalding [37], the "Four Environement Model" of Mehta and Tarbell [38], have been proposed so far.

Based on the same IEM equation, with a random exchange time instead of a mean one, a new closure in the MIL model (Modèle Intermittent Lagrangien) has been proposed recently by Gonzalez and Borghi [39].

The turbulence chemistry interaction can be expressed with the extension of (29) in the non homogeneous case by two models: the MIl model, which assumes a very high global activation energy, the PEUL model, wich solves (29) for several fluid particles. In complex turbulent flow fields, the use of lagrangian models is very similar to the simulation of a pdf equation with a Monte Carlo technique.

In the following section, the implementation of the PEUL model and some results concerning methane jet flames are presented.

3.   RECENT RESULTS WITH THE PEUL MODEL

The high price that has to be paid, in term of computing time, with the pdf models allows to think that an interesting compromise may again exist for engineering purposes in getting an approximate shape of the multidimensional pdf $\tilde{P}(\psi_1,...,\psi_N)$. On the other hand, the use of Monte Carlo method for computing the pdf equation, where "particles" are followed within the flame, shows the great interest of lagrangian concepts in order to physically describe the pdf. These two remarks explain the development, these last years, of approximate Probabilistic Eulerian Lagrangian models (PEUL).

The LNH is particulary interested in the development of an approximate PEUL model for industrial devices, following the studies and propositions of Borghi [40]. It may be a good compromise to take into account a finite rate reaction with reasonable computer time. This model is implemented in a finite difference code HADES, for 2D or axisymetric turbulent reactive flows with a k - ε model. Details about the numerical method (fractional step) can be found in [43]. The following results are obtained with a simple reaction step.

$$CH_4 + \frac{7}{4}O_2 \rightarrow \frac{1}{2}(CO + CO_2) + 2H_2O \qquad (30)$$

3.1. BASIC PRINCIPLE FOR THE PEUL APPROACH

The PEUL model considers particles only in the vicinity of each point of an eulerian computation code, and assumes that particles entering the neighbouring cell have only a limited number of possible states. For a diffusion jet flame, in the vicinity of any given point of the flow field, we can represent the life of these samples of particles in the plane (Y_i, Φ) where Y_i is the mass fraction of the species i in the particle.

This approach can be connected to the pdf formulation.

- If local homogeneity is assumed, the pdf transport equation evolves only in the composition space as:

$$\bar{\rho}\,\frac{\partial \widetilde{P}}{\partial t} = -\sum_{i=1}^{N}\frac{\partial}{\partial \Psi_i}\left(\widetilde{S}_i + \widetilde{D}_i\right) \tag{31}$$

where

$$\widetilde{S}_i = \rho\,\dot{\omega}_i\left(\Psi_1,\ldots,\Psi_N\right)\widetilde{P} \tag{32}$$

$$\widetilde{D}_i = \frac{\widetilde{Y}_i - \Psi_i}{\tau_{ex}}\,\widetilde{P} \tag{33}$$

Pope [42] has shown that in particle-model form, the same equation becomes:

$$\rho^{(n)}\,\frac{dy_i^{(n)}}{dt} = \frac{S_i\left(y_1^{(n)},\ldots,y_N^{(n)}\right) + D_i\left(y_1^{(n)},\ldots,y_N^{(n)}\right)}{\widetilde{P}\left(y_1^{(n)},\ldots,y_N^{(n)}\right)}\quad \forall n \tag{34}$$

or

$$\rho^{(n)}\,y_i^{(n)}(t) = \rho^{(n)}\,y_i^{(n)}(0) + \int_0^t \frac{S_i\left(y_1^{(n)},\ldots,y_N^{(n)}\right) + D_i\left(y_1^{(n)},\ldots,y_N^{(n)}\right)}{\widetilde{P}\left(y_1^{(n)},\ldots,y_N^{(n)}\right)}\,d\xi \tag{35}$$

Where $y^{(n)}$ is a sample of the discretized probability space, or a concentration of a fluid particle. The equation (34) is a trajectory in the composition space, which is similar to eq. (29) when the Dopazo O'Brien closure is used.

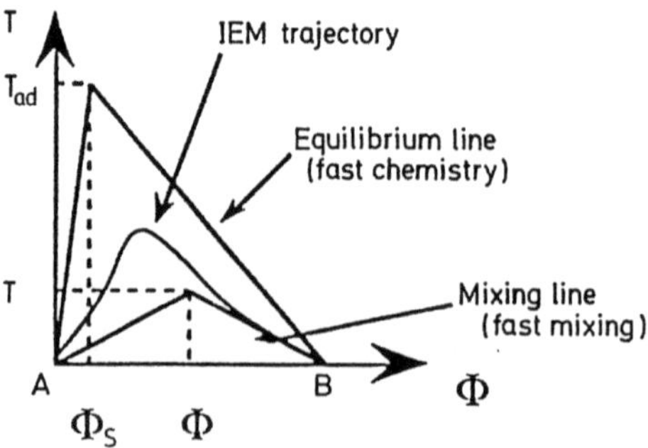

Fig. 5: Trajectories of particles from pure A jet and from pure B jet F is the final state $\left(\frac{dT}{dt} = 0\right)$

In the particular (Y_i, Φ) plane, (34) becomes:

$$\begin{cases} \dfrac{dY_i}{dt} = \dfrac{\widetilde{Y}_i - Y_i}{\tau_{ex}} + \dot{\omega}_i \\[2ex] \dfrac{d\Phi}{dt} = \dfrac{\widetilde{\Phi} - \Phi}{\tau_{ex}} \end{cases} \tag{36}$$

The points corresponding to the fuel jet an the external air are considered as initial conditions to solve (36). These trajectories can be plotted at the same time in each of the planes (Y_i, Φ), with different i; for instance Figure 7 represent the plane (T, Φ).

The initial conditions and the Damköhler number, $D_i(y_i^{(n)})$ determine the trajectories. In a well-stirred reactor this trajectory is the support of the pdf. The limiting situations $D_i(y_i^{(n)}) \gg 1$ and $D_i(y_i^{(n)}) \ll 1$ give the allowed domain for the pdf. These trajectories depending on the magnitude of τ_{ex} with respect to the chemical times involved in with the $\dot{\omega}_i$, are lying between the fast chemistry, and the fast mixing trajectories. A mean reaction rate can be computed as: (37)

$$\widetilde{\dot{\omega}}_i = \int_0^1 \dot{\omega}_i \left(Y_1^T(\Psi), \dots, Y_N^T(\Psi) \right) \widetilde{P}(\Psi) \, d\Psi \tag{37}$$

if the pdf $\widetilde{P}(\Psi)$ of the inert Φ is known. In the HADES PEUL model $\widetilde{P}(\Psi)$ is assumed as a beta function eq. (7).

- In the non homogeneous case, the main assumption is that a number N_{tr} of these two trajectories, with different Damköhler numbers (and/or different "initial conditions"), is enough to represent the chemical state of the fluid particles at one point in the turbulent flow field. A weight α_k is affected respectively to the trajectories k. The joint pdf $\widetilde{P}(\Psi_1, \dots, \Psi_N)$ can be supported by these trajectories:

$$\widetilde{P}(\Psi_1, \dots, \Psi_N) = \widetilde{P}(\Psi) \sum_{k=1}^{N_{tr}} \alpha_k \prod_{i=1}^{N} \delta\left(\Psi_i - Y_i^{T^k}(\Psi) \right) \tag{38}$$

(2) becomes:

$$\widetilde{\dot{\omega}}_i = \sum_{k=1}^{N_{tr}} \alpha_k \int_0^1 \dot{\omega} \left(Y_i^{T^k}(\Phi) \right) \widetilde{P}(\Phi) \, d\Phi \tag{39}$$

Where $Y_i^{T^k}(\Phi)$ represents these k trajectories in the plane (Y_i, Φ). The mean reaction rate $\tilde{\omega}$, given by (32), can be used in eulerian equations like (2) in order to compute the $\tilde{Y}_i$ and determine a new field of $\bar{\rho}$. So, the joint use of the lagrangian equation (29), together with eulerian equations (1) close the problem.

3.2. THE HADES PEUL MODEL

In the HADES PEUL computer code, each mesh cell is characterized by the mean inert scalar, the rms value of its fluctuation and the mean mass fraction of oxygen $\tilde{Y}_{O_2}$. The chemical system state is discretized in three possible evolutions (six trajectories), wich correspond to different Damköhler numbers.

- Da $(y_i^{(n)}) \gg 1$ Equilibrium:

$$\dot{\omega}_i^{(eq)}(\Phi) = \frac{dy_i^{(eq)}}{d\Phi} \left(\frac{\tilde{\Phi} - \Phi}{\tau_{ex}} \right) - \frac{\tilde{Y}_i - y_i^{(eq)}}{\tau_{ex}} \tag{40}$$

- Da $(y_i^{(n)}) \ll 1$ Pure mixing:

$$\dot{\omega}_i^{(mix)}(\Phi) = 0 \tag{41}$$

- Chemistry/Mixing competition:

$$\begin{cases} y_i^{(C/M)}(t) = y_i^{(C/M)}(0) + \int_0^t \left(\dot{\omega}_i - \frac{\tilde{Y}_i - y_i^{(C/M)}}{\tau_{ex}} \right) d\xi \\[2ex] \Phi(t) = \int_0^t \left(\frac{\tilde{\Phi} - \Phi}{\tau_{ex}} \right) d\xi \end{cases} \tag{42}$$

Depending on the sign of:

$$\Delta Y_{O_2} = \tilde{Y}_{O_2} - \int_0^1 Y_{O_2}^{(C/M)}(\Phi) \, \tilde{P}(\Phi) \, d\Phi \tag{43}$$

a weight α is computed for each evolution, so that the mean mass fraction of O_2 computed in the lagrangian form, is equal to the mean eulerian mass fraction given by (2), so:

$$\text{if} \quad \Delta Y_{O_2} > 0 \rightarrow \tilde{\omega}_i = \alpha \, \tilde{\omega}_i^{(C/M)} + (1 - \alpha) \, \tilde{\omega}_i^{(eq)} \quad \forall i \tag{44}$$

or

$$\text{if} \quad \Delta Y_{O_2} < 0 \rightarrow \tilde{\omega}_i = \alpha \, \tilde{\omega}_i^{(C/M)} \quad \forall i \tag{45}$$

The Chemistry/Mixing competition trajectories have to be integrated very carefully. In the HADES PEUL program, the implicit predictor corrector Mac-Cormack Method is used, and, the grid on the Φ axis is a function of $\gamma = \Delta t/\tau_{ex}$, where Δt is the appropriate step time.

This approach gives quite quantitative interesting results for jet flames, as we will see in 3.2.2. The use of a PEUL method requires less computer time than a pdf model if the chemical scheme is not too complex. For the Monte Carlo method, the reconstruction of the pdf at each point requires a sufficient number of particles of the order of one thousand, while in the PEUL approach a small number of trajectories have to be integrated. In the case of premixed combustion, PEUL may be developed similarly, see [41].

3.3. RESULTS

Comparisons are now reported between computation and measurements (respectively solid line and stars on the following figures) of mean temperature and concentrations of major species for two turbulent diffusion flames.

3.3.1. Computation of a jet flame in open air

In the first case, a natural gas turbulent diffusion flame in air is studied (see. Lockwood [18]). The particular interest of this flame is the finite oxygen concentration on the jet axis upstream of the main combustion zone; its presence is likely due to kinetics effects, and can not be represented with the fast chemistry assumption. This flame is stabilized by a pilot reactive flame of hydrogen. In the computation with the PEUL model, this pilot flame is not taken into account. Since the initial flow field is obtained with the equilibrium assumption, this approximation is not crucial, but it leads to too high levels of $\tilde{Y}_{O_2}$ near the burner exit.

For a Reynolds number of the order of 15 000, the flame is stabilized on the burner nozzle, Figure 6-9 show axial and radial profiles of O_2, CO and temperature.

For a 30 000 Reynolds number, a lifted flame is obtained. We have compared our results with Warnartz and Rogg computations [17]. Figure 10 shows computed axial profiles of mean temperature, Figure 11 of mean mass fraction of O2.

102

3.3.2. Computation of a jet flame with an air coflow

The second validation case of the model is related to the recent measurements of non premixed turbulent methane jet flames performed by Masri, Bilger and Dibble [44]. A central jet of methane, 7,2 mm in diameter is surrounded by a stoichiometric premixed and fully burned annulus of C2H2 - H2 and air with C/H ratio adjusted to that of methane. The outer diameter of the annulus is 18 mm. The coflowing air stream velocity and the pilot burnt gas velocity are 15 and 24 m.s.-1 respectively. Increasing the central jet velocity decreases the turbulent mixing time scale; local flame extinction can then occur when the chemical kinetics are unable to keep up with the turbulent mixing processes. The flame is reignited further downstream as the strain rate is decreased. This flame is a good and severe test for the turbulent combustion models.

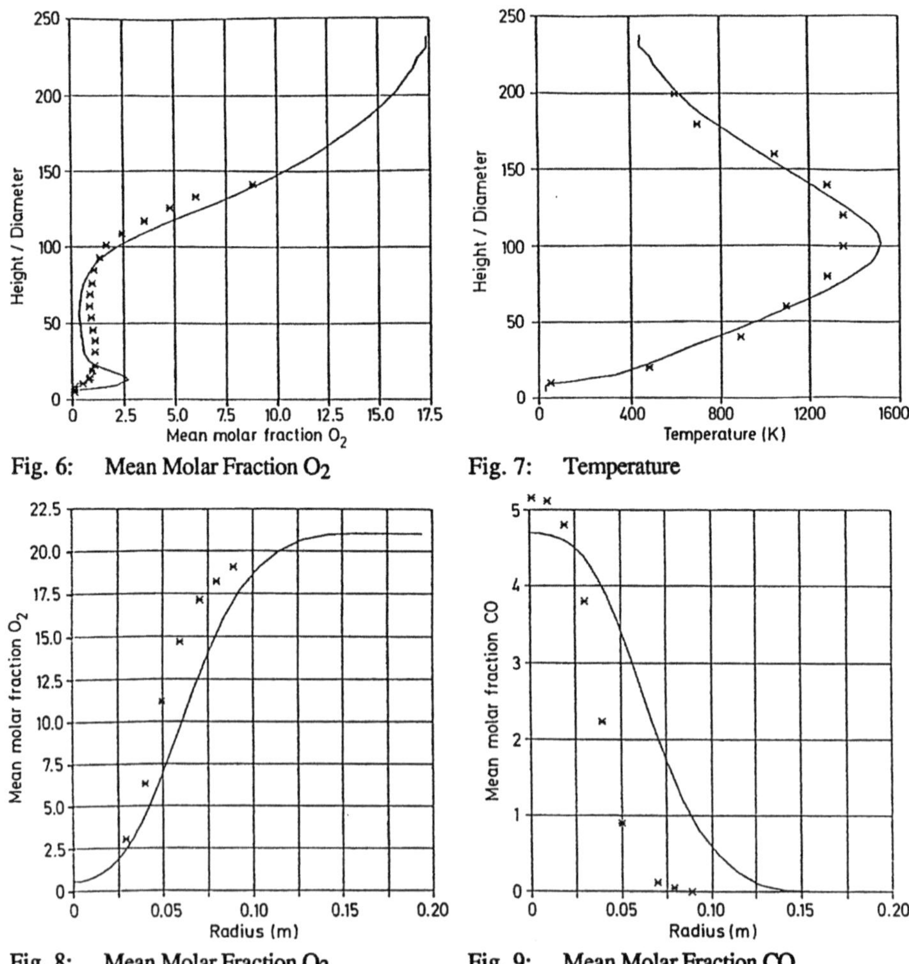

Fig. 6:   Mean Molar Fraction O_2

Fig. 7:   Temperature

Fig. 8:   Mean Molar Fraction O_2

Fig. 9:   Mean Molar Fraction CO

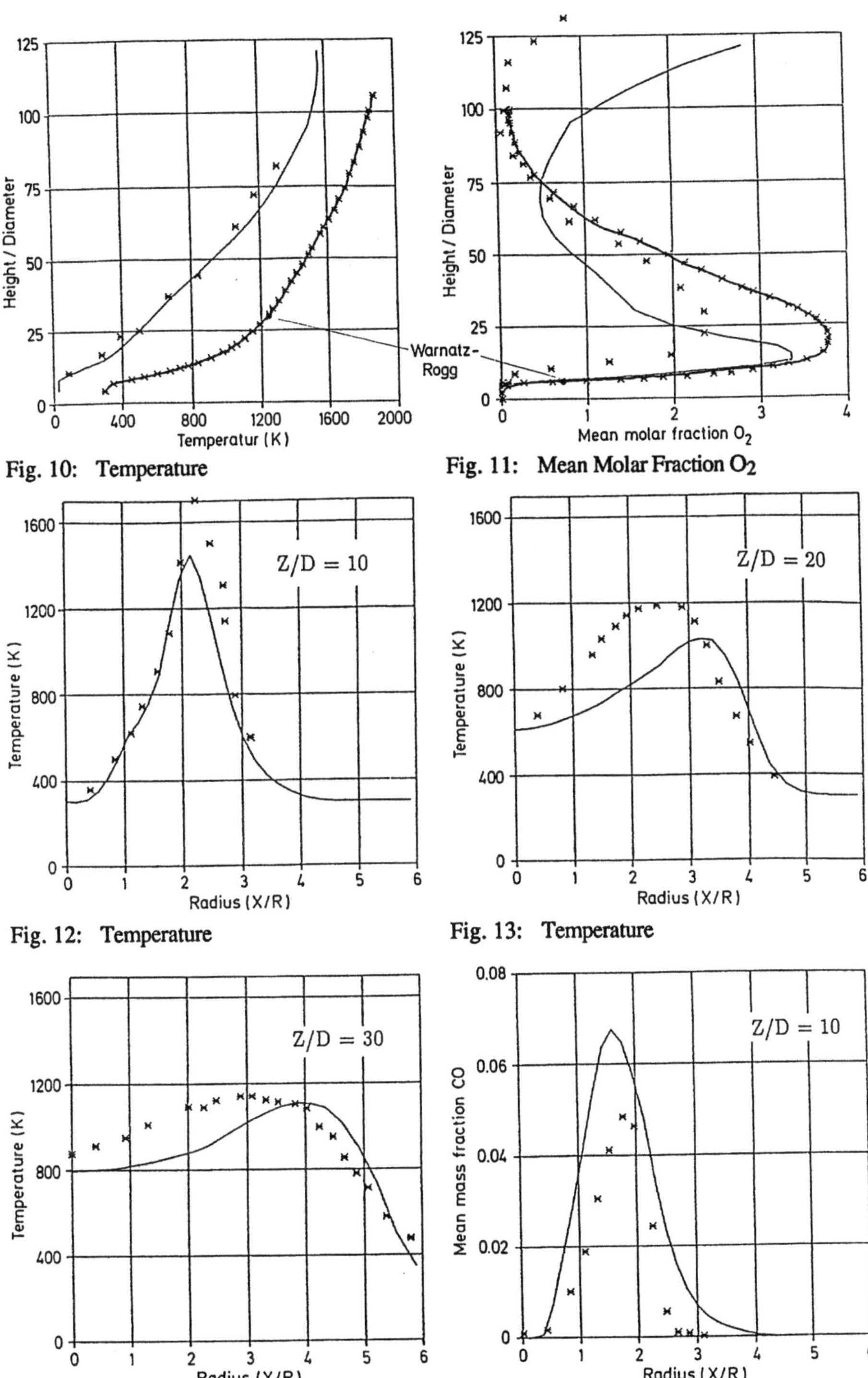

Fig. 10: Temperature

Fig. 11: Mean Molar Fraction O_2

Fig. 12: Temperature

Fig. 13: Temperature

Fig. 14: Temperature

Fig. 15: Mean Mass Fraction CO

104

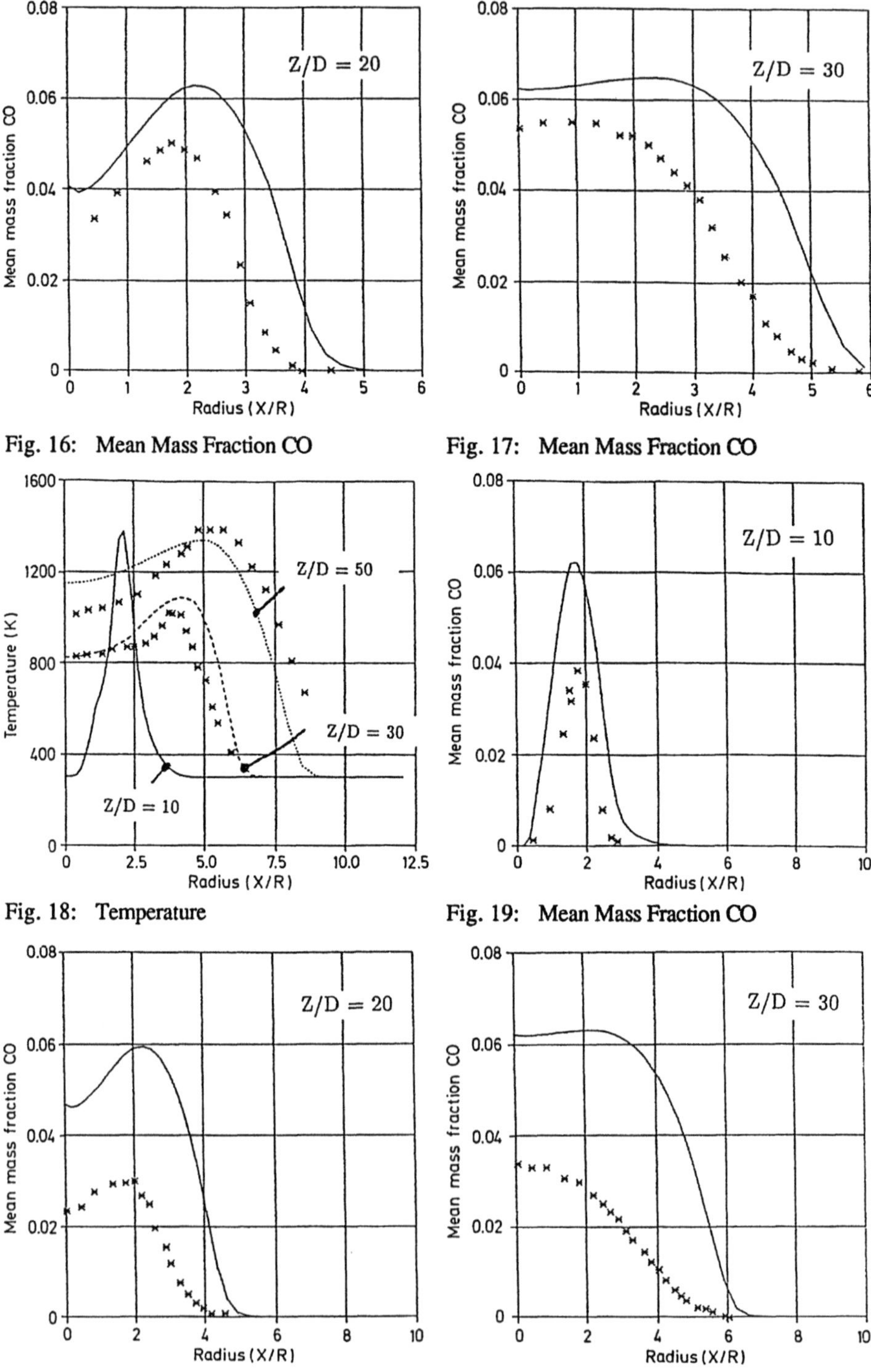

Fig. 16:   Mean Mass Fraction CO

Fig. 17:   Mean Mass Fraction CO

Fig. 18:   Temperature

Fig. 19:   Mean Mass Fraction CO

Fig. 20:   Mean Mass Fraction CO

Fig. 21:   Mean Mass Fraction CO

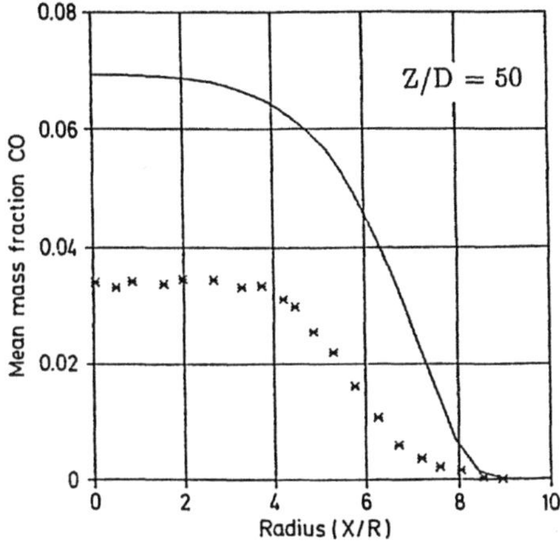

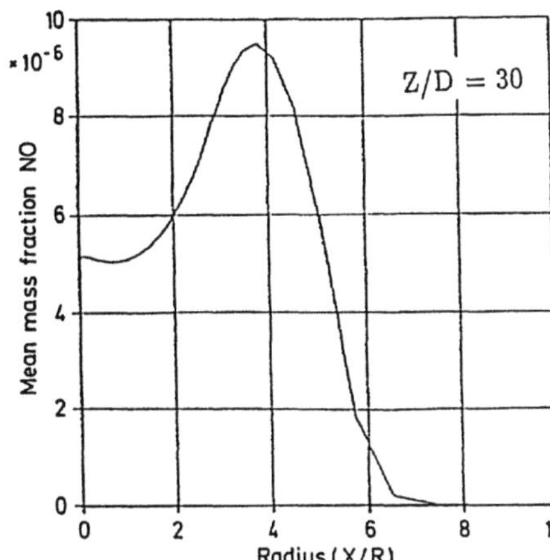

Fig. 22:   Mean Mass Fraction CO   Fig. 23:   Mean Mass Fraction CO

Two flame are computed:

- For $V_{CH_4} = 41m.s^{-1}$ the flame is stabilized at the burner exit. The Figures 12-17 show radial profiles of mean temperature and CO at different heights in the flame,

- For $V_{CH_4} = 48m.s^{-1}$ extinction and reignition occur around 30 diameters and reignition around 40 diameters, which is clearly expressed by the radial profile of temperature (Figure 18.) Figure 19 - 22 display radial profiles of CO. Taken into account the NO formation with the Zeldovich mechanism, NO levels can be predicted (on Figure 23, a good order of magnitude is observed.)

3.3.3. Conclusion

The agreement between numerical and experimental results is quite good. The study of these two flames to assess the capability of our program to predict the good interaction between the turbulent flow and the combustion phenomena is very promising. Even if only one chemical time is taken into account, the qualitative behaviour can be reproduced without the modelling of the radical pool.

We now focus our studies on two points: first, a Reynolds stress model for density varying turbulent flows is developed in our lab [45], and, secondly, the introduction of reduced schemes with radical species is in progress.

4.   AN ATTEMPT FOR TAKING INTO ACCOUNT RADIATION

4.1. THE BASIC EQUATION AND PROBLEM

When radiation is allowed to play, the mean energy equation in the very basic set has to include a radiation flux. The basic expression of the instantaneous radiative source term is:

$$\frac{\partial}{\partial x_1} q_1^r = \frac{\partial}{\partial x_1} \int \int_{4\pi} \omega_1 \ell d^2 \Omega \tag{46}$$

ℓ is the intensity of the radiation in the direction ω_1, integrated for all wavelengths, and the integration over solid angle must be all over the space.

The intensity obeys a transfer equation, which (see for instance [46]), is written as followed:

$$\omega_1 \frac{\partial \ell}{\partial x_1} = \int_v K_a^v \ell_v \, dv - \int_v K_a^v B_v \, dv + \int_v K_s [\ldots] \, dv \tag{47}$$

- K_a^v: the monochromatic absorption factor,
- B_v: the emission function of the Black Body,
- K_s^v a scattering coefficient.

When neglecting the scattering phenomena, and assuming a grey gas, this equation involves ℓ itself and no more ℓ_v, ($\ell = \int \ell_v \, d_v$).

$$\omega_1 \frac{\partial \ell}{\partial x_1} = K_a \left(\ell - \frac{\sigma T^4}{\pi} \right) \tag{48}$$

The grey gas assumption is usually very crude, except perhaps when the radiation is mainly governed by soot. If it be adopted, the difficulties of using (48) instead of (46) are tremendously increased.

In a turbulent flame, the mean radiative flux is needed, i.e., the mean value of $\frac{\partial q_1^r}{\partial x_1}$. Because the presence of the double integral in (46), the computation of the mean of q_1^r requires the joint pdf of ℓ at any point of the space: the radiation phenomena are non local phenomena, with quite large length scales. So it is very difficult to compute, in the general case, the mean value of q_1^r. However the problem is simpler in the two limiting cases of an optically thin medium, or of an optically thick medium.

In the first case, one can write directly:

$$\frac{\partial}{\partial x_l}\bar{q}_l^r = -4\sigma \overline{K_a(T,Y)\,T^4} \tag{49}$$

and so the mean value remains local, although it needs to take into account the correlation between the fluctuations of K_a and the ones of T.

In the second case, one has:

$$\frac{\partial}{\partial x_l}\bar{q}^r_l = -4\sigma\frac{\partial}{\partial x_l}\overline{\left(\frac{T^3}{K_a}\frac{\partial T}{\partial x_l}\right)} \tag{50}$$

Here the mean values is not local, but involves only two close points, because the gradient plays. It is often assumed that small scale fluctuations (at the end of the turbulence spectrum) are very poorly correlated with the large scale fluctuations, and in this case eq. (50) may be simplified in:

$$\frac{\partial}{\partial x_l}\bar{q}^r_l = -4\sigma\frac{\partial}{\partial x_l}\left(\frac{\overline{T^3}}{K_a}\frac{\partial \tilde{T}}{\partial x_l}\right) \tag{51}$$

The problem then, in order to be able to introduce $\dfrac{\partial\,\bar{q}_l^r}{\partial x_l}$ in the mean energy equation, is to compute the correlations $\overline{K_a\,T^4}$ and $\overline{T^3/K_a}$. Symmetrically, it is necessary also to compute the influence of the radiation terms on the fluctuations of temperature and mass fractions.

4.2. SOME CHARACTERISTIC SCALES

Before any calculation, it is very often useful to know the order of magnitude of the relevant characteristic scale of the phenomena we are interested in, this can be done simply.

If L is the characteristic length scale of the turbulent flame; for an optically thin medium ($K_a L \ll 1$), the time scale associated to the radiative heating (or cooling) is:

$$\tau_r = \rho\, C_p/K_a\,\sigma\, T^3.$$

For an optically thick medium ($Ka\ L \gg 1$), it is: $\tau_r = \rho\, C_p/K_a\ L^2/\sigma\ T^3$.

In the turbulent flame, this time scale is to be compared with the integral time scale of the turbulence, τ_t, which is very roughly of the order of L/U, where U is a reference velocity. The influence of the radiation will depend on the temperature level in the flame, the size, the velocity and the absorption coefficient.

For instance, for a flame with L = 1m, Ka = .2m-1, T = 2000K, $\tau_r \approx$ s; and for L=10m the medium is thicker $\tau_r \approx$ 11 s.

τ_r will be smaller only for higher temperatures. The turbulence time τ_r is usually, in furnaces, less than 1s.

4.3. A PROPOSITION OF USING THE PEUL MODEL WITH RADIATION

The approximate method PEUL can handle without any great difficulty the radiation in the optically thin limit.

The lagrangian equations corresponding to this case are:

$$
\begin{cases}
\dfrac{dY_i}{dt} = \dfrac{\widetilde{Y}_i - Y_i}{\tau_{ex}} + \dot{\omega}_i \\[2ex]
\dfrac{d\Phi}{dt} = \dfrac{\widetilde{\Phi} - \Phi}{\tau_{ex}} \\[2ex]
\dfrac{dh}{dt} = \dfrac{\widetilde{h} - h}{\tau_{ex}} - 4\sigma K_a(T,Y)\, T^4
\end{cases}
\tag{52}
$$

These equations can be integrated, with $\widetilde{Y}_i$, $\widetilde{h}$, $\widetilde{\Phi}$ and τ_{ex} given, from initial points, absorption can be included in the instantaneous radiative term. The temperature is determined by the relation:

$$
h = \int_{To}^{T} \sum_{i=1}^{N} Y_i\, C_{pi}(T)\, dT + \sum_{i=1}^{N} Y_i\, Q_i^0
\tag{53}
$$

In the vicinity of a given point, the trajectories from A, and B will have the shape of Figure 25, to be compared to the adiabatic case of Figure 5. Figure 24 shows the corresponding trajectories in the plane (h,Φ). The curves $T^r(\Phi)$ that are obtained lie between the curve corresponding to the adiabatic equilibrium, and the curve that would correspond to the equilibrium with radiation losses. The nondimensional parameters τ_r/τ_t and τ_c/τ_t will control the position of $T^r(\Phi)$.

Once the curve $T^r(\Phi)$ and $Y_i^r(\Phi)$ are obtained, as previously, the mean reaction rate $\widetilde{\omega}$ and the mean radiative flux $\bar{q}_t^r$ can be computed. In particular:

$$\frac{\partial}{\partial x_1}\overline{q}_1^r = -\int_0^1 4\sigma K_a T^L(\Phi)^4 \widetilde{P}(\Phi)\frac{\overline{\rho}}{\rho(\Phi)}\,d\Phi \tag{54}$$

Indeed, if we neglect the pressure fluctuations (for low Mach number flames) it can be shown that

$$\rho \overline{P}(\Phi) = \overline{\rho}\widetilde{P}(\Phi) \tag{55}$$

(see [2])

The mean reaction rate and the mean radiation flux can be used in the balance equations for the mean mass fractions and the mean energy, forming a closed set of equations. In this model, the soot content of the flame, which plays usually an important role for radiation, can be taken into account: the soot volume fraction can be represented by one of the components of the vector $\underline{Y}$. A chemical model for soot formation is then needed (but this is not easy).

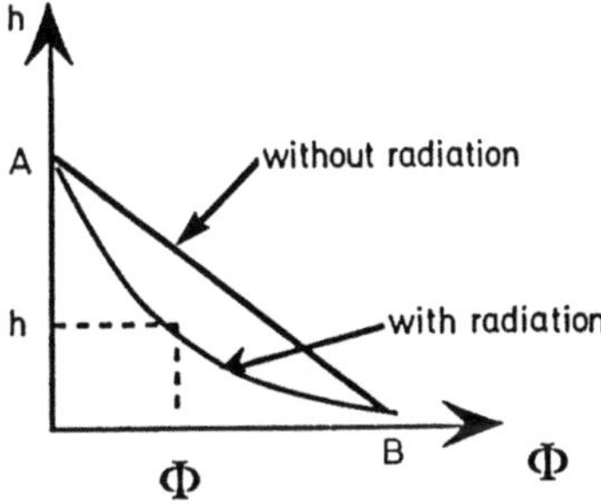

Fig. 24: Trajectories in the plane (h,Φ)

Fig. 25: Trajectories of particles from A and B

The PEUL method appears then very easily extended in this case. But this constitutes a proposition only; indeed, we have not done yet these calculations...

5.   CONCLUSION

A review of existing methods for the numerical computation of non premixed turbulent flames is presented. The state of the art is such that the fluctuations of temperature and main species can be predicted in the models, even for the case of non infinitely fast chemistry. These fluctuations have a great influence on the mean temperature field, and on the mean species fields. The comparisons with experimental results are very encouraging, and show that a noticeable progress has been done. Simple pollutant species, like CO and NO_x, can be predicted now with the right order of magnitude, in some cases. But problems are remaining if we want to include slightly more detailed chemical schemes, which are necessary in order to enlarge the validity domain of the current predictions.

Details are given for the approximate PEUL model which takes into account one finite reaction rate. Its capability to predict jet flames behaviours, particularly lift off phenomena is encouraging.

It seems possible, without a big amount of computation, to extend the PEUL method for including the radiation losses, when the flame can be assumed to be an optically thin medium. Taking into account the more general case of non thin media seems far more complicated, because the joint pdf of fluctuations at many different points of the flame is needed.

REFERENCES

1. Faeth: Prog. Energy. Comb. Sci 9 (1983).

2. Borghi: Turbulent Combustion Modelling; Prog. Energy. Combust. Sci. 14 (1988) 245-292

3. Toor: Mass Transfer in a Dilute Turbulent and Non Premixed Turbulent Systems with Rapid Irreversible Reactions and Equal Diffusivities. A.I.Ch.E.J. 8 (1962) 70

4. Lin, O'Brien: J.F. Mech. 64 part 1 195-206

5. Viollet, Gabillard, Mechitoua: Modélisation Tridimensionnelle des Plasmas en Ecoulement. Revues Phys. Appl. 25 (1990) 843-857

6. Bilger: Turbulent Jet Diffusion Flames. Prog. Energy Combust. Sci. 1 (1976) 87-109

7. Spalding: Chem. Engin. Sci. 95 (1971) 26

8. Lockwood, Naguib: The Prediction of the Fluctuations in the Properties of Free Round-Jet Turbulent Diffusion Flame. Combustion and Flame 24 (1975) 109-124

9. Rhodes, Harsha: AIAA Paper No 72-68 (1972).

10. Drake: Probability Density Functions and Correlations of Temperature and Molecular Concentrations in Turbulent Diffusion Flames. AIAA 81-0103 (1981).

11. Magnussen: Combustion Institute (1976).

12. Mao, Toor: A Diffusion Model for Reactions with Turbulent Mixing. A.I.Ch.E.J. 16 49 49-52.

13. Gibson, Libby: Combustion Sci. Tech. 8 (1972) 29-35.

14. Peters: Laminar Diffusion Flamelet Models in non Premixed Turbulent Combustion; Prog. Energy. Combust. 10 (1984) 319-339.

15. Marble, Broadwell: The Coherent Flame Model For Turbulent Chemical Reactions. TRW Defense and Space Systems Group Redondo, CA Jan (1977).

16. Liew, Bray, Moss J.B.: A Flamelet Model of Turbulent no Premixed Combustion. Comb. Sci. Tech. 27 (1981) 69-73.

17. Rogg, Warnatz: Turbulent Non-Premixed Combustion In Partially Premixed Flamelets with Detailed Chemistry. 21 Symp. on Comb. (1986) 1533-1541.

18. Hassan, Moneib, Lockwood: Fluctuating Temperature and Mean Concentration Measurements in a Vertical Turbulent Jet Diffusion Flame. XXXIV 9-12 Italian Flame Days (1980).

19. Broadwell, Dimotakis: Implications of Recent Experimental Results for Modelling Reactions in Turbulent Flows. AIAA J. 24 6 (1986) 885-889.

20. Lundgren: Distribution Functions in the Statistical Theory of Turbulence. The Physics of fluids 10 5 (1967).

21. Dopazo, O'Brien: An Approach to the Autoignition of a Turbulent Mixture. Acta Astronautica 1 (1974) 1239-1255.

22. Pope: T.S.F. 2 (1979) 31.

23. Chen, Kollmann: Dibble, Pdf Modelling of Turbulent Nonpremixed Methane Jet Flames. SAND 89-8403.

24. Frost: Model of a Turbulent Diffusion Flame Jet. Fluid Mech. Soviet Reasearch 4 2 (1975) 124 - 133.

25. Bonniot, Borghi: Joint Probability Density Function in Turbulent Combustion. Acta Astronaut. 6 (1979) 309-307.

26. Pope: A Monte Carlo Method for the P.D.F. Equation of Turbulent Reactive Flow. Combustion Science and Technologie 25 (1981) 159-174.

27. Curl: Dispersed phase mixing; I. Theory and effects of simple reactors. A.I.Ch.E. J. 9 (1963) 185.

28. Spielman, Levenspiel: A Monte Carlo Treatment for Reacting and Coalescing Dispersed Phase Systems. Chem. Engineer. Sci. 20 (1965) 247-254.

29. Valino, Dopazo: A Binomial Sampling Model For Scalar Turbulent Mixing. to be published

30. Janicka, Kolbe, Kollmann: Closure of the Transport Equation for The Probability Density Function of Turbulent Scalar Fields. J.Non-Equilib.Thermo. 4 (1979) 47-66

31. Pope: An Improved Turbulent Mixing Model. Combustion Science and Technology 28 (1982) 131-135.

32. Chen, Kollmann: Application of reduced Chemical Mechanisms for Prediction of Turbulent Nonpremixed Jet Flame Methane. Sand. (1990) 90-8447

33. Pope: Pdf Method for Turbulent Reactive Flows. Prog. Energy Combust. Scien. 11 (1984) 119-192.

34. Chen, Dibble, Kollmann: Application of reduced Chemical Mechanisms for Prediction of Turbulent Nonpremixed Jet Flame Methane. Sand. (1990) 90-8447

35. Aubry, Villermaux: J. Chemical Engineering Science 30 (1975) 457.

36. Vervisch, Garréton: Modélisation de Flames de Diffusion Turbulentes avec Prise en Compte de l'Interaction entre les Effets Thermochimiques et Turbulents. EDF DER LNH HE-44/91.04

37. Spalding: A General Theory of Turbulent Combustion. J. Energy 2 (1977) 16-32.

38. Metha, Tarbell: A Four Environment Model of Mixing and Chemical Reaction Part II comparaison with experiments, A.I.Ch.E. J. 65 125-133.

39. Gonzales, Borghi: A Lagrangian Intemittent Model for Turbulent Combustion Theorical Basis and Comparaisons with Experiments. Post Conference Proceedings of T.S.F. 7 Springer-Verlag

40. Pourbaix, Borghi: Lagrangian models for Turbulent Reacting Flows. T.S.F. 4 Springer-Verlag (1983).

41. Pourbaix, Borghi: Numerical Investigation of a Model of Turbulent Combustion of Hydrocarbons. AIAA 19 th Aero. SCI. Meet. (1981) 12-15.

42. Pope: The Relationship Between the Probability Approach and Partical Models for Reaction in Homogeneous Turbulence. Combustion and flame 35 (1979) 41 -45.

43. Simonin, Bechard: Modélisation des Ecoulements Compressibles. E.D.F. HE 44/89-12.

44. Masri, Bilger, Dibble: Turbulent Nonpremixed Flames of Methane Near Extinction. Combustion and flame 73 (1988) 261-258.

45. Kanniche, Baron, Viollet: 7 th T.S.F. 12.3 Stanford (1979).

46. Oran, Boris: Numerical Simulation of Reactive Flow. Elsevier New York Amsterdam London.

MODELLING OF THERMALLY RADIATING DIFFUSION FLAMES WITH DETAILED CHEMISTRY AND TRANSPORT

Y. Liu and B. Rogg
University of Cambridge, Dept. of Engineering,
Trumpington Street, Cambridge CB2 1PZ, UK

ABSTRACT

Strained laminar counterflow-diffusion flames of CO-H_2-O_2-N_2 mixtures in air are numerically simulated using detailed models for thermal radiation, chemistry and molecular transport. The overall model is validated by comparison of the numerical results with experimental data available in the literature. It is shown that at medium to low values of the strain rate radiation has a particularly strong impact on the flame structure.

INTRODUCTION

In high-temperature combustion, radiation from flames represents an important mode of energy transport. For instance, in gas-turbine combustion chambers, where combustion takes place in the non-premixed mode, a large proportion of the total heat flux from the flame to the liner walls is contributed by radiation, especially by radiation emitted from particles produced in the fuel-rich regions of the flame, thereby causing serious problems with regard to the liner durability. Another example is given by unwanted fires, where flame radiation may exert a strong influence on fire detection, ignition and flame spread. Thus, confident numerical predictions of radiating flames are desirable for design of equipment, diagnosis of malfunction problems and development of equipment modifications for improvement of performance. As a consequence, combustion *and* radiation transfer models are needed, that allow accurate and confident predictions of the temperature and the species concentrations throughout the flame. The detailed prediction of these quantities in *laminar* flames requires detailed models for radiation transfer, chemistry and molecular transport to be taken into account.

In recent years laminar counterflow diffusion flames have become a powerful tool for analysing experimentally, numerically and theoretically a variety of combustion phenomena in laminar non-premixed flows. Furthermore, such flames can be used as the basis to model non-premixed turbulent combustion in the laminar-flamelet regime, see e.g. Peters [1]. From an experimental point of view, counterflow geometries are attractive because they allow strain rate and flow-residence time to be varied from one experiment to the other by simply adjusting the flow rates of the opposed jets. From a numerical and theoretical point of view, these spatially two-dimensional flames are attractive because a similarity formulation exists which allows them to be described one-dimensionally in the similarity space. As a consequence, in the similarity space these flames are amenable to theoretical, say asymptotic, methods, and in numerical simulations detailed models for molecular transport and chemical kinetics can be employed.

Up to date the influences of radiation on the structure of laminar counterflow diffusion flames have received inadequate study. Radiation has been taken into account in only a few studies, such as in the work by Sohrab *et al* [2] who investigated diffusion-flame extinction with thermal radiant loss from the flame zone using asymptotic methods, or in the work by Müller [3], who studied the influence of radiative thermal heat loss on the thermal NO production in laminar $CO/H_2/N_2$ - air diffusion flames.

In the present paper a simplified radiation model is derived from the fundamental integro-differential equation governing radiation transport. Since for the flames considered herein soot formation is not relevant, only thermal radiation of the combustion products H_2O and CO_2 is considered. In contrast to Müller [3], herein boundary conditions are adopted that allow comparison of the numerical predictions to experimental data available in the literature [4]. Specifically, $CO/H_2/N_2$ - air counterflow diffusion flames stabilized in the forward stagnation-point region of the cylindrical Tsuji burner have been modelled. A schematic of this geometry is shown in Figure 1. In the numerical simulations a detailed reaction mechanism and detailed models of molecular transport are taken into account. The influence of radiation heat loss on the flame structure is examined for a wide range of values of the strain rate.

FORMULATION

Boundary-layer equations are adopted to describe the variations of density, velocity, pressure, energy and species mass fractions in a laminar, chemically reacting, low-Mach-number stagnation-point diffusion flame. Specifically, the flow field is modelled in the Tsuji counterflow configuration, a schematic of which is shown in Figure 1. An ideal-gas mixture comprising N chemical species is considered, and Dufour effects, diffusion caused by pressure gradients, and external forces are neglected. The resulting governing equations can be cast into a similarity form, see e.g. Rogg [5], viz.,

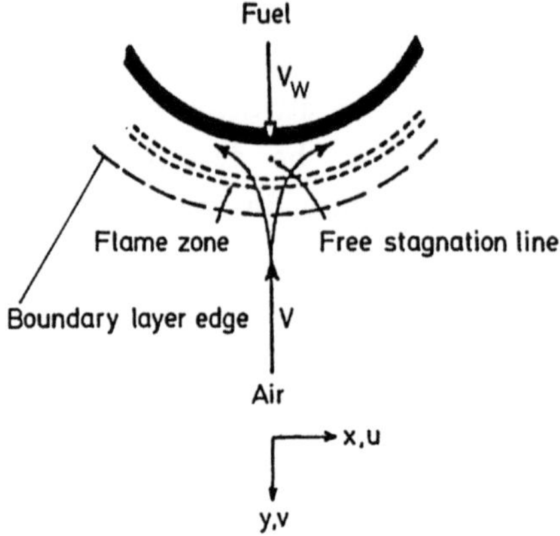

Fig. 1: Schematic of the Tsuji stagnation-point flow geometry.

$$L(f') = \frac{1}{j+1}\left(f'^2 - \frac{\rho_o}{\rho}\right) - \left(\frac{\rho\mu}{\rho_o\mu_o}f''\right)', \tag{1}$$

$$L(T) = -\frac{1}{c_p}\left(\frac{\rho\lambda}{\rho_o\mu_o}T'\right)' + T'\sum_{i=1}^{N}\frac{c_{pi}}{c_p}\frac{\rho Y_i V_i}{\sqrt{\rho_o\mu_o a(j+1)}}$$
$$+\sum_{i=1}^{N}\frac{h_i}{c_p}\frac{w_i}{\rho_a(j+1)} + \frac{1}{c_p\rho(j+1)}\frac{\partial q_R}{\partial y}, \tag{2}$$

$$L(Y_i) = \frac{(\rho Y_i V_i)'}{\sqrt{\rho_o\mu_o\, a(j+1)}} - \frac{w_i}{\rho_a(j+1)}, \quad i = 1,...,N, \tag{3}$$

where the accumulative-convective operator **L** is given by:

$$L(\phi) \equiv f\phi' - \frac{1}{a(j+1)}\frac{\partial\phi}{\partial t},$$

with ϕ representing any of the variables f', T or Y_i, $i = 1,...,N$. Subsequently the subscripts w and e will be used to identify conditions or quantities at the surface of the burner and the boundary-layer edge, respectively. In eqs. (1)-(3) and below, the prime denotes a partial derivative with respect to the similarity variable η, $\eta = \sqrt{(j+1)a/(\rho_e\mu_e)}\int_0^y \rho dy$, with t held fixed; y denotes the physical distance variable. The dependent variables in eqs. (1)-(3) are the stream function derivative f', the temperature T and the mass fractions Y_i, $i=1,...,N$;

upon specifying the pressure, the density ρ is obtained from the ideal-gas equation of state. The stream function f is obtained by integration of $\partial f/\partial \eta = f'$; a is the strain rate which is taken as constant. The transport coefficients, λ and μ, as well as the thermodynamic properties, c_p, c_{pi} and h_i, are defined as usual; V_i denotes the diffusion velocity of species i in the y direction, i=1,...,N. The relationships employed herein for the evaluation of c_p, c_{pi}, h_i and V_i were discussed previously [5]. In the energy equation (2), the term involving q_R is the y component of the radiant-heat-flux vector. Note that as a consequence of the boundary-layer approximation, the x derivative of q_R does not appear in eq. (2). The w_i appearing in eqs. (2) and (3) is the rate of production of species i; it is given by:

$$w_i = W_i \sum_{k=1}^{M} \left(v''_{i,k} - v'_{i,k}\right) \left(AT^\alpha\right)_k \exp\left(-\frac{E_k}{R^0 T}\right) \prod_{j=1}^{N} \left(\frac{\rho Y_j}{M_j}\right)^{v'_{j,k}},$$

(4)

i=1,...,N. In eq. (4) it is understood that forward and backward chemical reactions are treated separately; M is the number of elementary steps contained in the kinetic mechanism; W_i is the molecular weight of species i, i=1,...,N, and $v'_{i,k}$ and $v''_{i,k}$ are its stoichiometric coefficients in reaction k, k=1,...,M, representing there reactant and product, respectively; $(AT^\alpha)_k$ and E_k are the pre-exponential factor in the specific reaction-rate constant and the activation energy of reaction k, respectively; R^0 denotes the universal gas constant.

The reaction mechanism and the rate data as recommended by Warnatz [6,7] are listed in Table 1. The mechanism comprises 77 elementary steps among the 16 chemical species CO, O_2, CO_2, H_2O, H_2O_2, HO_2, H, OH, O, H_2, CHO, CH_2O, N, NO, NO_2 and N_2. The formation of NO and NO_2 has been modelled on the basis of Zel'dovich mechanism, which corresponds to steps 68-77 of Table 1.

Table 1: Kinetic mechanism of elementary reactions for CO-H_2-O_2-N_2 systems. Units are cm, mol s, kJ and K; third-body efficiencies were taken from Warnatz [6,7].

Nº			Reaction			A	α	E
1.	O_2	$+H$	$\rightarrow OH$	$+O$		2.00×10^{14}	0.00	70.30
2.	OH	$+O$	$\rightarrow O_2$	$+H$		1.57×10^{13}	0.00	3.52
3	H_2	$+O$	$\rightarrow OH$	$+H$		5.06×10^{4}	2.67	26.30
4.	OH	$+H$	$\rightarrow H_2$	$+O$		2.22×10^{4}	2.67	18.29
5.	H_2	$+OH$	$\rightarrow H_2O$	$+H$		1.00×10^{8}	1.60	13.80
6.	H_2O	$+H$	$\rightarrow H_2$	$+OH$		4.31×10^{8}	1.60	76.46

No.							A	n	E
7.	OH	$+ OH$		$\rightarrow H_2O$	$+ O$		1.50×10^9	1.14	0.42
8.	H_2O	$+ O$		$\rightarrow OH$	$+ OH$		1.47×10^{10}	1.14	71.09
9.	H	$+ H$	$+ M'$	$\rightarrow H_2$	$+ M'$		1.80×10^{18}	-1.00	0.00
10.	H_2	$+ M'$		$\rightarrow H$	$+ H$	$+ M'$	7.26×10^{18}	-1.00	436.81
11.	H	$+ OH$	$+ M'$	$\rightarrow H_2O$	$+ M'$		2.20×10^{22}	-2.00	0.00
12.	H_2O	$+ M'$		$\rightarrow H$	$+ OH$	$+ M'$	3.83×10^{23}	-2.00	499.48
13.	O	$+ O$	$+ M'$	$\rightarrow O_2$	$+ M'$		2.90×10^{17}	-1.00	0.00
14	O_2	$+ M'$		$\rightarrow O$	$+ O$	$+ M'$	6.55×10^{18}	-1.00	495.58
15.	H	$+ O_2$	$+ M'$	$\rightarrow H_2O$	$+ M'$		2.30×10^{18}	-0.80	0.00
16.	HO_2	$+ M'$		$\rightarrow H$	$+ O_2$	$+ M'$	3.19×10^{18}	-0.80	195.39
17.	HO_2	$+ H$		$\rightarrow OH$	$+ OH$		1.50×10^{14}	0.00	4.20
18.	OH	$+ OH$		$\rightarrow HO_2$	$+ H$		1.50×10^{13}	0.00	170.84
19.	HO_2	$+ H$		$\rightarrow H_2$	$+ O_2$		2.50×10^{13}	0.00	2.90
20.	H_2	$+ O_2$		$\rightarrow HO_2$	$+ H$		7.27×10^{13}	0.00	244.33
21.	HO_2	$+ H$		$\rightarrow H_2O$	$+ O$		3.00×10^{13}	0.00	7.20
22.	HO_2	$+ O$		$\rightarrow HO_2$	$+ H$		2.95×10^{13}	0.00	244.51
23.	HO_2	$+ O$		$\rightarrow OH$	$+ O_2$		1.80×10^{13}	0.00	-1.70
24.	OH	$+ O_2$		$\rightarrow HO_2$	$+ O$		2.30×10^{13}	0.00	231.71
25.	HO_2	$+ OH$		$\rightarrow H_2O$	$+ O_2$		6.00×10^{13}	0.00	0.00
26.	H_2O	$+ O_2$		$\rightarrow HO_2$	OH		7.52×10^{14}	0.00	304.09
27.	HO_2	$+ HO_2$		$\rightarrow H_2O_2$	$+ O_2$		2.50×10^{11}	0.00	-5.20
28.	OH	$+ OH$	$+ M'$	$\rightarrow H_2O_2$	$+ M'$		3.25×10^{22}	-2.00	0.00
29.	H_2O_2	$+ M'$		$\rightarrow OH$	$+ OH$	$+ M'$	1.69×10^{24}	-2.00	202.29
30.	H_2O_2	$+ H$		$\rightarrow H_2$	$+ HO_2$		1.70×10^{12}	0.00	15.70
31.	H_2	$+ HO2$		$\rightarrow H_2O_2$	$+ H$		1.32×10^{12}	0.00	83.59
32.	H_2O_2	$+ H$		$\rightarrow H_2O$	$+ OH$		1.00×10^{13}	0.00	15.00
33.	H_2O	$+ OH$		$\rightarrow H_2O_2$	$+ H$		3.34×10^{12}	0.00	312.19
34.	H_2O_2	$+ O$		$\rightarrow OH$	$+ HO_2$		2.80×10^{13}	0.00	26.80
35.	OH	$+ HO_2$		$\rightarrow H_2O_2$	$+ O$		9.51×10^{12}	0.00	86.67
36.	H_2O_2	$+ OH$		$\rightarrow H_2O$	$+ HO_2$		5.40×10^{12}	0.00	4.20
37.	H_2O	$+ HO_2$		$\rightarrow H_2O_2$	$+ OH$		1.80×10^{13}	0.00	134.75
38.	CO	$+ OH$		$\rightarrow CO_2$	$+ H$		4.40×10^6	1.50	-3.10
39.	CO_2	$+ H$		$\rightarrow CO$	$+ OH$		4.97×10^8	1.50	89.74
40.	CO	$+ HO_2$		$\rightarrow CO_2$	$+ OH$		1.50×10^{14}	0.00	98.70
41.	CO_2	$+ OH$		$\rightarrow CO$	$+ HO_2$		1.70×10^{15}	0.00	358.18
42.	CO	$+ O$	$+ M'$	$\rightarrow CO_2$	$+ M'$		7.10×10^{13}	0.00	-19.00
43.	CO_2	$+ M'$		$\rightarrow CO$	$+ O$	$+ M'$	1.42×10^{16}	0.00	502.65
44.	CO	$+ O_2$		$\rightarrow CO_2$	$+ O$		2.50×10^{12}	0.00	200.00
45.	CO_2	$+ O$		$\rightarrow CO$	$+ O_2$		2.21×10^{13}	0.00	226.07
46.	CHO	$+ H$		$\rightarrow CO$	$+ H_2$		2.00×10^{14}	0.00	0.00

47.	CO	$+H_2$		$\rightarrow$ CHO	$+H$		1.30×10^{15}	0.00	376.49
48.	CHO	$+O$		$\rightarrow$ CO	$+OH$		3.00×10^{13}	0.00	0.00
49.	CO	$+OH$		$\rightarrow$ CHO	$+O$		8.55×10^{13}	0.00	368.48
50.	CHO	$+O$		$\rightarrow$ CO_2	$+H$		3.00×10^{13}	0.00	0.00
51.	CO_2	$+H$		$\rightarrow$ CHO	$+O$		9.66×10^{15}	0.00	6.32
52.	CHO	$+OH$		$\rightarrow$ CO	$+H_2O$		1.00×10^{14}	0.00	0.00
53.	CO	$+H_2O$		$\rightarrow$ CHO	$+OH$		2.80×10^{15}	0.00	439.16
54.	CHO	$+O_2$		$\rightarrow$ CO	$+HO_2$		3.00×10^{12}	0.00	0.00
55.	CO	$+HO_2$		$\rightarrow$ CHO	$+O_2$		6.70×10^{12}	0.00	135.07
56.	CHO	$+M'$		$\rightarrow$ CO	$+H$	$+M'$	7.10×10^{14}	0.00	70.30
57.	CO	$+H$	$+M'$	$\rightarrow$ CHO	$+M'$		1.14×10^{15}	0.00	9.98
58.	CH_2O	$+H$		$\rightarrow$ CHO	$+H_2$		2.50×10^{13}	0.00	16.70
59.	CHO	$+H_2$		$\rightarrow$ CH_2O	$+H$		1.99×10^{12}	0.00	78.01
60.	CH_2O	$+O$		$\rightarrow$ CHO	$+OH$		3.50×10^{13}	0.00	14.60
61.	CHO	$+OH$		$\rightarrow$ CH_2O	$+O$		1.22×10^{12}	0.00	67.90
62.	CH_2O	$+OH$		$\rightarrow$ CHO	$+H_2O$		3.00×10^{13}	0.00	5.00
63.	CHO	$+H_2O$		$\rightarrow$ CH_2O	$+OH$		1.03×10^{13}	0.00	128.97
64.	CH_2O	$+HO_2$		$\rightarrow$ CHO	$+H_2O_2$		1.00×10^{12}	0.00	33.50
65.	CHO	$+H_2O_2$		$\rightarrow$ CH_2O	$+HO_2$		1.03×10^{11}	0.00	26.92
66.	CH_2O	$+M'$		$\rightarrow$ CHO	$+H$	$+M'$	1.40×10^{17}	0.00	320.00
67.	CHO	$+H$	$+M'$	$\rightarrow$ CH_2O	$+M'$		2.76×10^{15}	0.00	-55.51
68.	O	$+N_2$		$\rightarrow$ N	$+NO$		1.90×10^{14}	0.00	319.03
69.	N	$+NO$		$\rightarrow$ O	$+N_2$		4.23×10^{13}	0.00	4.248
70.	NO	$+H$		$\rightarrow$ N	$+OH$		1.30×10^{14}	0.00	205.85
71.	N	$+OH$		$\rightarrow$ N	$+OH$		4.79×10^{13}	0.00	5.226
72.	NO	$+O$		$\rightarrow$ N	$+O_2$		2.40×10^{9}	1.00	161.67
73.	N	$+O_2$		$\rightarrow$ NO	$+O$		1.13×10^{10}	1.00	27.82
74.	NO	$+HO_2$		$\rightarrow$ NO_2	$+OH$		3.00×10^{12}	0.50	10.04
75.	NO_2	$+OH$		$\rightarrow$ NO	$+HO_2$		1.20×10^{13}	0.50	47.22
76.	NO	$+OH$		$\rightarrow$ NO_2	$+H$		5.90×10^{12}	0.00	129.49
77.	NO_2	$+H$		$\rightarrow$ NO	$+OH$		2.35×10^{14}	0.00	0.048

THE RADIATION MODEL

Since we are interested in steady solutions to the governing equations, time-dependencies will be neglected in the derivation of the radiation model. For a monochromatic beam passing through a non-scattering medium with absorption coefficient k_λ, the radiation transfer equation can be written as [8]

$$\cos\theta \frac{dI_\lambda}{dy} = k_\lambda \left[\frac{e_{b\lambda}(y)}{\pi} - I_\lambda(y,\cos\theta) \right], \tag{5}$$

120

where I_λ is the intensity of the beam at wavelength λ, θ the angle between the direction of the radiative path and the coordinate y, and $e_{b\lambda}$ the spectral blackbody emissive power given by Planck's law. Nontrivial algebra shows that, if the assumption of the optically thin limit is made, the monochromatic radiation flux in differential form can be written as

$$\frac{dq_{R\lambda}}{dy} = k_\lambda \left[4e_{b\lambda}(T) - 2(B_{\lambda,w} + B_{\lambda,e}) \right],$$

(6)

where B_λ denotes the monochromatic radiation intensity multiplied by π. The expression finally to be substituted into the energy equation is obtained from eq. (6) by adopting the grey-medium approximation and by integrating eq. (6) over all wave lengths, viz.,

$$\frac{dq_R}{dy} = 2k_P(2\sigma T^4 - B_w - B_e).$$

(7)

In eq. (7), k_P denotes the Planck mean absorption coefficient, σ is the Stefan-Boltzmann constant, and B is given by $B = \varepsilon \sigma T^4$, where ε denotes the mean emissivity. At the boundary-layer edge ε is that of the incoming air stream, at the burner surface ε represents the emissivities of both the burner-surface and the gas mixture. The Planck mean absorption coefficient accounts for the absorption and emission from the gaseous species CO_2 and H_2O, and is expressed as

$$k_P = p[X_{CO_2} k_{P,CO_2}(T) + X_{H_2O} k_{P,H_2O}(T)].$$

(8)

Here p is the pressure, and $k_{p,i}$ and X_i denote the mean absorption coefficient and mole fraction, respectively, of species i. Data given by Hubbard and Tien [9] are used to approximate the $k_{p,i}$'s as polynomials in T.

THE BOUNDARY CONDITIONS

Specification of the problem is completed by imposing boundary conditions at the cylinder surface and at the edge of the boundary-layer. For the temperature T, the mass fractions Y_i and the stream-function derivative f' herein straightforward steady-state boundary conditions are adopted; the time derivatives of these quantities have been retained in the governing equations solely in order to allow a time-dependent approach to a steady-state solution. The radiative fluxes at the boundaries are described as $B_w = \varepsilon_w \sigma T_w^4$ and $B_w = \varepsilon_e \sigma T^4$; ε_w and ε_e represent the emissivities of the boundaries which are taken as constants.

THE NUMERICAL METHOD

The numerical method employed herein was discussed previously [10,11] and, therefore, only a short outline is given here. The governing equations are discretized on a mesh of grid points, and the resulting system of nonlinear equations is then solved by Newton's method. In applying Newton's method a damping strategy is employed which allows the Jacobian to be re-evaluated only periodically. Adaptive gridding is implemented into the numerical procedure. It is worthwhile to note that the employed adaptive-gridding strategy leads to CPU-time savings in the order of a factor of 20 compared with calculations performed on, say, an equidistant grid.

RESULTS AND DISCUSSION

For the results shown in Figs. 2 to 7 boundary conditions were chosen to match those of the flames investigated experimentally by Drake and Blint [4]. Thus, on a molar basis the fuel composition was taken as 40% CO, 30% H_2 and 30% N_2, and the air composition as 79% N_2 and 21% O_2. The temperature at the burner surface was taken as 395 K and 465 K, respectively, that at the boundary-layer edge as 298 K.

Shown in Figure 2 are temperature profiles for strain rates of 70 s^{-1} and 180 s^{-1}. The symbols in this Figure represent experimental results by Drake and Blint [4], who have also carried out numerical simulations (not shown here), however, without taking effects of radiation into account. Although these authors use a different kinetic mechanism, their numerical results virtually agree with ours obtained without radiation model (solid lines). Thus, it can be concluded, firstly, that the chemistry of CO-H_2-O_2-N_2 systems is sufficiently well understood, and, secondly, that the difference of the experimentally and numerically determined flame position at low values of the strain rate must be contributed to deficiencies in the experimental technique or the flow-field model. Arguments along these lines can be found in Drake and Blint [4], and, therefore, need not to be repeated here. It is interesting to note that the numerical simulations with radiation model predict absolute temperatures closer to the experimental data than without radiation model.

Shown in Figure 3 is the computed maximum flame temperature T_{max} as a function of 1000 times the reciprocal of the strain-rate; circles represent results without radiation model, i.e., $q_R=0$, triangles results with radiation model; the solid lines are interpolated, the dashed lines extrapolated. As it is to be expected

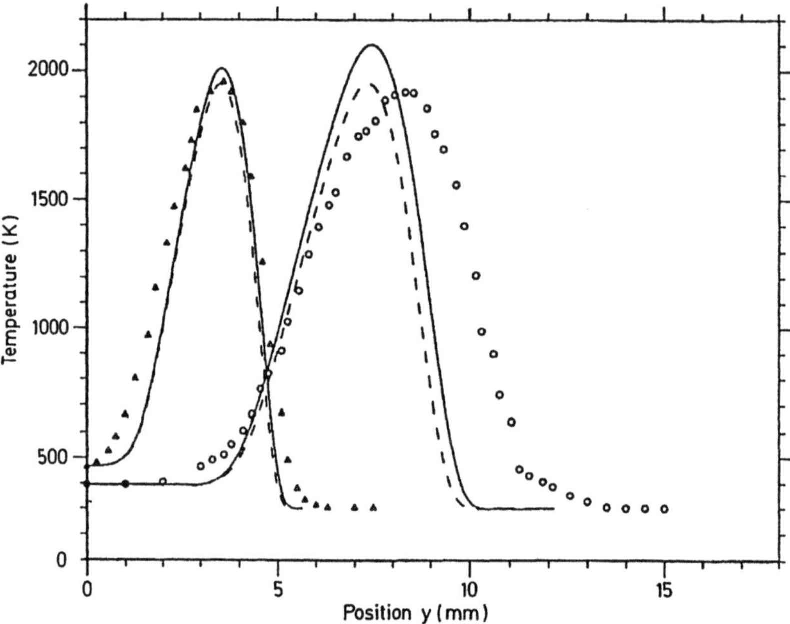

Fig. 2: Temperature profiles for a=70 s-1 (right) and a=180 s-1 (left). Symbols: experimental results taken from Drake and Blint [4]; solid lines: numerical results without radiation model; dashed lines: numerical results with radiation model.

on physical grounds, for the same strain rate lower temperatures are predicted if radiation is taken into account. It is interesting to note that with increasing rate of strain the effects of radiation become less and less pronounced until at sufficiently high rates of strain a blow-out extinction occurs. Furthermore it is seen from Figure 4 that with increasing a flames become thinner and move towards the burner surface. These phenomena were observed earlier by Dixon-Lewis and Missaghi [12] for non-radiating counterflow-diffusion flames of hydrogen-nitrogen mixtures in air.

It is seen from Figs. 2 and 3 that the smaller the strain rate, the stronger is the effect of radiation heat loss on the flame structure. This finding suggests that in addition to blow-off extinction, radiation heat loss may lead to extinction even at low strain rate. The physical reason for the possible existence of a radiative extinction limit is that with decreasing rate of strain convection becomes less important and, simultaneously, the flame becomes thicker. As a consequence, radiation heat loss reduces the flame temperature substantially, thereby eventually leading to extinction.

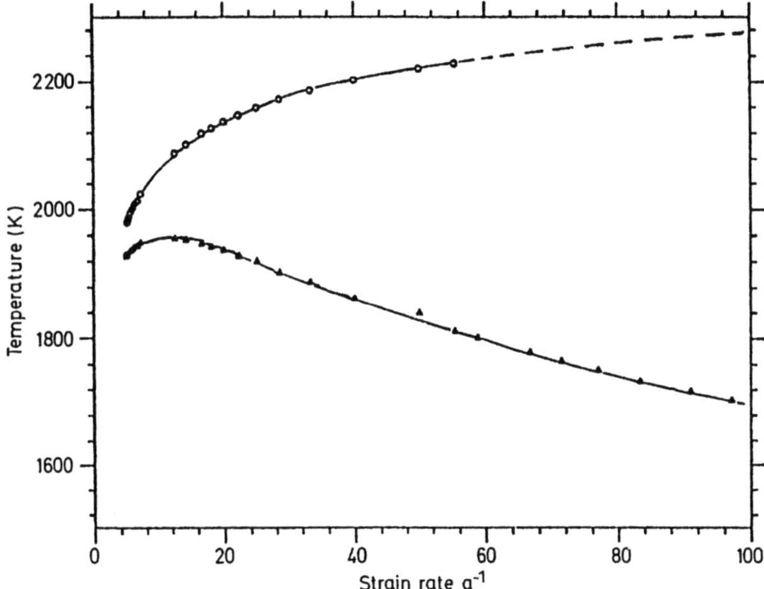

Fig. 3: Computed maximum temperature T_{max} as a function of 1000 times the reciprocal of the strain-rate; Circles: numerical results without radiation model, i.e., $q_R=0$; Triangles: numerical results with radiation model; solid lines: interpolated; dashed lines: extrapolated.

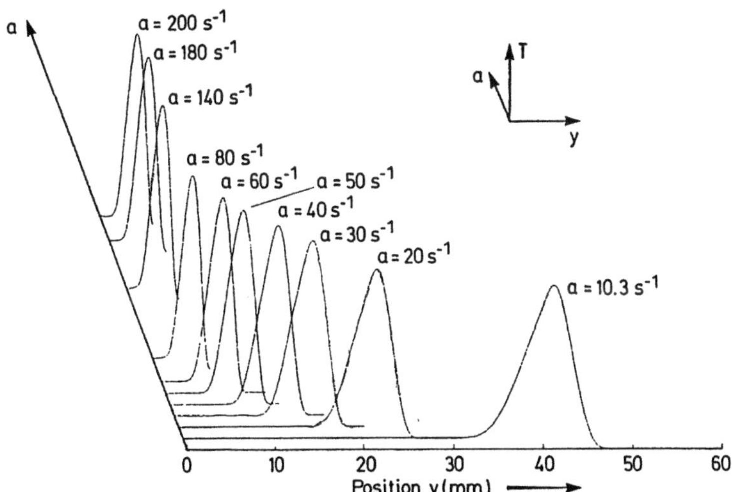

Fig. 4: Computed temperature profile obtained with taking the radiation model into account at various values of the strain rate.

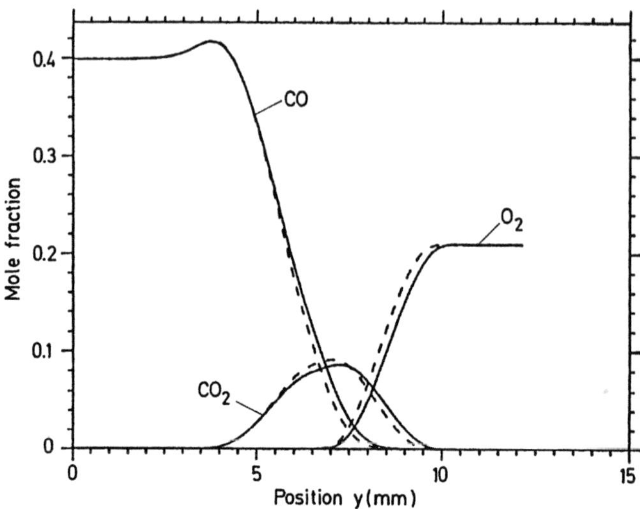

Fig. 5: Calculated major-species mole-fraction profiles for $a=70$ s^{-1}. Solid lines: without radiation model; dashed lines: with radiation model.

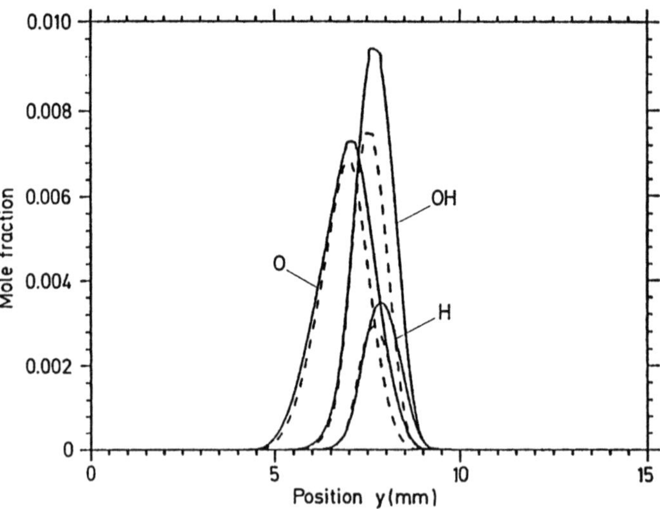

Fig. 6: Calculated mole fraction profiles of the radicals in the H_2-O_2 radical pool for $a=70$ s^{-1}. Solid lines: without radiation model; dashed lines: with radiation model.

Shown in Figs. 5 to 7 are numerically predicted mole-fraction profiles for a strain rate of 70 s^{-1}. Solid lines in these Figs. represent numerical results obtained without radiation model, dashed lines results obtained with radiation model. Shown in Figure 5 are the concentrations of the major species CO, O_2 and CO_2, in Figure 6 those of the species of the hydrogen-oxygen radical pool,

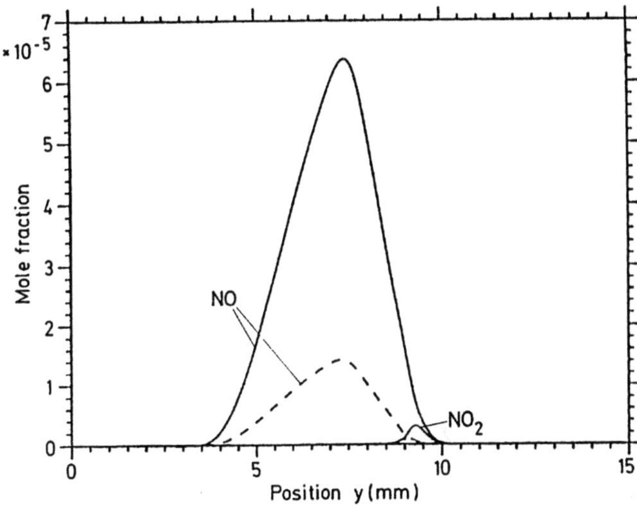

Fig. 7: Calculated mole fractions of NO and NO_2 for $a=70$ s^{-1}. Solid lines: without radiation model; dashed lines: with radiation model.

H, OH and O. It is seen that the effect of radiation on the concentrations of CO, O_2, CO_2, O and H is rather modest, whereas the effect on OH is quite pronounced. Shown in Figure 7 are the predicted concentrations of NO and NO_2. It is seen that radiation has a strong effect on the production of NO_x. This is to be expected on physical grounds, since radiation losses lower the flame temperature and, thereby, the production of NO_x. These findings confirm numerical results obtained by Müller [2] who has investigated effects of radiation on the formation of NO_x in a similar geometry, however, with boundary conditions not suitable for comparisons with the experimental data and with a less detailed kinetic mechanism.

CONCLUSIONS

Strained counterflow diffusion flames have been modelled and simulated using detailed models of radiation transport, molecular transport and chemistry. From the fundamental integro-differential equation governing radiation transport a simplified radiation model has been derived. Numerical results agree well with experimental data. In particular, temperatures predicted with the radiation model are closer to the experimentally observed temperature than temperatures predicted without radiation model. The results show that the smaller the strain rate, the stronger the effect of radiation heat loss on the laminar diffusion flame. These findings suggest that radiation heat loss can possibly lead to extinction at a sufficiently low value of the strain rate. Furthermore, it was found that at a low strain rate radiation losses strongly reduce the production of NO_x through their impact on the flame temperature.

ACKNOWLEDGEMENT

The authors would like to thank Prof. K.N.C. Bray for helpful discussions.

REFERENCES

1. Peters N.: Laminar Diffusion Flamelet Models in Non-Premixed Turbulent Combustion. Prog. Energy Combust. Sci. 10 (1984) 319-339.

2. Sohrab S.H., Liñán A., Williams F.A.: Asymptotic Theory of Diffusion Flame Extinction with Radiant Loss from the Flame Zone. Combust. Sci. and Tech. 27 (1982) 143-154.

3. Müller U.C.: Der Einfluß von Strahlungsverlusten auf die thermische NO-Bildung in laminaren CO-H_2-Diffusionsflammen, Diplomarbeit, RWTH Aachen (1989).

4. Drake, M.C., Blint, R.J.: Thermal NOx in Stretched Laminar Opposed-Flow Diffusion Flames with $CO/H_2/N_2$ Fuel. Combust. Flame 76 (1989) 151-167.

5. Rogg B.: Response and Flamelet Structure of Stretched Premixed Methane-Air Flames. Combust. Flame 73 (1988) 45-65.

6. Warnatz J.: Rate Coefficients in the C/H/O/Systems, in W.C. Gardiner, Jr. (Ed.). Combustion Chemistry, Springer, New York (1984)197-360.

7. Warnatz J.: private communication (1988).

8. Siegel R., Howell J.R.: Thermal Radiation Heat Transfer. 2nd ed., Hemisphere Publishing Corporation, New York (1981).

9. Hubbard G.L., Tien C.L.: Infrared Mean Absorption Coefficients of Luminous Flames and Smoke. Journal of Heat Transfer 100 (1978) 235-239

10. Rogg B.: Numerical Analysis of Strained Premixed CH_4-Air Flames with Detailed Chemistry. In Numerical and Applied Mathematics. Edited by W.F. Ames, Baltzer (1989) 159-167.

11. Rogg B.: Numerical Modelling and Computation of Reactive Stagnation-Point Flows. In Computers and Experiments in Fluid Flow. Edited by G.M. Carlomagno and C.A. Brebbia, Springer Verlag, Berlin-Heidelberg, (1989) 75-85.

12. Dixon-Lewis G., Missaghi M.: Structure and Extinction Limits of Counterflow Diffusion Flames of Hydrogen-Nitrogen Mixtures in Air. 22nd Symposium (Int.) on Comb., The Combustion Institute, Pittsburgh (1988) 1461-1470.

CALCULATIONS OF VELOCITY, COMPOSITION AND TEMPERATURE FIELDS IN A MODEL SWIRL COMBUSTOR WITH A REYNOLDS-STRESS TRANSPORT TURBULENCE MODEL

W.P. Jones
Imperial College of Science Technology and Medicine
Dept. of Chemical Engineering and Chemical Technology
London SW7 2BY

and
A. Pascau
Fluid Mechanics Group
Centro Politecnico Superior, Universidad de Zaragoza
Maria de Luna 3,
50015 Zaragoza

ABSTRACT

A Reynolds-Stress Transport Turbulence model is applied to calculate the flow, species and temperature fields in a model swirl combustor. Favre-averaging has been used to derive the equations, although weighted and unweighted averages are presented. Based on the results obtained some weaknesses of the model are pointed out.

INTRODUCTION

Industrial combustors currently use swirl as a means of providing a low-velocity region where the flame stabilizes. The mixing and combustion takes place in the shear layer formed by the fuel jet and the surrounding dilution air jet. Good mixing in this layer is important to ensure an appropriate temperature distribution and a complete combustion.

It is already well known that the swirling motion produces a stabilising effect on the flow [1]. This is related to the pressure gradient experienced by a fluid particle in its outward/inward motion. A particle with a given normal acceleration will tend to move to equilibrium regions in which the pressure gradient in radial direction is balanced with the centrifugal force, i.e., $\partial P/\partial r = \rho\, w^2/r$.

The radial motion of fluid particles in flows which satisfy Rayleigh's criterion $[\partial/\partial r(\rho\, w^2\, r^2) > 0]$ is restrained as they undergo a pressure gradient which drives them back to their original position. The stabilization of swirling flows is a feature which standard two-equation models fail to grasp. The k-ε model in particular contains no mechanism to account for the stabilization of the flow.

Variable density introduces a new participant in the set-up of the flow field. In addition to the pressure gradient, density gradients have a large stabilization effect. As a result of the density-pressure coupling, turbulence is reduced in regions where there is a positive gradient of density in radial direction. This is in fact the case in reacting flows in which products (low density particles) tend to move towards the centreline, whereas excess air (higher density particles) move outwardly, effectively setting up a positive density gradient.

The appraisal of a turbulence model has to be based on its capability of mimicking the features of reacting flows commented above. The ability of two-equation models to simulate complex swirling flows, both inert and reacting ones, is at least questionable. In inert flows, overprediction of the azimuthal turbulent momentum flux in radial direction $\overline{\rho v'w'}$ causes the flow to tend to a solid-body-rotation type of motion in regions where measurements show a combined forced-free vortex [2]. A Reynolds-Stress Transport model has been used to predict these flows successfully, at least at the mean flow level [3]. In inert, variable-density flows, the situation is similar and the same RST model has shown its superiority over the k-ε model [4]. Comparable results have been reported elsewhere [5]. On these grounds, RST models are an obvious alternative to two-equation models in reacting, swirling flows.

The purpose of this paper is to present a complete calculation of a complex, reacting, swirling flow performed with a RST model. Firstly, a brief overview of the turbulence and combustion models will be given. After this, the numerical discretization and time-advancing procedure to obtain a steady state solution will be described. To finish, an assessment of the model by comparing its predictions with the measurements of Wilhelmi [6] will be presented.

TURBULENCE AND COMBUSTION MODELS

Favre-averaging has been used throughout to obtain the mean flow equations. The instantaneous variable ϕ is split into a mean and fluctuating part ϕ'' in such a way that $\overline{\rho\phi''} = 0$ but $\overline{\phi''} \neq 0$. The equations are to be presented in Cartesian tensor form although they have in fact been derived and used in cylindrical polar coordinates.

GOVERNING EQUATIONS.

The Favre-averaged (density-weighted) equations which govern the mass and momentum transport in a turbulent, reacting flow are

130

Continuity

$$\frac{\partial \rho}{\partial t} + \frac{\partial \overline{\rho} \, \tilde{U}_j}{\partial x_j} = 0$$

Momentum

$$\frac{\partial \overline{\rho} \, \tilde{U}_i}{\partial t} + \frac{\partial \overline{\rho} \, \tilde{U}_i \tilde{U}_j}{\partial x_j} = - \frac{\partial \overline{P}}{\partial x_i} + \frac{\partial}{\partial x_j} \left(\overline{\tau}_{ij} - \overline{\rho} \, \widetilde{u''_i u''_j} \right)$$

Where

$$\overline{\tau}_{ij} = \mu \left(\frac{\partial \tilde{U}_i}{\partial x_j} + \frac{\partial \tilde{U}_j}{\partial x_i} \right) + \mu \left(\frac{\partial \overline{u''}_i}{\partial x_j} + \frac{\partial \overline{u''}_j}{\partial x_i} \right) - \frac{2}{3} \mu \left(\frac{\partial \tilde{U}_1}{\partial x_1} \right) \delta_{ij} - \frac{2}{3} \mu \left(\frac{\partial \overline{u''}_1}{\partial x_1} \right) \delta_{ij}$$

Scalar

$$\frac{\partial \overline{\rho} \, \tilde{F}}{\partial t} + \frac{\partial \overline{\rho} \, \tilde{F} \, \tilde{U}_j}{\partial x_j} = - \frac{\partial}{\partial x_j} \left(\rho \, D \frac{\partial \tilde{F}}{\partial x_j} - \overline{\rho} \, \widetilde{f'' u''_j} \right)$$

Use has been made of the fact that $\overline{U_i} = \tilde{U}_i + \overline{u''}_i$.

REYNOLDS STRESS TRANSPORT EQUATION MODEL

The model used in this work was proposed by Jones and Musonge [7]. Details can also be found in Pascau [4]. The time dependent form of the Reynolds stress transport equations is written as

$$\frac{\partial \overline{\rho} \, \widetilde{u''_i u''_j}}{\partial t} + \frac{\partial \overline{\rho} \, \tilde{U}_1 \widetilde{u''_i u''_j}}{\partial x_1} = \frac{\partial}{\partial x_1} \left\{ c'_s \overline{\rho} \, \frac{\tilde{k}}{\tilde{\varepsilon}} \, \widetilde{u''_1 u''_m} \frac{\partial \widetilde{u''_i u''_j}}{\partial x_1} \right\} -$$

$$- \overline{\rho} \left\{ \widetilde{u''_i u''_1} \frac{\partial \tilde{U}_j}{\partial x_1} + \widetilde{u''_j u''_1} \frac{\partial \tilde{U}_i}{\partial x_1} \right\} - \frac{2}{3} \delta_{ij} \overline{\rho} \, \tilde{\varepsilon}$$

$$- c_1 \overline{\rho} \, \frac{\tilde{\varepsilon}}{\tilde{k}} \left(\widetilde{u''_i u''_j} - \frac{2}{3} \delta_{ij} \tilde{k} \right) + c_2 \, \delta_{ij} \overline{\rho} \, \widetilde{u''_1 u''_m} \frac{\partial \tilde{U}_1}{\partial x_m} +$$

$$+ c_3 \left(\overline{\rho} \, \widetilde{u''_1 u''_j} \frac{\partial \tilde{U}_i}{\partial x_1} + \overline{\rho} \, \widetilde{u''_1 u''_i} \frac{\partial \tilde{U}_j}{\partial x_1} \right) +$$

$$+ c_4 \, \overline{\rho} \, \tilde{k} \left(\frac{\partial \tilde{U}_i}{\partial x_j} + \frac{\partial \tilde{U}_j}{\partial x_i} \right) + c_5 \left(\overline{\rho} \, \widetilde{u''_1 u''_j} \frac{\partial \tilde{U}_1}{\partial x_i} + \overline{\rho} \, \widetilde{u''_1 u''_i} \frac{\partial \tilde{U}_1}{\partial x_j} \right)$$

$$+ c_6 \, \overline{\rho} \, \tilde{k} \, \delta_{ij} \frac{\partial \tilde{U}_1}{\partial x_1} - \left(\overline{u''}_i \frac{\partial \overline{p}}{\partial x_j} + \overline{u''}_j \frac{\partial \overline{p}}{\partial x_i} \right)$$

Its complete derivation can be found in Jones and Musonge [7]. It contains the same terms as the earlier model of Launder *et al* [8], although with different constants (Table 1). It is cast in Favre-averaged form. The last term in brackets in the momentum equation appears only in variable density flows if Favre-averaged is used. It represents the effect of the pressure gradients on low- and high-density gas packets.

Table 1

$$C'_s = 0.22 \qquad C_1 = 3 \qquad C_2 = -0.44 \qquad C_3 = 0.46 \qquad C_4 = -0.23$$

$$C_5 = 0.2 \qquad C_6 = 1 \qquad C'_\varepsilon = 0.18 \qquad C_{\varepsilon 1} = 1.9 \qquad C_{\varepsilon 2} = 1 \qquad C_{\varepsilon 3} = 1$$

The transport equation for the unweighted mean of the Favre-fluctuating velocity $\overline{u''}$ used in the present work is

$$\frac{\partial \overline{\rho}\,\overline{u''_i}}{\partial t} + \frac{\partial \overline{\rho}\,\tilde{U}_1\overline{u''_i}}{\partial x_1} = -\overline{\rho}\,\overline{u''_1}\frac{\partial \tilde{U}_i}{\partial x_1} + \overline{u''_i u''_1}\frac{\partial \overline{\rho}}{\partial x_1} + \overline{\rho}\frac{\partial}{\partial x_1}\left(c' f\frac{\tilde{k}}{\tilde{\varepsilon}}\,\overline{u''_1 u''_m}\frac{\partial \overline{u''_i}}{\partial x_m}\right) -$$

$$- c^* f_1\frac{\tilde{\varepsilon}}{\tilde{k}}\,\overline{\rho}\,\overline{u''_i} - 2c_{f_2}\overline{\rho}\,b_{i1}\tilde{k}\frac{\partial \overline{\rho}}{\partial x_1} + c_{f_3}\overline{\rho}\,\overline{u''_1}\frac{\partial \tilde{U}_i}{\partial x_1} +$$

$$+ c_{f_4}\overline{\rho}\,\overline{u''_1}\frac{\partial \tilde{U}_1}{\partial x_i} + c_{f_5}\overline{\rho}\,\overline{u''_i}\frac{\partial \tilde{U}_1}{\partial x_1} + \overline{\left(\frac{\rho'}{\rho}\right)}\frac{\partial \overline{P}}{\partial x_i} - \overline{\rho}\,\overline{u''_i}\frac{\partial \tilde{U}_1}{\partial x_1}$$

with constants given in Table 2.

The above equation is the same as that for the scalar flux when there is no reaction taking place. In inert mixing, $\overline{u''_i} = C\overline{\rho}\,\widetilde{u''_i f''}$, C being a constant, and both equations should transform into each other term by term.

The equation for the dissipation rate of turbulent kinetic energy is

$$\frac{\partial \overline{\rho}\,\tilde{\varepsilon}}{\partial t} + \frac{\partial \overline{\rho}\,\tilde{U}_1\tilde{\varepsilon}}{\partial x_1} = \frac{\partial}{\partial x_1}\left(c'_\varepsilon \overline{\rho}\frac{\tilde{k}}{\tilde{\varepsilon}}\,\overline{u''_1 u''_m}\frac{\partial \tilde{\varepsilon}}{\partial x_m}\right) -$$

$$- c_{\varepsilon_1}\frac{\tilde{\varepsilon}}{\tilde{k}}\,\overline{\rho}\,\overline{u''_1 u''_m}\frac{\partial \tilde{U}_1}{\partial x_m} - c_{\varepsilon_2}\frac{\tilde{\varepsilon}^2}{\tilde{k}}\,\overline{\rho} - c_{\varepsilon_3}\frac{\tilde{\varepsilon}}{\tilde{k}}\,\overline{u''_1}\frac{\partial \overline{\rho}}{\partial x_1}$$

Last term is related to the extra-production term in the $\tilde{k}$-equation. For the value of the constants, *see* Table 1.

SCALAR FLUX MODEL

In inert mixing, the mean density can be obtained directly from the mixture fraction as they are algebraically related. In reacting flows this algebraic

132

relationship no longer holds and a new way of obtaining the mean density from the mixture fraction is required. In the present work, a two-parameter β function has been adopted for the probability density function of the mixture fraction. This pdf is uniquely determined by two moments, mean and variance, for which a transport equation is solved. The β-pdf is given as

$$p(f) = \frac{f^{a-1}(1-f)^{b-1}}{\int_0^1 f^{a-1}(1-f)^{b-1}\,df}$$

$$a = \tilde{f}\left\{\frac{\tilde{f}(1-\tilde{f})}{\widetilde{f''^2}} - 1\right\}\;;\; b = \frac{(1-\tilde{f})\,a}{\tilde{f}}$$

The mean density is obtained from

$$\bar{\rho} = \left\{\int_0^1 \frac{P(f)}{\rho(f)}\,df\right\}^{-1}$$

and some terms in the $\overline{u''_i}$ and $\widetilde{u''_i f''}$-equations are

$$\overline{\left(\frac{\rho'}{\rho}\right)} = 1 - \bar{\rho}^2 \int_0^1 \frac{P(f)}{\rho^2(f)}\,df$$

$$\overline{f''} = \bar{\rho}\int_0^1 \frac{f}{\rho(f)}\,P(f)\,df - \tilde{f}$$

The scalar fluxes are obtained by solving their transport equations to maintain the same level of modelling as that for the velocity field. The time dependent form of these equations is

$$\frac{\partial \bar{\rho}\,\widetilde{u''_i f''}}{\partial t} + \frac{\partial \bar{\rho}\,\tilde{U}_1\,\widetilde{u''_i f''}}{\partial x_1} = -\bar{\rho}\left(\widetilde{u''_1 f''}\frac{\partial \tilde{U}_i}{\partial x_1} + \widetilde{u''_i u''_1}\frac{\partial \tilde{F}}{\partial x_1}\right) +$$

$$+ \frac{\partial}{\partial x_1}\left(c'_f\bar{\rho}\,\frac{\tilde{k}}{\tilde{\varepsilon}}\,\widetilde{u''_1 u''_m}\frac{\partial \widetilde{u''_i f''}}{\partial x_m}\right) -$$

$$- c^*_{f1}\frac{\tilde{\varepsilon}}{\tilde{k}}\,\bar{\rho}\,\widetilde{u''_i f''} + 2\,c_{f2}\bar{\rho}\,b_{i1}\,\tilde{k}\,\frac{\partial \tilde{F}}{\partial x_j} + c_{f3}\bar{\rho}\,\widetilde{u''_1 f''}\frac{\partial \tilde{U}_i}{\partial x_1} +$$

$$+ c_{f4}\,\bar{\rho}\,\widetilde{u''_1 f''}\frac{\partial \tilde{U}_1}{\partial x_i} + c_{f5}\bar{\rho}\,\widetilde{u''_i f''}\frac{\partial \tilde{U}_1}{\partial x_1} - \overline{f''}\frac{\partial \bar{P}}{\partial x_i}$$

$$b_{i1} = 2\,\frac{\widetilde{u''_i u''_1}}{\tilde{k}} - \frac{1}{3}\delta_{ij}\quad\text{and}$$

$$c^*_{f1} = \frac{3}{1+c'_{f1}\sqrt{b^2}}\qquad b^2 = b_{ij}\,b_{ji}\qquad b_{ij} = \frac{\widetilde{u''_i u''_j}}{2\tilde{k}} - \frac{1}{3}\delta_{ij}$$

Its similitude with the equation for $\overline{u''_i}$ can be noticed. The model constants are given in Table 2.

Table 2

$C'_{f1} = 3$	$C_{f2} = 0.12$	$C_{f3} = 1.09$	$C_{f4} = 0.51$	$C_{f5} = 1$	$C'_f = 0.5$
$C'_v = 0.5$	$C'_{Df} = 0.41$	$C_{D1} = 2$	$C_{D2} = 1.8$	$C_{D3} = 1.7$	
$C_{D4} = 1.4$	$C_{D5} = 1$				

The equations for the variance and the scalar dissipation are written as

$$\frac{\partial \overline{\rho}\,\widetilde{f''^2}}{\partial t} + \frac{\partial \overline{\rho}\,\widetilde{U}_1 \widetilde{f''^2}}{\partial x_1} = -2\,\overline{\rho}\,\overline{u''_1 f''}\,\frac{\partial \widetilde{F}}{\partial x_1} - 2\,\overline{\rho}\,\widetilde{\varepsilon}_f +$$

$$+ \frac{\partial}{\partial x_1}\left(c'_v\,\overline{\rho}\,\frac{\widetilde{k}}{\widetilde{\varepsilon}}\,\overline{u''_1 u''_m}\,\frac{\partial \widetilde{f''^2}}{\partial x_m} \right)$$

$$\frac{\partial \overline{\rho}\,\widetilde{\varepsilon}_f}{\partial t} + \frac{\partial \overline{\rho}\,\widetilde{U}_1 \widetilde{\varepsilon}_f}{\partial x_1} = \frac{\partial}{\partial x_1}\left(C'_{Df}\,\overline{\rho}\,\frac{\widetilde{k}}{\widetilde{\varepsilon}}\,\overline{u''_1 u''_m}\,\frac{\partial \widetilde{\varepsilon}_f}{\partial x_m} \right) -$$

$$- C_{D1}\,\frac{\widetilde{\varepsilon}_f^2}{\widetilde{f''^2}} - C_{D2}\,\frac{\widetilde{\varepsilon}}{2\widetilde{k}}\,\widetilde{\varepsilon}_f - C_{D3}\,\frac{\widetilde{\varepsilon}}{\widetilde{k}}\,\overline{u''_1 f''}\,\frac{\partial \widetilde{F}}{\partial x_1} -$$

$$- C_{D4}\,\frac{\widetilde{\varepsilon}_f}{\widetilde{k}}\,\overline{u''_1 u''_m}\,\frac{\partial \widetilde{U}_1}{\partial x_m} - C_{D5}\,\frac{\widetilde{\varepsilon}_f}{\widetilde{k}}\,\overline{u''_1}\,\frac{\partial \overline{P}}{\partial x_1}$$

The constants are given in Table 2.

Composition and temperature have beeen related to the mixture fraction with a flamelet approach. A look-up table giving temperature and composition for different values of the mixture fraction was used to calculate mean properties of the mixture. This table was obtained in a stagnation-flow-type laminar flame with a typical strain rate of 300 s^{-1}.

NUMERICAL PROCEDURE

The location of the variables is shown in Figure 1. A staggered grid arrangement has been used throughout, wherein the velocity is located on the scalar control volume faces. All the quantities related to the velocities are stored at the same position as the corresponding velocity. Hybrid differencing has been used, whereby the convection discretization uses upwind differencing when the local Peclet number is greater than two and central differencing otherwise.

The pressure-velocity coupling is treated with a linearised, pseudo-temporal, implicit, two-step method. The solution advances from time t^n to time t^{n+1} in two steps:

Step 1: Using the pressure at time t^n, the momentum equations are solved for. The solution of this equation yields an intermediate solution for the velocity field which generally will not satisfy continuity.

Step 2: A pressure correction is calculated and added to the pressure so as to satisfy continuity. This pressure correction is calculated with the Poisson equation:

$$\nabla^2 \delta \, P^{n+1} = \frac{\nabla \, U^n}{\Delta t}$$

and the velocity is corrected with

$$U^{n+1} = U^n - \Delta t \, \nabla \, \delta \, P^{n+1}$$

A TDMA algorithm is used for solving the algebraic equations arising at each time step. Alternate direction sweeps were carried out, with typically one sweep in every direction for all the variables except for the pressure correction for which five sweeps are employed.

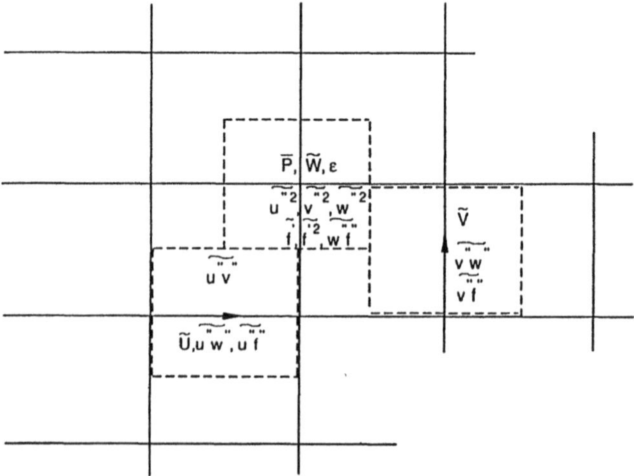

Fig. 1: Staggered grid arrangement and variables location

RESULTS AND DISCUSSION

The model was applied to calculate the flow, composition and temperature fields in a model swirl combustor measured by Wilhelmi. In the experiments, propane was used as fuel. The equivalence ratio was 0.67. Due to the geometry of the combustor - strong swirl combined with sudden expansion of large area ratio- a large toroidal recirculation zone was present which occupied about 80% of the cross-sectional area of the combustor. High turbulence levels were found in the interjet shear layer — where large mean velocity gradients exist — and just downstream of the mean recirculation zone. Burning of the mixture took place on the inside of the shear layer where hot products met the forward fuel jets. Steep temperature gradients were measured along this layer.

The grid used was 52 x 50 points in radial and axial direction respectively. As there were no measurements available for the density an estimation of the angular momentum flux was not possible. For this reason, the $\tilde{W}$ velocity in the swirler was set to $0.9 \times \tilde{U}$, the value used in previous runs of the isothermal flow in the same configuration. The fuel jet velocity was set to 12 m/s to provide the same propane flow rate measured by Wilhelmi.

A zero-gradient condition was used at the exit for all the variables except for the axial velocity for which the measured profile was employed. The velocity at the exit plane was corrected at each time step to conserve overall mass. When the steady state solution presented was eventually reached, the difference in velocities at the last station amounted to about 8%. Typical runs of 2000 time steps were required for steady conditions to prevail.

Figure 2a) illustrates the evolution of the axial velocity. The spread of the swirling jet is reflected in the outward shift of the $\tilde{U}$ velocity maximum. This expansion enhances the creation of a low pressure area over a large part of the radius which, consequently, draws fluid from downstream positions. The measured central recirculation zone is wider and shorter than that predicted, though from x/D=1, the discrepancies between the velocities are restricted to the region near the centreline. The unweighted axial velocity is also plotted (dashed line). The relation between both is simply $\overline{U} = \tilde{U} + \overline{u''}$. As $\overline{u''}$ is apparently negative throughout the flow, $\overline{U}$ is everywhere less than $\tilde{U}$. A maximum value of $\overline{u''}$ of about 5 m/s is found in the shear layer between the corner recirculation zone and the expanding jet.

V velocity profiles are consistent with the slower predicted spread of the jet. At x/D=0.4 where the measurements show that the jet has already impinged on the

a

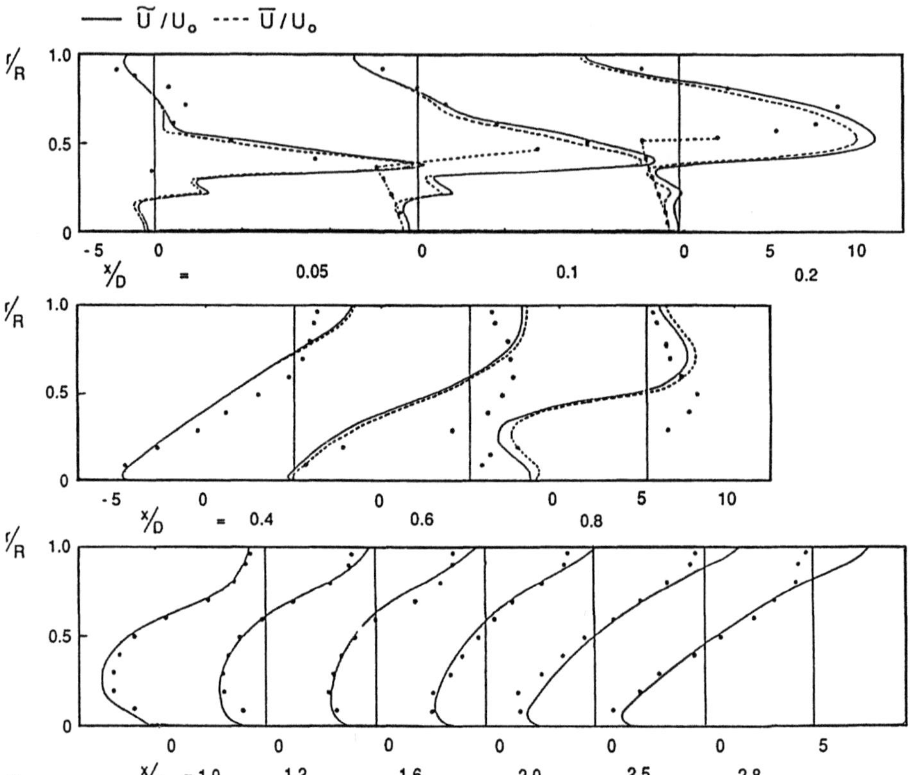

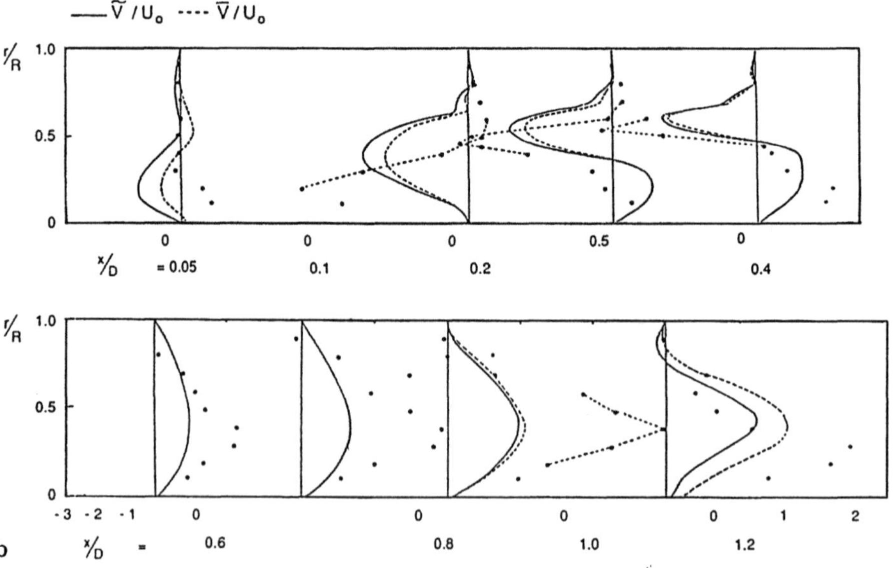

Fig. 2.a): Mean Axial Velocity Profiles

b

Fig. 2.b): Mean Radial Velocity Profiles

wall, the predicted jet is stil expanding, impoinging further downstream. The unweighted Favre-fluctuating velocity $\overline{v''}$ is negative everywhere with maximum magnitude of about 2.5 m/s (Figure 2b)).

The behaviour of the circumferential velocity is displayed in Figure 3. Several discrepancies arise when comparing the predictions and measurements. Firstly, the initial peak corresponding to the swirling jet is markedly overpredicted. Downstream of the first station the predictions have little resemblance with the data, the former showing a large peak in the inner part of the flow not present in the latter. From x/D=0.6 onwards, the agreement is closer but as the flow progresses downstream the $\widetilde{W}$ velocity near the wall evolves away from the measurements and the angular velocity of the forced vortex predicted around the centreline is slightly lower. Maximum absolute values of 3 m/s for $\overline{w''}$ are found at x/D=0.4.

A comparison of the values of the unweighted Favre-fluctuating velocities is instructive and helpful to an understanding of the combined effect of the non-uniform density field and the swirling motion. The fact that $\overline{v''}$ is negative everywhere implies that $\overline{\rho' v''}$ is positive throughout. Thus, positive/negative values of ρ' are associated with positive/negative values of v''. This would happen in a situation in which low-density fluid tended to move towards the centreline while higher-density fluid moved outwardly. In fact, this is the outcome of the density-rotation coupling: products (low density fluid) will be mainly found near the centreline with the outer part being filled with mostly reactants (high density fluid). Data on $\overline{w''}$ are consistent with this interpretation: a particle with low density will have relatively high rotating velocity as it needs to balance the pressure gradient in radial direction. The term which balances $\partial P/\partial r$ is $\rho\, W/r^2$; if ρ is small, W has to be large in relative terms for a particle to be stable at a certain position. It goes a little against intuition to find also positive values of $\overline{\rho' u''}$. One would expect low-density packets to accelerate because of mass conservation and, hence, to have larger axial velocities. Further thought can however explain away this apparent contradiction. Let us imagine a flow in which two packets of fluid with different densities move axially driven **only** by the pressure gradient, i.e., the axial momentum transport equation reduces to

$$\rho\, U\, \frac{\partial U}{\partial x} = -\frac{\partial P}{\partial x}$$

For the same pressure gradient, a particle with small ρ will undergo a reduction in velocity greater than the one with higher density. This argument implies that the correlation between ρ' and u'' will be positive.

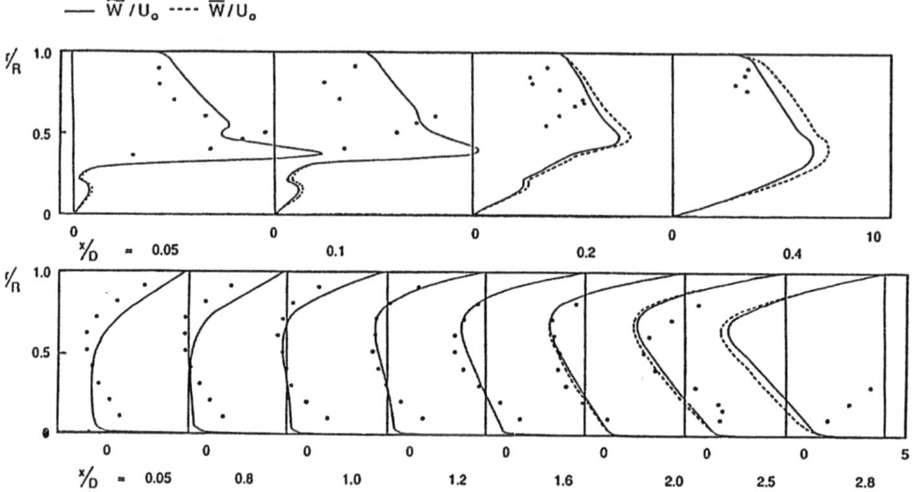

Fig. 3: Mean Circumferential Velocity Profiles

Values of the normal stresses are presented in Figure 4. It is immediately apparent that the values are overpredicted throughout, by as much as 400% at some stations. As an example at x/D=0.2 the normalised peak in $\widetilde{u''^2}$ in the inner shear layer of the swirling jet is about 28 whereas the measured value is slightly less than 10. The explanation of the discrepancy is not evident. It might be partly due to the overprediction of the unweighted Favre fluctuating velocity acting through the term $- \overline{u''} \dfrac{\partial \overline{P}}{\partial x}$. Against this explanation is the fact that, although the predicted swirling jet does not expand as the measurements show, the gradients are well predicted which suggests that the levels of shear stress are adequate. Any poor prediction of $\overline{u''}$ should also be reflected in $\widetilde{u''v'}$ via the term $- \overline{u''} \dfrac{\partial \overline{P}}{\partial r}$. The differences between the unweighted and weighted Reynolds stresses are much smaller than between the latter and measurements. At the centreline the relations between both averages reduce to

$$\overline{u'^2} = \widetilde{u''^2} - 2C_s \frac{\widetilde{k}}{\widetilde{\varepsilon}} \widetilde{u''^2} \frac{\partial \overline{u''}}{\partial x} - (\overline{u''})^2$$

$$\overline{v'^2} = \widetilde{v''^2} - 2C_s \frac{\widetilde{k}}{\widetilde{\varepsilon}} \widetilde{v''^2} \frac{\partial \overline{v''}}{\partial r}$$

$$\overline{w'^2} = \widetilde{w''^2} - 2C_s \frac{\widetilde{k}}{\widetilde{\varepsilon}} \widetilde{w''^2} \frac{\overline{v''}}{r}$$

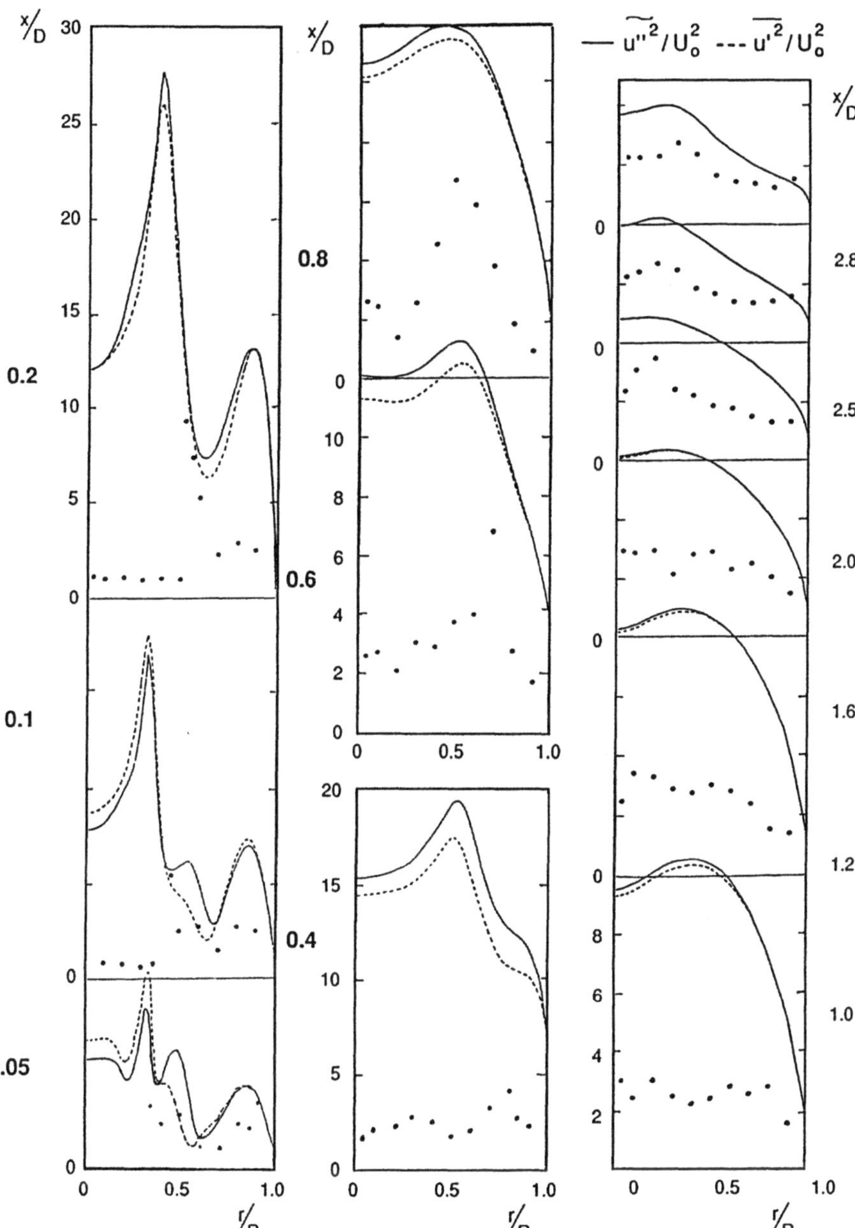

Fig. 4.a): Axial Turbulent Stresses

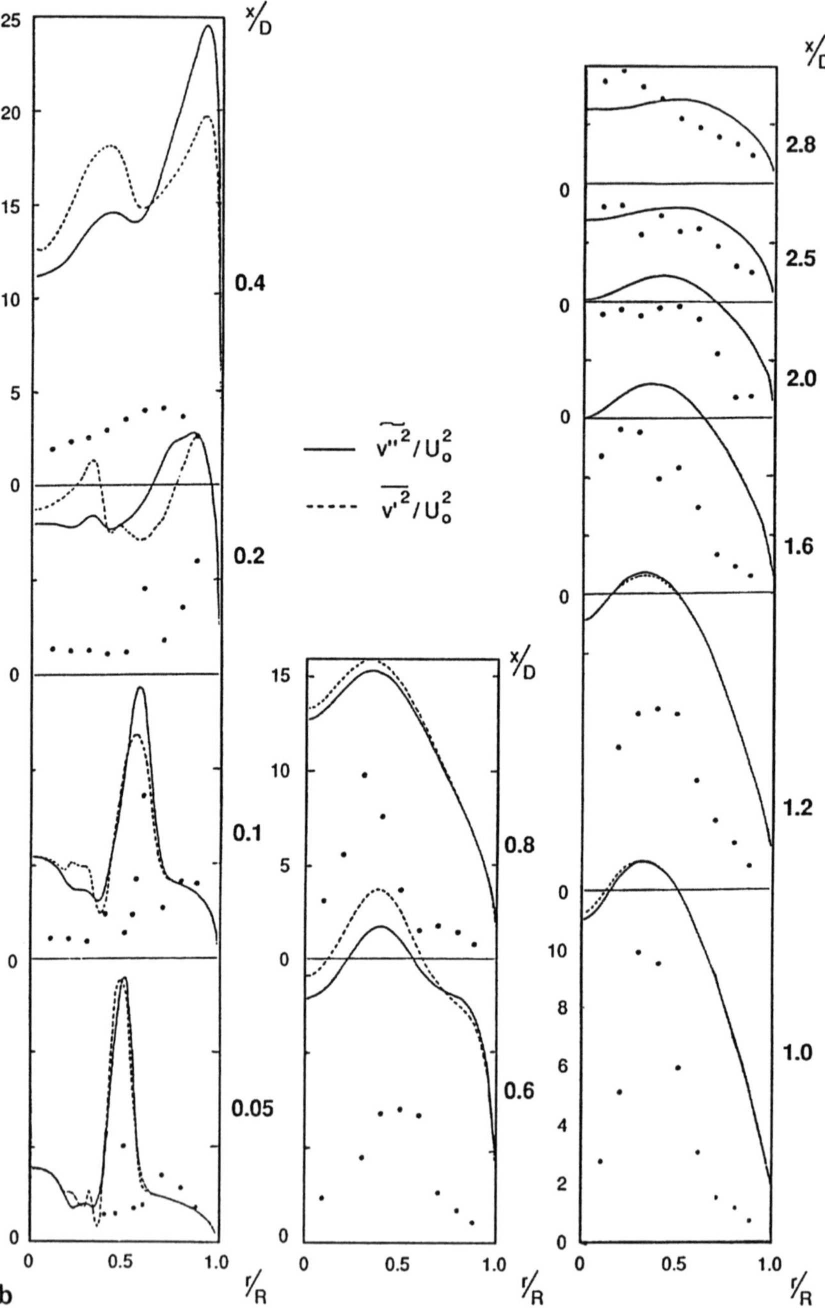

$$\widetilde{v''^2}/U_o^2 \qquad \overline{v'^2}/U_o^2$$

Fig. 4.b): Radial Turbulent Stresses

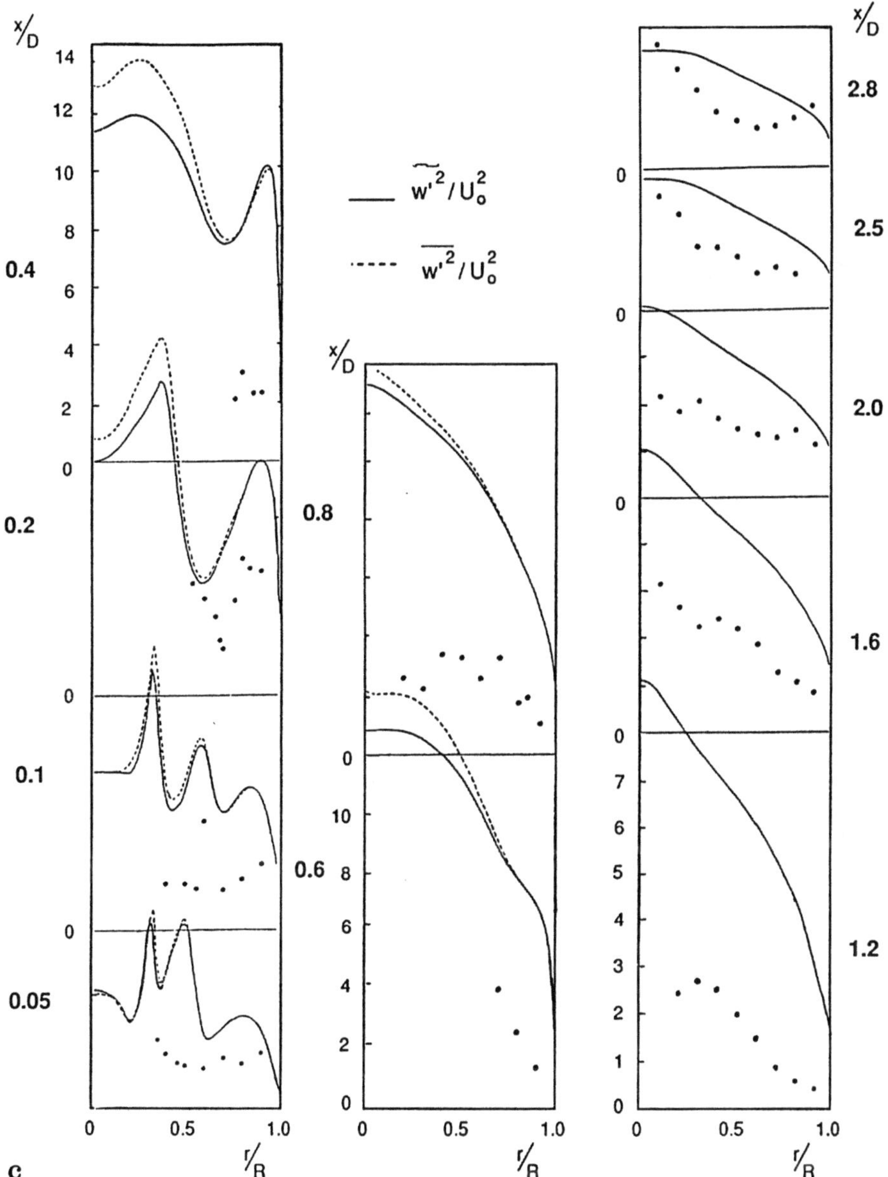

Fig. 4.c): Circumferential Turbulent Stresses

Near to the centreline, $\overline{v''}$ varies linearly with the radius, that is $\partial\,\overline{v''}/\partial r = \overline{v''}/r$. As $\widetilde{v'^2} = \widetilde{w''^2}$, the equality of unweighted normal stresses is satisfied by the model, i.e. $\overline{v'^2} = \overline{w'^2}$ at the centreline. The diffusion terms in the above equations act predominantly in the vicinity of the centreline causing $\overline{v'^2}$ and $\overline{w'^2}$ to be greater than its weighted counterparts. This is so because $\overline{v''}$ is negative and linear around the centreline, i.e., $\overline{v''} = -Cr$, C being a positive constant. Thus, at the centreline

$$\overline{v'^2} = \widetilde{v''^2}\left(1 + 2\,C\,C_s\frac{\widetilde{k}}{\widetilde{\varepsilon}}\right) \Rightarrow \overline{v'^2} > \widetilde{v''^2}$$

$$\overline{w'^2} = \widetilde{w''^2}\left(1 + 2\,C\,C_s\frac{\widetilde{k}}{\widetilde{\varepsilon}}\right) \Rightarrow \overline{w'^2} > \widetilde{w''^2}$$

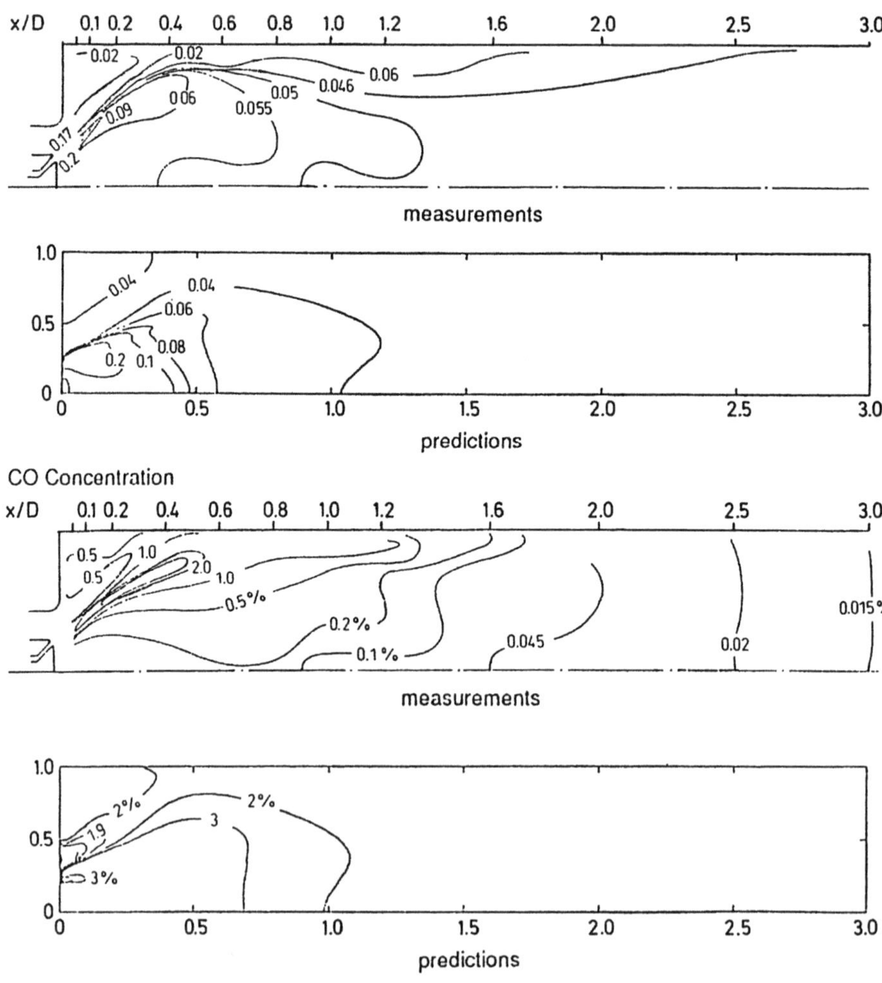

Fig. 5: Scalars Contours

The radial normal stresses at x/D=0.4 present the maximum difference. At that station, the jet is about to impinge on the wall and $\overline{v''}$ is near its peak, making $\overline{v'}^2$ in the outer maximum 25% smaller than $\widetilde{v''}^2$. From x/D=1 up to the end of the test zone,where the fluctuations in density are small, both averages coincide.

Figure 5 shows contours of scalars. The burning takes place on the inside of the conical shear layer where the recirculated hot products meet the annular forward jets. Peak temperature values of 1700 °C were predicted in a tiny region in the

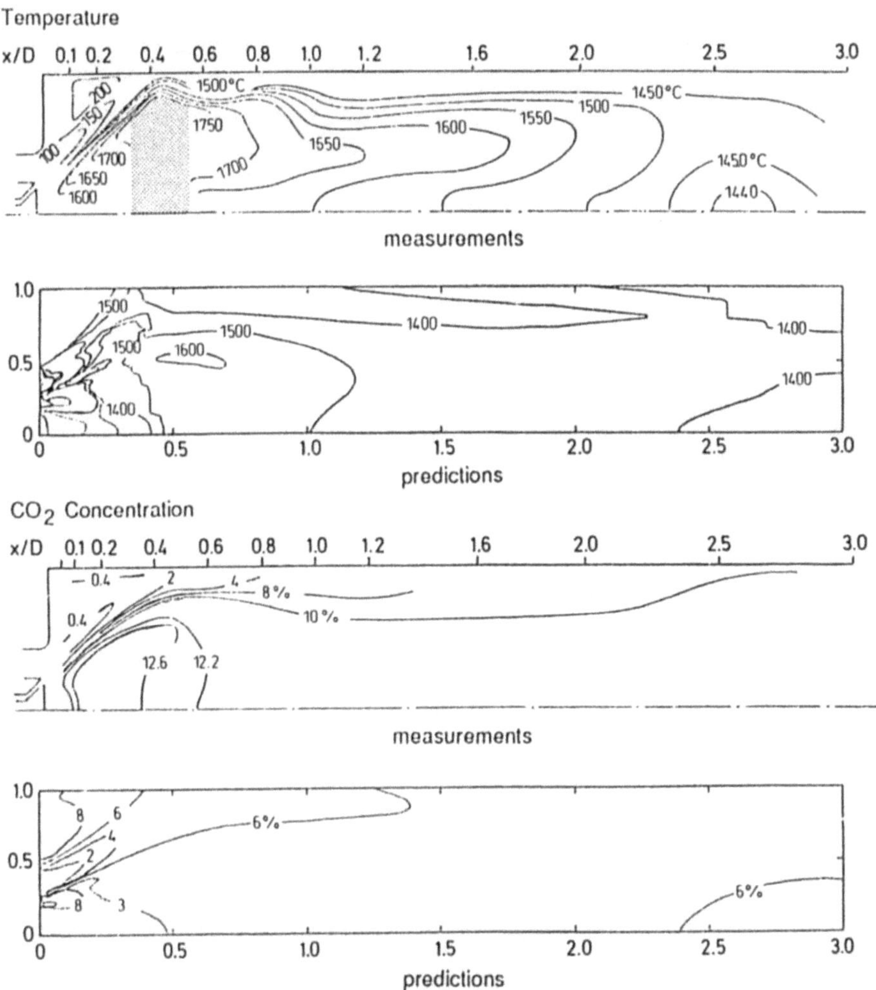

inner part of the expanding swirling jet. Predictions of mixture fraction show that the mixing is very rapid in the shear layer between the jets. The reported peak values for hydrogen and carbon monoxide are 2.3% and 4%, positioned in the direction of the expanding jet. Measured values are respectively 2.5% and 5% but maxima are in the central recirculation zone where hot products are recirculated. In the measurements, a large region with CO_2 between 12% and 13% in the upstream end of the recirculation zone with insignificant levels of hydrogen and carbon monoxide indicate complete combustion. As a contrast, the predicted maximum of CO_2 in the recirculation zone is less than 10% with significant amounts of hydrogen and CO. Both predictions and measurements show comparatively high level of CO along the combustor walls up to x/D=2. As the mixture fraction contours are reasonably well reproduced, the incorrectly predicted value and localisation of the maxima of CO_2 and CO must be due to the flamelet model adopted. Increasing the strain rate in the laminar flame calculations has a marked effect in the values of CO and CO_2. The strain rate adopted in this predictions gave maximum values of 9% for CO_2. In this situation, it is impossible to obtain predicted values of up to 12% as the measurements show. No attempt was made to use another flamelet table or to employ different flamelet relationships at different points depending on the local rate of strain. This is currently being investigated.

CONCLUSIONS

Several discrepancies with the measured profiles highlight some shortcomings of the model. The large values of the unweighted Favre-fluctuating velocity may be one of the causes for the abnormally high levels of normal stresses. The expansion of the swirling jet was not well predicted, but once the jet impinged on the wall and the axial velocity recovered downstream of the recirculation zone the discrepancies were restricted to a very small region near the centreline. Wilhelmi also compared his measurements with calculations carried out with the k-ε model. Although the axial velocity fields are very similar, the k-ε model wrongly predicted the azimuthal velocity to tend to a solid-body-rotation profile very quickly. Another significant difference was found in the scalar field, the k-ε model predicting a much slower mixing, resulting in the species contours being markedly different from the measurements.

REFERENCES

1. Rayleigh: On the Dynamics of Revolving Fluids. Sci. Papers. Cambridge University Press 6 (1916) 447-453.

2. Sloan D.G., Smith P.J., Smoot L.D.: Modelling of Swirl in Turbulent Flow Systems. Progress in Energy and Combustion Science 12 (1986)163-250.

3. Jones W.P., Pascau A.: Calculation of Confined Swirling Flows with a Second Moment Closure. Transactions of the ASME, Journal of Fluids Engineering 111 (1989) 248-225.

4. Pascau A.: The Application of a Second-Order Closure to Inert and Reacting Swirling Flows. Ph.D. Dissertation, University of London (1989).

5. Hogg S., Leschziner M.: Computation of Strongly-Swirling, Reacting Flow in a Model Combustor with Second-Moment Closure. 7th Symposium on Turbulent Shear Flows, Stanford University (1989).

6. Wilhelmi J.: Axisymmetric Swirl Stabilised Combustion. Ph.D. Dissertation, University of London (1984).

7. Jones W.P., Musouge P.: Closure of the Reynolds Stress and Scalar Flux Equations. Physics of Fluids 31 (1988) 3588-3614.

8. Launder B.E., Reece G.J., Rodi W.: Progress in the Development of a Reynolds Stress Turbulence Closure. Journal of Fluid Mechanics 68 (1975) 537-566.

NUMERICAL PREDICTION OF FLAME IMAGE

J.C.F. Pereira, P. Coelho, J.M.P. Rocha and M.G. Carvalho
Instituto Superior Técnico
Technical University of Lisbon
Av. Rovisco Pais
1096 Lisboa Codex
Portugal

ABSTRACT

The paper reports the prediction of a turbulent propane free flame. The numerical procedure includes turbulent combustion and radiation models. A numerical procedure is outlined to predict the digitised image from a CCD-camera. A qualitative good agreement between the predicted and digitised images was obtained.

1. INTRODUCTION

The optimum operation of a furnace depends upon the flame characteristics which influences the efficiency, pollutant levels release and the overall safety of the installations. There are currently great interests in the vision systems in order to allow acquisitions of images within furnaces. Most of furnaces have already been installed with normal CCD cameras that detect in the visible spectrum range the indoor furnace and flame.

A research group at Instituto Superior Técnico is currently concerned with the development of a vision system to detect flame location, flame length and flame colour brightness. The digitised images may assist the overall control strategy aiming to improve the furnaces control.

Although radiation transfer in the visible range has been studied for many decades, the authors are not aware of any studies related with the attempt to predict a turbulent flame brightness image. Thus, the main objective of the present work is to outline an engineering predictive tool to simulate a CCD camera image. Several assumptions have to be made. Among them, it is assumed that the radiation intensity is mainly due to continuous radiation from soot in the visible range. Other assumptions rely on the standard numerical modelling procedures to calculate mean properties in turbulent flames.

The next section describes the mathematical model including turbulence, combustion and radiation models. This is followed in the third section by the description of the numerical procedure used to predict the flame picture. The paper ends with a summary of the study.

2. MATHEMATICAL MODEL

THE MEAN FLOW CONSERVATION EQUATIONS

The incompressible, time-averaged form of the conservation equations, for mass, momentum and scalar quantities for a stationary turbulent high Reynolds number flow with chemical reaction are

Continuity

$$\frac{\partial \rho U_i}{\partial x_i} = 0 \; ,$$
(1)

Momentum

$$\frac{\partial}{\partial x_j}\left(\rho \, U_j U_i\right) = -\frac{\partial P}{\partial x_i} - \frac{\partial}{\partial x_j}\left(\rho \left\langle u_i u_j \right\rangle\right),$$
(2)

Scalar

$$\frac{\partial}{\partial x_j}\left(\rho U_j \Theta\right) = -\frac{\partial}{\partial x_j}\left(\left\langle \rho u_i \theta \right\rangle\right) + S_\Theta$$
(3)

Θ stands for scalar properties as mean temperature field, enthalpy and mixture fraction. In the above equation S_Θ represents a source/sink of the scalar quantity.

THE TURBULENCE MODEL

The mean flow equations are closed by the k-ε eddy viscosity/diffusivity model (EVM) which comprises transport equations for the turbulent kinetic energy k, its dissipation rate ε and constitutive relations for the Reynolds stresses and, turbulent fluxes of heat and mass as

$$- \rho \left\langle u_i u_j \right\rangle = \mu_t \left(\frac{\partial U_i}{\partial x_j} + \frac{\partial U_j}{\partial x_i}\right) - \rho \, \frac{2}{3} \, \delta_{ij} \, k$$
(4)

and

$$- \rho \langle u_i \theta \rangle = \kappa_{ij} \frac{\partial \Theta}{\partial x_j}, \qquad \kappa_{ij} = \rho \delta_{ij} \kappa_t \tag{5}$$

$$\mu_t = \rho \, C_\mu \frac{k^2}{\varepsilon} \tag{6}$$

$$\frac{\partial}{\partial x_j} (\rho \, U_j k) = \frac{\partial}{\partial x_j}\left[\frac{\mu_t}{\sigma_k}\frac{\partial k}{\partial x_j}\right] + P - \rho \, \varepsilon \tag{7}$$

$$\frac{\partial}{\partial x_j} (\rho \, U_j \varepsilon) = \frac{\partial}{\partial x_j}\left[\frac{\mu_t}{\sigma_\varepsilon}\frac{\partial \varepsilon}{\partial x_j}\right] + C_{\varepsilon 1}\frac{P}{\tau_m} - C_{\varepsilon 2}\rho \, \frac{\varepsilon}{\tau_m} \tag{8}$$

$$P \equiv - \rho \, \langle u_i u_j \rangle \frac{\partial U_i}{\partial x_j} \, ,$$

where the eddy viscosity μ_t is given by $\mu_t = \rho \, C_\mu \dfrac{k^2}{\varepsilon}$ and $\tau_m = \dfrac{k}{\varepsilon}$ is the characteristic time-scale of the turbulent mechanical field. The set of constants is standard [1]. The turbulent eddy - diffusivity $\kappa_t = \dfrac{\mu_t}{\sigma_t}$ is calculated considering the turbulent Prandtl number $\sigma_t = 0.7$.

The calculation of fluctuations of the scalar quantity is made considering a transport equation for the fluctuations mean variance $\langle \theta^2 \rangle$ as

$$\frac{\partial \rho \, U_j \langle \theta^2 \rangle}{\partial x_j} = \frac{\partial}{\partial x_j}\left[\frac{\mu_t}{\sigma_{\theta 2}}\frac{\partial \langle \theta^2 \rangle}{\partial x_j}\right] + 2 \, P_\theta - 2 \rho \, \varepsilon_\theta \, , \tag{9}$$

$$P_\theta = - \rho \, \langle u_i \theta \rangle \frac{\partial \Theta}{\partial x_i} \, .$$

This quantity is required for the calculation of the pdf of mixture fraction at all points of the flow domain as described below.

In almost all of the previous studies the dissipation rate of the scalar variance ε_θ has been calculated assuming that the ratio of the characteristic time-scale of turbulent mechanical and scalar field is constant as

$$2 \, \varepsilon_\theta = \frac{1}{R}\frac{\varepsilon}{k}\langle \theta^2 \rangle \, , \tag{10}$$

and prescribing a constant value for R. It has been shown elsewhere [2] that under buoyancy effects (or chemical reaction) the value of R is not constant and rather different from 0.5 (a value accepted for passive scalar dispersion). In the present study, calculations have been made with R equal to 0.55. A more general and physically meaningful route to calculate ε_θ is to consider a transport equation for the quantity. Extensions of the model proposed by Newman *et al.* [3] have been applied to the calculation of scalar dispersion in turbulent flows [4,5]. The present model considers a similar equation,

$$\frac{\partial \rho\, U_j \varepsilon_\theta}{\partial x_j} = \frac{\partial}{\partial x_j}\left[\frac{\mu_t}{\sigma_{\varepsilon\theta}}\frac{\partial \varepsilon_\theta}{\partial x_j}\right] + C_{P1}\frac{\varepsilon_\theta}{<\theta^2>}P_\theta + C_{P2}\frac{\varepsilon_\theta}{k}P - C_{D1}\rho\frac{\varepsilon_\theta^2}{<\theta^2>} - C_{D2}\rho\frac{\varepsilon_\theta}{\tau_m} \qquad (11)$$

$$\text{I} \qquad\qquad \text{II} \qquad \text{III} \qquad \text{IV} \qquad \text{V}$$

The terms (II) and (III) represent production of the quantity due to mean scalar and velocity gradients; terms (IV) and (V) are dissipative and term (I) represents the diffusion of process of the property. The model constant values are the same as in a previous study [5] except for the dissipating terms constants which were reduced by about 20%.

COMBUSTION MODEL

The instantaneous scalar field was obtained using the laminar flamelet [6] model for propane-air diffusion flame data. The instantaneous temperature and density fields were obtained from constitutive relations as

$$h \equiv \int_0^T \sum_{\text{all } j} m_j\, C_{pj}(T)\, dT + m_{fu}\, H \qquad (12)$$

where C_{pj} is the constant-pressure specific heat of a species j. Density is determined from the equation of state j

$$\rho = p\left(RT \sum_{\text{all } j} m_j/M_j\right)^{-1} \qquad (13)$$

where p is pressure, R_0 is the universal gas constant, and M_j is the molecular weight of species j.

The stochastic behaviour of the turbulence is introduced by means of a pdf of the mixture fraction, supposed to obey a clipped normal distribution as described elsewhere [7]. The mean values of density, temperature and concentration of the species C_3H_8, O_2, N_2, CO_2, CO, OH, O, H_2O, H_2 and H were obtained following the averaging procedure.

$$\overline{\phi_i} = \int_{-\infty}^{+\infty} \text{pdf(f)}\ \phi_i\ df \quad .$$

(14)

To obtain mean temperature one has to functionalise temperature with mixture fraction. For an adiabatic flame the flamelet data allow to obtain an expression for temperature functions of mixture fraction. The integration procedure (14) is straightforward. However, if flame radiation is considered the enthalpy loss presents additional difficulties to link temperature with mixture fraction. Figure 1 illustrates the procedure adopted in this study. For a given point, corresponding to a mean fraction mixture, the calculated enthalpy is smaller than the adiabatic enthalpy. Thus the instantaneous enthalpy over the mixture fraction range (0-1) is considered either to be obtained by the displaced curve shown in the Figure 1 or by two straight lines, connecting the calculated enthalpy to their extreme values. The first outlined procedure leads to higher mean temperature values than the second one in points corresponding to low mixture fraction values. The curve displacement over the entire mixture fraction domain was adopted in the present study.

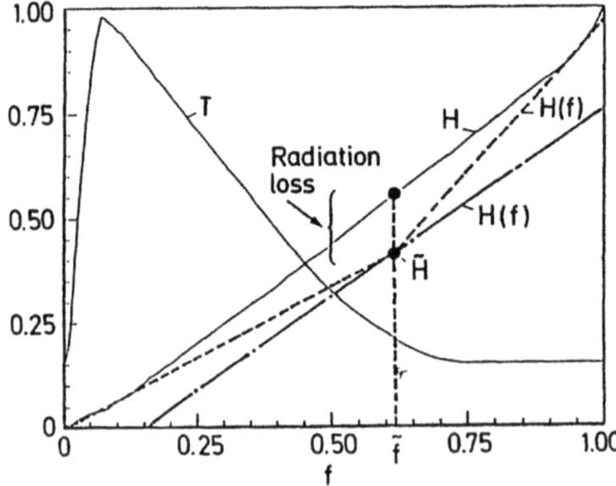

Fig. 1: Laminar Flamelet Model Curves.

SOOT MODEL

The distinctive feature of a propane flame is its significant soot content. The proportion of the total carbon content of the fuel which converts to soot is too small to influence significantly the overall flame heat release distribution.

Rather, soot is of concern because its presence greatly augments the radiation heat transfer. Most reasonably operated and maintained modern burners ensure complete combustion of soot. So the primary function of the soot will be good characterizing of the optical behavior of the flame. In other flames, say oil-flames the soot content is so great that those of industrial dimensions approach the black body limit. In consequence the accurate prediction of the local soot concentration is a prerequisite for the good calculation of the radiation transfer.

This is indeed fortunate since the mechanisms of soot formation are far from being established in the simplest laboratory flames [8].

Soot concentration was predicted by solving the transport equation

$$\frac{\partial}{\partial x_j}(\rho\, U_j\, m_s) = \frac{\partial}{\partial x_j}\left[\frac{\mu_t}{\sigma_{ms}}\frac{\partial m_s}{\partial x_j}\right] + P_{s+} - P_{s-} \tag{15}$$

A simple global expression similar to that used by Khan and Greeves [9] is chosen to characterize soot production:

$$P_{s+} = C_f P_{fu}\,\phi^n\,\exp\left(-E/T_g\right)\ . \tag{16}$$

C_f is a function which ideally depends on an easily definable fuel property such as the C/H ratio. The value of this constant was taken in accordance with [10]. Although the work of Khan and Greeves [9] was performed in connection with diesel engines the authors find that their values of 3 and 40200 cal/mol for n and E are useful. Soot production is essentially zero for equivalence ratio, ϕ, less than that corresponding to the incipient sooting limit [11] and for ϕ in excess of a value corresponding roughly to the upper flammability limit. Following Khan and Greeves the upper and lower limits have been set to 2 and 8 respectively.

In one sense the determination of the soot burning rate poses a much less demanding modelling problem since the particle sizes are so small that near-particle diffusion cannot possibly be a controlling process. Rather, the combustion rate will be controlled by the rate of mixing of the particle-bearing vortices with adjacent oxygen bearing material. A simple method to estimate this rate has been proposed by Magnussen and Hjertager [12] who, following conventional turbulence concepts [13], presume that the mixing rate is proportional to the magnitude of the time soot concentrations, and the timescale of the large scale turbulence motion ε/k. Their expression for the soot consumption rate is:

$$P_{s-} = A\, m_s\left(\varepsilon/k\right) \tag{17}$$

where A is a model constant assigned the value 4 based on numerical experimentation. This relation will not be satisfactory in regions where the reaction rate is limited by oxygen deficiency in which case Magnussen and Hjertager propose:

$$P_{s-} = A \left(\frac{m_{ox}}{m_s S_s + m_{fu} S_{fu}} \right) m_s \left(\frac{\varepsilon}{k} \right)$$
(18)

where S_s and S_{fu} are the soot and fuel stoichiometric ratios. The alternative giving the smallest reaction rate is to be used.

RADIATION MODEL

Radiation heat transfer plays a dominant role in most industrial furnaces. An adequate treatment of thermal radiation is essential to develop a mathematical model of combustion system. A review of the fundamentals of radiation heat transfer and the recent progress in its modelling in combustion systems was recently presented [14]. The Discrete Transfer method [15] which combines the virtues of the zonal, Monte Carlo and discrete-ordinate methods, has been applied yielding accurate results in one and two-dimensional geometries [16]. The Discrete Transfer method has been applied with success to full scale industrial furnaces [17, 18]. This method, applied in the present study, is briefly described next.

The fundamental equations for the transfer of thermal radiation may be expressed as:

$$\frac{dI}{ds} = -(k_a + k_s) I + k_a \frac{E_g}{\pi} + \frac{k_s}{4\pi} \int_{4\pi} P(\overline{\Omega}, \overline{\Omega}') I(\overline{\Omega}') d\Omega'$$
(19)

where I is the radiant intensity in the direction of Ω, s is distance in the Ω direction, $E_g \equiv \sigma T_g^4$ is the black body emissive power of the gas at temperature $T_{g',a}$ and k_a, k_s are the gas absorption and scattering coefficients, and $P(\Omega, \Omega')$ is the probability that incident radiation in the direction Ω' will be scattered into the increment of solid angle $d\Omega$ about Ω. For conciseness we define an extinction coefficient $k_c \equiv k_a + k_s$, an elemental optical depth $ds^* = k_c ds$, and a modified emissive power. $E^* \equiv 1/k_c (k_a E_g + (k_s/4) \int 4\pi P(\Omega, \Omega') I(\Omega') d\Omega')$, in which case the transfer equation may be reexpressed as:

$$\frac{dI}{ds^*} = -I + \frac{E^*}{\pi}$$
(20)

IMAGE BRIGHTNESS PREDICTION

The basic assumptions of the image/brightness numerical model are the following:

- A CCD-camera only detects radiation in the visible range of the spectrum.

- In the propane flame under consideration the radiation in the visible range is mainly due to soot.

- If the 3D-temperature and species concentration fields are known inside the flame, the absorption coefficient can be related to wave length via the Planck mean assumption theoretical description.

- Integration of the radiation transfer equation along discrete rays from every point inside the flame to the location point where the virtual CCD camera is located will provide the prediction of the flame shape brightness and image.

The numerical procedure proposed in this work to yield flame prediction image/brightness is the following:

1. To transform the predicted 2D-axisymmetric temperature and soot concentration into a 3D cylindrical space.

2. To locate the virtual CCD-camera at a point in a 3D reference space with origin at the flame region, see Figure 2.

3. To trace from each point inside the 3D cylindrical space a ray into the CCD-camera.

   The radiation transfer equation is integrated from the point to the limiting surface boundary, see Figure 3.

$$\mathrm{I}e^{s^*} = \frac{E^*}{\pi}\, e^{s^*} + \text{constant} \tag{21}$$

   giving

$$\mathrm{I}_{n+1} = \mathrm{I}_n\, e^{-\delta s^*} + \frac{E^*}{\pi}\left(1 - e^{-\delta s^*}\right) \tag{22}$$

where, I_n and I_{n+1} are respectively the values of intensity entering and leaving any control volume, n, that the ray has crossed, δs^* is the optical length within the volume.

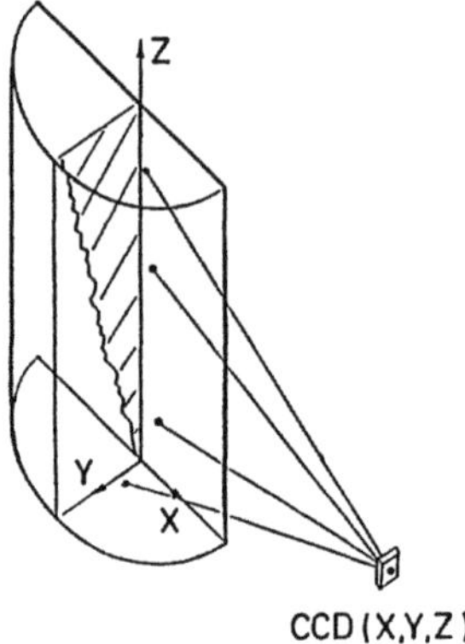

Fig. 2:   Flow Geometry and Reference Space.

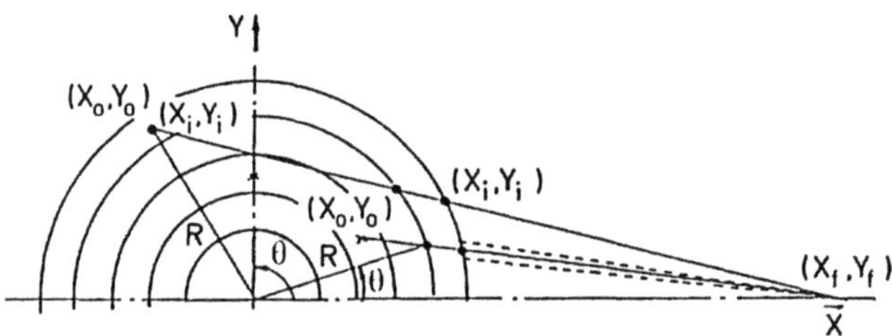

Fig. 3:   Integration Surface.

4. The surface of the cylinder (viewed from the CCD virtual point) is divided into a number of (m x n) "pixels". At each point, the radiation intensity is the sum of all the intensities of the rays that have crossed this surface element. The CCD-camera receives the integration in a solid angle of the surface radiation intensity. An image/brightness is predicted in (m x n) number of "pixels". The present calculations were obtained considering 40 x 40 "pixels".

The problem of computing the absorption coefficient of soot clouds can be attached from a fundamental point of view by means of the well known Lorenz-Mie theory. This theory solves the problem of the scattering of a plane wave by a spherical, homogeneous, isotropic particle of arbitrary particle diameter by giving the solution of Maxwell's field equation with appropriate boundary conditions [19,20]. To interpret soot data, the complete Lorenz-Mie theory is rarely used because of its reputed difficulty and also because it could be argued that soot particles are not spherical, homogeneous or isotropic [21]. In multidimensional radiative transfer analysis use of Mie codes is impractical.

Because of this, it is desirable to have simple approximations. An extensively used soot absorption coefficient is [14]:

$$K_\lambda = A\, f_v / \lambda \tag{23}$$

where f_v is the volume fraction of soot particles and the value A is a coefficient dependent of the type of fuel. Hottel and Sarofim [22] have suggested the value of 7 for typical soot particles observed in combustion chambers. Siegel [23] has analysed the available experimental data for several flames yielding values of A coefficient for different types of fuel. Based on that study, the present work uses the value of 4.0, typical of propane flames.

The absorption coefficient can be related to the spectral absorption coefficient by the following definition

$$\overline{K}_{\lambda 1,\,\lambda 2} = \frac{\displaystyle\int_{\lambda 1}^{\lambda 2} E_{b\lambda}\, K_\lambda\, d\lambda}{\displaystyle\int_{\lambda 1}^{\lambda 2} E_{b\lambda}\, d\lambda} \tag{24}$$

where $\lambda 1$, $\lambda 2$ represent the limits of the spectrum of electromagnetic radiation.

3. NUMERICAL MODEL

The numerical model is based on a finite-volume discretization of the governing equations, employing a staggered grid arrangement for the mean-velocity components relatively to scalar properties. An elliptic solver especially adapted for free flows and in particular for free jets was used in the computations [24]. Boundary conditions are imposed in all the boundaries. Far enough from the jet centreline ambient pressure is prescribed and entrainment flow velocities are implicitly calculated as any other quantity in the computational domain. The SIMPLE algorithm was chosen for the pressure-correction scheme and the Strong Implicit Procedure is employed for the solution of the algebraic set of equations.

Solution grid-dependence tests were performed and a computational domain comprising a 240 x 42 control volumes was considered satisfactory.

4. RESULTS

Figure 4 (a,b) shows radial profiles of mean temperature across several axial stations of jet development. The calculations were made using both a constant

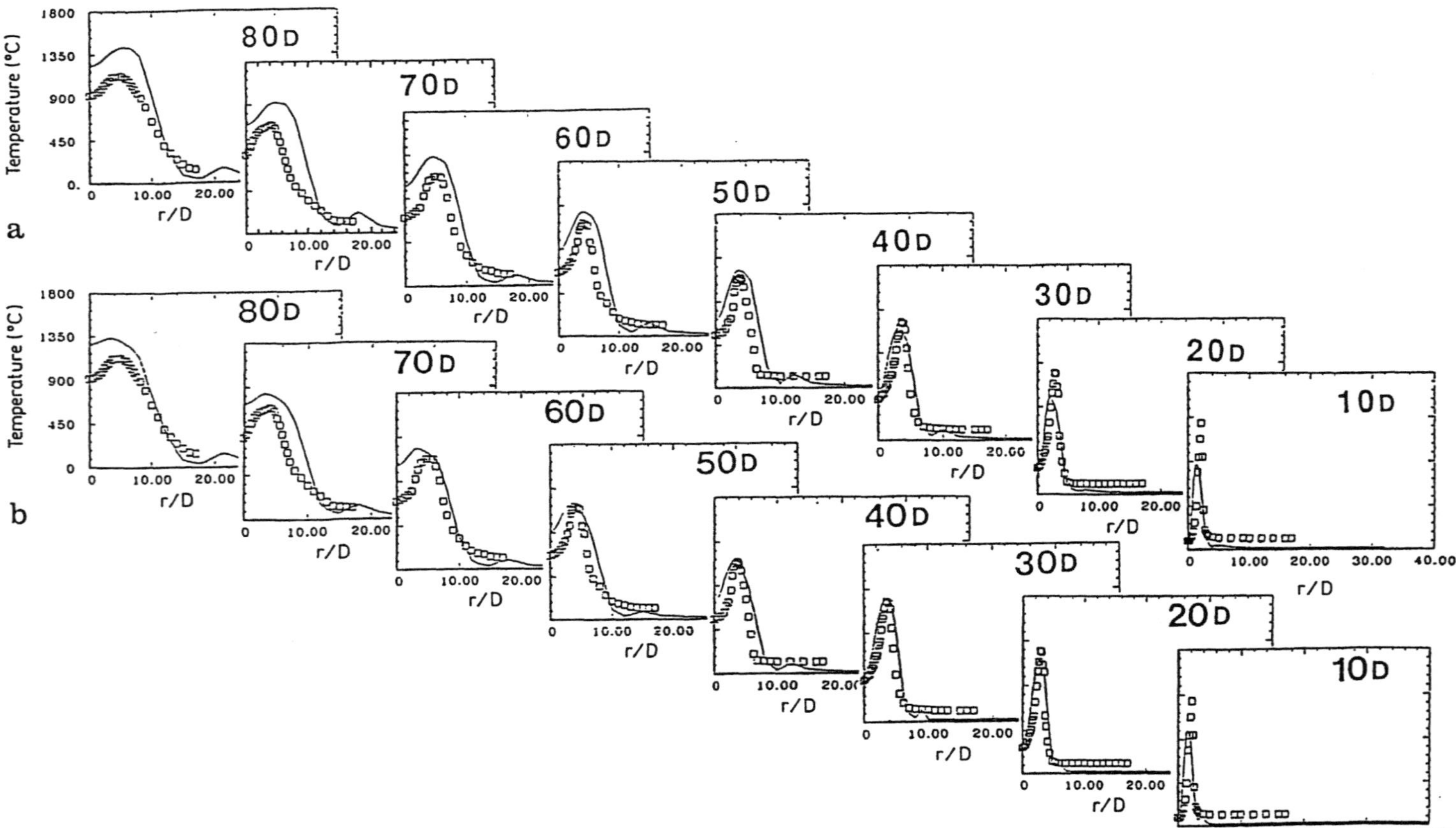

Fig. 4: Radial Profiles of Mean Temperature. ——— Predictions: a) R = Constant,
b) ε_θ - Equation; Symbols: Measurements.

time-scale ratio and a transport equation for the dissipation rate of mixture fraction variance. Globally, an acceptable agreement is observed between measured and predicted profiles. Though, it is depicted that the model responses too slowly to the temperature peak growth in the flame front as well as overpredicts the measurements at last section. This reveals a certain inertial behaviour which can be linked with deficiencies of the combustion model. The use of a transport equation for the dissipation rate of mixture fluctuations yields slightly better description of the temperature field which is an indication that further research is needed in modelling of the turbulence thermal field in reacting flows. Figure 5 a) shows a three-dimensional view of the flame brightness (Costeira *et al.* [25]) obtained by digitising a flame image. Figure 5 b) shows the corresponding flame brightness prediction obtained with the method previously outlined. A qualitative good agreement between the two Figures can be observed.

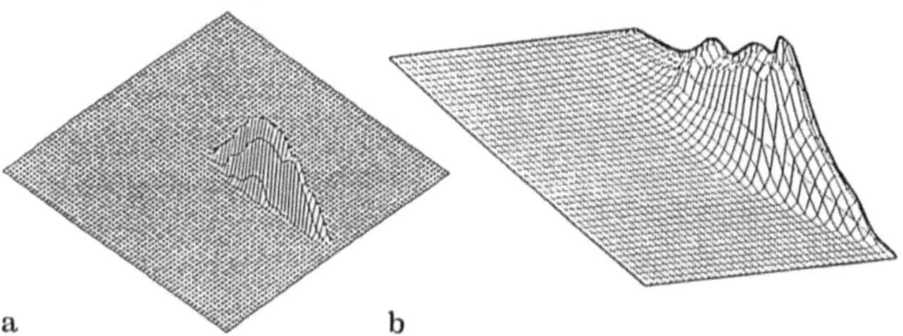

Fig. 5:   3D Plot of Flame Brightness. a) Digitised, b) Predicted

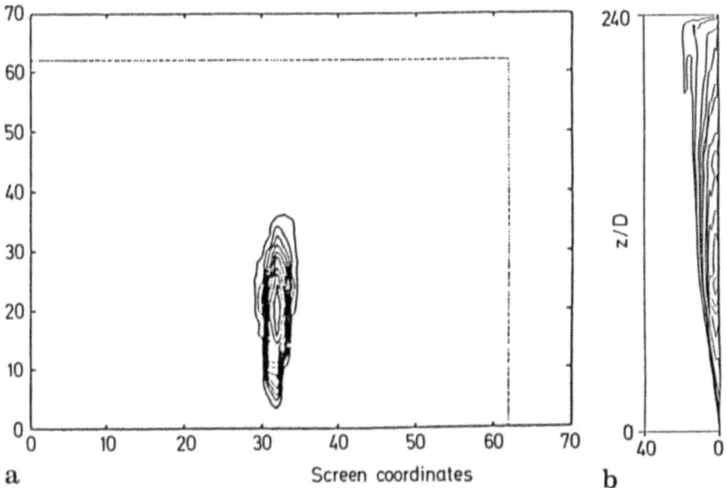

Fig. 6:   Isocontours of Flame Brightness. a) Digitised, b) Predicted.

Figure 6 a) and b) shows the brightness isocontours. Despite the different scales used in the prediction and measurement graphical representation, the digitised measured data indicate the maximum intensity located at the flame half-length while the prediction displays it at a shorter distance. This can be attributed to the soot concentration prediction.

5. CONCLUDING REMARKS

An engineering predictive tool was proposed to simulate a flame image brightness in the visible range. The basic assumption is that in luminous flames the radiation intensity in the corresponding visible wave length range $[3.9 - 7.8] \times 10^{-7}$ comes from soot. Thus, the contribution from CH, OH, and C_2 has been neglected as well as transitions of electrons from one configuration to another. The proposed model would be improved with the mentioned contribution; however, this is out of the scope of an engineering calculation procedure.

Turbulence/combustion models still require development for the calculation of turbulent combustion, especially the representation of the scalar field. The adopted transport equation for scalar dissipation rate has slightly improved the calculations.

ACKNOWLEDGEMENTS

The authors are deeply grateful to Ms. Sofia Erse Arruda and Ms. Rita Maia for typing and helping to prepare this manuscript.

REFERENCES

1. Launder B.E., Spalding D.B.: The Numerical Computation of Turbulent Flows. Comp. Meth. Appl. Mech. Eng. 3 (1974) 269-289.

2. Shabbir A., George W.K.: Energy Balance Measurements in an Axisymmetric Turbulent Buoyant Plume in a Neutral Environment. Sixth Symposium on Turbulent Shear Flows, Toulouse (1987).

3. Newman G.R., Launder B.E., Lumley J.L.: Modelling the Behaviour of Homogeneous Scalar Turbulence. J. Fluid Mech. 11 (1981) 217-232.

4. Durão D.F.G., Pereira J.C.F., Rocha J.M.P.: Evaluation of k-ε Turbulence Model for Predicting Buoyant Free Round Jets. 26th AIChE National Heat Transfer Conference, Edited by R.K. Shah, ASME HTD 107 (1989) 99-107.

5. Pereira J.C.F., Rocha J.M.P.: Numerical Calculation of Heat Transfer in a Turbulent Separated-Reattaching Shear Flow, to be presented in 3th Brazilian Thermal Science Meeting, ENCIT -90 (1990).

6. Liew S.K., Moss J.B., Bray K.N.C.: Flamelet Modelling of Chemical Closures in Turbulent Propane Air Combustion. Report TP-84-81, School of Mech. Engng., Cranfield Institute of Technology (1984).

7. Lockwood F.C., Naguib A.S.: The Prediction of the Fluctuations in the Properties of Free, Round-Jet, Turbulent Diffusion Flames. J. Combustion and Flame 24 (1975) 109-124.

8. Wagner H.G.: Soot Formation in Combustion. in Proc. 17th Int. Combustion Symp. Combustion Institute, Pittsburgh (1978).

9. Khan I.M., Greeves G.A.: A Method for Calculating the Formation and Combustion of Soot in Diesel Engines: Heat Transfer in Flames. Edited by Afgan and Beer (1974) 391-402.

10. Abbas A.S., Koussa S.S., Lockwood F.C.: The Prediction of a Variety of Heavy Oil Flames. in Proc. ASME Winter A. Mtg, Washington (Special Session on Two Phase Combustion Liquid Fuels) Nov. (1981).

11. Glassman I., Yaccarino P.: The Temperature Effect in Sooting Diffusion Flames. in Proc. 18th Int. Combustion Symp., Combustion Institute, Pittsburgh (1981).

12. Magnussen B.F., Hjertager B.H.: On Mathematical Modelling of Turbulent Combustion with Special Emphasis on Soot Formation and Combustion. in Proc. 16th Int. Combustion Symp., Combustion Institute, Pittsburgh (1976).

13. Tennekes H., Lumley J.L.: A First Course in Turbulence. MIT Press, Cambridge (1973).

14. Viskanta R., Menguç M.P.: Radiation Heat Transfer in Combustion Systems. Prog. Energy Combust. Sci. 13 (1987) 97-160.

15. Lockwood F.C., Shah N.G.: A New Radiation Solution Method for Incorporation in General Combustion Prediction Procedures. 18th Symposium (Int.) on Combustion (1981).

16. Shah N.G.: A New Method of Computation of Radiant Heat Transfer in Combustion Chambers. in PhD. Thesis, Imperial College, London (1979).

17. Carvalho M.G., Durão D.F.G., Pereira J.C.F.: Prediction of the Flow, Reaction and Heat Transfer in an Oxy-Fuel Glass Furnace. International Journal of Engineering Computations 4(1) (1987) 23-34.

18. Carvalho M.G., Oliveira P., Semião V.: A Three-Dimensional Modelling of an Industrial Glass Furnace. J. of the Institute of Energy (1988)143-156.

19. Van der Hulst H.C.: Light Scattering by Small Particles. John Wiley & Sons, Inc., New York, 470 (1957)

20. Bohren C.F., Huffman D.R.: Absorption and Scattering of Light by Small Particles. John Wiley & Sons, Inc., New York (1983)530.

21. Felske J.D., Hsu P.F., Ku J.C.: The Effect of Soot Particle Optical Inhomogeneity and Agglomeration on the Analysis of Light Scattering Measurements in Flames. J. Quant. Spectrosc. Radiat. Trans. (1986) 35-447.

22. Hottel H.C., Sarofim A.F.: Radiative Heat Transfer. McGraw Hill, New York (1967).

23. Siegel R.: Radiative Behaviour of Gas Layer Seeded with. NASA TND-8278, Washington, D.C. (1976).

24. Durão D.F.G., Pereira J.C.F.: Calculation of Isothermal Three--Dimensional Free Multi-Jet Flow. Applied Mathematical Modelling (1991), 338-350.

25. Costeira J.P., Fernandes E.C., Heitor M.V., Sentieiro J., Simões J.P., Vitor J.A.: A Vision System for Analysis and Classification of Industrial Flames. in Proc. 6th CIM-Europe Annual Conference, 15-17 May, Lisboa, Portugal, Eds. L. Faria and W. van Puymbroeck, Springer (1990).

THE ROLE OF RADIATION IN MICROGRAVITY FLAME STRUCTURES AND BEHAVIOUR

J. Buckmaster
University of Illinois
Urbana, IL 61801, USA

and

P.D. Ronney
Princeton University
Princeton, NJ 08544, USA

ABSTRACT

This paper is concerned with heat losses from flames by radiation. In particular, it discusses the importance of these losses and their role in some unusual phenomena and structures that have been observed in combustion under microgravity conditions. These are 'self-extinguishing flames' (SEFs) and 'flame-balls'/'flame-strings'.

INTRODUCTION

Band radiation losses are not normally thought of as being of great significance in controlling the propagation of a deflagration wave. There is good reason for this — the radiation from the combustion product gases will be small compared to the volumetric heat production from reaction except at extremely small flame-speeds, speeds not normally observed. Simple theories [1] that account for radiation losses can predict flammability limits: that is, the impact of the losses grows stronger as the fraction of deficient reactant in the mixture is reduced (reducing the flame-speed) until quenching occurs. But the predicted flame-speeds at the limit are much lower than those usually observed in the laboratory [2]. This is because, under normal laboratory conditions, non-radiative losses dominate radiation losses. In particular, no configuration can eliminate buoyancy effects, which are known to have a strong influence on flammability limits [3][*] . Certainly if convective velocities induced by buoyancy are larger than the limit flame-speed that would be defined by radiation losses alone, this speed will bear little relationship to the observed limit speed. In

---

[*]   In upward propagation the bubble of burnt gas can rise at speeds much greater than the burning velocity. In downward propagation in a standard flamability tube, differential buoyancy-induced velocities due to wall cooling are significantly larger than radiation-limited burning velocities.

162

general, conditions where radiative losses can be expected to influence deflagrations are not accessible at earth gravity.

MICROGRAVITY COMBUSTION

The situation is quite different in the absence of gravity, for it is then possible to devise a configuration in which radiation has no competitor[**] and so plays an important role. Recent ignition experiments in a large chamber at μg confirm this expectation [2]. That radiation is energetically significant is clear from the examination of the thermal decay in the burnt gas, Figure 1. The measured decay compares well with theoretical predictions. Moreover, evidence that these losses are responsible for limit behavior is provided by examining variations of the flame-speed with equivalence ratio or dilution, Figure 2 [4]. Theory [1] predicts an infinite slope at the limit, and the curves in Figure 2 appear to exhibit this trend.

Indirect evidence of the importance of radiation comes from consideration of 'self-extinguishing flames' (SEFs). Ignition of a mixture at a point leads to spherical propagating flames which are highly curved and experience stretch, both of which can significantly affect the flame-temperature and speed [5]. The nature of the effect depends on the Lewis number, defined as the ratio of the thermal diffusivity of the bulk mixture to the mass diffusivity of the scarce reactant. For a lean CH_4-air mixture (Le < 1) a small radius flame travels faster than one of large radius so that a flame-radius — time plot has a slope that decreases with time in the neighborhood of the origin, Figure 3 [6]. For lean C_3H_8-air mixtures (Le > 1) the reverse is true. Thus for mixtures for which Le < 1, radiation effects become more important as the radius of the flame increases and the flame-speed decreases. In particular, outwardly propagating flames can extinguish at some finite radius at which radiation losses have, in relative terms, become large enough (SEFs). Figure 4 [4] shows experimental evidence for this, with extinction occurring at about 4 cms for dilution with 63.2% CO_2, for example. This kind of behavior can be duplicated by a simple asymptotic analysis of equations whose ingredients include a flame-sheet model for the kinetics, Le<1, and volumetric heat losses ($\mathcal{L}$) [7]. Variations in $\mathcal{L}$

(**) There is an analogy with surface tension effects in ordinary fluid flow involving a material interface. For earth-bound experiments at all but very small scales gravity effects dominate those of surface tension which is, accordingly, unimportant. But under microgravity conditions surface tension has no competitor and plays a profound role.

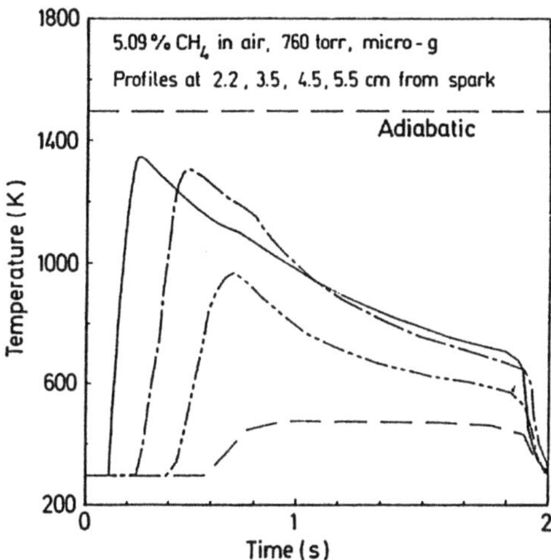

Fig. 1:   Measured thermal decay in a methane-air flame at microgravity. Here the losses are sufficient to cause extinction when the flame radius is about 4.7 cm. Comparisons with theoretical predictions in [2] establish that the decay is due to radiation.

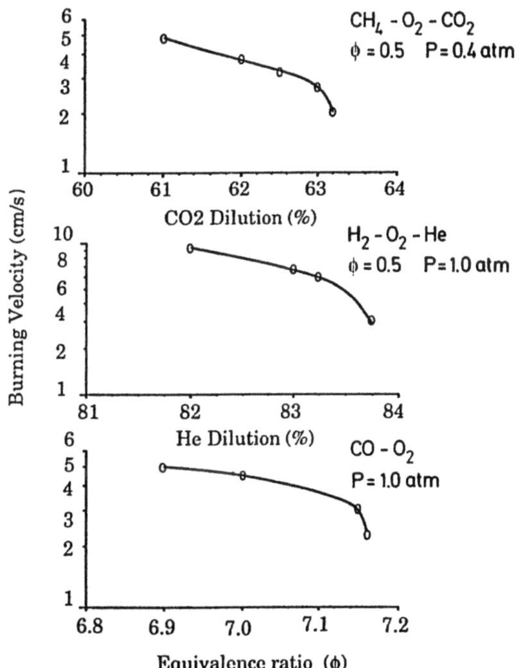

Fig. 2:   Burning velocity vs. dilution from [4]. The theories of [1] predict infinite slope at extinction.

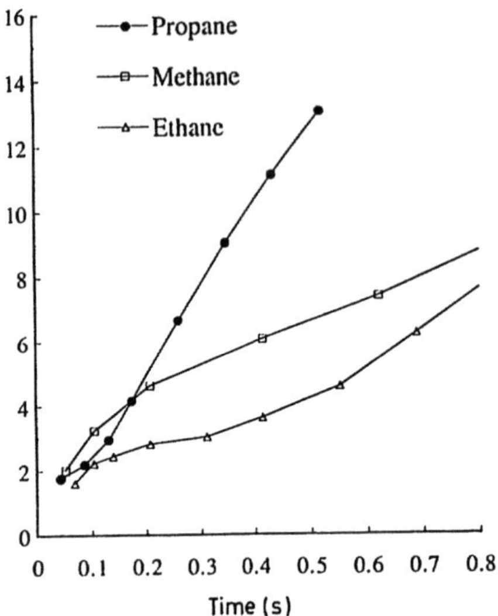

Fig. 3: Observed flame radius (in cms) vs. time for different mixtures, from [6]. For methane the flame slows as it grows, for propane it accelerates.

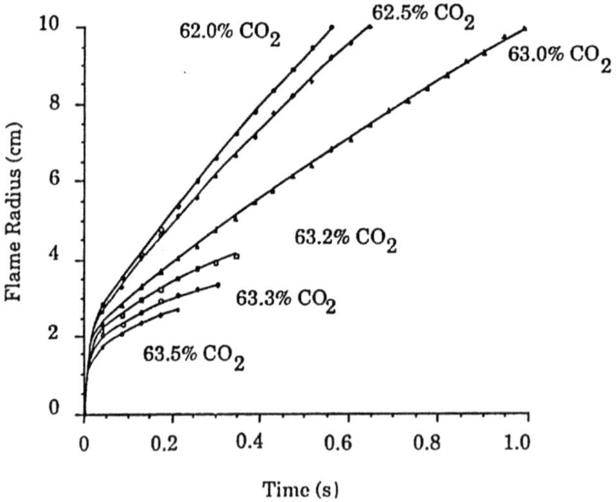

Fig. 4: Dynamic behavior of methane-oxygen flames diluted with carbon-dioxide, from [4]. Equivalence ratio = 0.5, pressure = 1 atm. The Lewis number is approximately 0.74. Extinction is observed for dilutions of 63.2% and greater.

correspond to variations in the percentage of CO_2, and Figure 5 shows how the theory predicts flames of different final radii (extinction radii) as ℓ is varied.

A noteworthy characteristic of SEFs is that the extinction radius depends strongly on the energy deposited in the mixture by the igniting spark. Memory of the initial conditions is retained even after the flame has released chemical energy which is orders of magnitude greater than the ignition energy. Figure 6 shows this effect in a lean-limit CH_4-air mixture [8]. A 72.1 milliJoule (mJ) spark produces a SEF which extinguishes after generating 3600 times the ignition energy, whereas a 78 mJ spark produces a flame which does not extinguish. This effect is not duplicated by the analysis of [7] but can be mimicked by numerical simulations in which the chemistry is modeled by one-step Arrhenius kinetics (with Le<1 and volumetric heat losses, as before), Figure 7 [9].

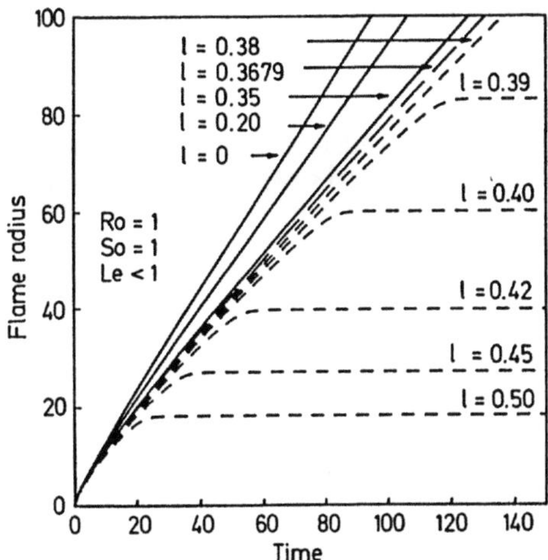

Fig. 5: Flame history for different values of the heat loss parameter ℓ, from [7]. Extinction occurs at different radii provided the heat-loss is large enough.

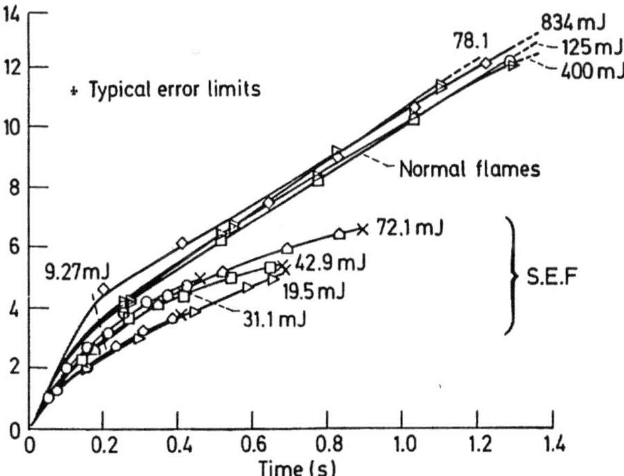

Fig. 6: Variations of the observed flame radius (in cms) with time, from [8]. A small increase in ignition energy from 72mJ to 78mJ creates a normal flame, rather than a self-extinguishing flame.

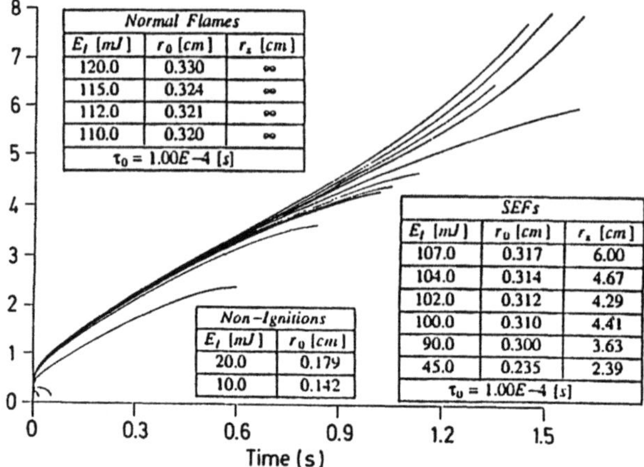

Fig. 7: Variations of the calculated flame radius (in cms) with time, from [9]. These results show the same sensitivity of final flame radius to initial energy deposition as Fig. 6. For the 1-step kinetics of [9], extinction is defined when the maximum temperature attains a prescribed minimum value.

Radiation also appears to be responsible for the existence of 'flame-balls', stationary spherical structures approximately 1.5 cms in diameters. These can be generated by point ignition in near-limit mixtures for which Le $\leq$ 0.5. The initial flame fragments because of the well-known cellular instability, the cells separate from each other, and then form spheres. The balls were first observed in 2.2 sec drop-tower experiments but, very recently, have been seen for up to 15 secs in NASA's KC-135 during Keplerian trajectories [10]. A stationary spherical structure implies that the mass-averaged velocity is zero everywhere, so that the only fluxes (apart from radiation) are diffusional. A premixed flame in which diffusion alone drives the supply of mixture to the burning zone, and the dispersion of heat and products from the burning zone, is singular.

The theory of flame-balls appears to be on a secure footing [11]. If we adopt a flame-sheet model to account for the kinetics and assume constant transport coefficients then, in the burned gas,

$$\underline{r \leq r^*} \qquad T = T_b, \ Y=0; \tag{1}$$

and in the unburned gas,

$$\underline{r \leq r^*} \qquad T = T_f + (T_b - T_f)\frac{r^*}{r}, \quad Y = Y_f\left(1-\frac{r^*}{r}\right). \tag{2}$$

The flame-temperature T_b is given by the formula

$$T_b = T_f + \frac{(T_a - T_f)}{Le}. \tag{3}$$

Here T_f is the remote temperature, Y_f the remote (farfield) mass fraction of mixture, T_a is the adiabatic flame-temperature, and r* is the flame-radius. The thermal source strength of the flame-sheet is $(T_b-T_f)/r*$, a measure of the strength of the reaction, an assigned quantity in the flame-sheet model. This determines r*.

The solution (1)-(3) was identified many years ago by Zeldovich [*], but is unstable. If one considers infinitesimal one-dimensional unsteady perturbations these grow exponentially, signalling either collapse of the flame or expansion out into the surrounding mixture. When radiative losses are incorporated however, stable solutions can be generated. More precisely, heat losses that are too large will quench the flame, but for losses less than the quenching value there are two possible steady solutions, Figure 8, and a portion of the upper branch is stable.

We can not resist ending this brief discussion with a reference to 'flame-strings', although a detailed discussion must wait for a different forum. During the KC-135 experiments, g-jitter (small fluctuations in the gravitational acceleration) generate a relative motion between a flame-ball and the surrounding mixture. In this way a flame-ball can be stretched out to form a cylinder 5-8 cms long, a 'flame-string'. Flame-strings are unstable and collapse into a number of discrete flame-balls.

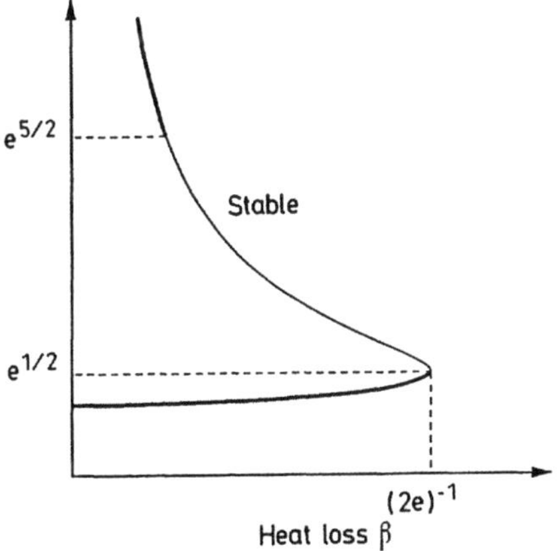

Fig. 8: Non-dimensional flame-ball radius vs. a heat loss parameter defined by radiative losses from the burnt gas, from [11b]. Solutions on a portion of the upper branch are stable, and so are candidates for the observed structures. Note that in the absence of heat losses there are no stable solutions.

---

[*] He was motivated by 'flame-caps', the lg counterpart of flame-balls [12].

It is possible to carry out a simple stability analysis that throws some light on this behavior by assuming that the undisturbed string is steady and 'long and slender' so that it can be treated as quasi-one-dimensional [13]. In its simplest form the analysis predicts the variations of growth rate with kR_s shown in Figure 9. Here R_s is the radius of the undisturbed string, and k is the wavenumber, characterizing the disturbance, viz, the disturbed radius is

$$R = R_s + \delta\, e^{\lambda t + ikx} \tag{4}$$

where x is axial distance. Short waves are stable but sufficiently long waves are unstable, and this is consistent with the experimental observations. Various perturbations to the basic model can be considered (e.g., small variations of the string radius) and these have the effect of shifting the wave-number for neutral stability from the value of about 0.6 seen in Figure 9.

All of these observations and results provide unambiguous direct and indirect evidence of the vital role that radiation can play in microgravity combustion.

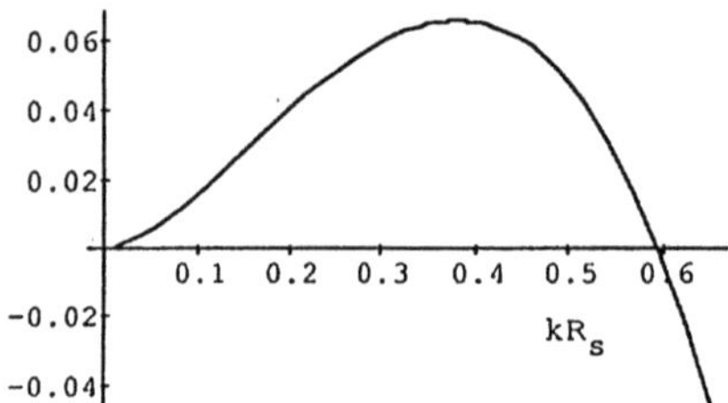

Fig. 9: Variations of growth rate (the ordinate is proportional to λ) with wave-number for flame-strings. The peristaltic instability for sufficiently long waves causes the string to break up into a number of flame-balls.

ACKNOWLEDGEMENTS

The work of J.B. is supported by AFOSR and NASA; that of P.D.R. by NASA-Lewis, NSF (PYI program), and the Gas Research Institute.

REFERENCES

1. Spalding D.B.: Proc. Roy. Soc. (London) A240, 83 (1957); Buckmaster J.: Combustion and Flame 26 (1976) 151; Joulin G., Clavin P.: Acta Astronautica 3 (1976) 223.

2. Ronney P.D.: Twenty-Second Symposium (International) on Combustion. The Combustion Institute, Pittsburgh, PA (1988), 1615.

3. Levy A.: Proc. Roy. Soc. (London) A283 (1965) 134; Buckmaster J.D., Mikolaitis D.: Combustion and Flame 45 (1982) 109; Jarosinski J., Strehlow R.A., Azarbarzin A.: Nineteenth Symposium (International) on Combustion. The Combustion Institute, Pittsburgh, PA (1982) 1549.

4. Abbud-Madrid A., Ronney P.D.: Effects of Radiative and Diffusive Transport on Premixed Flames Near Flammability Limits. Twenty-Third Symposium (International) on Combustion, to appear.

5. Buckmaster J.D., Ludford G.S.: Theory of Laminar Flames. Cambridge University Press (1982); Williams F.A.: Combustion Theory, 2nd edition, Benjamin-Cummins (1985).

6. Ronney P.D.: Acta Astronautica 12 (1985) 915.

7. Ronney P.D., Sivashinsky G.I.: SIAM J. on Appl. Math. 49 (1989) 1029.

8. Ronney P.D.: Combustion and Flame 62 (1985) 120.

9. Farmer J.N., Ronney P.D.: A Numerical Study of Unsteady Nonadiabatic Flames. Combustion Science and Technology, in press.

10. Ronney P.D.: Near-Limit Flame Structures at Low Lewis Number. Combustion and Flame, in press; Whaling K.N., Abbud-Madrid A., Ronney P.D.: Structure and Stability of Near-Limit Flames with Low Lewis Number. Abstract of a presentation at the 1990 Fall Technical Meeting, Western States Section. The Combustion Institute, October 15-16, La Jolla, CA.

11. Zeldovich Ya. B., Barenblatt G.I., Librovich V.B., Makhviladze G.M.: The Mathematical Theory of Combustion and Explosions. Consultants Bureau, New York (1985) 327; Buckmaster J., Joulin G., Ronney P.D.: Combustion and Flame 79 (1990) 381; Buckmaster J., Joulin G., Ronney P.D.: Structure and Stability of Non-Adiabatic Flame Balls, II. Effects of Far-Field Losses. Combustion and Flame, in press; Buckmaster J., Joulin G.: Flame-Balls Stabilized by Suspension in Fluid with a Steady Linear Ambient Velocity Distribution. Submitted; Lee C.J., Buckmaster J.: The Structure and Stability of Flame-Balls: A NEF Analysis. Submitted.

12. Lewis B., von Elbe G.: Combustion, Flame, and Explosions of Gases. Academic Press, New York (1961) 314; Weeratunga S., Buckmaster J., Johnson R.E.: Combustion and Flame 79 (1990) 100.

13. Buckmaster J.: Manuscript in preparation.

CHAPTER 3

PARTICLE PHASE — GAS PHASE RADIATIVE AND CONVECTIVE INTERACTION

SIMULATION OF PARTICLE MULTIPLE SCATTERING AND APPLICATIONS TO PARTICLE DIAGNOSIS

G. Gouesbet, B. Maheu and J.N. Le Toulouzan
Laboratoire d'Energétique des Systèmes et Procédés
INSA de Rouen, UA CNRS 230, Coria, BP08,
76131, Mont-Saint-Aignan Cedex, France

ABSTRACT

A review of the work carried out in Rouen during the last ten years on the topic of light multiple scattering in particulate media is provided. Theoretical issues concern analytical radiative transfer computations by a four-flux model and Monte-Carlo simulations. One of the practical outcome of this theoretical effort is the development of a Visible Infra-red Double Extinction technique to simultaneously measure particle sizes and number-densities in high optical thickness particulate media. Validation of this technique is carried out by studying standard media from multiple scattering to dependent scattering situations with particle proximity effects. Prospective ideas are also discussed, some of them hopefully providing new opportunities for further research.

1. INTRODUCTION

Much effort has been devoted since a long time to the study of quasi-elastic scattering of light due to a fundamental interest but also due to the existence of interesting applications, such as the development of optical particle diagnosis that we shall consider in this paper. However, applications have been usually restricted to situations where the optical thickness of the medium under study is not too high, severely limiting the range of potential applications. In a lot of industrial, environmental, and fundamental situations, optical thickness may be high enough to put multiple scattering and even dependent scattering phenomena on the stage. For instance, in interstellar clouds and dusty atmospheres, big optical thicknesses result from big geometrical thicknesses which overcome the fact that particle number-densities are small. Conversely, big optical thicknesses may be obtained for small geometrical thicknesses and big number-densities, for instance in coal and spray furnaces, fluidization and sedimentation, particle pneumatic transport in conduits. Therefore, there was a need for the development of optical particle diagnosis techniques in such cases.

Nowadays, optical particle diagnosis has become synonymous of laser source. Therefore, before designing an experimental set-up for measurements, we must address a theoretical problem. Given a particulate medium illuminated by a laser beam, we want to know something inside the medium (particle sizes, number-densities) by measuring something outside of it. Although, this is actually an inverse problem usually of tremendous difficulty, we shall be fortunate enough to find a simple solution to it.

The paper is organized as follows. We first discuss simple theoretical tools for multiple scattering computations and simulations. Although these tools do not specifically concern the case of laser beam sources, we keep in mind that this laser case corresponds to our original motivation for optical diagnosis. Section 2 is devoted to Monte-Carlo techniques and Section 3 to four-flux models. Relying on a result of the four-flux models, we describe the Visible Infra-red Double Extinction (VIDE-) technique in Section 4. Section 5 is a conclusion.

We now mention that this paper is not strictly speaking a scientific paper providing the reader with new results, the presented material being published or going to be published elsewhere. It rather looks as a general guideline from which the reader may reach other more technical papers giving required details. We shall avoid mathematical discussions to focus on offered opportunities.

2.   MONTE-CARLO TECHNIQUE

2.1. GENERALITIES

The newcomer wanting to start working in the field of multiple scattering of light should begin with the very pedagogic book by Chandrasekhar [1] and pursue with the treatise by van de Hulst [2].

Being interested with particulate scattering, we then realize that we need a connection between the radiative transfer formulation and scattering properties of volume elements containing the scatter centers. This connection is established by using a single scattering theory for individual particles and by lifting the single particle scattering properties to equivalent volume elements properties (Gouesbet et al [3]). In some cases such as when a cloud of spherical, homogeneous, isotropic, non-magnetic particles is illuminated by a plane wave, the basic single scattering theory to use is the classical Lorenz-Mie Theory (LMT). When the connection is established, it is possible to go back to Chandrasekhar and van de Hulst and to rely on their radiative transfer formulation. Multiple scattering by particles is then easily handled in principle by combining single scattering theory and radiative transfer.

However, if the incoming light is a shaped beam like a laser beam, two new difficulties arise. The first is that the interaction between a laser beam and a particle cannot always be described accurately enough by LMT which assumes an incident plane wave. A more sophisticated single scattering theory is required, namely GLMT (Generalized Lorenz-Mie Theory). Although GLMT is more difficult by at least "one order of magnitude" than LMT, such a theory does exist (see for instance Refs. [4]-[6] from which earlier references dating back to 1980 may be found). Implementation of GLMT in a multiple scattering study however still remains to be done and would probably provide a very decent opportunity for new research. This problem is not considered any more in this paper. The second new difficulty is the loss of symmetry implied by the use of a laser beam source, leading to a radiative transfer problem of excessive complexity. Although such problems have been attacked by some researchers (see for instance Crosbie and Dougherty [7]), we are probably very far of owning a general enough solution.

Consequently, the most reasonable approach appears to be the use of a Monte-Carlo code. Monte-Carlo simulations have a bad reputation in terms of computer time requirements. This reputation is no more justified due to the efficiency of modern mainframe computers, workstations and even personal computers. On the other hand, they own the advantage of a dramatic flexibility to be adapted to arbitrary light sources, arbitrary geometry, arbitrary location of detectors, etc. We now describe our code MUSCAT (MUltiple SCATttering) for Monte-Carlo simulations. Such simulations being actually not new, some emphasis will be given to the case of laser beam multiple scattering which is of current interest nowadays. Other details are available from Maheu et al [8] and will appear later in subsequent papers.

2.2. PRINCIPLE

We designed the Monte-Carlo program MUSCAT to handle multiple scattering under the following assumptions:

i)   the scattering medium is homogeneous and isotropic (with respect to the scattering properties) with arbitrary geometry,

ii) the scattering process is quasi-elastic without any account for polarization nor for dependent scattering, but no restrictive assumption holds for the phase function which can range from the simplest isotropic one up to the most realistic Mie one,

iii) the source is monochromatic with arbitrary geometry and indicatrix,

iv) the detectors have arbitrary view angles, sizes and locations inside or outside the scattering medium.

General information about Monte-Carlo techniques can be found in classic books (for instance Cashwell and Everett [9]). In the case of multiple scattering we here deal with, the Monte Carlo method essentially consists in numerically simulating the life of each photon from its emission by the source up to its absorption inside- or escape from- the scattering medium. The process has to be repeated as many times as required by the accuracy of the statistics. The algorithm of MUSCAT is shown by Figure 1 where it is easy to recognize the history of a photon which is first emitted by the source, then possibly encounters successive scattering events and eventually dies or escapes.

Random numbers are used, first to choose a photon inside the source (I) and then to determine:

(i)   if it is scattered before escaping;
(ii) where a scattering event takes place;
(iii) whether the photon is absorbed or scattered;
(iv) which direction it travels after scattering.

The code then stores the escape position and direction of the photon (S).

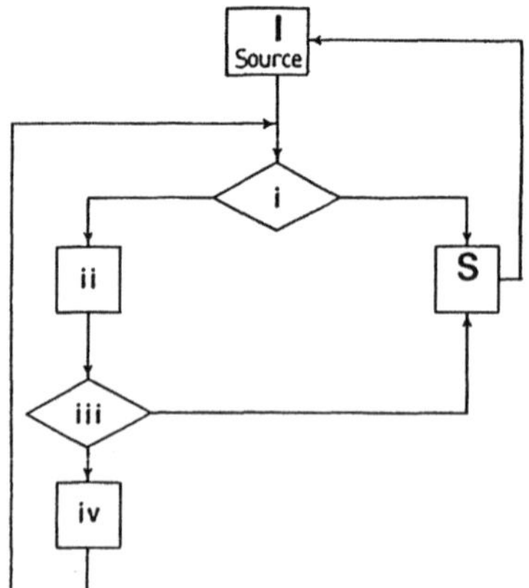

Fig. 1:   Algorithm of MUSCAT code: (I) Source; (i) Scattering/ /Escaping ?; (ii) Scattering Location ?; (iii) Absorption/Scattering ?; (iv) Scattering direction ?; (S) Outgoing Storage.

The random numbers are weighted by p.d.f.'s to allow a statistically exact simulation of the actual physical processes. Of course, the accuracy of the numerical simulation closely depends on the physical correctness of the p.d.f.'s and on the statistical validity of the random numbers. Hence, numerical simulations like the results of MUSCAT need careful validations before we give them plain confidence.

2.3. VALIDATION

Numerical simulations must be carefully checked and validated because the results of computer programs may be misleading much more easily than experimental data. Such validations have been achieved for the code MUSCAT. They range from obvious statistical and physical tests to comparisons with experimental data. For instance, we checked the uniformity of the random generator of MUSCAT and the independence of three successive (pseudo)random numbers. Also, exhaustive comparisons have been made between MUSCAT and the exact analytical doubling method [2] for the plane-parallel geometry (collimated and diffuse illumination). Several papers by a group of the university of Firenze [10]-[11] allowed successful comparisons with another numerical simulation and with experimental data. The experiment by this group is the scattering of a laser beam by a cylindrical cell filled with a suspension of small spheres (Figure 2). The transmittance of the medium is measured for various sphere sizes, optical depths, aperture angles, and distances from the scattering medium.

Again, a good agreement was found between MUSCAT and this well-controlled laboratory experiment.

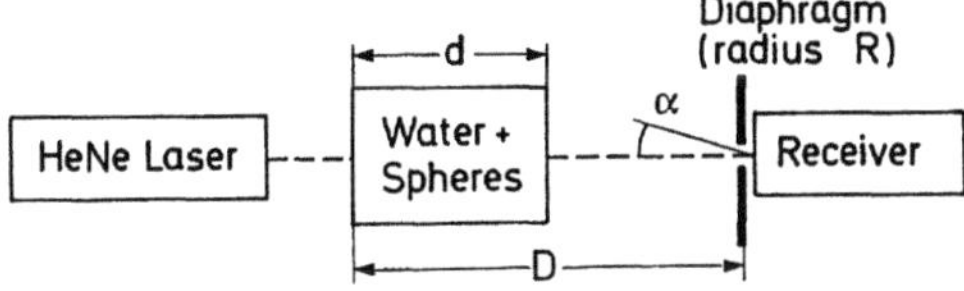

Fig. 2: Experiment by Bruscaglioni et al [11]. The HeNe laser beam propagates through a suspension in water of latex spheres; d=10 cm is the width of the turbid medium; D is the (variable) distance between the receiver and the point where the beam enters into the medium; α is the (variable) field of view semiaperture of the receiver. Diameter of the laser beam at the input plane of the scattering cell is 2 mm. R is the radius of the diaphragm at the input area of the receiver.

2.4. EXAMPLES OF RESULTS

A striking feature of Monte-Carlo simulations is their versatility and ability to handle the complexity of concrete geometries where the usual physicist becomes a little bit overwhelmed. A contract with the LCPC (Laboratoire Central des Ponts et Chaussées, Paris) gives a good example of a numerical simulation of a concrete problem. The aim is to predict the visual environment of a driver by foggy weather. For instance, Figure 3 shows the light coming from the headlights of a car into the eyes of another driver. The curves give the radiance as a function of the distance for three aperture angles: 1° (direction of the headlights); 10° (halo around the car); 90° (ambient light). In this example, the subroutine simulating the light source uses a tabulation of the actual (very

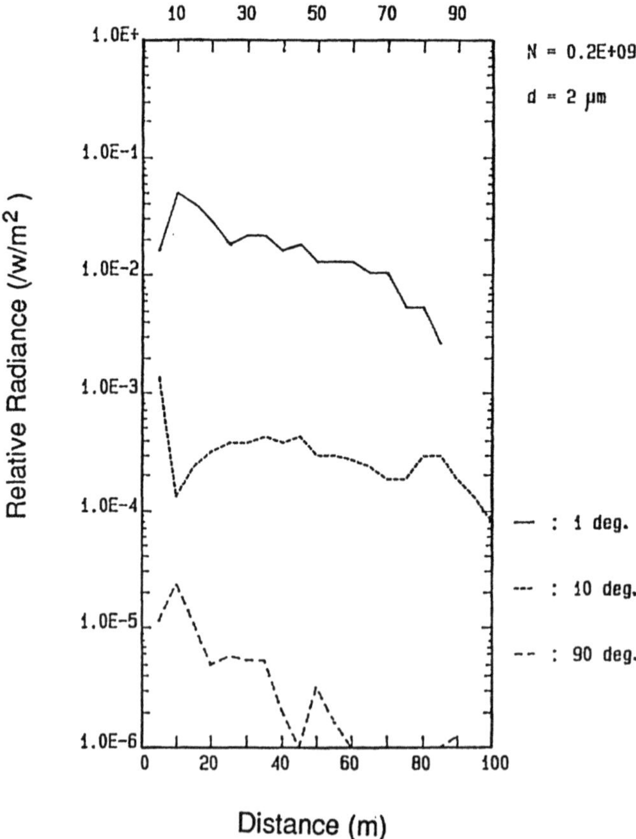

Fig. 3: Radiance onto the eye of a driver vs fog width.
Source: car headlights (tabulated)
Fog: 2.10^8 droplets/m^3 (diameter 2 μm)
Detector: Eye of a driver (3 fields of view)

unsymmetrical) headlight of a car and different road surface properties can be simulated (absorbing surface/Lambert surface/wet reflecting road). Going further in the same field of physics, other applications have been considered: visibility of targets on the side of the road, backscattering from the frontlights towards the car driver, etc...

Due to the flexibility of the Monte-Carlo method, MUSCAT was used also for helping in designing diagnosis set-ups inside turbines (vapor embedding water droplets) or for atmosphere control near open-air wires (dust in the air). Last but not least, we are currently developing extensive tabulations of the four-flux model by means of MUSCAT (see section 3).

An attractive feature of a Monte-Carlo code is that adapted subroutines may be inserted to account for a great wealth of sources (plane wave, laser beams, cylindrical beams, car lights,...), media (various geometries and boundary conditions) and detectors (aperture angle, size, number of detectors...). Actual developments carry on account for polydispersity of the scatterers and polychromaticity of the light.

3.   FOUR-FLUX MODEL

N-flux models for radiative transfer modelling also have a very long tradition, seemingly dating back to a 2-flux model by Schuster [12]. We also mention the celebrated Kubelka and Munk theory [13] designed for studying luminous properties in paints. The reader should furthermore refer to the easy and useful book by Kortüm [14]. The basic idea is to discretize the radiative flux in N parts, to write down radiative balance equations for each flux leading to a system of coupled ordinary differential equations, and to solve this set. Here again, the link between old usual N-flux models and present case when scattering is produced by macroscopic particles is obtained by relying on a single scattering theory, like LMT. The four-flux model we shall discuss in this section is very simple, leading to a set of differential equations which may be solved on the corner of the desk, and corresponding numerical computations can be most of the time carried out with pocket computers. Also, in numerous cases, the loss of accuracy with respect to a full rigorous radiative transfer theory is not significant.

We consider the standard problem sketched in Figure 4. The scattering slab contains homogeneously distributed particles with diameter d, number-densities N, and complex refractive index m. It is illuminated by an incident

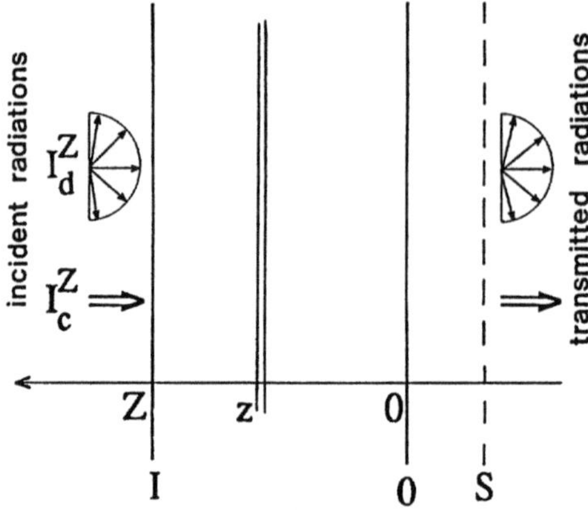

Fig. 4:   Scattering slab is limited by planes I (in) and 0 (out) located at z=Z and z=0, respectively. The background surface S is parallel to the slab and located at a negative z. Incident radiation is constituted of (1) a collimated beam of intensity I_c^Z with infinite lateral extension hitting the slab perpendicularly and (2) a semi-isotropic diffuse radiation of intensity I_d^Z.

radiation which is the mixing of a collimated and a diffuse flux. At a given location z in the slab, we analyze the radiative energy as a mixing of four fluxes, two collimated fluxes travelling in opposite directions and two diffuse fluxes also travelling in opposite directions. For each of these fluxes, knowing volume elements scattering properties by using a single scattering theory, we write a radiative balance equation. The set of equations is solved accounting for radiative boundary conditions. The outcome of the analysis is a set of analytical expressions for the various reflectances and transmittances involved in the problem. Details are available from Maheu et al [15]. By specifying restrictive assumptions in the general framework, we may recover previously published simpler models, such as three-flux models described by Körtum [14] and two-flux models given by Ryde, Kubelka and Munk, Sagan and Pollack and Schuster (see Maheu and Gouesbet [16]).

To achieve computations in this framework, everything is known from single scattering theory but for a parameter, the so-called average crossing parameter $\langle \varepsilon \rangle$, which is a pure representative of the radiative transfer ingredients involved in the problem. It is defined by saying that, when the diffuse light crosses a length dz in the slab, the average path length which is actually travelled over is $\langle \varepsilon \rangle$dz. For a collimated beam, this parameter is clearly equal to 1, and it is readily shown to be equal to 2 for a semi-isotropic diffuse flux [14]. However, it is not known in general. A first approach is then to carry out four-flux computations with the extreme values 1 and 2, and to deduce from the obtained differences a typical uncertainty in the results. A better solution is to provide tabulations of the average crossing parameter, as briefly discussed by Maheu et al [8] and Wang et al [17]. Such extensive tabulations will be published elsewhere in a forthcoming paper, providing an essential ingredient for systematic and accurate use of four-flux models.

Our original aim in designing four-flux models has been to provide us with a simple approach for further generalization to the case when the slab is illuminated by a Gaussian beam. We however failed in designing this generalization, may be due to a lack of intensive enough effort. The energetic reader might attempt to attack again this problem whose interest is warranted. Nevertheless, we were content enough in recognizing that the results we obtained in four-flux models for the evaluation of collimated-collimated transmittances remained valid for laser beam illumination. This remark provided us with a simple theoretical result to develop a measurement technique described in the next section.

4. VIDE-TECHNIQUE

We consider again the particulate slab sketched in Figure 4.

Assume that this slab is illuminated by two laser beams at two different well-chosen wave-lengths, and that the transmittances of the slab at these two wave-lengths are measured. From four-flux models, we find that these transmittances are simply expressed by a Beer-Lambert law exponential decrease, depending on the diameter d and on the number-density N of the particles. If the complex refractive index of the particle material is known at both wave-lengths, then it is an exercise to show that the measurements of both transmittances permit to determine the unknown elements d and N. For particle diameters in the typical range 10 µm-100 µm, a good choice for the wave-lengths is 632.8 nm (Visible He-Ne laser) and 337 µm (Infra-red, HCN laser). Hence, the technique is named VIDE: Visible Infra-red Double Extinction.

First experiments we carried out by using the VIDE-technique are reported in Gouesbet and Ledoux [18], Gougeon et al [19], Gouesbet et al [20], from which other references may be found. Our conclusion concerning the validity of the VIDE-technique was positive but spoiled by the poor quality of the multiple scattering medium under study, namely a falling cloud of particles (coal or glass) in a rotating and vibrating cylinder.

To pursue the work more accurately, it has then been necessary to produce high quality standard media for multiple scattering (and even dependent scattering) studies. Our efforts to reach this goal are discussed in Thioye et al [21] and Guidt et al [22], eventually leading to an accurate validation of the technique [23]. In this last study, optical thickness could be as big as 9 corresponding to transmittance as small as 10^{-4}. A by-product of this work was to experimentally define a limit of Beer-Lambert law for collimated beams in multiple scattering media and to settle down a very simple criterion for the appearance of proximity effects. This experimental criterion states that proximity effects arise when the optical diameter of the particles defined from single scattering theory is about half of the average distance between particle centers. Qualitatively, these proximity effects lead to a decrease of efficient number-densities perceived by the light and therefore to an increase of transmittances (which may be by orders of magnitude) with respect to the Beer-Lambert law theoretical ones. Diameter measurements however are not significantly affected.

Our conclusion in Guidt et al [23] was that VIDE-technique appears to be suitable and accurate enough for simultaneous measurements of sizes and number-densities of particles in single, multiple and even dependent scattering situations. Change of diameter ranges requires another choice of the wave-lengths. The technique can also deal with non-spherical particles like coal ones. We nevertheless point out that obtained data are averaged over a line of sight and also over the particle diameter probability density function. But preliminary theoretical works (unpublished) suggest that polydispersity information might be obtained by using a three-wave technique. Also, for cylindrical symmetry problems, we expect that local measurements might be feasible by using an Abel inversion which could lead to the development of VIDE-tomography. The principle of the technique would not be essentially affected in combustion systems in so far as wave-length windows of transparency exist. Finally, for more opaque media in which transmittances would be too weak to be measurable, information might be retrieved from reflectance measurements [16].

5. CONCLUSION

This paper provides a brief review of the work carried out in our team in the field of multiple scattering of light. Emphasis is stressed on the case when the medium under study is illuminated by a laser beam. Simple theoretical tools have been discussed, namely Monte-Carlo techniques and four-flux models. On the experimental side, we reported on the existence of an optical technique for simultaneously measuring particle diameters and number-densities in big optical thickness media. Accurate experimental works in designing and calibrating instruments require the use of standard media under multiple scattering situations. The existence of such media has also been reported. We took the opportunity of this review to settle down some ideas for further works, both theoretical and experimental. The field of laser beam multiple scattering is a domain of fundamental interest leading to applications for instance in optical particle diagnosis. We hope that this paper will encourage other researchers in entering this field of research where much work remains to be done.

REFERENCES

1. Chandrasekhar S.: Radiative transfer. Oxford U.P., London (1950).

2. Van de Hulst H.C.: Multiple light scattering, tables, formulas and applications. Academic Press, New York 1, 2(1980).

3. Gouesbet G., Gréhan G., Maheu B.: Single scattering characteristics of volume elements in coal clouds. Appl. Opt. 22 (1983) 2038-2050.

4. Gouesbet G., Maheu B., Gréhan G.: Light scattering from a sphere arbitrarily located in a Gaussian beam, using a Bromwich formulation. J. Optical Society of America A, 5 (1988) 1427-1443.

5. Maheu B., Gouesbet G., Gréhan G.: A concise presentation of the Generalized Lorenz-Mie Theory for arbitrary location of the scatterer in an arbitrary incident profile. J. Optics (Paris) 19 (1988) 59-67.

6. Gouesbet G., Gréhan G., Maheu B.: A localized interpretation to compute all the coefficients g_n^m in the generalized Lorenz-Mie Theory. J. Opt. Soc. Am. A, 7 (1990) 998-1007.

7. Crosbie A.L., Dougherty R.L.: Two-dimensional isotropic scattering in a finite thick cylindrical medium exposed to a laser beam. J. Quant. Spectrosc. Radiat. Transfer 27 (1982) 149.

8. Maheu B., Briton J.P., Gouesbet G.: Four-flux model and a Monte-Carlo code: comparisons between two simple and complementary tools for multiple scattering calculations. Appl. Opt. 28 (1989) 22-24.

9. Cashwell E.D., Everett C.J.: The Monte-Carlo methods for random walk problems. Pergamon Press (1959).

10. Battistelli E., Bruscaglioni P., Lo Porto L., Zaccanti G.: Separation and analysis of forward scattered power in laboratory measurements of light beam transmittance through a turbid medium. Appl. Opt. 25 (1986) 420-430.

11. Bruscaglioni P., Zaccanti G., Ismaelli A., Pili P.: Comparison between measured and calculated contributions of multiply scattered radiation to the transmittance of a light beam through a turbid medium. Radio Science 22 (1987) 899-905.

12. Schuster A.: Radiation through a foggy atmosphere. Astrophysics J. 21 (1905) 1.

13. Kubelka P., Munk F.: Ein Beitrag zur Optik der Farbanstriche. Zeit. Tech. Phys. 11a (1931) 593.

14. Kortüm G.: Reflectance spectroscopy. Springer, New-York (1969).

15. Maheu B., Le Toulouzan J.N., Gouesbet G.: Four-flux models to solve the scattering transfer equation in terms of Lorenz-Mie parameters. Appl. Opt. 23 (1984) 3353-3362.

16. Maheu B., Gouesbet G.: Four-flux models to solve the scattering transfer equation: special cases. Appl. Opt. 25 (1986) 1122-1128.

17. Yi Ping Wang, Zheng Sen Wu, Kuan Fang Ren: Four-flux model with adjusted average crossing parameter to solve the scattering transfer equation. Appl. Opt. 28 (1989) 24-26.

18. Gouesbet G., Ledoux M.: Supermicronic and submicronic optical sizing, including a discussion of densely laden flows. Special issue of optical engineering on particle sizing for spray analysis. Optical Engineering 23 (1984) 631-640.

19. Gougeon P., Le Toulouzan J.N., Gouesbet G., Thenard C.: Optical diagnosis in multiple scattering media using a visible/infra-red double extinction technique. J. of Physics E 20 (1987) 1235-1242, selected for inclusion in J. of Engineering Optics 1 (1988) 95-102.

20. Gouesbet G., Gougeon P., Le Toulouzan J.N., Thioye M., Guidt J.B.: VIDE-technique in densely laden flows, new progress. Part. Part. Syst. Charac. 5 (1988) 51-56.

21. Thioye M., Le Toulquzan J.N., Gouesbet G., Gougeon P., Chabot F.: Réalisation d'échantillons pour l'étude de la diffusion de la lumière. J. of Aerosol Science 19 (1988) 55-64.

22. Guidt J.B., Le Toulouzan J.N., Thioye M., Gouesbet G.: Standard media for particle size and number-density measurements and calibrations under single, multiple and dependent situations, Part. Part. Syst. Charac. 7 (1990) 36-43.

23. Guidt J.B., Gouesbet G., Le Toulouzan J.N.: An accurate validation of visible infra-red double extinction simultaneous measurements of particle sizes and number-densities by using densely laden standard media. Appl. Opt. 29 (1990) 1011-1022.

SENSITIVITY OF RADIATIVE TRANSFER TO REFRACTIVE INDEX OF FLY ASH IN HIGH TEMPERATURE ENVIRONMENT

E. Lecossais
Laboratoire de Thermodynamique, URA CNR 230/CORIA
BP 118, 76134 Mont Saint Aignan Cedex, France

and

Ziad-G. Habib
TECHNODES SA
Centre de Recherche Industrielle et Technique
Group Ciments Français, 78931 Guerville Cedex, France

ABSTRACT

Measurements and calculations of radiative energy emitted and scattered by a hot suspension of calibrated alumina and two industrial samples of fly ash are presented. Excellent agreement was observed for alumina when using the available data of refractive index. For ash, the results are very sensitive to both carbon content and size distribution. The results suggest that carbon, even in low proportion, is a major constituent of fly ash regarding radiative heat transfer in pulverized coal furnaces. On the other hand, when scattering is significant, very reliable information concerning particle concentration and size distribution is required in order to predict accurate radiative fluxes.

INTRODUCTION

Radiative properties of fly ash at moderate thermal emission wavelengths (1-5 μm) are of great importance since fly ash is a major contribute to heat transfer in pulverized coal furnaces. The corresponding published values of the refractive index n-ik show a wide variation especially for the imaginary part k [1-3]. The difference of the origin of ash samples cannot be considered as a unique explanation of this wide variation, i.e., $10^{-1} < k < 10^{-6}$, depending on wavelength and composition. Carbon free ash is a mixture of mostly oxides for which resonnance bands are beyond 5 μm. This suggests that, most probably, the presence of unburnt carbon is responsible of the measured high values of k in the 1-5 μm range. It may seem surprising when considering that carbon content of *in situ* fly ash is less than 10%, the carbon-affected values are many orders of magnitude higher than what is expected for pure oxides. In fact, a significant amount of unburnt carbon in ash may probably consist of individual particles of size going down to the submicron range. Hence, both size and density of unburnt carbon, lower than those of oxides, make that carbon, based on number density, can easily reach significant amounts.

A question that one can ask is to which extent radiative heat transfer is sensitive to the refractive index of material having an imaginary part as low as 10^{-6}? and can the influence of residual carbon, when present, be quantified? Answering to these questions is a tedious job especially that radiative transfer depends strongly on the knowledge of the other important parameters, such as particle size distribution and particle concentration.

The approach used in the present work is experimental and theoretical, both complementing each other. Ash samples of different composition and carbon content were injected in a drop tube furnace and spectral radiative energy (1-5 µm) escaping from a small hole was measured and compared to the result of a n-flux computer code modeling the experiment. Special care has been undertaken in determining the input parameters of the code, whether experimental or calculated. These are: particle concentration, particle size distribution, wall and particle temperature.

EXPERIMENTAL

The experiments have been performed by using two different ash samples collected in the ash trays of a 3 MW coal fired experimental furnace (Station d'Essais de Mazingarbe, CERCHAR, France). A third sample consisting of pure alumina particles was used in order to calibrate the code.

In order to avoid residual carbon combustion and infrared emission bands of active gas like carbon dioxide and water vapor, the ash samples where injected in hot nitrogen (1100 K) flowing at a mean velocity of 1 m/s through a hot walls laboratory scale cylindrical furnace (50 mm i.d. and 1 m long). This leads to a residence time of 1 s. The furnace operates vertically (drop tube furnace) and is electrically heated providing a fixed wall temperature gradient. Two diametrically disposed optical access holes of 1 cm diameter are located at mid-distance from top and bottom of the furnace. This allows emission measurements as well as crossing of the particle suspension by an infrared optical reference beam for transmission measurements.

Figure 1 shows the wall temperature profile, $T_w(z)$, z=0 and z=100 cm are the top and bottom of the furnace respectively. The continuous line corresponds to the gradient injected in calculation. Symbols on Figure 1 are the temperature readings given by seven Pt-Pt/10% Rh thermocouples disposed along the furnace wall and used for temperature control and regulation. Operation with a wall gradient was chosen so that a black body situation can be avoided. In

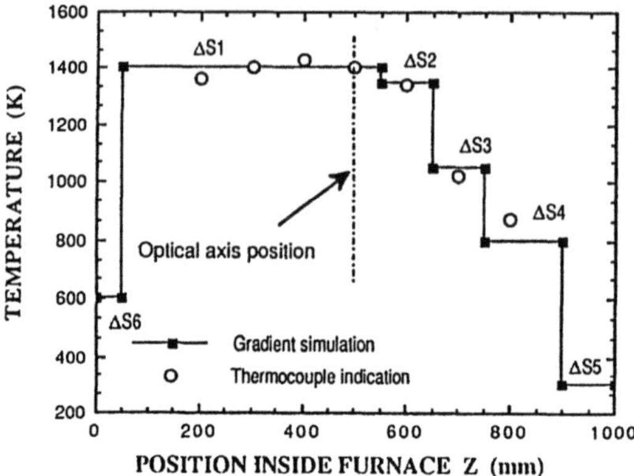

Fig. 1:   Temperature gradient inside furnace and cylindrical surfaces used in solving the surface exchange equations.

fact, with a uniform wall temperature, the design of the furnace is very similar to a black body configuration, i.e., a radiative output surface (1 cm diameter) much less than the internal heated surface (~1500 cm^2). The latter situation will decrease the sensitivity of both measurements and calculations.

A top view of the optical set up is given in Figure 2. It has the classical configuration for emission/transmission optical measurements. An infrared fiber optics links the collection lens with a classical f/6 spectrometer operating with a 150 groves/mm grating. The fiber optics exit replaces the entrance slit of the spectrometer (400 μm diameter). An InSb detector is placed at the exit of the spectrometer (no exit slit) because of the relatively low signal level. As a consequence, the spectral resolution of the spectrometer is poor: around 10% of wavelength. However, this is not of much importance since we are not interested in a spectroscopic analysis of the radiative fluxes.

During measurements, the furnace operated with a slight depression because of the absence of optical access windows. These do disturb the emission measurements by adding a uncontrolled signal due to reflection of radiation, especially on the back window. However, uncontrolled signal due to multi-reflection inside the furnace could not be entirely eliminated and was of about 5% of the measured signal.

Transmission measurements were performed in the 1-5 μm spectral region. Those lead to the determination of the extinction coefficient of the ash particles suspension K_e and of the particle concentration N_0 (*see details in the next sections*). Emission measurements were performed in the same spectral range

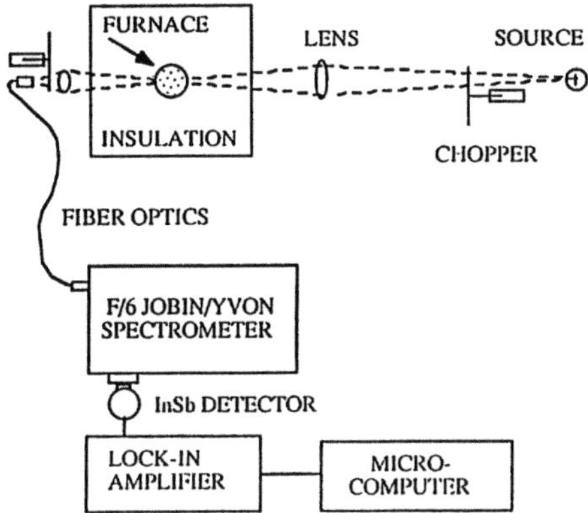

Fig. 2:   Top view of the optical setup.

then converted in terms of spectral intensity by using a calibration source (commercial black body).

THEORETICAL BACKGROUND

The theoretical part of the study consists of building an n-flux numerical code capable of predicting the normal spectral intensity impinging on a surface element of a cylindrical furnace containing a suspension of particles. The code must provide the solution of the radiative heat transfer equation in an emitting, absorbing and scattering medium. With the variable y being the direction of propagation of radiation, this equation is given by,

$$\frac{dI}{dy} = -(k_a + k_s)\, I + k_a I_{bp} + \frac{k_s}{4\pi} \int_{4\pi} I'(\Omega')\, P(\Omega' \rightarrow \Omega)\, d\Omega \qquad (1)$$

where k_a and k_s are respectively the absorption and scattering coefficients of the particles. The sum $k_a + k_s$ is known as the extinction coefficient k_e. I_{bp} is the black body intensity at particle temperature T_p. The integral term represents the increase of intensity by in-axis scattering of radiation. $I'(\Omega')$ is the intensity incident in a direction centered on the solid angle Ω' and scattered in the y direction, centered on the solid angle Ω. $P(\Omega' \rightarrow \Omega)$ is the phase function of the particles.

For polydispersions where particle suspension is characterized by the particle size distribution, psd, the scattering coefficient is given by,

$$k_s = \int_0^\infty \pi\, a^2 Q_s\,(m,a,\lambda)\, dN(a) \qquad (2)$$

190

where a is the particle radius, λ the wavelength, m=n-ik the refractive index. dN(a) is the particle size distribution and Q_s is the scattering efficiency. The absorption coefficient is derived by a similar relation. Scattering and absorption efficiencies can be derived from the Mie theory for spheres [4].

The solution of eq. (1) in an isothermal homogeneous particle suspension of width Y is given by,

$$I(Y) = I(0)\, e^{k_e\, Y} + \frac{k_a}{k_e}\left(1 - e^{-k_e\, Y}\right) I_{b,p}$$

$$+ \frac{k_s}{4\pi} \int_0^Y \left[\int_{4\pi} I'(\Omega')\, P(\Omega' \to \Omega)\, d\Omega\right] e^{k_e\,(y-Y)}\, dy \tag{3}$$

ke is the extinction coefficient I(0) is the value of intensity for y=0, i.e., background intensity.

The right hand first and second terms of eq. (2) stand for transmission and emission respectively. They can be calculated analytically providing the knowledge of the background intensity I(0), particle temperature T_p and extinction and scattering coefficients k_e and k_s. The estimation of the integral term which stands for in axis scattering is the main difficulty of the equation of transfer. This term cannot be determined analytically unless large simplifications are introduced and which sometimes are based on unsupported assumptions, e.g., isotropic scattering or balance between in and out scattering. Otherwise, the integral term can be numerically estimated when all physical variables of the problem are known. These are the same as for the first second terms of eq. (2) cited above, to which are added the phase function of the particles and the boundary conditions over the 4π steradians at any location y of the integral. Obviously, this is a drastic work to perform which can be undertaken only in controlled laboratory conditions.

THE NUMERICAL CODE

Equation of Transfer. The transmission and emission terms of Eq. (3) being analytical, the code only transforms the integral term into a discrete sum over the variable y and also over the total solid angle 4π, i.e.,

$$I_s = \sum_y \left[k_s \sum_{i=1}^N \frac{1}{4\pi}\int_{\Omega_i} I'(\Omega')\, P(\Omega' \to \Omega)\, d\Omega\right] e^{k_e\,(y-Y)}\, \Delta y \tag{4}$$

Summation over y is not of major difficulty. In opposite, angular discretization is of great importance. As it is shown in eq. (4), the $4\,\pi$ steradians are subdivided into N discrete solid angles Ω_i for which a continuous integral term is remaining. By considering very narrow solid angles, we can reasonably assume that the intensity inside Ω_i is uniform so that summation over solid angles in eq. (4) becomes,

$$I'_s = \frac{1}{4\pi} \sum_{i=1}^{N} I'(\Omega') <P_i> \Omega_i \tag{5}$$

where $<P_i>$ is the mean phase function averaged over Ω_i such as,

$$<P_i> = \frac{1}{\Omega_i} \int_{\Omega_i} P(\Omega' \to \Omega)\, d\Omega \tag{6}$$

Whatever angular discretization is chosen, it is submitted to two important conditions: solid angles have to be complementary in a way that the norm of the phase function is conserved, i.e.,

$$\frac{1}{4\pi} \sum_{i=1}^{N} \Omega_i = 1 \tag{7}$$

and

$$\frac{1}{4\pi} \sum_{i=1}^{N} <P_i> \Omega_i = 1 \tag{8}$$

Coordinate Systems. Figure 3 shows the coordinate systems corresponding to the geometrical configuration of the problem. The coordinate system OXYZ is cylindrical, centered on the axis of a cylinder representing the furnace of radius R and length L. The location of the scattering volume in this system is given by (0,y,0), OY being the direction of interest of eq. (3).

A spherical coordinate system $O'X_3X_1X_2$ is centered on the scattering volume so that $O'X_3//OZ$ and $O'X_2//OY$. The intensity $I'(\Omega)$ is incident on the scattering volume following the O'Z direction (*see* Figure 3). The scattering angle is hence defined as $\rho=(O'Z,O'X_2)$. O'Z is one axis of a spherical coordinate system O'XYZ also centered on the scattering volume. It is used in order to calculate the phase function of the particles by using the Mie theory for spheres according to Kerker formalism [4,5]. The directions of O'X and O'Y are not shown in Figure 3 since they are not of great importance, the phase function is calculated for unpolarized radiation for which only the O'Z direction is important.

Finally a fourth spherical coordinate system $O'X_1X_2X_3$ is used in order to relate the geometrical variables of each of the three other systems by using simple transformation relations.

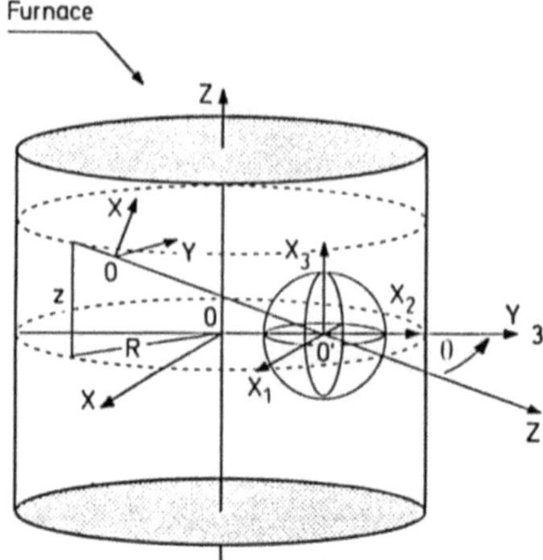

(OXYZ):   Cylindrical coordinate system
      associated to furnace
$(O'X_3X_1X_2)$: Spherical coordinate system
      associated to scattering volume
(O XYZ):   Cylindrical coordinate system
      associated Kerker formalism

Fig. 3: Geometrical configuration and coordinate systems associated with the numerical code. The sphere represents the 4π steradians of the integral term in eq. (3).

Angular Discretization. The scattering angle being defined as $\theta=(O'Z,O'X2)$, the angular discretization is made in $O'X_3X_1X_2$ by using the spherical coordinate and complementary angle θ_2, i.e.,

$$\theta_2 = (O'X_2,ZO') = \pi - \theta \tag{9}$$

Basically, in order to cover the entire 4π steradians, θ_2 has to vary in the $[0,\pi]$ interval and the azimuthal angle $\phi_2 = (O'X_3,O'X_1)$ in the $[0,2\pi]$ interval. The axisymmetrical feature of the furnace regarding $O'Z$ and that of the phase function regarding $O'X_2$ enables to reduce the ϕ_2 interval to $[0,\pi]$. Hence, angular discretization is limited to 2π steradians defined in $O'X_3X_1X_2$ as $0 \leq \phi_2 \leq \pi$ and $0 \leq \theta_2 \leq \pi$. Numerical results of eq. (4) are then multiplied by a factor 2.

The meridians and azimuths, i.e., the interval $[0,\pi]$ both for θ_2 and ϕ_2, are divided into 2n equal arcs, n being the order of the angular discretization. The 2π steradians are hence subdivided into a total number of $N=4n^2$ solid angles. Incrementation from 1 to $4n^2$ is made following θ_2 so that the direction of the first solid angle Ω_1 is given by $\theta_{2,1} = \phi_{2,1} = \pi/4n$. All other solid angles are derived from Ω_1. The boundaries of each solid angle are precisely known and are given by,

$$\overset{m}{\phi}_{2,i} = \phi_{2,i} - \frac{\pi}{4n} \qquad\qquad \overset{m}{\phi}_{2,i} = \theta_{2,i} - \frac{\pi}{4n}$$

$$\text{and} \tag{10}$$

$$\overset{M}{\phi}_{2,i} = \phi_{2,i} + \frac{\pi}{4n} \qquad\qquad \overset{M}{\theta}_{2,i} = \theta_{2,i} + \frac{\pi}{4n}$$

superscripts m,M stand for minimum and maximum respectively.

Figure 4 is an example of angular discretization for n=2, i.e., a total number of solid angles of N=16 (per 2π steradians). The sphere of Figure 4 represents the 4π total solid angle centered on the scattering volume localized at O'. A value of n=16, leading to a total of <u>2048 solid angles for the 4π steradians</u> was fixed for this study. Consequently, the assumption of uniform intensity within a single solid angle in deriving eq. (4) is quite reasonable.

Mean Phase function. The basic phase function used in calculation corresponds to unpolarized radiation. In that case $P(\Omega' \rightarrow \Omega)$ depends only on the scattering angle θ. On the other hand, $P(\Omega' \rightarrow \Omega)$ is also function of particle size, i.e., of radius a. The mean phase function, averaged over particle size and solid angles is determined by using the following equation,

$$<P_i> = \frac{1}{\left(\cos\overset{m}{\theta}_{2,i} - \cos\overset{M}{\theta}_{2,i}\right)} \int_{\overset{m}{\theta}_{2,i}}^{\overset{M}{\theta}_{2,i}} \left[\frac{1}{N_0} \int_0^\infty P(\theta,a)\, dN(a) \right] \sin\theta\, d\theta \tag{11}$$

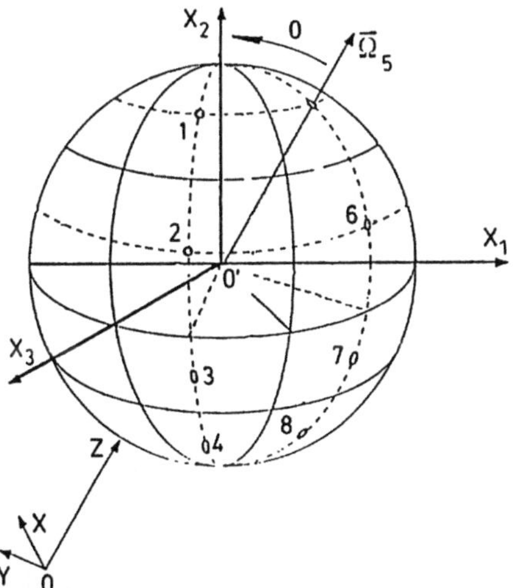

Fig. 4: Example of the angular discretization used in the code. The direction OX_2 is parallel to OY in Fig. 3.

where N_0 (m^{-3}) is the total number density of particles. We may easily verify that the norm of the phase function is conserved, i.e.,

$$2n \times \frac{1}{4\pi} \sum_{i=1}^{2n} <P_i> \Omega_i = 1 \tag{12}$$

Wall Radiosity. In presence of a wall temperature gradient $T_w(z)$ inside the furnace, wall radiosity $J_0(z,\Omega')$ is also function of position. Knowing the real temperature profile and providing the wall hemispherical spectral emissivity, wall radiosity profile can be determined by solving the surface to surface radiation exchange equations, based here on six uniform temperature surfaces ΔS_i (*see* Figure 1). The furnace wall, a mixture of alumina and silica, was assumed to be diffuse-gray having an hemispherical emissivity varying with temperature following $\varepsilon(T)=1.471 - 2.67 \ 10^{-3} \ T + 2.2 \ 10^{-6} \ T^2 - 6.764 \ 10^{-10} \ T^3$, where $300 \leq T \leq 1400$ K, $\varepsilon(300) = 0.85$ and $\varepsilon(1400) = 0.2$ based on literature [6-7].

By inverting Planck's function, the calculated radiosities lead to the determination of a so-called spectral radiosity temperature profile for each of the six surfaces, these are given in Figure 5. Radiosity temperatures were then injected in calculation. These results were validated where possible (for the three hottest surfaces) by using an optical pyrometer operating at 0.6 µm (dots on Figure 5). The agreement is excellent, consequently these profiles were assumed valid in the infrared.

PRELIMINARY RESULTS

Phase Function. Figure 6 shows a comparison between exact and averaged phase functions. The curve in dashed line is the exact function given by means of the Mie theory for spheres and corresponding to a particle diameter of 37 µm. The curve in continuous line is the size averaged phase function over a ZOLD distribution (Zero Order Log-Normal) of 37 µm outer diameter. One can note that the sauter diameter d_{32} is representative of the particle size distribution regarding the phase function. The curve in broken line is the phase function averaged over both size distribution and solid angles according to eq. (12) and corresponding to the angular discretization given in Figure 6. A very good agreement is noticed between the three functions.

Simple/Multiple Scattering boundaries. The term $I'(\Omega')$ in eq. (3) is the main unknown of the problem. It represents the radiative energy incident on the particle volume sample and to be scattered in axis. It is formally identical to

I(Y) for the same equation since it is also solution of the equation of transfer, i.e., $J(0,\Omega')$ is the wall radiosity per unit solid angle, I'_{bp} is black body intensity at particles temperature, those located on the trajectory $[0,\psi]$ separating the wall surface element from the particle volume sample. Hence $I''(\Omega'')$ will be formally identical to $I'(\Omega')$. When assuming simple scattering, i.e., a photon is scattered only once before escaping from the particle suspension, the integral term in the above equation can be neglected. However, particle concentration boundary between simple and multiple scattering is always critical to define [5]. In the present study, no hypothetical assumption is made but calculations are performed by using the numerical code in order to determine the upper limit of concentration beyond which simple scattering assumption is not valid. The basic result of such a code is "black body situation". When particles, gas and furnace wall are at the same temperature, the intensity at any location in the furnace is that of black body at this same given temperature. This is also the case of the intensity escaping from a small hole performed in the furnace wall.

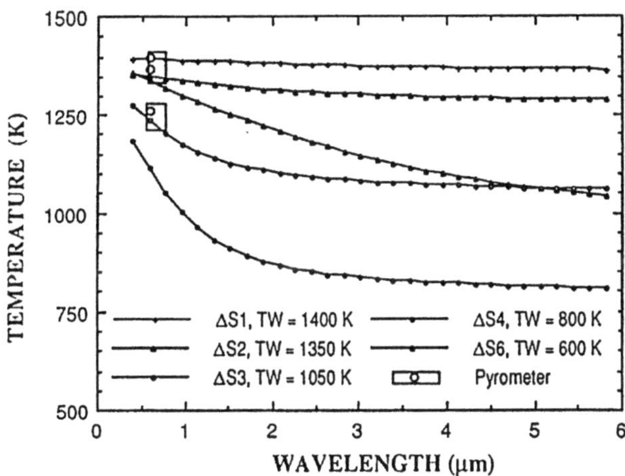

Fig. 5: Spectral radiosity temperatures of ΔS_i injected in the code.

$$I'(\psi,\Omega') = J(0,\Omega')\, e^{-k_e\psi} + \frac{k_a}{k_e}(1 - e^{-k_e\psi})\, I_{bp}$$

$$+ \frac{k_s}{4\pi} \int_0^{\psi} \left[\int_{4\pi} I''(\Omega'')\, P(\Omega'' \to \Omega')\, d\Omega' \right] e^{k_e(\xi-\psi)}\, d\xi \tag{14}$$

Figure 7 shows the variation versus particle concentration of the calculated normalized intensity (to black body) escaping from the furnace when a single temperature of 1400K is considered. Here also a ZOLD of 37 μm sauter diameter and 1.3 standard deviation have been used and representing relatively well the

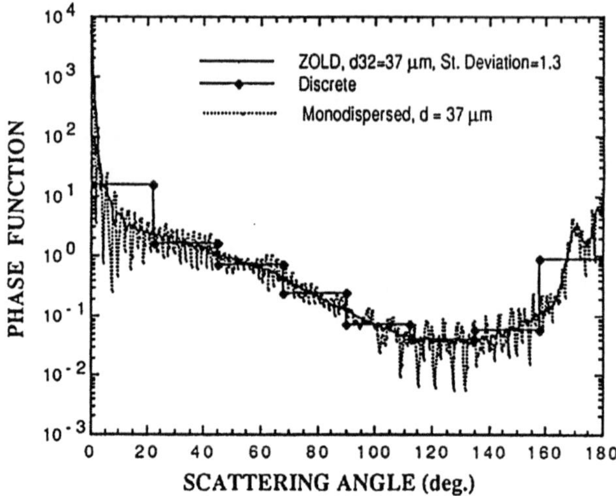

Fig. 6: Example of discrete averaged phase function used in calculations.

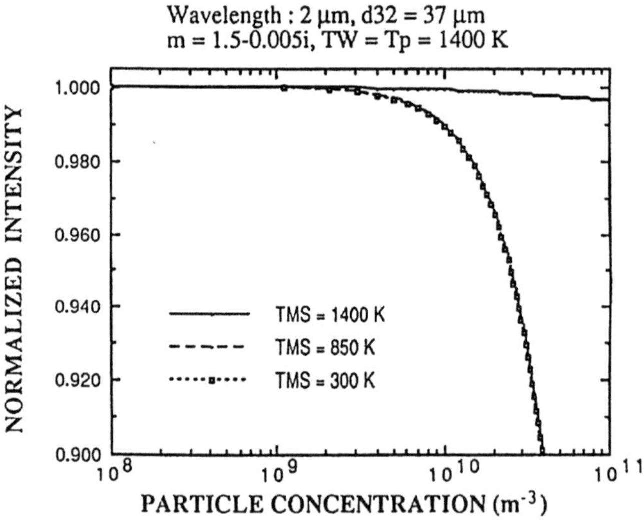

Fig. 7: Validation of the code using black body configuration.

coarsest ash sample used in the experiments. A complex refractive index of 1.5-.005i was used. The integral term in eq. (14) was replaced by a black body intensity corresponding to different temperatures TMS (Multiple Scattering Temperature). Notice that, as long as particle concentration does not exceed 10^{10} m^{-3}, and for TMS varying between 300 and 1400 K, neglecting the integral term (i.e., assuming simple scattering) leads to a numerical error less than 2-3%. Experiments have been designed in order not to exceed the 10^{10} order of magnitude in particle concentration, allowing for the integral term in eq. (14) to be neglected.

RESULTS AND DISCUSSION

Ash Samples Preparation. Ash samples were first passed through 160 μm sieve in order to clean them from large aggregates. Then they were passed through a 20 μm sieve in order to eliminate as maximum as possible of submicron particles. The presence of submicron particles in large amounts leaded to unstable injection and formation of slags inside the furnace. Characteristics of ash samples are given in Table 1.

Table 1:   Composition of ash and calibration sample (% mass)

	Rietspruit	Freyming	Calibration
SiO_2	49.7	41.4	-
Al_2O_3	21.5	23.5	100
Fe_2O_3	10.83	11.92	-
MgO	2.53	4.5	-
CaO	5.84	3.38	-
K_2O	1.88	1.63	-
SO_3	2.58	·2.03	-
TiO_2	.94	.45	-
C	3.1	7	-

Size distribution of both ash samples and alumina was determined prior to injection by using a commercially available Malvern 2600 D instrument. The results which exhibit a large percentage of fines are given in Figure 8. The sauter diameter d_{32} was of 15 and 37 μm for Rietspruit and Freyming ash respectively, and of 18 μm for alumina.

Extinction Coefficient and Particle Concentration. The feed rate of particles inside the furnace was varied between 200 g/h for alumina and 120g/h for the coarsest ash (Freyming). These feed rates were calculated in order not to exceed 10^{10} particles per cm^3 following the result of Figure 7.

As preliminary calculations, the sensitivity of extinction coefficient k_e to refractive index m=n-ik was first tested. This is shown on Figure 9 where the real part n was maintained constant (n=1.5) and the imaginary part k was

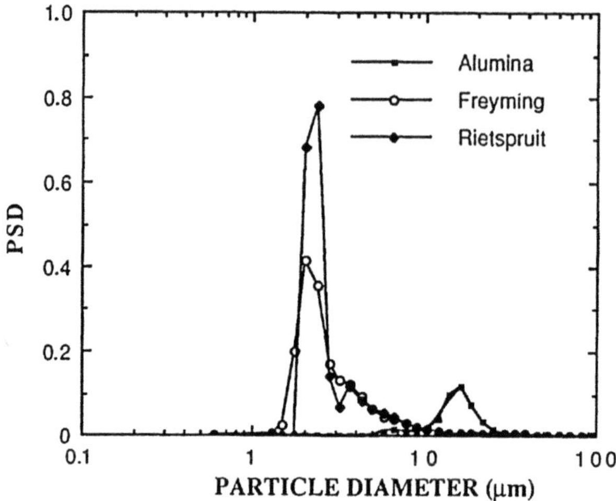

Fig. 8:   Particle size distribution, psd, of the three samples used in experiments.

$n = 1.5$, WAVELENGTH = 3 μm, No = 1E+10 part/m3

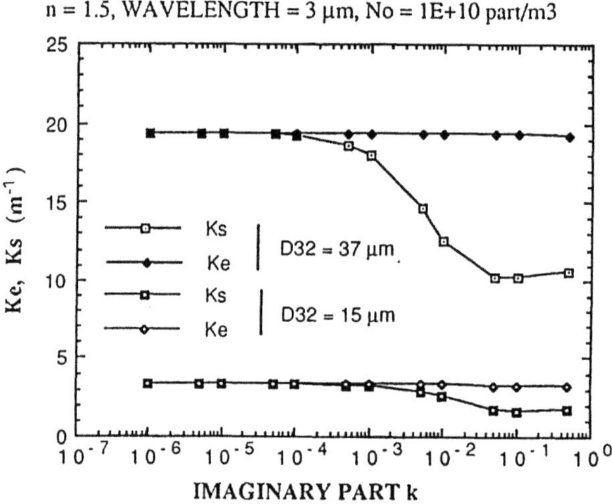

Fig. 9:   Variation of extinction and scattering coefficients versus imaginary part of refractive index for two ZOLD psd of a same standard deviation of 1.3.

varied between 10^{-7} and 1. Obviously, k_e is independent of k. The same calculations versus n lead to a similar result. This is due to the fact that the size parameters we are dealing with are within the Mie region. It is an important result since it suggests that, knowing the size distribution of particles, dN(a), the particle concentration No can be determined from transmission measurements according to eq. (2). Following this procedure, particle concentration of the three samples was measured and injected in calculations.

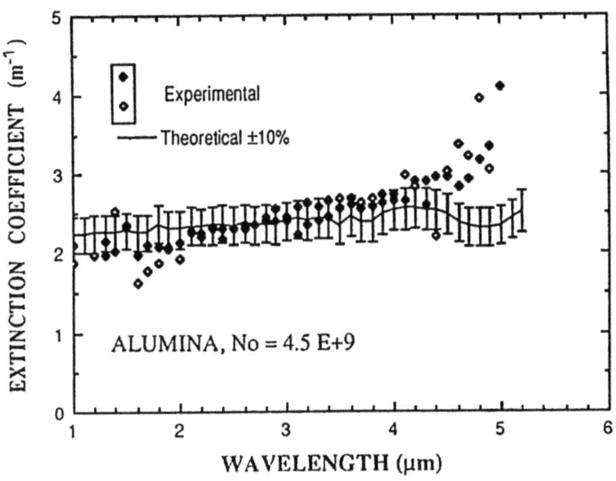

Fig. 10: Comparison between experimental and theoretical extinction coefficient for alumina.

Figure 10 shows the variation of the experimental extinction coefficient of alumina versus wavelength. The theoretical fitting of the spectra was achieved by using eq. (2) associated with the refractive index available in the literature [8-9]. Particle concentration was varied until a good agreement was reached within 10% (reproducibility of measurements). The similar procedure was applied to both Rietspruit and Freyming ashes which exhibit a same quasi-constant variation versus wavelength. The results are grouped in Table 2.

Table 2: Measured extinction Coefficients and Concentrations

	$d_{32}(\mu m)$	$k_e(m^{-1})$	$N_0(m^{-3})$	
Alumina	18	2.3	4.45	10^9
Rietspruit	15	1.7	1.7	10^{10}
Freyming	37	0.8	1.89	10^9

Particle Temperature. Basically, direct determination of particle temperature in the present experimental conditions is unfeasible. Thermal emission of particles is mixed with a component due to scattering of wall radiation by the particle suspension according to eq. (3). A reasonable assumption is that particles, for which maximum size is of 60 µm, are in thermal equilibrium with the surrounding nitrogen flow. Radial and axial profile of gas temperature along the furnace was measured by using a 50 µm Pt-Pt 10%Rh thermocouple. These show that a mean gas temperature of 1100 K reasonably represents the thermal condition of both gas and particles inside the furnace.

Calibration of the Code Input Parameters. As it was mentioned earlier, the input parameters of the code DIRAC are: particle concentration, N_0, particle size distribution, $dN(a)$, particle temperature T_p and wall radiosity $J_0(z,\Omega')$. It is the latter parameter which is affected by the highest uncertainty due to the several simplifications made in determining the coefficients of the surface to surface exchange equations. Consequently, only this parameter was varied during the calibration procedure performed by using the alumina intensity spectra.

Figure 11 shows the calibration curve of the code. A typical experimental intensity spectrum of alumina (averaged over several spectra) was taken as a reference. Given the spectral refractive index of alumina, the spectral radiosity profiles were adjusted so that the best agreement is achieved with the theoretical spectra. A satisfactory result is that Figure 11 was obtained with only a slight adjustment of the radiosity temperature T_R(50 K down shift). The excellent agreement between experimental and calculated spectra for alumina is a demonstration that the input parameters of the code are of confidence. The discrepancy observed beyond 4 μm is due to the presence of the atmospheric CO_2 absorption band where the signal to noise ratio was low.

Intensity Spectra of Ash Samples. Wall radiosity and particle temperature have been validated with alumina, these were maintained as input parameters for the other two ash samples. Providing the particle size distribution by the Malvern instrument and particle concentration from extinction measurement

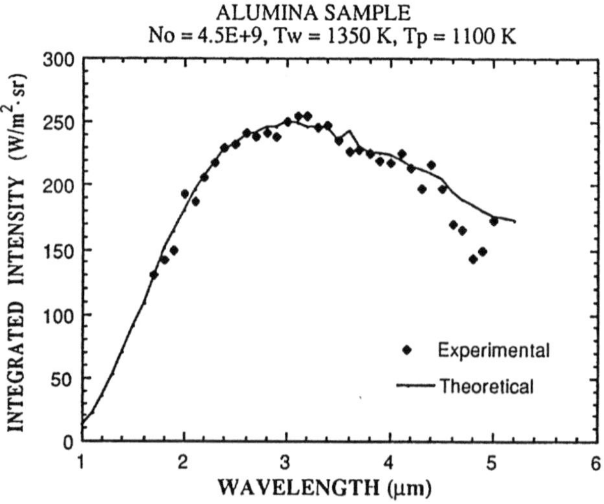

Fig. 11: Alumina calibration curve of the numerical code.

(Table 2), the number of unknowns in order to predict the intensity spectra is reduced to one, i.e., the complex refractive index m=n-ik. As sensitivity calculations have shown, scattering and mainly emission is not very sensitive to the real part of the refractive index. Consequently, a constant value of n=1.5 was injected in calculations. Therefore, the work is reduced to the search of a best fit value of the imaginary part k.

The method used here consists of finding an upper and lower limits of variation of k versus wavelength. The use of a spectral point by point iterative procedure is not justified due first to the poor spectral resolution of the optical apparatus and second to the uncertainty of measurements compared to variation of the theoretical calculation with variation of k. A 10-15% uncertainty covers an interval of variation of one order of magnitude for k. This is shown on Figure 12 for Rietspruit and on Figure 13 for Freyming. The experimental points are plotted together with an upper and lower limit of k=10^{-3} and 10^{-1} respectively.

As one can notice, the curves of Figures 12 and 13 are far from being concluding as it was the case of alumina. The only consistent observation that can be formulated is that the imaginary part k has 10^{-3} as a lower limit. This value, which is much higher than what is expected for ash, is true for both samples. On the other hand, the value of 10^{-1} represents more an average than an upper limit. The upper limit could easily go up to k=1.

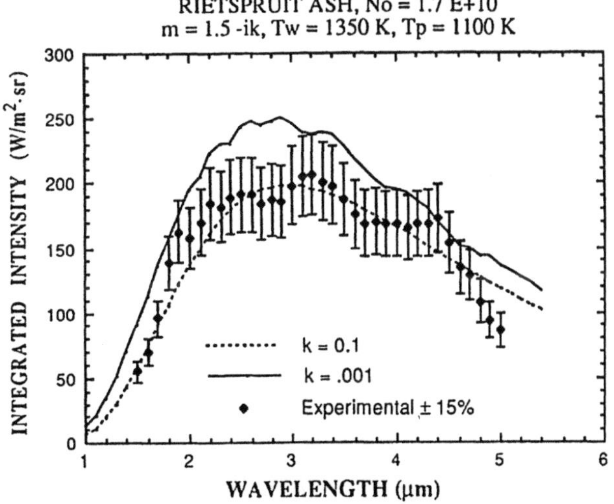

Fig. 12: Experimental intensity spectrum for Rietspruit ash compared to calculated spectra by using two different values of refractive index.

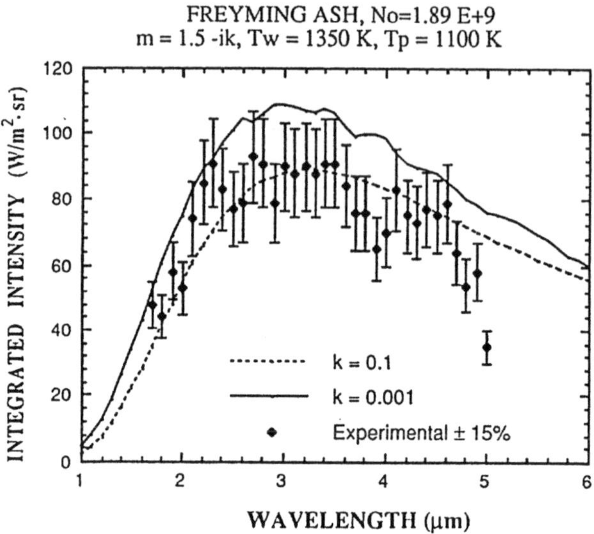

Fig. 13: Experimental intensity spectrum for Freyming ash compared to calculated spectra by using two different values of refractive index.

Even though we believe that, based on alumina calibration, the above results are basically self-consistent, it is probable that they are affected by a relatively high error. In fact, among the input data of the numerical code, it is particle size distribution which is not in common between alumina and ash. Alumina sample consists of commercial calibrated particles such as 90% are between 10 and 20 μm (*see* Figure 8). The Malvern instrument is indeed sensitive in this size range and the psd provided by this instrument is of confidence. We believe that, the excellent agreement between theoretical and measured spectra for alumina is mainly due to the good knowledge of the particle size distribution.

In opposite, both ash samples present a high percentage (~90%) of particles located around 2 μm, which is the lower limit of sensitivity of the Malvern instrument. In fact, the psd's histograms seem to be truncated at 2 μm. Submicron particles are most probably present in the ash samples without being detected. In that case, particle concentrations determined from extinction measurements are erroneously low for both ashes. This is the most probable explanation of the relatively high values of k determined from Figures 13 and 14.

On the other hand, sensitivity calculations show that, the underestimation which may affect k does not exceed one order of magnitude, unless a significant down shifting of both size distributions is performed, which does not seem realistic.

CONCLUSION

The most reliable conclusion corresponds to alumina for which radiative transfer is exclusively dominated by scattering due to its imaginary part of refractive index, k, which is less or equal to 10^{-4} in the 1-5 μm range. The corresponding values available in the literature are quite satisfactory in predicting the experimental intensity spectra. This suggests that values of refractive index of pure constituents of ash may be taken as basic data in order to predict the optical behaviour of a given ash, as in Goodwin's approach [8] (dispersion equations).

For the two industrial samples of fly ash, and following the analysis of the preceding section, values of k for which calculation and experience match is sensitive to carbon content. An average value such as $10^{-2} < k < 10\text{-}1$ may be representative of the radiative behaviour of both ash samples. However, the reliability of this result is questionable due to the uncertainty affecting the knowledge of particle size distribution, mainly because of the presence of small and submicron particles. Unburnt carbon, which is potentially localized in the submicron range, is responsible of increasing the average value of the imaginary part of refractive index of fly ash. Consequently, reliable prediction of radiative properties of *in situ* ash requires the knowledge of not only the overall size distribution but also the one corresponding to unburnt carbon. It is only then that one can estimate with confidence the optical properties of an *in situ* ash suspension.

REFERENCES

1. Gupta R.P., Wall T.F.: The Optical Properties of Fly Ash in Coal Fired Furnaces. Combustion and Flame 61 (1985) 145-151.

2. Cowen S.J., Ensor D.S., Sparks L.E.: The Relationship of Fly Ash Light Absorption to Smoke Plume Opacity. Atmospheric Environment 15 10/11 (1981) 2091-2096.

3. Pollack J.B., Toon O.B., Khare B.N.: Optical Properties of some Terrestrial Rocks and Glasses. Icarus 19 (1973) 372-389.

4. Van de Hulst H.C.: Scattering of Light by Small Particles. Dover Publications Ltd, NY (1981).

5. Kerker M.: The scattering of Light and Other Electromagnetic Radiations. Academic Press, NY (1969).

6. Incropera F.P., De Witt D.P., Fundamentals of Heat and Mass Transfer. Second edition, John Wiley & Sons, NY (1985).

7. Siegel R., Howell J.R.: Thermal Radiation Heat Transfer. Second edition, Hemisphere Publishing Corporation NY (1981).

8. Goodwin D.G.: Infrared Optical Constants of Coal Slags. Ph.D. Thesis, Stanford University, HTGL Rept. T-255 (1986).

9. Godwin D.G., Mitchener M.: Fly ash Radiative Properties and Effects on Radiative Heat Transfer in Coal-fired Systems. Int. J. Heat, Mass Transfer 32(4) (1989) 627-638.

MODELLING OPTICAL PROPERTIES OF AIR CO_2-H_2O- SOOT MIXTURES FOR ENGINEERING CALCULATIONS

S. Pasini, N. Pintus
ENEL - Centro Richerche Termiche e Nucleari - Pisa

and

L. Castellano, S. Lattanzi
MATEC - Milano

ABSTRACT

Computer codes based on Narrow Band Models were developed in order to calculate total and spectral emissivities of gas-soot mixtures.

Comparisons with numerous published results are presented.

The agreement is good in the low temperature range (<1500 K) when parameters from AFGL are utilized; for higher temperatures the calculated values have a different behaviour, especially for long path-lengths.

1. INTRODUCTION

Recent developments in the numerical simulation of radiation heat transfer in furnaces, e.g. [1], [2], show that consideration of spectral variations in the optical properties may substantially improve theoretical predictions.

For engineering calculations on energy systems involving combustion of hydrocarbon fuels, the subdivision of the Thermal Radiation Spectrum into wavenumber intervals whose width lies in the range

$$50 \text{ cm}^{-1} \leq \delta w \leq 1000 \text{ cm}^{-1} \tag{1}$$

should ensure a good degree of accuracy.

This paper deals with the POSMLT and POSMHT computer codes developed at ENEL-CRTN (Pisa, Italy) to compute transmissivities and emissivities of Air-H_2O-CO_2-Soot mixtures for any discretization of the Thermal Radiation Spectrum.

The physical and mathematical backgrounds of these codes are based on the statistical Narrow-Band Models described in [1].

A detailed review of the main features of this approach was made in [3].

In the present paper, emphasis is placed on the comparison of the total emissivities evaluated by means of the above-mentioned computer codes and the results obtained from Wide-Band Models, the Weighted Sum of Gray Gas Models and Hottel's Chart.

2. MAIN FEATURES OF THE POSMLT, POSMHT COMPUTER CODES

The POSMLT computer code works in the following range of low temperatures

$$300 \text{ K} \leq \text{T} \leq 1500 \text{ K} \tag{2}$$

and makes use of the band-model parameters $\bar{k}$, δ^{-1} for H_2O and CO_2 published by Hartmann and co-workers [4], and by Soufiani and co-workers [1] in the recent years.

The present version of POSMHT works in the following range of temperatures

$$1500 \text{ K} \leq \text{T} \leq 1800 \text{ K} \tag{3}$$

and makes use of the band-model parameters given in [5].

The correlations recommended in [1], [4], [6] have been adopted for the mean half-width of spectral lines $\bar{\gamma}_L$.

To make up for the lack of spectral data for the following zone

$$0 \leq w \leq 150 \text{ cm}^{-1} \tag{4}$$

of water vapor, the contribution of the isolated pure rotational band to the total emissivity, was evaluated from the empirical correlation

$$\varepsilon_{RB} = (1 + \gamma(w_E)) \, \overline{\varepsilon_{RB}} \tag{5}$$

where $\overline{\varepsilon_{RB}}$ indicates the contribution of the sub-region

$$150 \text{ cm}^{-1} \leq w \leq 975 \text{ cm}^{-1} \tag{6}$$

and w_E is the parameter (Equivalent Band Center) introduced by Charalampoulos and Felske [7].

The function $\gamma(w_E)$ is defined as:

$$\gamma(w_E) = \begin{cases} \dfrac{w_E - 150}{975} & \text{for } 300 \text{ K} \leq T \leq 1200 \text{ K} \\[2mm] \dfrac{w_E - 150}{w_e} & \text{for } 1200 \text{ K} \leq T \leq 1500 \text{ K} \\[2mm] \dfrac{w_E - 150}{(w_E + 975)} & \text{for } T > 1500 \end{cases} \qquad (7)$$

The absorption coefficient of the soot was taken to be inversely proportional to the wavelength

$$k_S = c_0 f_v / \lambda \qquad (8)$$

where f_v is the volume fraction and c_0 is a constant which was set equal to 5.0, [8].

3. APPLICATIONS AND RESULTS

Several sets of numerical experiments were performed. The results presented here deal with the total emissivities of single pure substances and homogeneous Air-H_2O-CO_2-Soot mixtures.

The aim of this study was to check the agreement of the optical properties evaluated by means of our codes with the total properties commonly used for the less expensive engineering calculations.

The limited numbers of results reported below are nevertheless sufficiently representative of the type of research being carried out and of the main indications which have transpired so far.

Figures 1-4 summarize the total emissivity of water vapour and CO_2 in relation to temperature. The choice of the values of the optical paths (PL) considered is based on the requirements of a comparison with carefully documented data in the current technical literature.

As for water vapour (Figures 1-2), we find that the results of the model used in our tests are virtually always systematically situated along the line of average values of the results obtained by Leckner and Schmidt [9]. We also find there is an abrupt break at the point of transition from the interval of the low temperatures to that of the high temperatures (>1500 K); this discontinuity tends to decrease as the optical path increases. At the present state of the art, the exact cause of such behaviour is not clear; indeed, strangely enough,

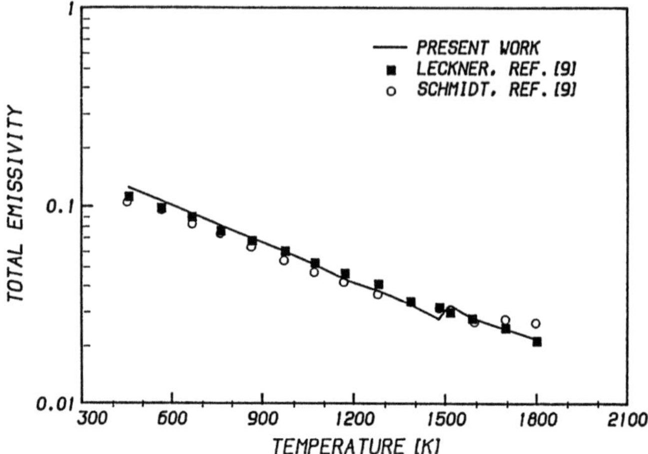

Fig. 1: H_2O Total emissivity, PL=1.5 (bar x cm), P=1 (bar), X_{H_2O}=1.

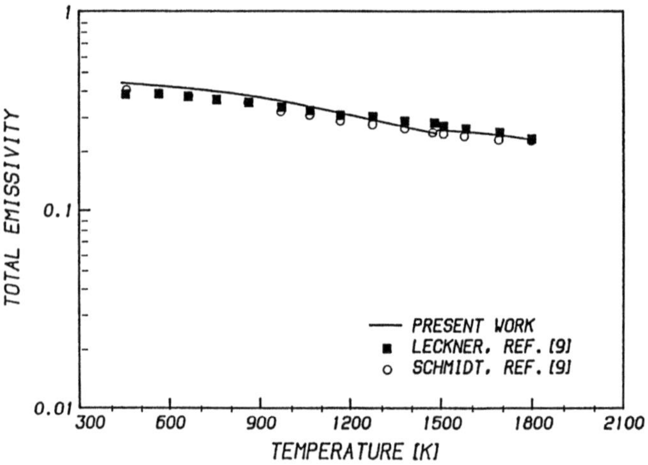

Fig. 2: H_2O Total emissivity, PL=40 (bar x cm), P=1 (bar), X_{H_2O}=1.

Figures 3 and 4 show that, in the case of CO_2, the same discontinuity tends to increase as the optical path increases.

When we examine Figures 5-8, concerning mixtures of Air H_2O and CO_2 in the typical composition that is produced by burning oil, we see that the agreement between the total emissivities observed in the course of the present research and those obtained with the "Gray-Gas Model" and with the "Band Absorptance Formulation" in [10] can be considered satisfactory for the range

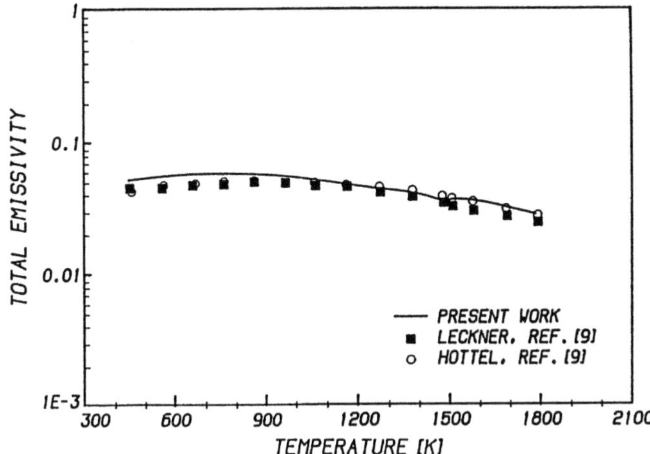

Fig. 3: CO$_2$ Total emissivity, PL=1 (bar x cm), P=1 (bar), X$_{CO_2}$=1.

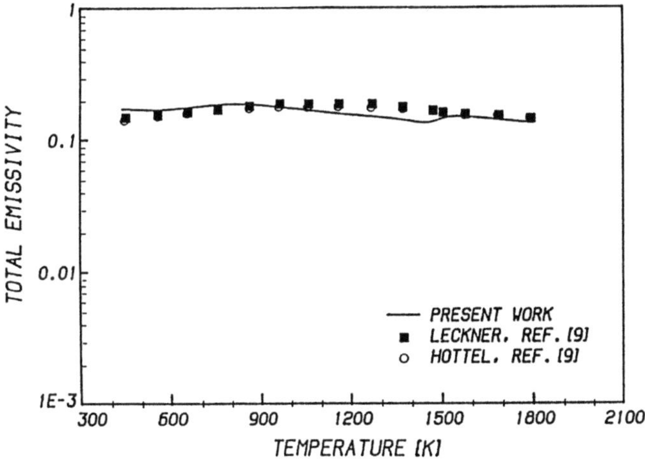

Fig. 4: CO$_2$ Total emissivity, PL=100 (bar x cm), P=1 (bar), X$_{CO_2}$=1.

of temperatures considered, but only up to the point where the value of the optical path is PL=0.5 m.atm.

Whereas, for greater optical paths, we observe that the values supplied by the Narrow-Band model for temperatures higher than 1500 K begin to be systematically under-evaluated.

This behaviour seems to be similar to that of CO$_2$ in Figures 3-4.

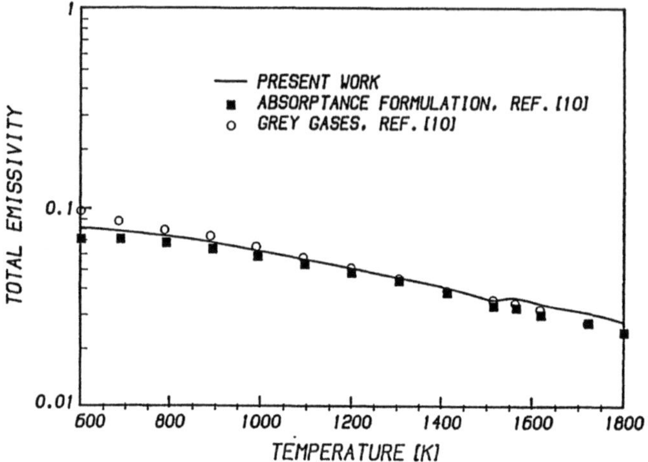

Fig. 5: Total emissivity of a mixture AIR-H_2O-CO_2, PL=0.05 (m x atm), X_{H_2O}=X_{CO_2}=0.1.

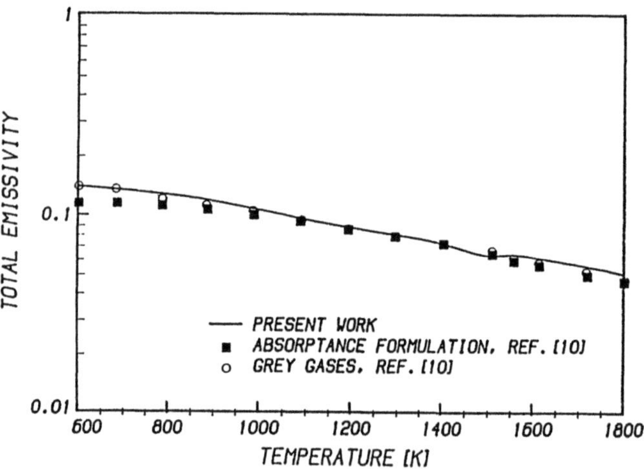

Fig. 6: Total emissivity of a mixture AIR-H_2O-CO_2, PL=0.15 (m x atm), X_{H_2O}=X_{CO_2}=0.1.

Lastly, Figures 9-10 show that the disagreement is greatly attenuated in the presence of a certain amount of soot.

4. CONCLUSION

The increasing need to improve instruments for simulating and testing combustion processes has stimulated research is shown by the development of

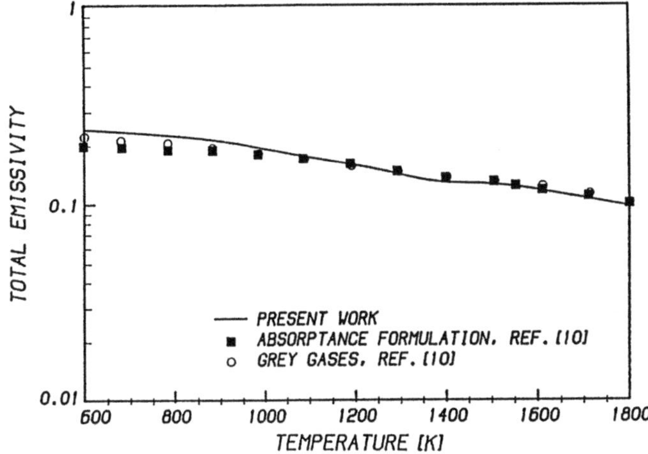

Fig. 7: Total emissivity of a mixture AIR-H_2O-CO_2, PL=0.5 (m x atm), X_{H_2O}=X_{CO_2}=0.1.

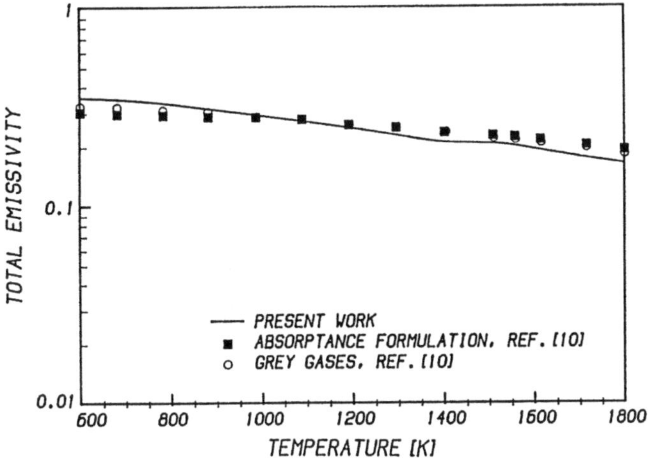

Fig. 8: Total emissivity of a mixture AIR-H_2O-CO_2, PL=1.5 (m x atm), X_{H_2O}=X_{CO_2}=0.1.

numerical models for the calculation of radiative thermal exchange, as well as for the assessment of the properties of mixtures of optically active gases.

Several conclusions can be drawn from the above discussions and results:

a) Relatively simple yet accurate calculations are possible using statistical Narrow-Band Models.

b) Numerous well-tested and reliable empirical correlations are available for the evaluation of parameters $\bar{\gamma}_L$.

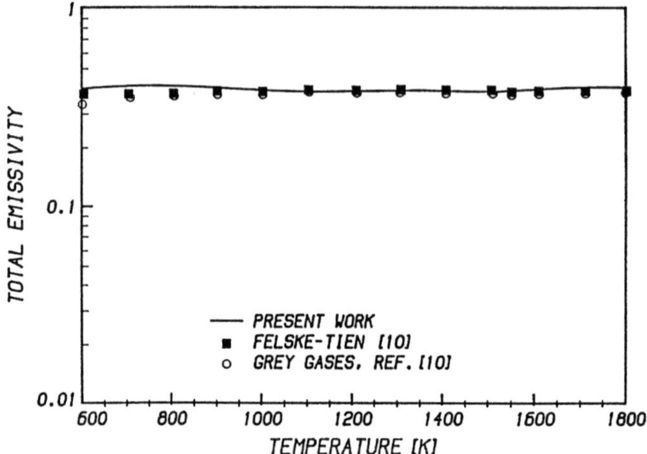

Fig. 9: Total emissivity of a mixture AIR-H_2O-CO_2-soot, PL=1.5 (m x atm), $X_{H_2O}=X_{CO_2}=0.1$, $f_v=1$ x 10^{-7}.

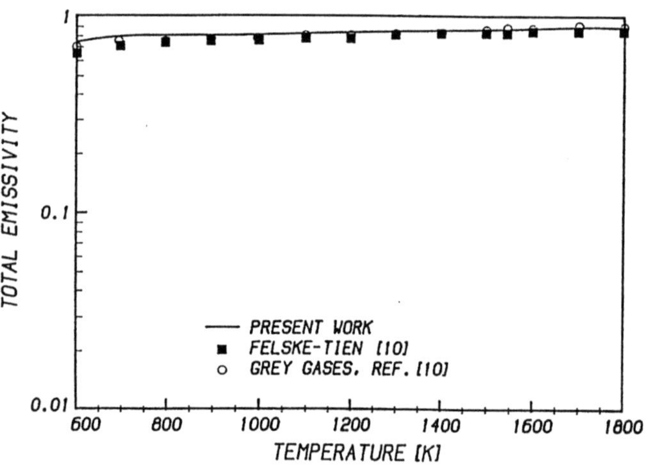

Fig. 10: Total emissivity of a mixture AIR-H_2O-CO_2-soot, PL=1.5 (m x atm), $X_{H_2O}=X_{CO_2}=0.1$, $f_v= 1$ x 10^{-6}.

c) Parameter $\bar{k}$, δ, relevant to the main bands of Carbon Dioxide and Water Vapor, are available for a sufficient number of temperatures between 300 K and 1800 K. These parameters are available for groups of sub-bands of variable extension between 5 cm^{-1} and 50 cm^{-1}.

d) Such intervals are comparable to the resolution power of the instruments at present used for sophisticated measurements in boilers. On the other hand, the discretization is still too fine to be utilized as it is in radiative heat transfer codes. However, Narrow-Band Models can be used to calculate optical models of bigger groups of lines; for instance, physical bands with relevant overlapping.

e) As far as the total properties are concerned, the numerical experiments performed show that:

e1) the model which transpires from the Exponential-Tailed-Inverse (line-intensity) Distribution provides the most accurate results;

e2) the agreement with the results of previous studies is satisfactory in the range of low temperatures;

e3) a transitional zone is clearly evident around the 1500 K temperature mark;

e4) in the range of high-temperatures, the present Narrow-Band approach over-estimates the existing data for pressure-path-length of less than 0.5 (m.atm); beyond this value an increasing under-estimation has been observed;

e5) more exhaustive studies are needed in order to understand if the validity of the models falls within this range or if the available spectral data is insufficiently accurate;

e6) the disagreement is less evident in the presence of soot.

REFERENCES

1. Soufiani A., Hartmann J.M., Taine, J.: Validity of Band-Model Calculations for CO_2 and H_2O Applied to Radiative Properties and Conductive-Radiative Transfer. JQSRT, 33 (1985) 243-258.

2. Fiveland W.A., Jamaluddin A.S.: Three-Dimensional Spectral Radiative Heat Transfer Solutions by the Discrete-Ordinate Method. 1989 National Heat Transfer Conference, HTD in 106, Heat Transfer Phenomena in Radiation, Combustion Fires (1989).

3. Castellano L., Pasini S., Pintus N.: A review of Narrow-Band Models for Calculating Spectral Properties of Homogeneous and Non-Homogeneous Mixtures. Joint Meeting of the German and Italian Section of the Combustion Institute, Ravello (1989)

4. Hartmann J.M., Levi Di Leon R., Taine J.: Line-by-line and Narrow-Band Statistical Model Calculations for H_2O. JQSRT 32 (1984) 119-127.

5. Ludwig C.B., Malkmus W., Reardon J.E., Thompson A.L.: Handbook of Infrared Radiation from Combustion Gases edited by R. Goulard and A.L. Thompson NASA SP-3080 (1973).

6. Farag I.H.: Radiative Heat Transmission from Non-Luminous Gases. Computational Study of the Emissivities of Water Vapor and Carbon Dioxide. in Sci. D. Thesis, Chem. Engr. Dept., MIT (1974).

7. Charalampopoulos T.T., Felske J.D.: Total Band Absorptance, Emissivity, and Absorptivity of the Pure Rotational Band of Water Vapor. JQSRT 30 (1983) 89-96.

8. Hottel H.C.: Radiation from Carbon Dioxide, Water Vapor and Soot. (American Flame Committee), Fuels Res. Lab., Chem. Engr. Dept., MIT (1985).

9. Leckner B.: Spectral and Total Emissivity of Water Vapor and Carbon Dioxide. Comb. and Flame 19 (1972) 33-48.

10. Felske J.D., Charalampopoulos T.T.: Gray-Gas Weighting Coefficients for Arbitrary Gas-Soot Mixtures. Int. J. Heat Mass Transfer 25 (1982) 1849-1855.

THE RADIATIVE PROPERTIES OF SOOT AGGLOMERATES

P.A. Bonczyk and R.J. Hall
United Technologies Research Center
East Hartford, Connecticut USA

ABSTRACT

Soot particle sizing and density measurement by laser light scattering has been reformulated to take into account non-spherical shape effects. By assuming a morphology consisting of chains of monomer spheroids, and the concepts of fractal theory, it has been possible to set up extinction and multiangle scattering measurements which make it possible to determine the structural parameters of such clusters. This permits more realistic calculations of soot radiative properties, and puts soot sizing/density measurements on a better basis. Experiments performed in an ethylene-air slot burner yield credible cluster parameters using this approach.

INTRODUCTION

When soot particulates are present in combustion, they usually make greater contributions to radiative transfer than do molecular by-products. In sooting laboratory flames, 20-30% conversion of fuel enthalpy release to radiation has been reported [1] and soot radiation is important in practical contexts such as furnaces and gas turbines. In most analytical studies of the effect, the particulates are assumed to have spherical shapes, with calculated extinction, absorption, and scattering parameters given by the well-known Mie theory [2]. Further, most in-situ density and size determinations by laser light scattering, which provide input to such studies, also make this assumption. While there are certainly situations where the assumption of sphericity is valid [3,4], it is more typical to observe in ex-situ electron micrographs evidence of agglomeration of spherical monomers [5]. When, as is often the case, the inferred effective radius from light scattering (500-1000 Angstroms, say) greatly exceeds the monomer radius observed in micrographs (50 to 200 Angstroms, typically), one can reasonably assume that the flame particulates exist as aggregates. The theoretical tools now exist to make it possible to discard the spherical Mie theory in such situations. The concept of a mass fractal is essential to the development of the proper theory.

THEORY OF FRACTAL CLUSTER SCATTERING

Soot particulates seem to belong to a class of objects called fractals. These are objects which vary irregularly in a geometric sense from one to another, but have certain statistical similarities; they are said to be "self-similar" and scale invariant" over certain length scales. The concept of a mass fractal dimension makes it possible to concisely characterize, in a statistical sense, properties of the mass distribution such as the radius of gyration. Examples of mass fractal objects are gold colloids [6], soot [7,8], metallic smokes [9], and "fumed" silica or silica soot [10]. Growth models such as DLA (Diffusion Limited Aggregation) [11] or DLCA (Diffusion Limited Cluster Aggregation) [12] predict such structures, and information about the growth process can be inferred from the fractal dimension.

For fractal clusters with N primary spheroids, the radius of gyration R_g can be shown to be related to the primary spheroid radius a by [13, 14]

$$\sqrt{2}\, R_g = \alpha a N^{(1/D)} \tag{1}$$

where $\alpha \approx 1$ and D is the fractal dimension, typically non-integer. For most particulates, values of D around 2 are typical. This can be seen to be associated with a tenuous or sparse structure through the infra-cluster density relationship [13]

$$\rho \sim \frac{N}{R_g^3} \sim \frac{1}{a^3 N^{(3/D-1)}} \tag{2}$$

which is intended to apply on length scales intermediate between a and R_g. The lower the value of D, the more tenuous the structure is seen to be. An important formal restatement of eq. (2) is that the two-point density-density correlation function (the monomer pair distribution function)

$$\rho_2(\vec{r}) = \int \rho(\vec{r}')\, \rho(\vec{r}' + \vec{r})\, d^3\vec{r}' \tag{3}$$

has a power-law form for values of r in the stated range. Specifically,

$$\rho_2(r) = r^{-(3-D)} h(r/R_g) \tag{4}$$

for clusters embedded in three-dimensional space. In eq. (4), h is a cutoff function important for values of r approaching R_g. This is usually of exponential or Gaussian form [13-15].

Fractal cluster parameters can be obtained by visual, structural examination of electron micrograph images. However, with such ex-situ measurements of samples, there is always the question whether the in-situ properties are truly represented. For silica soot, to give an example, the fractal dimension obtained from in-situ light scattering is less than that obtained by examination of samples [16]. Also, a model is needed to compute the radiative coefficients when these values are in hand. Light scattering measurements provide a way of determining the structural parameters without the need to be concerned over sampling perturbation of the cluster structures, and, if a scattering model is available, a self-consistent treatment is possible.

For singly scattered waves, the scattering cross-section and density autocorrelation are Fourier Transform pairs [17]

$$\frac{d\sigma}{d\Omega}(\vec{q}) \sim \int \rho_2(\vec{r}) \, e^{i\vec{q}\cdot\vec{r}} d^3\vec{r} \tag{5}$$

where q is the change in the incident wave vector due to scattering,

$$q = 2K_i \sin(\theta/2) = \frac{4\pi}{\lambda} \sin(\theta/2) \tag{6}$$

where K_i is the incident wave vector, λ is the incident wavelength, and θ is the scattering angle. The reciprocal of q is a measure of the resolution or length scale probed by the scattering.

One important consequence of eq. (5) which will be utilized in the analysis and experiments to follow is that the forward scattering intensity is proportional to the squared mass of the particulate; thus, if the particle consists of N monomer spheroids, there will be an N^2 coherent enhancement of the forward cross-section relative to its monomer value. The asymptotic limits expected for the cross-section are [18]

$$\frac{d\sigma}{d\Omega} \sim 1 - q^2 R_g^2/3 + ... \qquad (R_g < 1/q) \tag{7a}$$

$$\sim q^{-D}(1 - q^2 a^2/3 + ...) \qquad (R_g > 1/q > a) \tag{7b}$$

$$\sim q^{-(6-D_s)} \qquad (a > 1/q) \tag{7c}$$

where for large q there is a surface fractal dimension D_s related to surface roughness. We will present data from another source which permits a determination of D_s.

It is possible to correct for multiple scattering (intraparticle) using the self-consistent field approximation [19]. Such multiple scattering would be expected to become important for very large and dense clusters (D>2), and for strongly scattering monomers. Inter-particle multiple scattering for large optical paths is also something that in principle would need to be corrected for. Neither multiple scattering effect has been thought to be important in the work to be described, however, for reasons which will be discussed.

If the primary monomers are in the Rayleigh range (the cluster itself need not be so) and consequently scattering weakly, then Rayleigh-Debye theory [13,15] gives an additional important result; namely, that the absorption coefficient α_a for the cluster is just the sum of the monomer absorption coefficients. This assumes that the monomers are non-conducting dielectrics. Thus, if there are n monodisperse N-clusters per unit volume

$$\alpha_a = nN \frac{8\rho^2}{\lambda} a^3 \operatorname{Im}\left(\frac{m^2 - 1}{m^2 + 2}\right) \tag{8}$$

where m is the complex index of refraction.

The absorption can be equated with extinction if the measurement is made at a wavelength where scattering is small. In general, however, it can be determined from the relationship

$$\alpha_a = \alpha_e - \int_{4\pi} \alpha_s(\vec{\Omega}_s) d\vec{\Omega}_s \tag{9}$$

where α_s is the scattering coefficient and $\vec{\Omega}_s$ is a unit vector in the direction of scattering.

Thus, multiangle scattering supplemented by an absorption coefficient measurement provides a way of determining all of the cluster structural and density parameters. The foregoing theory suggests the following approach, essentially that given in [15].

#1) Determine Rg from small q scattering

#2) Deduce the fractal dimension D from the asymptotic slope of the log scattering - log q plot if accessible experimentally, or least squares fit a model $\rho_2(r)$.

#3) Use an optical absorption measurement to deduce nNa^3.

#4) Make an absolute intensity or cross-section measurement in the near forward direction, using, for example, calibration by room air Rayleigh scattering. This will yield a value for nN^2a^6 from the relationship

$$n\frac{d\sigma}{d\Omega}(N) = nN^2 \frac{d\sigma}{d\Omega}(1) = K_i^4 a^6 \left|\frac{m^2 - 1}{m^2 + 2}\right|^2 N^2 n \tag{10}$$

Note that the cluster number density results from division of the square of the number derived in #3) by the number derived in #4), and that the result does not depend on the fractal dimension or on any assumptions about the aggregate shape. It would be valid for isolated monomers, assuming that they are in the Rayleigh range with respect to the incident light. The fact that the monomer radius a appears raised to relatively high powers means that this quantity may be subject to smaller measurement error than the other quantities; experimental errors or uncertainties in the complex index of refraction which goes into the overall constants in #3) and #4) will have a mitigated impact on the error in a because of this.

This theoretical-experimental approach assumes monodispersity in N and a. It is theoretically possible to take into account polydispersity, and this will be discussed in the section concerning analysis of the experimental data.

EXPERIMENT

Apparatus. The apparatus used to carry out the measurements is shown schematically in Fig. 1. The coherent INNOVA 90-4 argon ion laser has 1.7 Watts of nominal cw power at 5145 Å. It serves primarily as a source of intense light for soot particulate and molecular Rayleigh scattering. The Rayleigh scattering provides an accurately known intensity to which soot signals are referenced for the purpose of determining their required absolute intensities.

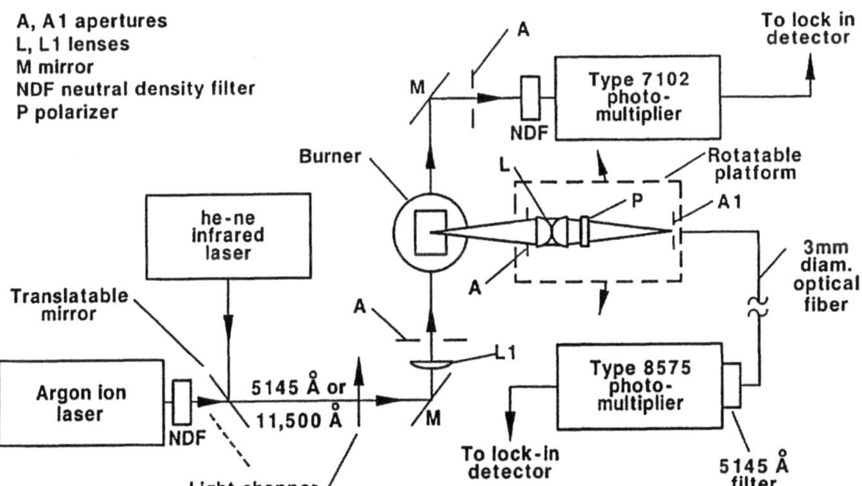

Fig. 1: Schematic diagram of the apparatus.

Since the Rayleigh signals are weak, the full power of the laser is required to observe them. On the other hand, the much stronger soot signals are observed with the laser attenuated to about 0.5 milliwatts of power by the NDF in Fig. 1. The ion laser is linearly polarized in a direction perpendicular to the plane of scattering, and the resulting light intensities are analysed in this same direction using the polarizer, P.

The Jodon HN-10G infrared He-Ne laser in Fig. 1 is used exclusively to measure soot extinction; it has a nominal cw output of 3 milliwatts at 11,500 Å. In the approach taken here, we require that the extinction be dominated by absorption, or equivalently that the near forward scattering contribution to the extinction be negligible. This is the case if $2\pi a/\lambda \ll 1$, where $2a$ is the diameter of the primary soot spherule. By using an infrared as opposed to a visible wavelength, as we have done, the preceding inequality is best approximated. It will be seen that this assumption of negligibly small relative scattering is borne out by the experimental data.

Scattering or extinction measurements are done alternately by properly inserting a translatable reflecting mirror in the path of one laser beam or the other. For scattering, the 5145 Å radiation is focused into the flame with a plano-convex lens, L_1, having a 40 cm focal length. The scattered light is collected by optics mounted to a rotatable platform, and then is sent via an optical fiber to a 10 Å wide narrow-band 5145 Å filter and photomultiplier. The resulting electrical signal, modulated at the 400 Hz frequency of the light chopper, is synchronously converted to a dc output by the lock-in detector. On the platform, the two 15 cm focal length lenses, L, are symmetrically located with respect to the burner center and, removed 30 cm from it, the 0.50 mm diameter aperture, A1. These optics and the lens, L_1, define, at 90°, a cylindrical sample volume having a 0.2 mm diameter and 0.5 mm length. Since the platform rotates in steps of 4° from 0 to 180°, the sample length is in fact a $1/\sin\theta$ function of the scattering angle, θ, which must be accounted for properly in the data reduction. For extinction measurements, the 5145 Å beam is blocked and the 11,500 Å beam is passed through the flame and detected by an infrared sensitive type 7102 photomultiplier; synchronous signal detection by a second lock-in detector is used in this case as well.

The rectangular burner in Fig. 1 has been described previously [20]. The diffusion flame is fueled by ethylene and air and is nearly two-dimensional. The gas flows to the burner and an end view of its sooting overventilated flame are given in [20].

Approach. In order to test a fractal description of the soot in our flame, measurement of the parameters n_s $(d\sigma_s/d\Omega)_s$ and α_e from scattering and extinction, respectively, are required.

These are defined and determined as follows. For soot, the scattered intensity, I_s, is given by $I_s=I_s^0\, n_s\, (d\sigma/d\Omega)_s\, \Omega l\varepsilon$, where I_s^0 is the incident 5145 Å laser intensity, n_s the soot number density, $(d\sigma/d/\Omega)_s$ the differential scattering cross-section, Ω the light collection solid angle, l the angular dependent optical sample length, and ε the light collection efficiency. In principle, the evaluation of n_s $(d\sigma/d\Omega)_s$ from experiment is straightforward once I_s is measured and I_s^0 and $\Omega l\varepsilon$ are assigned their respective numerical values. In practice, the precise evaluation of $\Omega l\varepsilon$ is not easy, especially since l is a function of scattering angle. Fortunately this difficulty may be overcome by making auxiliary Rayleigh scattering measurements. In this case, the scattered intensity, I_R, is given by $I_R = I_R^0\, n_R\, (d\sigma/d\Omega)_R\, \Omega l\varepsilon$, where I_R^0 is the incident laser intensity, n_R the number density of Rayleigh scatterers (room air at 20°C), $(d\sigma/d\Omega)_R$ the Rayleigh differential scattering cross-section, and $\Omega l\varepsilon$ the same as above. Using the equations above for I_s and I_R, it follows immediately that n_s $(d\sigma/d\Omega)_s = (I_s/I_R)$ $(I_R^0/I_s^0)\, n_R\, (d\sigma/d\Omega)_R$. Since n_R and $(d\sigma/d\Omega)_R$ are known [21], the left side of the preceding equation can be evaluated without reference to $\Omega l\varepsilon$. Further, since scattered intensities occur in the ratio, (I_s/I_R), it is not even necessary to calibrate the type 8575 photomultiplier in Fig. 1.

The parameter α_e above is the optical extinction coefficient. It is determined from $I=I_0\, \exp(-\alpha_e L)$, where I is the transmitted light intensity, I_0 the incident 11,500 Å laser intensity, and L the extinction path length. Since the flame in this work is only approximately two-dimensional, there is a variation of L with height. For the four measurement heights of 17, 20, 25 and 28 mm (.68, .8, 1., and 1.12 inches, respectively), the L's were 34.7, 32.3, 28.3 and 25.9 mm, respectively. The L's were determined by observing the appearance and disappearance of 90° scattered light as the burner was translated in a direction orthogonal to the viewing axis.

In addition to the procedures described above, it was necessary to correct for the attenuation of I_s as scattered light passed between the center of the flame and its boundary. Again, this correction varied with scattering angle. Details of how this correction was applied are omitted here except to point out that extinction measurements at 5145 Å were made for it, and that the corrections

222

ranged from 1/2 to 11% depending on viewing angle and measurement height. These values should be small enough to justify neglect of multiple scattering effects in the theory used to reduce the data. Fig. 2 shows the soot scattering at 5145 Å at the height of 1.12 inches above the burner surface.

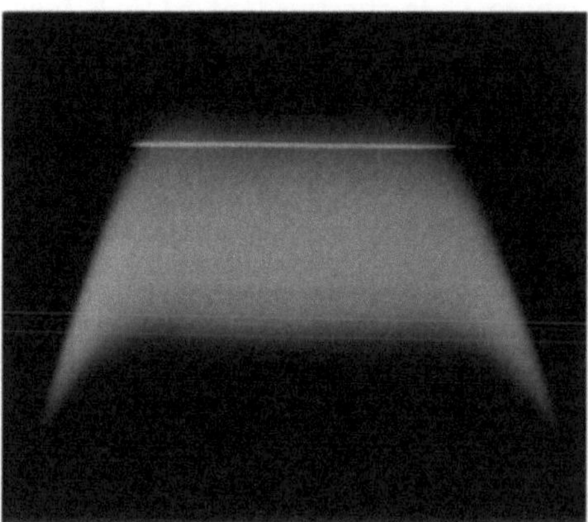

Fig. 2:   Scattering of 514.5 nm laser light by soot particulates at 1.12 in.

INTERPRETATION OF EXPERIMENTAL DATA

Summaries of the data taken in the ethylene-air slot burner are shown in Fig. 3 and Table 1. Referring to Fig. 3, which displays the log-log plots of $n(d\sigma/d\Omega)$ vs. q, it is clear that accurate extrapolation to exact forward scattering can be performed, and that, with the exception of the .68 in. data, which appears to reflect essentially Rayleigh scattering, the curvature in the near forward direction will lead to well-defined R_g. However, it is unclear whether the available optical source wavelength and particulate size have provided sufficient access to the asymptotic regime of eq. (7b), which would have permitted unambiguous determination of D. The high $R_g q$ range is limited, as seen. To obtain D, therefore, use will be made of least squares fitting model $\rho_2(r)$, as discussed, or to fitting a function like the Fisher-Burford form [22]. Table 1 presents, in addition to the extrapolated forward scattering data, the measured extinction coefficients at .5145 and 1.15 microns. Taking the Fig. 3 scattering data, and performing the angular integral over all scattering directions as in eq. (9), gives the integrated scattering information shown, from which it is possible to calculate the particle albedoes given. It is clear that scattering is relatively small even at .5145 micron, and is most certainly entirely negligible at 1.15 micron. The absorption coefficients at the two

Table 1
Data Summary

	0.68 in.	0.8 in.	1 in.	1.12 in.
$\displaystyle\lim_{q \to 0} n \frac{d\sigma}{d\Omega} \left(cm^{-1} sr^{-1}\right)$	2.05×10^{-4}	6.3×10^{-4}	1.05×10^{-3}	4.24×10^{-4}
$(0.5145\mu)\ \alpha_e\ (cm^{-1})$	0.118	0.144	0.139	0.068
$(1.15\mu)\ \alpha_e\ (cm^{-1})$	0.0507	0.059	0.056	0.0268
$\displaystyle (0.5145\ \mu)\ n \int_{4\pi} d\vec{\Omega}\ \frac{d\sigma}{d\Omega}\ (\vec{\Omega})\ \left(cm^{-1}\right)$	1.48×10^{-3}	3.76×10^{-3}	5.0×10^{-3}	1.77×10^{-3}
$(0.5145\ \mu)\ \omega\ (albedo)$	1.25×10^{-2}	2.6×10^{-2}	3.6×10^{-2}	2.6×10^{-2}

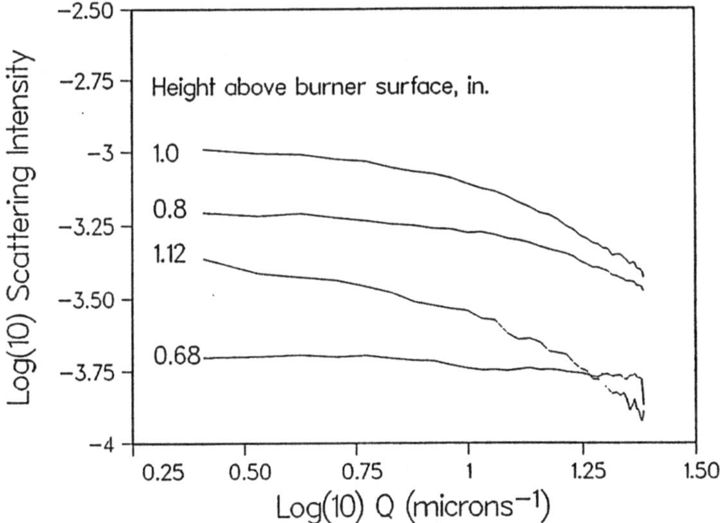

Fig. 3: Scattered intensity vs q for four different vertical positions in the flame.

wavelengths have roughly the expected inverse wavelength dependence with some index of refraction dispersion perhaps responsible for departures from this. In any event, our calculated results do not depend strongly on which of the two absorption coefficients we use, provided that we take into account the differing wavelengths.

Extraction of R_g and D from the data involved several different approaches. The first of these was a polynomial fit in q^2 to the multiangle scattering data. This gives R_g from the q^2 coefficient of the expansion. The second was least squares fitting the Fisher-Burford form

$$I(q)/I(0) = (1 + 2\ q^2 R_g^2/3D)^{-D/2} \tag{11}$$

which yielded R_g and D. The third was least squares fitting a model density function using eq. (4) and (5) and the relationship

$$\frac{d\sigma}{d\Omega}(\vec{q}) \sim \int \rho_2(\vec{r})\, e^{i\vec{q}\cdot\vec{r}}\, d^3\vec{r} \sim \int r^2 \rho_2(r)\, \frac{\sin(qr)}{(qr)}\, dr \tag{12}$$

with the r-integration performed numerically. Three cutoff functions were employed; these were the exponential form of Berry and Percival [13], the modified Gaussian form of Mountain and Mulholland [15], and the expression given by Hurd and Flower [16]. The .68 in. data were not analysed in this part of the data reduction because the clusters at that height seem to be so small that a fractal interpretation may be inappropriate, and attempting to fit these formulas to a Rayleigh object would be subject to error.

With regard to the R_g determination, all of the approaches gave reasonably similar results. The Fisher-Burford fits and those resulting from use of the Berry-Percival cutoff function also were quite close, with fitted fractal dimensions of 1.4-1.5. The Mountain-Mulholland and Hurd-Flower cutoffs gave inferred fractal dimensions of one or less. Attempts to fit a straight line to the high-q regimes of the Fig. 3 curves suggest values of D closer to 1.4-1.5, but the numbers cannot be specified with precision because of the narrow q range and noise in the scattering at the backward scattering angles. We do not know whether the peculiar inferred fractal dimensions from the Refs. [15, 16] cutoff functions are the result of inadequacies in the data or in the cutoffs themselves, but the dimensions are not consistent with our data. Figs. 4-6 show the results of fitting the Berry-Percival density function to our data, together with the best fit R_g and D. Because the latter values were supported by the Fisher-Burford fits, which were also excellent, and by the q^2 polynomial fitting, they were used in the subsequent data reduction. The average inferred D value is around 1.5; ex-situ structural examination of soot particles yields values of 1.5 to 1.7 [7, 8]. Structural examination of carbon black samples yields values of 1.46 and higher [23, 24]. For further data reduction, we pick a nominal value of D=1.49, equal to the Hurd-Flower silica soot value.

Before proceeding to the calculation of a, N, and n, it is worthwhile to discuss other data bearing on the question of soot fractal dimension. We know of no other multiangle laser scattering data on soot, but small angle x-ray scattering data (SAXS) have been reported [25]. If a log-log plot is made of the butadiene slot burner SAXS data reported there, one finds that the data cover mainly the Porod regime (eq. (7c)) with a slope of -4.06. This implies a surface fractal dimension of about 2, meaning that the monomer surfaces are relatively

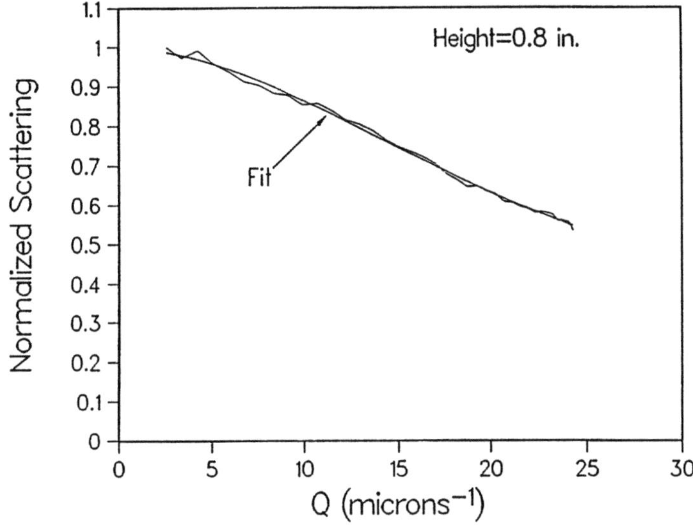

Fig. 4: Berry-Percival fit to scattered intensity vs q at 0.8 in. Height: R_g=0.069 μ; D=1.5532.

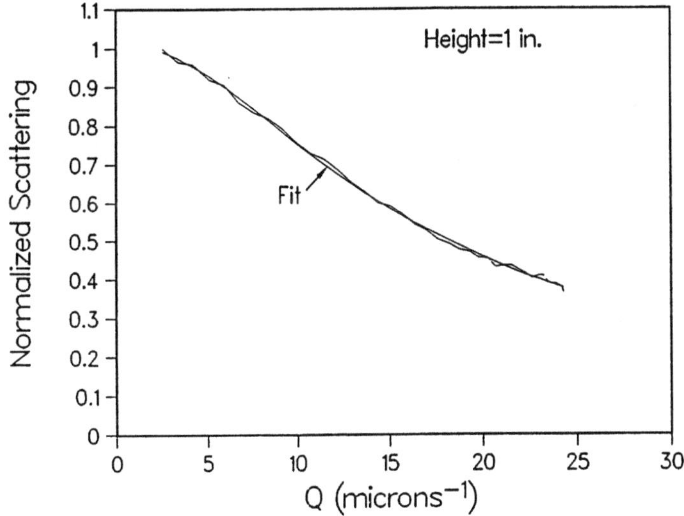

Fig. 5: Berry-Percival fit to scattered intensity vs q at 1.0 in. Height: R_g=0.1063 μ; D=1.5240.

smooth. This will be of relevance to soot oxidation rate analysis, for example. The smallest q values seem to cover the transition from the mass fractal to the Porod regime. The slope between the two smallest scattering angles is about -1.5; the count levels are very high, so noise is probably not a factor, but with so few points it is hard to draw any conclusions.

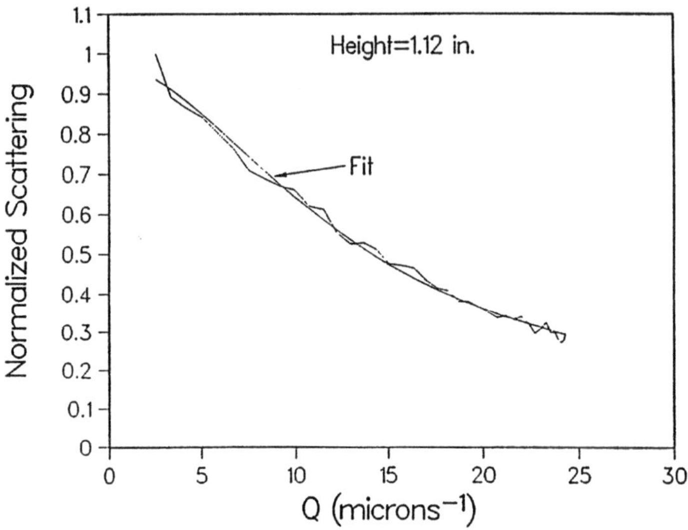

Fig. 6: Berry-Percival fit to scattered intensity vs q at 1.12 in. Height: R_g=0.1323 μ; D=1.4474.

With R_g and D determined, the parameters a, N, and n follow in a straightforward way if the index of refraction is known. We have chosen the Senftleben and Benedict [26] value of m=1.94 +.66 i. As will be discussed, there is considerable sensitivity of the predicted results to index, but a recent summary of the field [27] lends support to our choice. Table 2 summarizes the predictions as a function of height above the burner surface. The evolution of the cluster number density with height suggests agglomerating clusters, and this is borne out by the variation of the predicted N. Although we cannot verify the values of N so obtained, they certainly seem reasonable based on electron micrographs from other sooting flames [5]. The predicted values of the monomer diameter start at about 160 Angstroms at .8 in. and reach a value of about 110 Angstroms at the greatest height investigated. These are also plausible numbers, and, as will be shown, are consistent with samples taken near the top of the flame. The monotonic decrease of spherule radius seems to

Table 2
(D=1.49)

Height	N	2a (nm)	n (cm^{-3})	R_g (microns)	f_v
0.68 in.	-	-	4.19×10^{10}	-	1.405×10^{-6}
0.8 in.	43	15.7	1.887×10^{10}	0.0694	1.644×10^{-6}
1.0 in.	87	15	1.016×10^{10}	0.106	1.567×10^{-6}
1.12 in.	194	10.9	5.66×10^{9}	0.1323	7.43×10^{-7}

suggest oxidative erosion of the primary spheroids rather than breakup of the clusters, at least over the range investigated. This qualitative variation does not seem to be sensitive to uncertain parameters like index of refraction or fractal dimension.

To test the reasonableness of the inferred monomer sizes, we collected soot on a metal disc in the region of the luminous flame tip. This was scraped off, and examined under an electron microscope, as shown in Fig. 7. The scraping presumably resulted in compaction of the aggregates, preventing any estimate of N. However, the observed diameter of the primary spheroids, 40 to 100 Angstroms, is quite consistent with the value of 109 Angstroms measured at the largest height probed by laser. The physical collection was performed even higher in the flame, so somewhat smaller spheroids are expected. This encouraging agreement is probably the most significant result thus far; to our knowledge, there has been no other determination of soot cluster primary spheroid size by laser light scattering. The relative smallness of D and a, together with predicted N values that are not particularly large, means that intraparticle multiple scattering is indeed probably small [13, 19].

There is considerable sensitivity of certain of the inferred parameters to the fractal dimension, as exhibited in Table 3 for an assumed D=1.78. Comparing

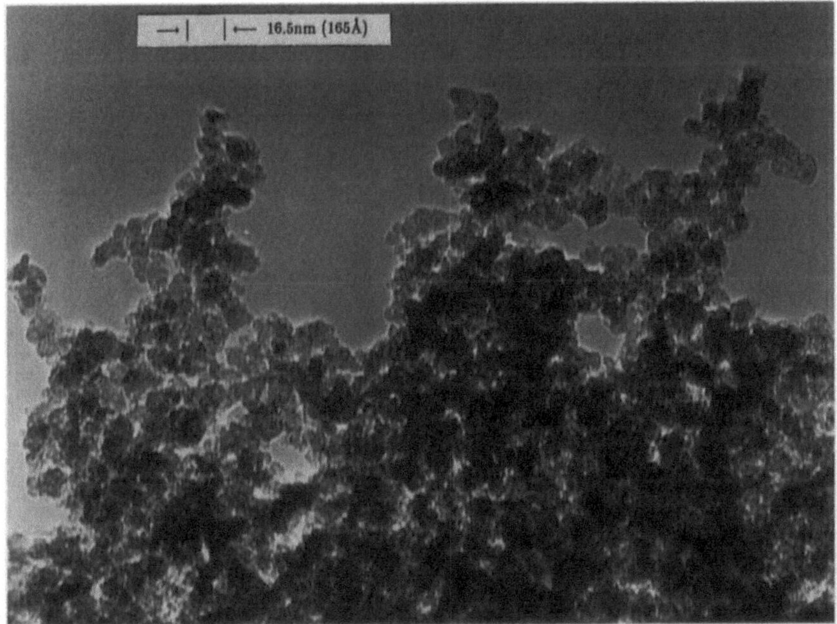

Fig. 7:   Transmission electron micrograph of soot collected at tip of ethylene/air flame.

Table 3
(D=1.78)

Height	N	2a (nm)	n (cm^{-3})	R_g (microns)	f_v
0.68 in.	-	-			
0.8 in.	262	8.6		Same as Table 2	
1.0 in.	736	7.4			
1.12 in.	2427	4.7			

with Table 2, the values of N are seen to be especially sensitive, and there is considerable sensitivity in the monomer radius. It cannot be said that these numbers are unreasonable, either, so it is clear that determination of the proper in-situ fractal dimension for soot aggregates should be a priority task. As we have stated, the limited high-q range of our data means that our inferred range of 1.4-1.5 is not definitive. Table 4 gives some indication of the predicted sensitivity to assumed index of refraction. Use of the Dalzell-Sarofim [28] value instead of the Senftleben-Benedict value is seen to result in factors of two changes in certain of the parameters. While the recent [27] summary would seem to lend weight to the latter number, the proper index of refraction for soot remains an important question.

Table 4
Height=1 in. (D=1.49)

N	2a (nm)	n (cm^{-3})	R_g (microns)	f_v
†87	15	1.016 x 10^{10}	.106	1.567 x 10^{-6}
‡47	22.6	4.622 x 10^9	.106	1.327 x 10^{-6}

†m = 1.94 + .66i [23]
‡m = 1.57 + .56i [25]

It is interesting to interpret our experimental data in terms of an "equivalent sphere" model for the scattering using conventional Mie scattering theory. This is normally done in one of two ways: one uses ratios of scattering intensities at two angles; the other is using extinction and scattering at one angle. The results of reducing the 1 in. data using these two approaches are shown in Figs. 8 and 9. In Fig. 8, the Mie scattering function is fitted to the measured intensities at 20 and 160 degrees, with the inferred hard sphere radius and radius of gyration. The predicted multiangle pattern from this sphere is then given by the triangles, and the overall agreement with what was actually observed is seen to be only fair. The agreement is better at the lower heights

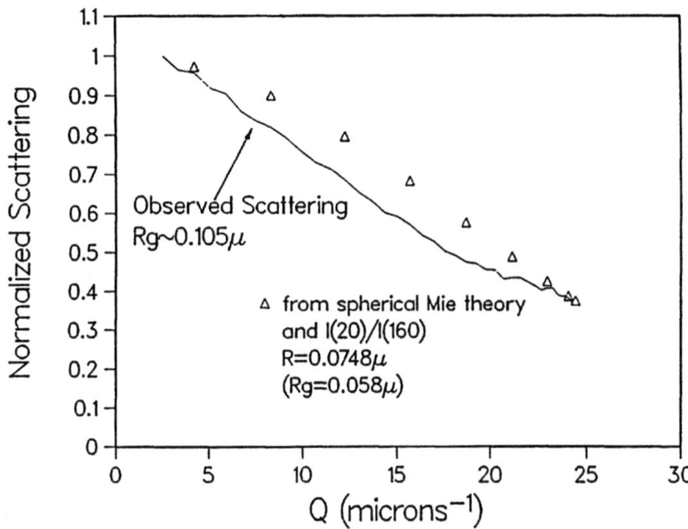

Fig. 8:   Mie fit to scattered intensity vs q at 1 in. height. Rg=0.105 μ is from Berry-Percival. R=0.0748 μ is from Mie theory and the ratio of scattered intensities at 20 and 160°. R_g=0.058 μ is the radius of gyration of the sphere of radius R.

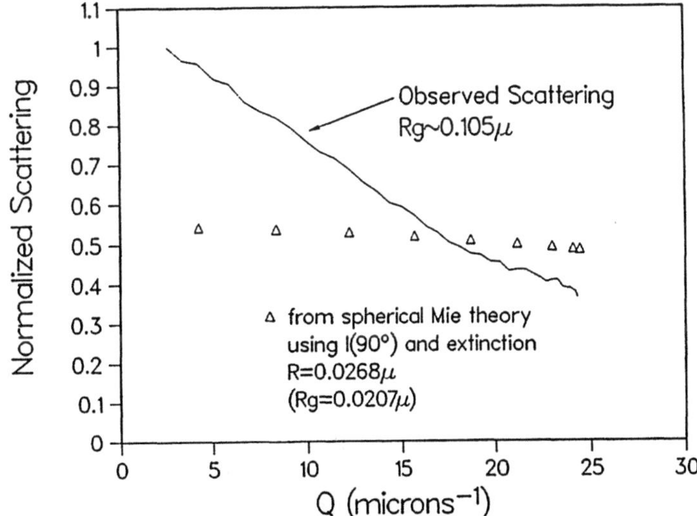

Fig. 9:   Alternate Mie fit to scattered intensity vs q at 1 in. height. R=0.0268 μ is from Mie theory and the ratio of the scattered intensity at 90° to extinction. Other radii are defined as in the Fig. 8 caption.

where the clusters are smaller, and much worse at 1.12 in. where the cluster is much bigger. Fig. 9 displays the results for fitting a Mie scattering function to the 90 degree scattering and the extinction, another approach in widespread use. The inferred sphere radius of gyration and the overall scattering pattern differ significantly from those observed.

It is also possible to make estimates of polydispersity effects on these results, and in principle, to estimate the width of the distribution from the data. We do not expect a high degree of polydispersity in the monomer sizes [29] and we have not included this effect here. Rather, we have assumed polydispersity in the N-distribution, and show how it changes the predicted cluster parameters. A scheme for estimating the width of the size distribution is also given. If a ZOLD (Zero Order Logarithmic Distribution) is assumed, then one has as parameters N, the average cluster number of spheroids, and σ_0, which gives the width of the distribution. It is not difficult to show that the following substitutions occur:

$$nN_a^3 \to n <N> a^3 \to n\overline{N}_a^3 \tag{13a}$$

$$nN^2 a^6 \to n <N^2> a^6 \to n\overline{N}^2 e^{\sigma_0^2} a^6 \tag{13b}$$

Since the measured R_g is a weighted average of the form

$$<R_g^2>(\text{measured}) = \frac{<N^2 R_g^2(N)>}{<N^2>} \sim \frac{<N^{2(1+1/D)}>}{<N^2>} \tag{14}$$

It can be shown that

$$aN^{1/D} \to a\overline{N}^{1/D} e^{\sigma_0^2(1/D^2 + 1.5/D)} \tag{15}$$

This still leaves the width of the distribution unknown, but if one picks a value it is then possible to carry through the data reduction as before without a new piece of experimental data. For soot, values of $\sigma_0 \approx .3$ appropriate to a diameter distribution are suggested in the literature [30, 31], but we emphasize that when we use such a number here it is in a loose sense intended only to give a feeling for the magnitude of the effect. Table 5 shows how the predicted parameters change (compare Table 2) for a predicted N-distribution σ_0 of .3. The changes are not drastic, particularly with regard to the monomer sizes.

To obtain σ_0 from the data, a new piece of experimental data is needed. This might in principle be provided by some of the higher order coefficients in the

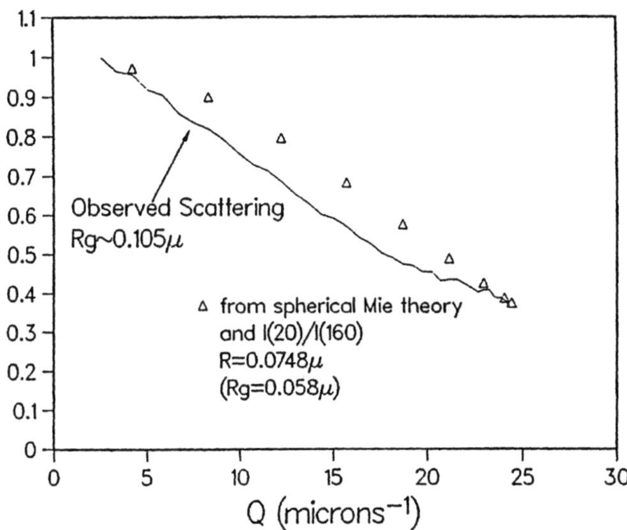

g. 8: Mie fit to scattered intensity vs q at 1 in. height. Rg=0.105 μ is from Berry-Percival. R=0.0748 μ is from Mie theory and the ratio of scattered intensities at 20 and 160°. R_g=0.058 μ is the radius of gyration of the sphere of radius R.

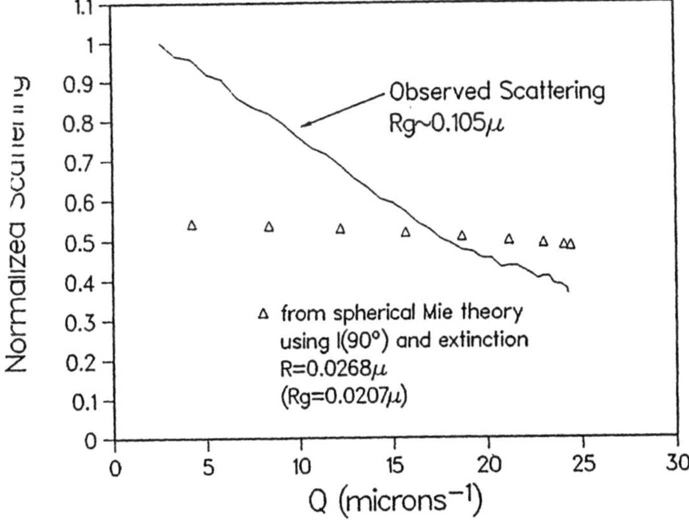

Fig. 9: Alternate Mie fit to scattered intensity vs q at 1 in. height. R=0.0268 μ is from Mie theory and the ratio of the scattered intensity at 90° to extinction. Other radii are defined as in the Fig. 8 caption.

ways. Doing so for the free molecular regime results in a predicted number density increase of about a factor of two, an increase in the monomer size by about a third, and a marked reduction in N by about a factor of five.

CONCLUSIONS

A theoretical analysis of fractal cluster scattering and absorption coefficients has suggested experimental approaches which make possible in-situ determination of soot cluster structural parameters. The approach has been tested in experiments performed in an ethylene-air slot burner and found to have considerable promise. Plausible values of cluster number density, monomer number, and monomer size are obtained from our measurements, although our fractal dimension determination is subject to some uncertainty. Given the sensitivity of certain cluster parameters to fractal dimension, its measurement should be a priority task, perhaps with neutron sources. This work presents an approach to revamping laser light scattering particulate sizing diagnostics for those situations where clusters are expected. Soot scattering is negligible relative to absorption in the burner investigated, and volume fraction is seen to be the size/density parameter governing radiative transfer. For soot radiative transfer calculations, there is no longer any need to employ unrealistic morphological models of soot particles.

ACKNOWLEDGMENT

The authors would like to thank Heidi Hollick and Kathi Wicks for their help in preparing the manuscript.

REFERENCES

1. Markstein G.H.: Relationship Between Smoke Point and Radiant Emission from Buoyant Turbulent and Laminar Diffusion Flames. in Twentieth Symposium (International) on Combustion, The Combustion Institute, Pittsburgh, PA (1984) 1055-1061.

2. Kerker M.: The Scattering of Light and Other Electromagnetic Radiation. Academic Press, New York (1969).

3. Bockhorn H, Fetting F., Heddrich A., Wannemacher G.: Investigation of the Surface Growth of Soot in Flat Low Pressure Hydrocarbon Oxygen Flames. in Twentieth Symposium (International) on Combustion, The Combustion Institute, Pittsburgh, PA (1984) 979-988

4. Harris S.J., Weiner A.M.: Determination of the Rate Constant for Soot Surface Growth. Combust. Sci. Technol. 32 (1983) 267-275.

5. Bonczyk P.A., Sangiovanni J.J.: Optical and Probe Measurements of Soot in a Burning Droplet Fuel Stream. Combust. Sci. Technol. 36 (1984) 135-147.

6. Weitz D.A., Huang J.S., Lin M.Y., Sung J.: Limits of the Fractal Dimension for Irreversible Kinetic Aggregation of Gold Colloids. Phys. Rev. Lett. 54 (1985) 1416-1419.

7. Megaridis C.M., Dobbins R.A.: Morphological Description of Flame-Generated Fractals. Combust. Sci. Technol. 71 (1990) 95-109.

8. Samson R.J., Mulholland G.W., Gentry J.W.: Structural Analysis of Soot Agglomerates. Langmuir 3 (1987) 272-281.

9. Forrest S.R., Witten Jr T.A.: Long-Range Correlations in Smoke-Particle Aggregates. J. Phys. A: Math. Gen 12 (1979) L109-L116.

10. Hurd A.J., Schaefer D.W., Martin J.G.: Surface and Mass Fractals in Vapor-Phase Aggregates. Phys. Rev. A 35 (1987) 2361-2364.

11. Witten T.A., Sander L.M.: Diffusion-Limited Aggregation, a Kinetic Critical Phenomenon. Phys. Rev. Lett. 47 (1981) 1400-1403.

12. Botet R., Jullien R., Kolb M.: Hierarchical Model for Irreversible Kinetic Cluster Formation. J. Phys. A: Math. Gen 17 (1984) L75-L79.

13. Berry M.V., Percival I.C.: Optics of Fractal Cluster Such as Smoke. Opt. Acta 33 (1986) 577-591.

14. Mountain R.D., Mulholland G.W.: Stochastic Dynamics Simulation of Particle Aggregation. in Kinetics of Aggregation and Gelation (F. Family and D.P. Landau Eds.) North-Holland, Amsterdam (1984) 83-86.

15. Mountain R.D., Mulholland G.W.: Light Scattering From Simulated Smoke Aggregates. Langmuir 4 (1988) 1321-1326.

16. Hurd A.J., Flower W.L.: In-Situ Growth and Structure of Fractal Silica Aggregates in a Flame. J. Colloid Interface Sci. 122 (1988) 178-192.

17. Berne B.J., Pecora R.: Dynamic Light Scattering Wiley, New York (1976).

18. Freltoft T., Kjems J.K., Sinha S.K.: Power-Lens Correlations and Finite-Size Effects in Silica Particle Aggregates Studies by Small-Angle Neutron Scattering. Phys. Rev. B 33 (1986) 269-275.

19. Nelson J.: Test of a Mean Field Theory for the Optics of Fractal Clusters. J. Mod. Opt. 36 (1989) 1031-1057.

20. Bonczyk P.A.: Suppression of Soot in Flames by Alkaline-Earth and Other Metal Additives. Combust. Sci. Technol. 59 (1988) 143-163.

21. Namer I., Schefer R.W., Chan M.: Interpretation of Rayleigh Scattering in a Flame. Lawrence Berkeley Laboratory, University of California, Report No. LBL-10655 (1980).

22. Fisher M.E., Burford R.J.: Theory of Critical-Point Scattering and Correlations. I. The Ising Model. Phys. Rev. 156 (1967) 583-622.

23. Bourrat X., Oberlin A.: Mass Fractal Analysis of Conducting Carbon Black Morphology. Carbon 26 (1988) 100-103.

24. Ehrburher-Dolle F., Tence M.: Determination of the Fractal Dimension of Carbon Black Aggregates. Carbon 28 (1990) 448-452.

25. England W.A.: An In-Situ X-Ray Small Angle Scattering Study of Soot Morphology in Flames. Combust. Sci. Technol. 46 (1986) 83-93.

26. Senftleben H., Benedict E.: Uber die Optischen Konstanten und die Strahlungsgesetze der Kohle. Ann. Phys. 54 (1918) 65-78.

27. Vaglieco B.M., Beretta F., D'Alessio A.: In-Situ Evaluation of the Soot Refractive Index in the UV-Visible From the Measurement of the Scattering and Extinction Coefficients in Rich Flames. Combust. Flame 79 (1990) 259-271.

28. Dalzell W.H., Sarofim A.F.: Optical Constants of Soot and Their Application to Heat-Flux Calculations. ASME J. Heat Trans. 91 (1969) 100-104.

29. Dobbins R.A.: Optical Cross-Sections of Flame Generated Aggregates. Fall Technical Meeting of Eastern Section of Combustion Institute, Albany, NY (30 October, 1989).

30. Prado G., Lahaye J.: Physical Aspects of Nucleation and Growth of Soot Particles. in Particulate Carbon Formation During Combustion (D.C. Siegla and G.W. Smith, Eds.), Plenum Press, New York, p. 143 (1981).

31. Wersborg B.L., Howard J.B., Williams, G.C.: Physical Mechanisms in Carbon Formation in Flames. Fourteenth Symposium (International) on Combustion, The Combustion Institute, Pittsburgh, PA (1973) 929-940.

32. Guinier A., Fournet G.: Small Angle Scattering of X-Rays, Wiley, New York (1955).

33. Friedlander S.K.: Smoke, Dust and Haze, Wiley, New York (1977) 175-208.

SPECTROPYROMETRIC DETERMINATIONS OF LOCAL PARTICLES AND GAS TEMPERATURES IN A PULVERIZED COAL FLAME

Jean Crabol
Laboratoire d'Energétique et d'Economie d'Energie
Université de Paris X-Nanterre
I.U.T. de Ville d'Avray, 1, Chemin Desvallières
92410 - Ville d'Avray - France

ABSTRACT

The method described is based on emission-absorption of thermal radiation in the near infrared. It allows the determination of local temperatures rather than an average temperature integrated along aiming line of the pyrometer. The particles and gas temperature radial fields in a semi-industrial pulverized coal flame have been established. These particles and gas temperature profiles exhibit the same shape but particles temperatures are about 100 C higher than gas temperatures.

NOMENCLATURE

a	albedo
$f(x)$	radial variation of transmissivity
$g(x)$	radial variation of absorptivity
ka_λ	absorption coefficient
ke_λ	extinction coefficient
L_λ	brightness of medium
$L^0_{\lambda T}$	brightness of blackbody
T	temperature
T_L	brightness temperature
T_g	gas temperature
T_p	particle temperature
α_λ	spectral absorptivity
ε_λ	spectral emissivity
λ	wavelength
τ_λ	monochromatic transmissivity

1. INTRODUCTION

When the energy problem raised in Europe during the seventies, France tried to develop all kinds of possible use of natural energy, other than petroleum. It seemed worthwhile to create new devices based on coal since it is available in large quantities inside the French soil.

The basic condition was to provide a challenging price of energy in terms of coal combustion. Large research programmes were then undertaken to create new types of installation that would allow to burn the coal in an easy way.

Moreover, in order to insure this challenge, an increase of the combustion efficiency was required. Combustion efficiency is slightly improved when using pulverized coal, but it requires new high efficiency burners.

In order to design such burners, a pilot furnace was installed in Mazingarbe (Northern France). This furnace was expected to be equipped with all metrology and control equipments, that would allow the monitoring of all kinds of burners aimed at the combustion of various sorts of coal.

More precisely, a measurement device allowing a thorough measurement of parameters connected to pulverized coal combustion inside the chamber was required.

Indeed, the knowledge of temperature field within the flame is a gateway to that of the chemical reaction mechanism and hence to combustion efficiency.

The measurement of these temperatures is often very difficult in the case of large scale industrial plants [1] and more specifically when the fuel is in the form of pulverized solid. The probes that are mainly used for this type of measurement are suction pyrometers. Their use is quite delicate and one should be cautious as to the results they provide.

In order to determine the temperature field, we used a method based on emission-absorption of thermal radiation in the near infrared between 1 and 5 μm [2]. It was first tested on an experimental set-up before being used on the pilot furnace. Its main advantage is to keep from perturbating the flow and, furthermore, it provides a more reliable interpretation on the measured values.

Optical methods have several advantages. They have no limitation in terms of high temperatures and they can follow the extremely rapid fluctuations connected to very turbulent flows. Furthermore, spectral measurements allow the distinction between solid and gaseous phases in two phase media. However they do not allow local measurements since they provide average values integrated along the aiming line of the apparatus. This is often a limiting effect, particularly in the case of industrial flames which are neither homogeneous nor isothermal.

Based on an endoscopic device, this present method allows local temperature measurements. A radial together with a longitudinal investigation of the flame yields a reliable assessment of the temperature field. In addition, in the case of pulverized coal combustion, both temperatures of solid particles and gaseous media were obtained since we can analyse the radiation of different species.

2. EXPERIMENTAL DEVICE

2.1. THE 1 MW FURNACE

First trials were conducted in a 1 MW experimental furnace at Cerchar (Centre d'Études et de Recherches du Charbonnage de France) [3]. This 1.4 m diameter and 5 m long cylindrical furnace is vertical. The burner is located on the top and is supplied with pulverized coal.

The access is possible through two opposite portholes located at a distance of 0.6 m from the furnace front plate. In order to keep a constant spatial resolution, the optics were settled and several water cooled tubes of different lengths were inserted into the flame, allowing to work with a limited thickness of the layer.

2.2. THE 3 MW FURNACE

First trials having shown the validity of the method, a similar set-up was installed on a 3 MW furnace at Cerchar, Mazingarbe, France [4]. This furnace is 1.5 m diameter, 7.7 m long and is horizontal. Since it is experimental, this furnace may be used to test various burners and various types of pulverized coal.

The walls of this furnace are made of 31 calorimetric loops that enable the determination of heat transfer between the wall and the flame. Each of these loops holds windows on each side.

Fig. 1:   View of the 1 MW furnace

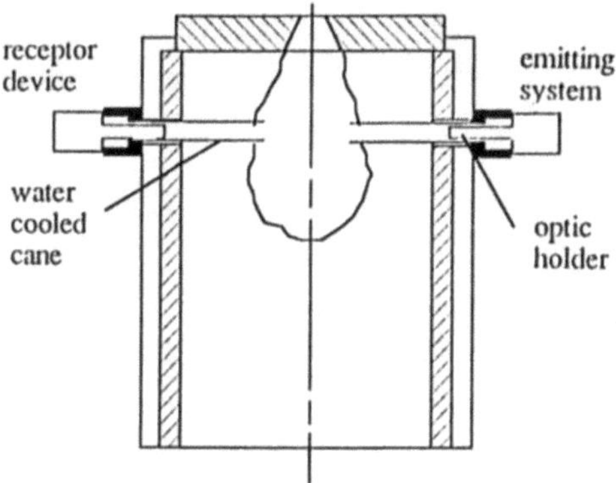

Fig. 2:   Sketch of the set-up in the 1 MW furnace.

The emitting part of the optical device may be moved in a longitudinal way by mean of a gage holder. The receptor device is installed on a mobil system moving on rails. Two sliding canes cooled by water are inserted into the furnace in order to confine the various flame thicknesses to be analysed.

Fig. 3:   View of the 3 MW furnace

2.3. MEASUREMENTS SET-UP

Measurements through transmission and emission were conducted by means of the optical set-up sketched on Figure 4. The emitter is a blackbody at a temperature up to 1200 C while the receptor is a monochromator with a diffraction grating equipped with an infrared detector. Four optical filters allow the separation of various orders in the diffraction grating for wavelengths ranging from 0.8 to 5 μm.

The signal emitted by the detector is amplified by means of a synchronous detection monitored by one of the modulators M_1 or M_2. For transmission

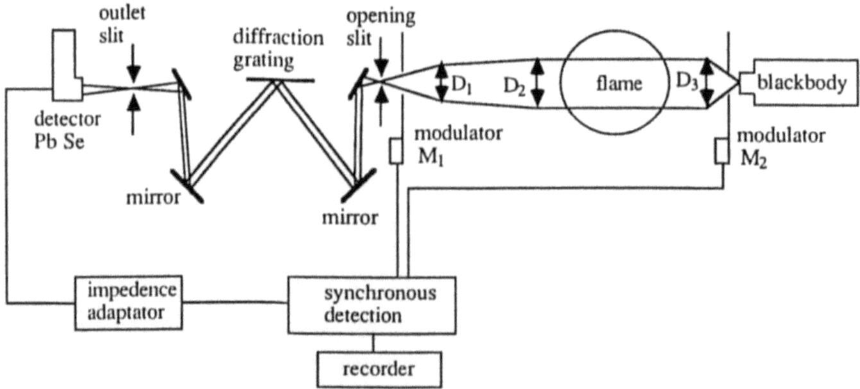

Fig. 4:   Sketch of the principle of the emission-absorption method

measurements, the modulator M_1 is at a standstill and the radiation flux emitted by the source is modulated by M_2 at a frequency of 650 Hz. In this case, the flux emitted by the source is solely transmitted by the flame and it is amplified. For emission measurements, the blackbody source is hidden and the flame radiation is solely amplified. Optics were calculated so that D_3 gives the image of the source on D_2, D_2 gives the image of D_3 on D_1 and D_1 gives the image of D_2 on the opening slit of the monochromator. In order to limit the achromatic defaults due to the fairly wide spectral field, these optics are doublets made of a fluorine and a saphir lens.

3. RADIATION OF TWO-PHASE MEDIA

This method is based on radiative properties of gaseous and two-phase media which will be recalled hereafter.

3.1. BRIGHTNESS TEMPERATURE

The brightness $L_{\lambda T}$ of a medium is given by the relationship:

$$L_{\lambda T} = \varepsilon_\lambda \overset{\circ}{L}_{\lambda T} = \overset{\circ}{L}_{\lambda T_L}$$

where λ, T, ε_λ are the wavelength, the temperature and the spectral emission coefficient respectively. $\overset{\circ}{L}_{\lambda T}$ denotes the brightness of blackbody at temperature T, while T_L is that corresponding to the same brightness. Therefore, T_L is named the brightness temperature of the medium. For values of T that would not be too high, namely as long as $\lambda T < 4500$ μm.K one may express:

$$1/T_L = 1/T - (\lambda/C_2) \ln \varepsilon_\lambda$$

where $C_2 = 14388$ μm.K. is the second Planck constant. Obviously when the medium radiates as a graybody, $1/T_L$ varies in a linear way as a function of λ. Hence, T is obtained experimentally by the origin ordinate of the $1/T_L = f(\lambda)$ line, and the emission coefficient by the slope of the line.

3.2. RADIATION OF AN ABSORBING AND DIFFUSING LAYER

Monochromatic transmission coefficient $(\tau_\lambda)_0^{x_0}$ of a given absorbing and diffusing layer of thickness x_0 is determined by the relationship [5]

$$(\tau_\lambda)_0^{x_0} = \exp\left(-\int_0^{x_0} k e_\lambda (x)\, dx\right)$$

where $ke_\lambda(x)$ is the extinction coefficient at the abscissa x. This coefficient takes into account both absorption and diffusion of the medium at this point. One must state that this coefficient is the same for a spectral radiation that propagates either in the $\overrightarrow{Ox_0}$ or $\overrightarrow{x_0O}$ directions. Thus:

$$(\tau_\lambda)_0^{x_0} = (\tau_\lambda)_{x_0}^0$$

Moreover, the absorptivity of this layer in the direction $\overrightarrow{Ox_0}$ is defined by the relationship:

$$(\alpha_\lambda)_0^{x_0} = \int_0^{x_0} ka_\lambda(x).(\tau_\lambda)_0^x \, dx$$

where ka_λ is the absorption coefficient at the abscissa x. Let us note that in the opposite direction $\overrightarrow{x_0O}$, the absorptivity is:

$$(\alpha_\lambda)_{x_0}^0 = \int_0^{x_0} ka_\lambda(x).(\tau_\lambda)_0^{x_0} \, dx$$

which is different from the previous one since one can easily note that:

$$(\tau_\lambda)_0^{x_0} = (\tau_\lambda)_0^x . (\tau_\lambda)_x^{x_0}$$

The brightness of the radiation emitted by this layer in the $\overrightarrow{Ox_0}$ direction from the boundary x_0 is defined by

$$(L_\lambda)_0^{x_0} = \int_0^{x_0} ka_\lambda(x) \, L_{\lambda T(x)}^0 (\tau_\lambda)_x^{x_0} \, dx$$

Likewise, brightness temperature emitted by this layer in the direction $\overrightarrow{x_0O}$ from boundary 0 is defined by relationship:

$$(L_\lambda)_{x_0}^0 = \int_0^{x_0} ka_\lambda(x) \, L_{\lambda T(x)}^0 (\tau_\lambda)_0^x \, dx$$

Like transmission coefficients, both brightnesses are usually different.

3.3. HETEROGENEOUS BUT ISOTHERMAL MEDIA

In the case of an isothermal medium at temperature T, $L_{\lambda T}^0$ is a constant along the axis $\overrightarrow{Ox_0}$ and therefore the integral can be expressed as follows:

$$(L_\lambda)_0^{x_0} = L_{\lambda T}^0 \int_0^{x_0} ka_\lambda(x) \, (\tau_\lambda)_x^{x_0} \, dx$$

Therefore it is possible to define an emission coefficient $(\varepsilon_\lambda)_0^{x_0}$ of the layer in the direction $\overrightarrow{Ox_0}$ in the following way:

$$(L_\lambda)_0^{x_0} = (\varepsilon_\lambda)_0^{x_0} \; L_{\lambda T}^0$$

Comparing to the proceeding relations, one obtains:

$$(\varepsilon_\lambda)_0^{x_0} = (\alpha_\lambda)_x^0$$

Likewise, one can obviously define an emission coefficient $(\varepsilon_\lambda)_{x_0}^0$ in the direction $\overrightarrow{x_0 O}$

$$\left(L_\lambda^0\right)_{x_0}^0 = (\varepsilon_\lambda)_{x_0}^0 \; L_{\lambda T}^0$$

Hence

$$(\varepsilon_\lambda)_{x_0}^0 = (\alpha_\lambda)_0^{x_0}$$

3.4. GENERAL PROBLEM: HETEROGENEOUS AND NON ISOTHERMAL MEDIA

One can always define the emission coefficients, which will be considered as average, by the same relations:

$$(\varepsilon_{\lambda m})_0^{x_0} = (\alpha_\lambda)_{x_0}^0$$

$$(\varepsilon_{\lambda m})_{x_0}^0 = (\alpha_\lambda)_0^{x_0}$$

Thus, the corresponding average temperatures T_m and T'_m can be defined

$$(L_\lambda)_0^{x_0} = (\varepsilon_{\lambda m})_0^{x_0} \; L_\lambda^0 \, _{T_m}$$

$$(L_\lambda)_{x_0}^0 = (\varepsilon_{\lambda m})_{x_0}^0 \; L_\lambda^0 \, _{T'_m}$$

In the case of a flame exhibiting a strong degree of transparency, i.e. τ_λ close to unity it is easy to show that $T_m = T'_m$ is the adiabatic temperature of the layer [6].

Obviously, in a small thickness flame this assumption is practically verified and temperature T_m is the temperature T of the quasi isothermal layer.

3.5. TWO PHASE MEDIA

In this case, when a gaseous flow contains particles of significant size, it is possible to show on the basis of Mie theory, that ke_λ and ka_λ are independent of the wavelength. In these conditions, the different coefficients that have been defined earlier, namely $(\tau_\lambda)_0^{x_o}$, $(\alpha_\lambda)_0^{x_o}$, $(\varepsilon_{\lambda m})_0^{x_o}$, as well as T_m are not wavelength dependent. More precisely, since $(\varepsilon_{\lambda m})_0^{x_o}$ is not depending on λ, any layer $\overrightarrow{Ox_0}$ may be regarded as graybody and the relationship

$$\frac{1}{(T_L)_0^{x_0}} = \frac{1}{(T_m)_0^{x_0}} - \left(\frac{\lambda}{C_2}\right) \ln (\varepsilon_\lambda \, m)_0^{x_0}$$

shows that the inverse of $(T_L)_0^{x_0}$ varies in a linear way with wavelength. Thus, one obtains $(T_m)_0^{x_0}$ and $(\varepsilon_{\lambda m})_0^{x_0}$ by plotting the experimental variation of $\dfrac{1}{(T_L)_0^{x_0}}$ as a function of λ.

4.   PRINCIPLE OF THE METHOD

4.1. MEASUREMENTS THROUGH TRANSMISSION

If an endoscope is available to confine several flame thicknesses, measurements allow to determine the transmitivity of these different thicknesses; that is

$$(\tau)_0^x = f(x)$$

One can obtain the value of the extinction coefficient at a given point x_0 through relationship

$$ke\,(x_0) = -\frac{1}{(\tau)_0^{x_0}}\left[\frac{d\,(\tau)_0^x}{dx}\right]_{x_0} = -\frac{1}{f(x_0)}\left(\frac{df}{dx}\right)_{x_0}$$

This coefficient, connected to the concentration, exhibits its evolution within the furnace.

4.2. MEASUREMENTS THROUGH EMISSION

In the same way, it is possible in this case to determine the variations of brightness temperatures as a function of the wavelength for various layers of the flame. Since the pulverized coal flame radiates as a graybody, the variation of $1/T_L$ as a function of λ is then a line whose slope allows to obtain $(\varepsilon)_x^0$ or $(\alpha)_0^x$ for a given thickness Ox. Variation of this absorptivity with thickness x

$$(\alpha)_0^x = g(x)$$

yields the non wavelength dependent value of the spectral absorption coefficient at the abscissa x_0.

$$ka\,(x_0) = \frac{1}{(\tau)_0^{x_0}}\left[\frac{d\,(\alpha)_0^x}{dx}\right]_{x_0} = \frac{1}{f(x_0)}\left(\frac{dg}{dx}\right)_{x_0}$$

Furthermore, it is possible to plot the variations of $\left(\frac{1}{T_L}\right)_0^x$ as a function of x for a given value of λ and to derive this function. Hence, values of $\left(\frac{1}{T_L}\right)_0^{x_0}$ and $\left[\frac{d}{dx}\left(\frac{1}{T_L}\right)_0^x\right]_{x_0}$ may be obtained.

4.3. CALCULATION OF PARTICLE TEMPERATURES

Values of $ke(x_0)$, $ka(x_0)$, $\left(\frac{1}{T_L}\right)_0^{x_0}$ and $\left[\frac{d}{dx}\left(\frac{1}{T_L}\right)_0^x\right]_{x_0}$ having been determined for a wavelength where only particles radiate, the temperature T_p of those located at x_0 is given by the expression:

$$\left(\frac{1}{T_p}\right)_{x_0} = \left(\frac{1}{T_L}\right)_0^{x_0} + \left(\frac{\lambda}{C_2}\right)\ln\left[\frac{ka(x_0)}{ke(x_0) - \frac{C_2}{\lambda}\left[\frac{d}{dx}\left(\frac{1}{T_L}\right)_0^x\right]_{x_0}}\right]$$

4.4. CALCULATION OF GAS TEMPERATURE

In the same manner, at a wavelength λ where both gas and particles radiate, it is possible to obtain gas temperature value. The global transmissivity $(\tau_{p+g})_0^x$ compared to $(\tau)_0^x$ previously obtained yields the value of gas transmissivity $(\tau_g)_0^x$

$$(\tau_{p+g})_0^x = (\tau)_0^x \cdot (\tau_g)_0^x$$

and, hence the value of the absorption coefficient ka_g of the gas at a location x_0

246

through relation

$$ka_g(x_0) = -\left[\frac{d}{dx}\left(\ln(\tau_g)_0^x\right)\right]_{x_0}$$

Using the variation of the inverse of the brightness temperature as a function of x, together with its derivative, one can obtain

$$\left(\frac{1}{T_g}\right)_{x_0} = \left(\frac{1}{T_L}\right)_0^{x_0} +$$

$$+ \left(\frac{\lambda}{C_2}\right)\ln\left[\frac{ka_g(x_0)}{(ke_p + ka_p)(x_0) - \frac{C_2}{\lambda}\left[\frac{d}{dx}\left(\frac{1}{T_L}\right)_0^x\right]_{x_0} - ka_p(x_0)\exp\left(-\frac{C_2}{\lambda}\left(\frac{1}{T_p(x_0)} - \frac{1}{T_L(x_0)}\right)\right)}\right]$$

5. EXPERIMENTAL RESULTS

5.1. MEASUREMENTS CONDUCTED IN THE 1 MW FURNACE

Figure 5 is an illustration of the variations of the extinction and absorption coefficients of the gases along the diameter of the furnace. More specifically, it shows the variations of concentration of particles (coal as well as solid combustion products) and carbon dioxide. One may observe an asymmetric behaviour of the combustion inside the chamber.

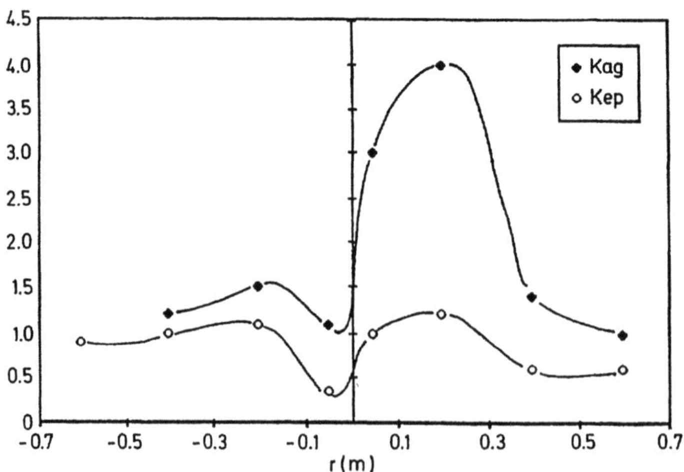

Fig. 5:   Variations of ka_p and ka_g along the diameter of the furnace

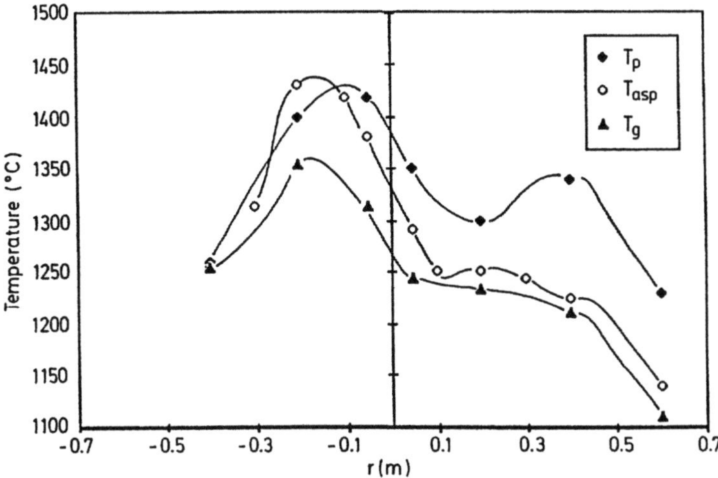

Fig. 6: Variations of temperature T_p of the particles and T_g of gases along the diameter of the furnace

Figure 6 exemplifies the radial variations of temperature obtained either for solid particles or for gases. One may notice that both curves exhibit a similar behaviour. However, the temperature of the gases is around 100 C less than that of particles.

Comparison of both curves to that obtained by means of a suction pyrometer Tasp shows a slight discrepancy. Although this latter pyrometer yields values that are quite close to gas temperatures, on the right hand side of the curve, those corresponding to the left hand side, where the stick is somewhat more deeply inserted in the flame, are slightly greater, due to its heating.

5.2. MEASUREMENTS CONDUCTED IN THE 3 MW FURNACE

Similar measurements were performed with the new set-up. However, since the apparatus were more sophisticated results were more accurate. Moreover the flame was more thoroughly investigated. Figure 7 exhibits the radial evolution of coefficients ke_p and ka_p for a given location of the aiming axis. In this case, the flame is relatively symmetrical.

Variations of the albedo

$$a = 1 - (ka_p/ke_p)$$

are depicted on the same Figure 7. One may notice that its value is slightly constant. We will describe in a next section of this paper an improvement of our method on the basis of this parameter.

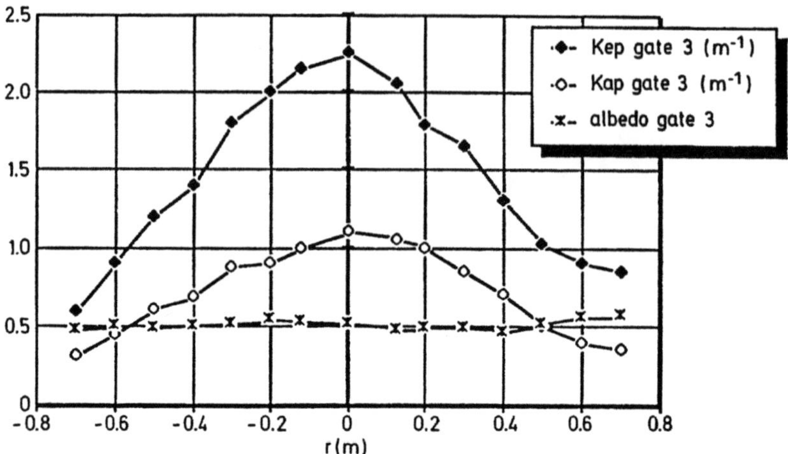

Fig. 7: Radial profiles of the extinction and absorption coefficients of particles and of the albedo.

Figure 8 depicts temperature measurements that were conducted at the same location. It shows temperature variations of both particles and gases as well as those obtained by means of the suction pyrometer.

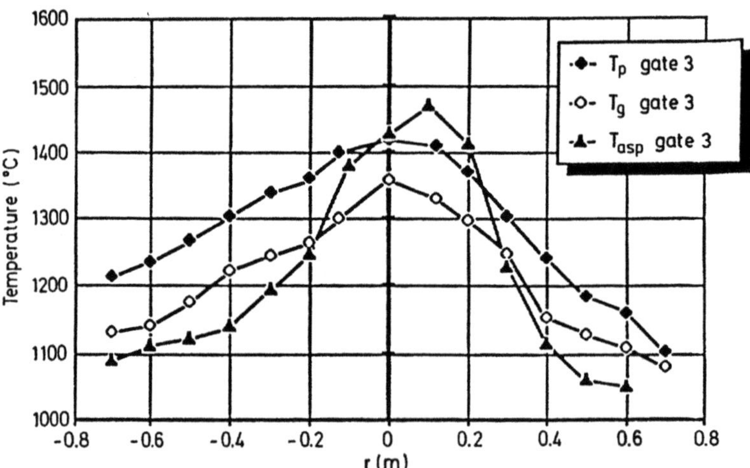

Fig. 8: Radial profiles of particles and gas temperatures and those of suction pyrometer.

Once again, gas temperature is 50 C to 100 C less than that of particles. Using the suction pyrometer yields a asymmetrical behaviour of the curve. This is not the case for the two other measurements.

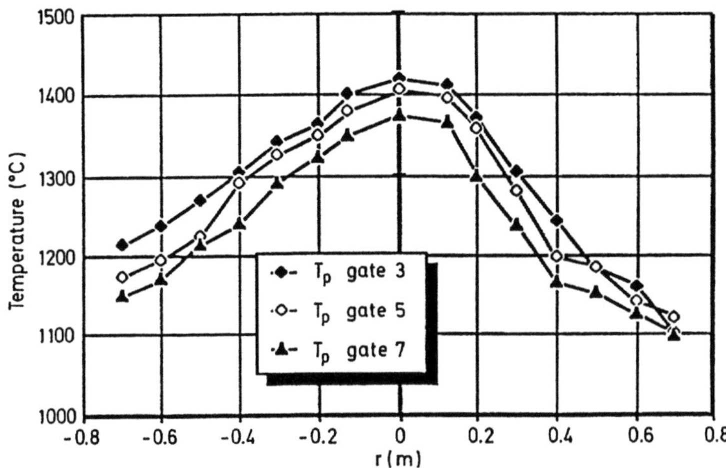

Fig. 9: Radial profiles of particle temperatures located at three different sections of the furnace

Figure 9 shows the radial variations of particle temperatures at three locations in the furnace. These temperatures decrease when the distance from the outlet of the burner increases. This was utterly predictable.

Many other results were obtained. However, the three preceding examples were most significant.

5.3. MEASUREMENT UNCERTAINTIES

Complexity of calculations may raise a problem as to their uncertainty. It could be shown that the uncertainty on particle temperature was generally less than 50 C. That of gas temperature is somewhat greater and may be up to 100 C in some particular cases.

6. IMPROVEMENTS OF THE METHOD

Since pure emission measurements are more reliable, the method was modified in the two ways described below.

6.1. PARTICLES TEMPERATURE MEASUREMENTS

According to the slightly constant value of the albedo a (around 0.5), measurement of ka_p is solely required and is obtained by emission measurements.

From relations

$$ke(x_0) = - \frac{1}{f(x_0)} \left(\frac{df}{dx} \right)_{x_0}$$

and

$$ka(x_0) = \frac{1}{f(x_0)} \left(\frac{dg}{dx} \right)_{x_0}$$

we have

$$\frac{ka}{ke} = 1 - a = 0.5 = - \frac{\left(\dfrac{df}{dx} \right)_{x_0}}{\left(\dfrac{dg}{dx} \right)_{x_0}}$$

Hence

$$f(x_0) = - 0.5 \, g(x_0) + K$$

where K is a constant. But for $x_0 = 0$,

$$f(x_0) = 0, \quad g(x_0) = 0$$

and

$$K = 0.5$$

Therefore, the value of the absorption coefficient is given by the relationship

$$ka(x_0) = - \frac{2}{1 - g(x_0)} \left(\frac{dg}{dx} \right)_{x_0}$$

Where $g(x_0)$ is obtained by emission measurements. Then

$$\left(\frac{1}{T_p} \right)_{x_0} = \left(\frac{1}{T_L} \right)_0^{x_0} + \frac{\lambda}{C_2} \ln \left[\frac{k_a(x_0)}{2k_a(x_0) - \dfrac{C_2}{\lambda} \left[\dfrac{d}{dx} \left(\dfrac{1}{T_L} \right)_0^{x} \right]_{x_0}} \right]$$

This method has been applied to the preceding measurements and has given the same results.

6.2. GAS TEMPERATURE MEASUREMENTS

Gas temperature measurements may be obtained by the study of the emission band head of CO_2[8] at $\lambda = 4.2 \ \mu m$. In emission, this band presents the behaviour of the Figure 10, with two band heads.

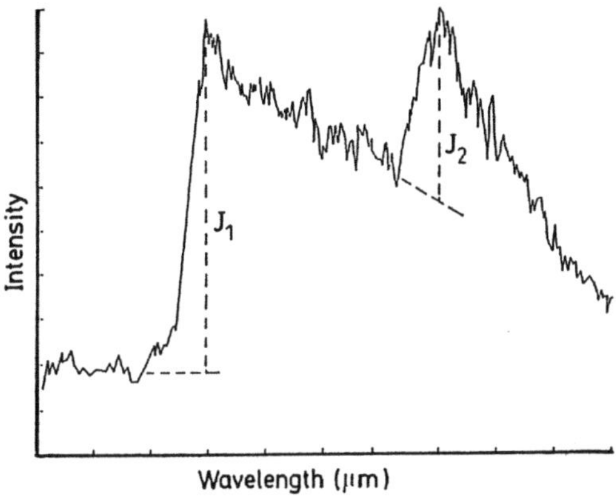

Fig. 10: Band heads of the emission of CO_2 at $\lambda = 4.2\ \mu m$

It is possible to show the relationship [9]

$$T = \frac{944}{\ln 2\ \dfrac{J_1}{J_2}}$$

where T is the gas temperature, J_1 and J_2 the intensities of two band heads.

The first trials of this method have shown a good agreement (Figure 11) between the spectroscopic measurements and those obtained with the suction pyrometer.

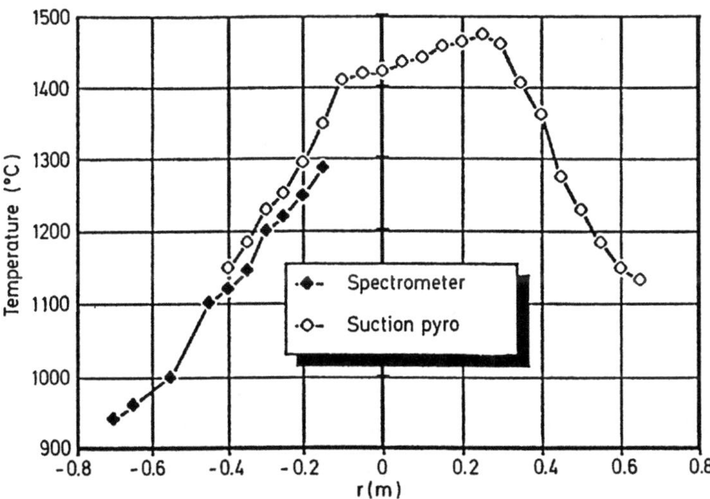

Fig. 11: Radial profiles of gas temperature obtained by spectroscopic measurements and by the suction pyrometer.

7. CONCLUSION

Particles and gas temperature measurements that have been described enabled the determination of thermal fields inside a semi-industrial furnace. These results allow an improvement in the control of the burner and they provide a better knowledge of pulverized coal combustion [10].

This emission-absorption method requires a great care in the positioning of the optical components. Also, since pure emission measurement is more reliable, two modifications of the method were carried out; one, in supposing a constant value of albedo for particle temperature measurements and the other, by studying the intensities of band heads of the emission of CO_2 for gas temperature measurements.

8. ACKNOWLEDGMENTS

This study was conducted at CERCHAR (centre d'Etudes et de Recherches du Charbonnage de France) and was under the grant of both "Ministère de la Recherche et de la Technologie" and "Groupement Scientifique Charbon".

9. REFERENCES

1. Gaydon A.G., Wolfhard H.G.: Flames: their structure, radiation and temperature. Chapman and Hall, London (1979).

2. Crabol J., Macé O., Potier B.: Etude spectropyrométrique des champs de températures dans des installations de type semi-industriel. Joint meeting of the French and Italian Sections of the Combustion Institut, Amalfi, Italie (1987).

3. Potier B.: Détermination des champs de températures et de concentrations dans une flamme de charbon pulvérisé de taille semi-industrielle. Thèse de Doctorat de l'Université d'Orléans, France (1986).

4. Macé O.: Etude des champs de températures dans des écoulements hétérogènes: Application aux flammes de charbon pulverisé et aux lits fluidisés circulants. Thèse de Doctorat de l'Université de Rouen, France (1989).

5. Crabol J.: Transfert de Chaleur — Tome II: Applications Industrielles, Masson, Paris (1990).

6. Crabol J., Potier B., Macé O.: Signification des mesures optiques de températures locales et moyennes dans un milieu diphasique et hétérogène — Application aux flammes semi-industrielles. Journée sur la Combustion, Chatenay-Malabry, France (1987).

7. Van de Hulst H.C.: Light scattering by small particles. Wilez, New York, (1957).

8. Hertzberg G.: Molecular spectra and molecular structure. D. Van Nostrand, New-York (1953).

9. Crabol J., Mace O.: Mesure de température de flamme par l'étude des têtes de bandes de l'émission du gaz carbonique. Journées d'études sur le rayonnement des gaz et des flammes, S.F.T., Paris (1989).

10. Crabol J., Potier B., Mace O.: Optical measurement of local temperatures in a semi-industrial pulverized coal flame by infrared spectroscopy. SPIE's. International symposium on optical and optoelectronic applied science and engineering, San Diego, U.S.A. (1990).

RADIATION TOMOGRAPHY OF SOOTING DIFFUSION FLAMES

R.J. Hall and P.A. Bonczyk

United Technologies Research Center
East Hartford, Connecticut USA

ABSTRACT

When soot is present in combustion, it usually dominates the radiative transfer process, and it is well known that very large conversions of chemical enthalphy to radiative loss can occur in sooting laboratory flames. Diagnosis of sooting flames, therefore, is important if the effects of soot radiation are to be accurately predicted. A sooting flame temperature measurement technique has been demonstrated that is based on emission/absorption tomography. The approach applies the algorithms of Fourier transform tomography to deconvolve local soot absorption coefficient and Planck function (temperature) from sets of parallel, line-of-sight measurements. The technique has the advantage that it is experimentally simple, and does not require involved data reduction. For soot-size particles, there is also no sensitivity of the inferred temperature to possibly uncertain medium parameters. Its main limitation seems to be that it will not work well for vanishingly small absorption, but this could be overcome in practice by seeding to some reasonable level of absorption, and then performing all work at the wavelength of a seed resonance. While in principle limited to optically thin flames, accurate corrections for moderate optical thickness can often be made. A self-consistent comparison of measured, global radiation from a sooting ethylene flame with a radiative transfer calculation based on measured temperature and soot absorption parameters has been performed.

INTRODUCTION

Temperature is an important indicator of the performance of combustion-driven systems, and considerable effort has always been devoted to the development of measurement techniques. Knowledge of temperature gives information about combustion efficiency, the rates of formation of undesired particulates and pollutant species, and allows one to predict radiant heat loading on surfaces. Optical measurement approaches have been historically favored because of minimal disturbance of the medium being probed. Physical probes could perturb the medium, can become coated with particulates, and need radiative corrections. Point techniques may be inconvenient in some situations, however, because a measurement is made at only one location at a time. It could be faster and easier to make a path-integrated measurement or a simultaneous set of them, along a line or lines which the temperature is varying. There is a special advantage, therefore, to any algorithm which can reconstruct this varying temperature from path-integrated measurements, because with it a two-dimensional slice of the medium can be mapped out rapidly. A further requirement of any viable technique is certainly simplicity of data reduction; retrieval of temperature should not involve extensive and time-consuming computer code use, nor require assumption of values for medium parameters that may not be well known.

We have obtained favorable results in sooting laboratory flames using emission/absorption tomography. The technique is simple to implement experimentally, and data reduction is relatively uncomplicated. The deduced temperatures are not sensitive to assumptions about values of any medium parameters, and seem to agree well with those obtained from advanced nonlinear optical diagnostics. Only relatively easily measured quantities, radiation intensity and absorption coefficient, are involved in the approach, which we now describe.

THEORY

The reconstruction of internal flame or plasma temperatures from external, path-integrated intensity measurements has long been a problem of interest. For a gas in thermodynamic equilibrium, the line-of-sight emission intensity is given by the path integral [1]

$$I_\lambda = \int \alpha_\lambda^{(a)}(x,y)\, B_\lambda(T(x,y))\, e^{-\int_{x,y}^{-} \alpha_\lambda^{(e)}(x',y')\,ds'}\, ds$$

$$+ I_\lambda^{(0)}\, e^{-\int_{-}^{-} \alpha_\lambda^{(e)}(x',y')\,ds'} \tag{1}$$

where $\alpha_\lambda^{(a)}$ and $\alpha_\lambda^{(e)}$ are the local absorption and extinction coefficients, $I_\lambda^{(0)}$ is the incident intensity, and $B_\lambda(T)$ is the Planck function

$$B_\lambda(T) = \frac{2\,hc^2}{\lambda^5 \left(\exp\left(\frac{hc}{\lambda\,kT}\right) - 1 \right)} \tag{2}$$

If the absorption is due to molecules or small particles in the Rayleigh scattering range, scattering will be negligible, and $\alpha_\lambda^{(e)} = \alpha_\lambda^{(a)} = \alpha_\lambda$.

In the work of Porter [1] and Kuhn and Tankin [2], axisymmetric media were divided into concentric zones and the internal emission coefficients deduced by algebraic reconstruction. The internal absorption coefficients were calculated by making intensity measurements with and without an external source or mirror in the experiment. It has been suggested more recently that Fourier transform tomography may be applicable to temperature measurement [3-6] and experimental demonstrations in laboratory flames have been reported [5-6].

The application of tomography to temperature measurements is based on the fact that the expression for line-of-sight radiant intensity (eq. (1)) for optical thinness is the path integral of a property field, which is essentially the local emission rate. By making lateral intensity measurements at multiple angles of incidence, and applying Fourier transform-based reconstructive algorithms, it is possible to retrieve the internal values of this property field (the product of Planck function and local absorption coefficient) in a slice through the medium. To derive temperature, it is also necessary to perform an auxiliary measurement in which the local absorption coefficient is derived, using the same reconstructive technique. When the medium is not optically thin, corrections for the self-absorption factor in eq. (1) are introduced; the details of how this is done follow. Strictly speaking, the line-of-sight radiant intensity is not a property field when self-absorption is significant.

We did not conduct a rigorous comparison of the tomographic reconstructive technique with other possible approaches, with the exception of some calculations based on "onion-peeling", which was found definitely to amplify experimental errors. As will be discussed, the Fourier transform method was found to give very good reconstruction of synthetic data, even for moderate optical thickness. The reconstructive algorithm is also simple and fast. We use the convolution method of Ramachandran [7]; in particular the computer algorithm given by Shepp and Logan [8]. Our application follows that given in Refs. [3-4]; the reader is referred there for further details.

Given a lateral intensity scan or projection $P(r,\theta)$ as given in Figure 1, the property field is reconstructed with the relationship

$$F(x,y) = \frac{a}{2N} \sum_{j=1}^{N} \sum_{k=1}^{M} P(r_k, \theta_j) \, \phi(x \cos \theta_j + y \sin \theta_j - r_k)$$

$$(3)$$

where a is the spacing between rays in a scan; M is the number of parallel ray paths in a scan; N is the number of angular scans ($\theta_j = (j-1)\pi/N$); and ϕ is a filter function given by

$$\phi(r_k) = -\frac{4}{\pi a^2 (4k^2 - 1)} \quad k = 0, \pm 1, \pm 2, \ldots$$

(Shepp — Logan Filter)

$$\overline{\phi(r_k)} = .4\phi(r_k) + .3\phi(r_{k+1}) + .3\phi(r_{k-1})$$

(Modified Shepp — Logan Filter)

(4)

As our experimental projections were relatively smooth, we saw no difference between the two filters. For the axisymmetric flames, all projections are the same, and only one experimental projection has to be made. We normally reconstructed on the x-axis at the input projection data radii, i.e. at $(r_k,0)$, as in Fig. 1. When the flame is not optically thin, it is necessary to account for the self-absorption factor in eq. (1). This was done in the following way: the deconvolved effective emission rate was corrected by the value of the self-absorption factor on the path leading from the point of reconstruction (x,0) to the edge of the flame in the ray direction (normal to origin) $\theta=0$. The known absorption coefficient profile makes it possible to do this calculation. We do not know what the rigorous basis for this procedure might be, or whether it is applicable to non-axisymmetric media. As has been discussed, it gives generally very good results in reconstruction of optically thick synthetic spectra. The question is moot for the results reported in this paper, because the flames were close to optically thin.

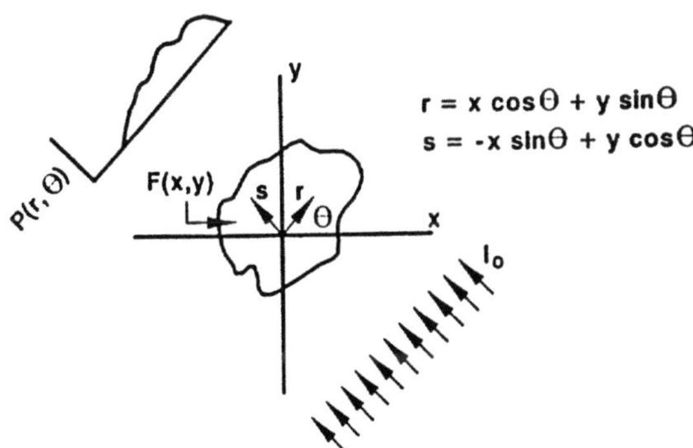

(Absorption) $P(r,0) = -\ln\left[I(r,0)/I_o\right] = \int \alpha_\lambda(x, y)ds$

(Emission) $P(r,\Theta) = \int \alpha_\lambda(x,y) B_\lambda(x,y) e^{-\int_{x,y} \alpha_\lambda(x',y')ds'} ds$

Fig. 1: Generation of projections from parallel path measurements at various angles of incidence (after [3,4]).

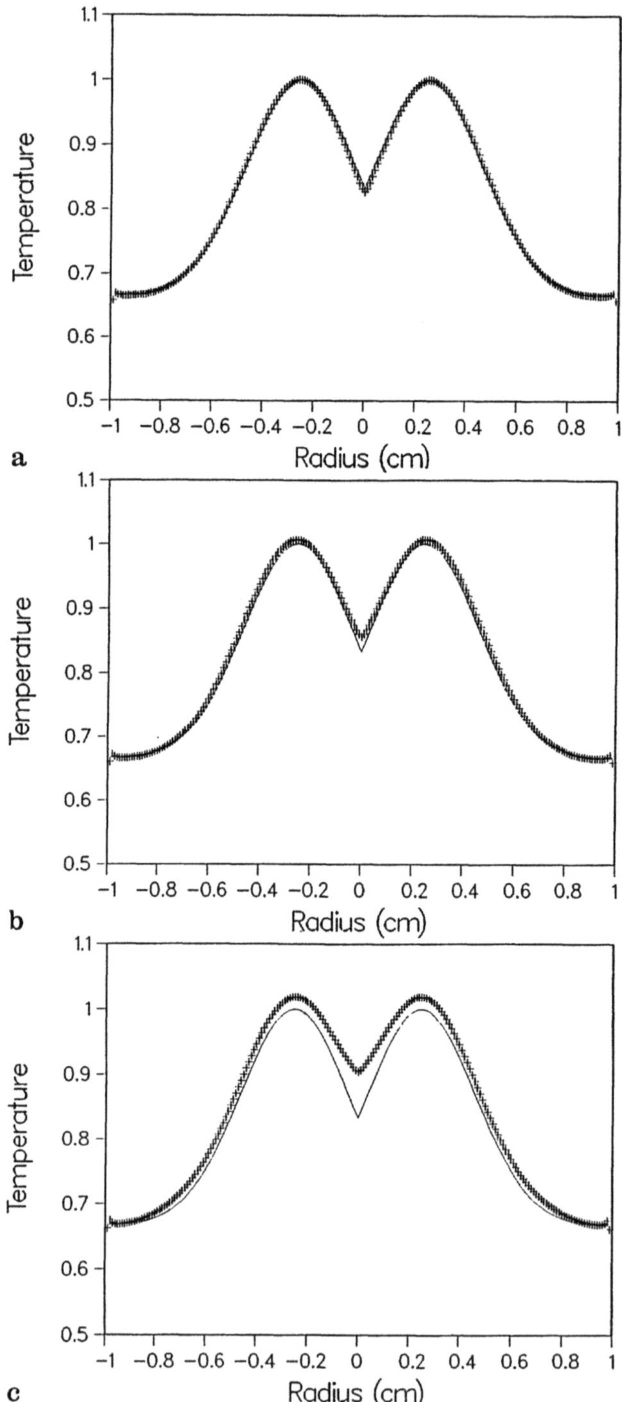

Fig. 2: Comparison of reconstructed temperature (++++) with true input temperature (———) for values of absorption coefficient of A) 1. per cm, B) 2. per cm, C) 3. per cm.

A number of computer experiments were performed to test the reconstructive capabilities of the algorithm. Axisymmetric temperature and absorption coefficient profiles were assumed, and synthetic projections were calculated numerically, including self-absorption. The projections were then input to the deconvolution code, and the ability to reconstruct the temperature field tested. For spatially uniform absorption coefficient profiles, the algorithm could retrieve the input temperatures up to absorption coefficient-radius products of at least two, which represents fairly significant optical thickness. For thickness much beyond this, errors started to appear. Sample calculations are given in Figures 2 (a-c). For the input "experimental" temperature profile (admittedly unphysical), the reconstructed temperature is shown for absorption coefficient-radius products of one, two and three. When the input temperature is also flat, the reconstruction was excellent for $\alpha R = 1.5$, which still represents significant optical thickness, but errors of a few percent in the reconstruction appeared at $\alpha R = 2$. The great sensitivity of Planck function to temperature (approximately T^4) had led to concern about dynamic range problems along paths with wide temperature variations, but this was not revealed by the computer experiments when the local absorption was uniform.

When the local absorption is highly non-uniform, the accuracy of the temperature reconstructions can, in certain cases, be unsatisfactory. Regions of very weak absorption or those that combine weak absorption and low temperature will not always give satisfactory reconstructions; examples of this will be shown in the experimental sooting flame results. A synthetic calculation is shown in Figure 3 for a spatially varying absorption coefficient.

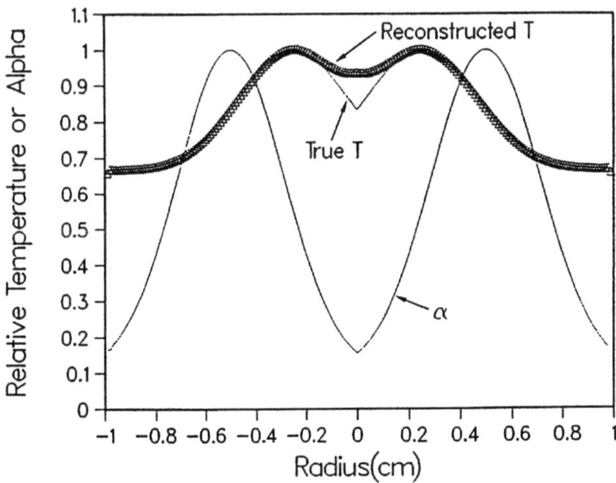

Fig. 3: Comparison of reconstructed and true temperatures for spatially varying absorption coefficient

Even though the overall optical thickness is quite large, the temperature reconstruction is good except in the regions of weak absorption around the centerline. A very strongly absorbing annulus can obscure the centerline region of a medium, and lead to inaccurate predictions there. On the whole, however, the synthetic reconstructions gave encouragement to the use of tomographic procedures for temperature measurement in flames. It is recommended that such deconvolutions of synthetic spectra for approximately expected temperatures and absorption coefficients be carried out by anyone interested in applying the technique.

EXPERIMENTAL ARRANGEMENT

The experimental apparatus used in our investigations is shown in Figure 4. The source employed was a tungsten filament lamp which radiates as a grey body. The axisymmetric flame was positioned on a translatable mount; at a given vertical height above the burner surface, the required projections were generated by moving the flame transverse to the optical axis. Transverse resolution is defined by the 100 micron vertical slit, and the vertical resolution by the 1.5 mm diameter optical fiber. As required for path-integrated measurements, the focal lengths of the lenses which bring the source radiation to a focus at the center of the flame, together with the vertical slit dimensions, are such as to create a large depth of field in the longitudinal or optical direction. This was verified by translating the flame in the longitudinal

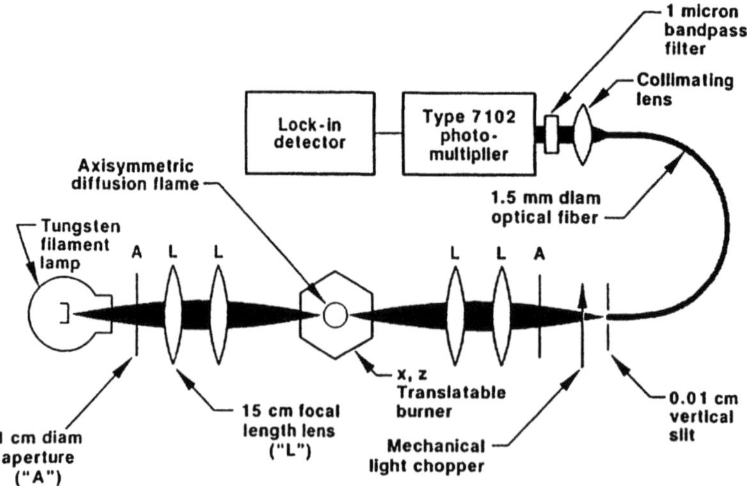

Fig. 4:   Schematic of experimental apparatus.

direction and observing no change in the intensity. (Conventional Kurlbaum can be performed by substituting a pin-hole for the vertical slit, giving fine longitudinal resolution). Because we wanted to avoid having to make estimates of soot particle scattering, all work was performed at a wavelength of one micron where the size parameters are such that scattering should be small, and measured extinction can be equated to absorption. Other wavelengths could be selected by substituting for the one micron fiber. The detection apparatus consisted of a photomultiplier and lock-in detector; the output signals were recorded on chart paper, and transferred manually to the computer for analysis.

The experimental procedure consists of making two projections at each vertical height; one with the source on (eq. (1)), and the other with it off (eq. (1) with $I_\lambda^0=0$). Because the flames we investigated were axisymmetric, all projections at different angles of incidence are the same, and only one angle of incidence need be considered experimentally. The difference between the two projections defines the source transmission profile, which is readily deconvolved to yield the local absorption coefficient. Deconvolution of the flame emission profile for optical thinness yields the local value of the product of absorption coefficient and Planck function. Knowledge of the previously derived absorption coefficient makes it possible to calculate the Planck function and thus the local temperature. It is highly desirable to derive temperature by measuring the Planck function because of the sensitive dependence of the latter on the former (approximately T^4). A 10% measurement error in the Planck function, for example, will be equivalent to roughly a 2% temperature error. It is also apparent that any error in absorption coefficient will generate an equivalent error in Planck function. The suggestion of Kuhn and Tankin [2] that temperature error in this type of measurement is approximately the fourth root of the absorption coefficient error seems reasonable. Because the latter is a quantity that can in principle be measured very accurately, there is the promise of precise temperature measurements. Accurate absorption measurements are unlikely to be made in the limits of extreme optical thinness or thickness; it is expected that regions of moderate absorption will yield the best results. Temperature measurements cannot be reliably made in regions containing little soot.

The Planck function measurements need to be placed on an absolute basis for temperature determination, and this is done by performing an optical pyrometer measurement of the tungsten filament temperature. The filament current can be set at an arbitrary value, and the emission measurements are put on an absolute scale in this way. If the lens transmission losses on the filament side of the flame are significant, as they were in our experiments, this needs to be accounted for in the calibration. Losses at windows need to be accounted for in a similar fashion.

In these first experiments, we generated the projections by a tedious procedure of translating the flame relative to the optical axis, and then transferring by hand the intensities from the recorder paper to the computer for data reduction. The technique could lend itself to the use of detector arrays with computer interfacing for much more rapid and accurate data acquisition and reduction, although Schlieren effects might have to be considered. Both the experiment and the analytical data reduction procedures are relatively simple, and do not require major investments in equipment or software. Real time temperature measurements should ultimately be possible.

EXPERIMENTAL RESULTS

Ethylene Flame. We first investigated an ethylene-air diffusion flame. The burner consists of a 1 cm diameter fuel tube surrounded by a honeycomb mesh to straighten the air flow. It was extensively investigated with CARS [9] and is not too dissimilar from the flame investigated by Santoro and co-workers, who used thermocouples and laser particle sizing [10-11]. The luminous tip height is 76 mm.

Figure 5 shows sample data from the ethylene flame at a height of about 4 cm above the burner surface. The filament current is set at an arbitrary value which keeps the source intensity on scale; the intensity scale has been calibrated using a pyrometer measurement as discussed in the experimental section. The actual recorded projections show slight asymmetries which have been removed by folding, about the centerline and averaging.

Subtracting the flame projection from the combined projection of the flame plus the source yields the source radiation transmission projection as shown. The path-integrated absorption is about 10%, which should be just adequate for accurate work. This is inverted by eqs. (3-4) to yield the soot absorption coefficient profile shown in Figure 6. Typically 100 adjacent intensities were input to the data reduction programs, with 50-100 angles of incidence assumed. The centerline ray was typically picked to be that midway between the emission annulus peaks. This could be well established to ±1 ray, and the results were not strongly sensitive to this sort of variation. Even though the soot probably

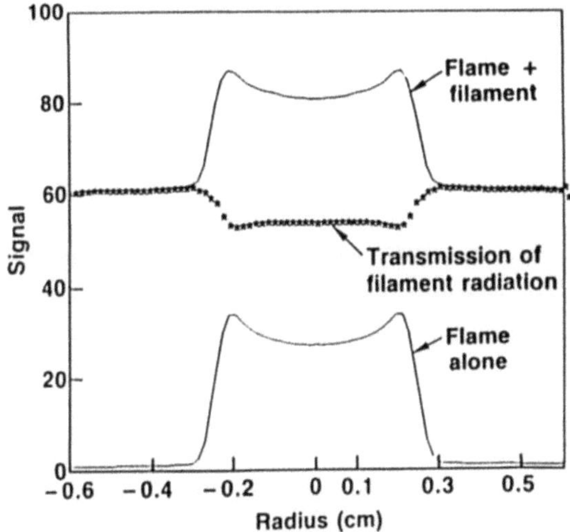

Fig. 5:   Line-of-sight projections in ethylene flame 4 cm above burner surface.

consists of chain-like aggregates which are not in the Rayleigh range, the Rayleigh relationship

$$\alpha_\lambda = C \frac{f_v}{\lambda} \tag{5}$$

between absorption coefficient and volume fraction f_v should still be valid. The theory of scattering by mass fractal aggregates [12-13] shows that the absorption coefficient of an aggregate of spheroids is, to a first approximation, simply the sum of the spheroid absorptions if the latter are in the Rayleigh range, and if multiple internal scattering effects are small.

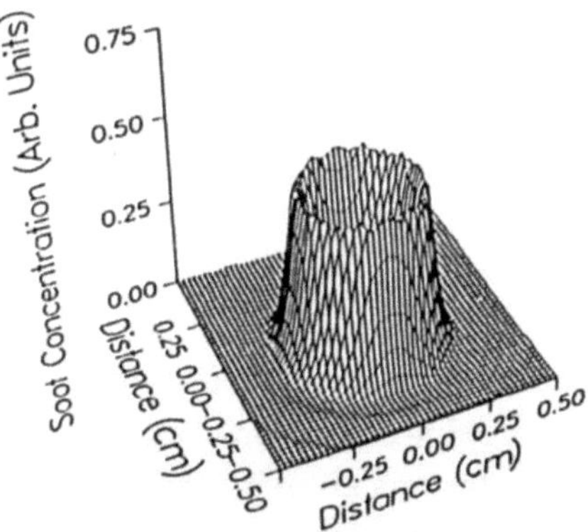

Fig. 6:   Reconstructed soot absorption coefficient/volume fraction profile at height of 4 cm. The approximate volume fraction can be obtained by dividing the absorption coefficient shown (in cm^{-1}) by 4.9 x 10^4.

The constant C is a function of the soot index of refraction; if one assumes the value used in Ref. [10], it has the value 4.9 x 10^4 for α_λ in cm^{-1} and λ in microns. Thus the peak absorption in Figure 6 corresponds to a soot volume fraction of about 10^{-5}. This is only slightly higher than the peak values reported in Ref. [10]. The temperatures deduced by deconvolving the flame projection in Figure 5 are shown in Figure 7. At the centerline, the temperature of ca. 1700 K agrees favorably with the 1650 K value reported for CARS [9]. It is perhaps 100 K higher than thermocouple measurements [10], but the flames are not identical. There is a discontinuity in the temperature at a radius of about .27 cm because the soot concentration drops abruptly to zero.

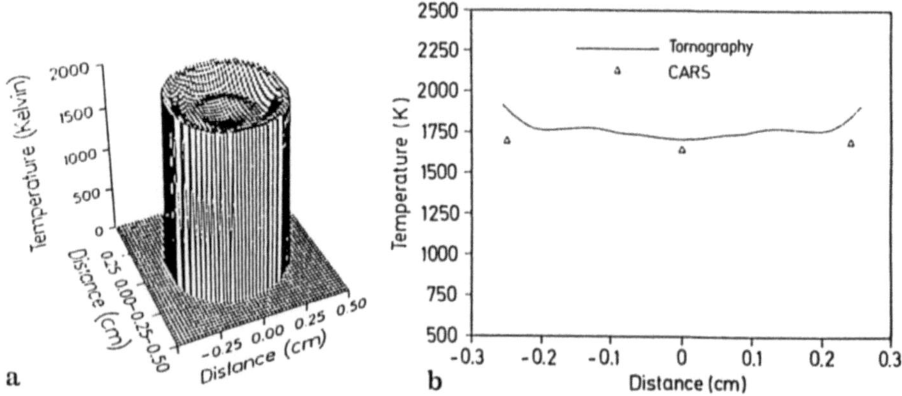

Fig. 7:   a) Reconstructed soot temperature at 4 cm height in ethylene flame; b) comparison of tomographic soot and CARS temperatures (Ref. [9]).

The calculated soot absorption profiles as a function of height graphically depict the evolution of the soot as shown in Figure 8. The soot forms in an annulus low in the flame. Eventually, it fills the center of the flame as the wings contract, and low absorption precludes measurement beyond 6.5 cm.

The annular shape of the soot distribution low in the flame presents a problem for temperature measurement, as depicted in Figure 9. The inferred local emission rate, which is the product of absorption coefficient and Planck function, is plotted for a vertical height of 1.5 cm. As seen, the emission in the center region of the flame is extremely weak, and it is futile to attempt to deduce temperature in such regions. We employed a restriction that the local emission rate be greater than 10% of the peak value at a given height to qualify for temperature determination. With this restriction, the 1.5 cm emission/absorption temperature distribution is as shown. A comparison of the temperature inferred in the sooting annulus at 1.5 cm with that reported in

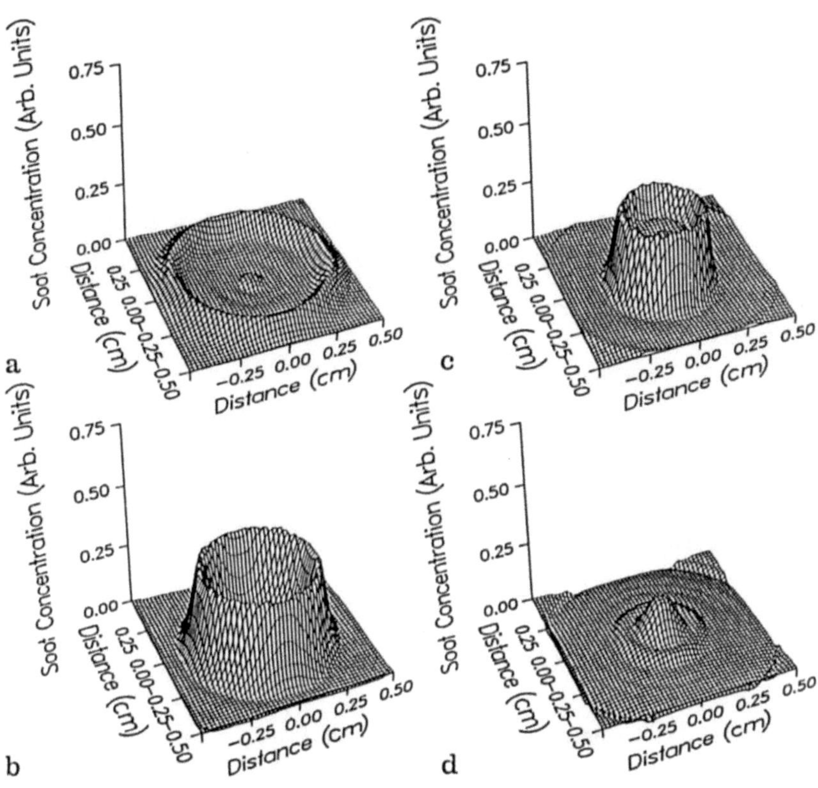

Fig. 8:   Evolution of absorption coefficient/soot volume fraction with height in ethylene flame: A) 1 cm; B) 2.5 cm; C) 5.0 cm; D) 6.5 cm.

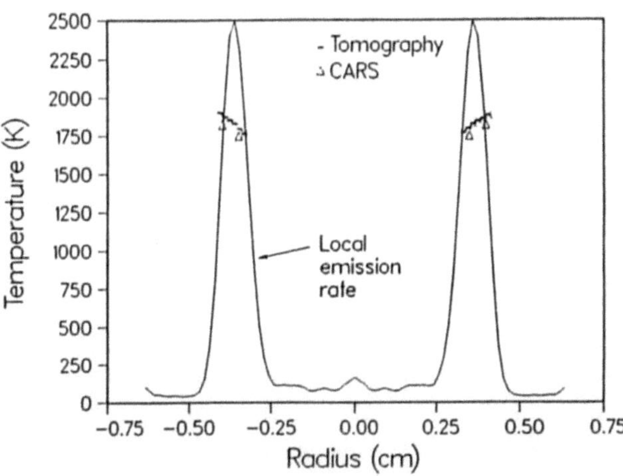

Fig. 9:   Comparison of CARS (Ref. [9]) and emission-absorption temperatures at 1.5 cm height in ethylene flame.

the same region using CARS [9] is also shown in Figure 9; the agreement is within a few percent. Higher in the flame, the soot occupies the center region, and the inferred soot temperatures tend to be relatively flat with values of about 1700 K, as shown in Figure 10 for a height of 5.5 cm. This qualitative behavior seems to be consistent with the thermocouple measurements of Ref. [10], if one restricts attention to the sooting regions. Because of insufficient absorption, we were not able to probe high enough in the flame to see any discernible fall-off of temperature.

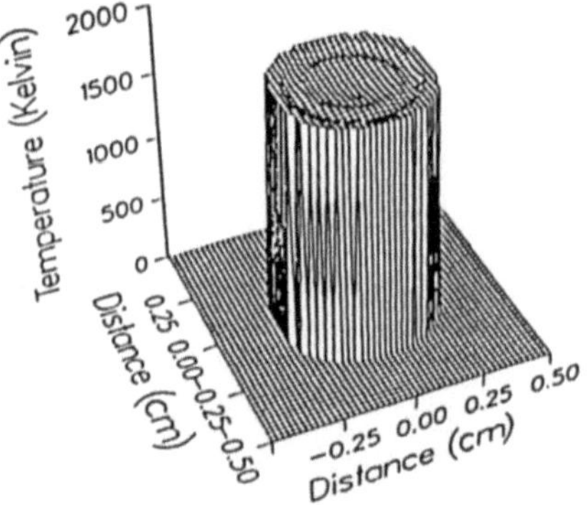

Fig. 10: Reconstructed soot temperature at 5.5 cm in ethylene flame

Substitution of a 100 μ pin-hole for the vertical slit in Figure 4 makes possible conventional Kurlbaum or soot reversal measurements with a longitudinal resolution of 1 mm. As discussed, one needs to make corrections for window and lens transmission losses on the source side of the flame. Further, one needs to account for absorption of the source radiation prior to the focal volume; the maps of absorption coefficient generated in these studies make this possible. Kurlbaum was performed at two heights, 32.5 mm and 40 mm, above the burner surface. At 32.5 mm, the line reversal measurement was 1848°, while the tomography prediction averaged over 1 mm around the centerline was about 1770 K. At 40 mm, the line reversal gave 1794 K, and the tomography predictions averaged over 1 mm gave about 1720 K.

As part of the experiments on the ethylene flame, we performed a measurement of the global, wavelength-integrated radiation. The measurement was performed in the manner of Markstein [14]; a conventional, calibrated radiometer was placed sufficiently far from the flame that the latter could be approximated as a point source. The measured radiative power was then multiplied by 4π divided by the radiometer acceptance angle; the result was 36.3 Watts. For the ethylene flow rate of 3.85 cc/sec, this corresponds to 20%

of the chemical enthalpy release, a result that is quite consistent with the results of Ref. [14]. For an optically thin medium with an absorption coefficient having the Rayleigh form, the theoretical radiated power should be [15].

$$\text{Radiated power} = \int d\lambda \int_{4\pi} d\Omega \int dV\, \alpha_\lambda B_\lambda(T)$$
$$= C' \int dV\, f_v T^5 \tag{6}$$

where C' has the value 2.97 x 10^{-10} (m.k.s) for the index of refraction of Ref. [10], and dV denotes a volume integral. Dispersion of the index has been neglected. Performing the above integration with our volume fractions and temperatures deduced from one micron measurements gave a value of 36.0 Watts. Of course, reasonable agreement is expected because our measurements are based on radiation at a particular wavelength, but the very precise agreement obtained suggest that the Rayleigh absorption law must be essentially valid, and that soot volume fraction is the concentration parameter needed for radiation prediction.

Iso-octane flame. We also employed the tomography technique in a highly sooting, iso-octane/air diffusion flame [16]. Figure 11 shows the deconvolved soot temperature at a height of 3 cm. Conventional Kurlbaum measurements were made at the centerline at heights of 2 and 3 cm. The soot absorption at 2 cm was weak and we could not measure a temperature tomographically there. However, the agreement at 3 cm was very good, as seen; Kurlbaum gave 1720K and tomography 1705 K.

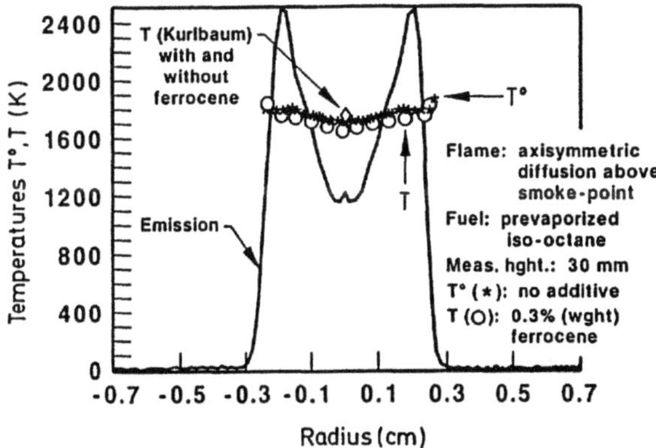

Fig. 11: Deconvolved soot temperature profiles in iso-octane/air diffusion flame at height of 3 cm. Kurlbaum measurement at centerline is indicated.

The deficiencies of the technique in the annular sooting zones low in the flames suggest that seeding to achieve a more uniform distribution of absorption might be a worthwhile idea. All work would be performed at the wavelength of a seed resonance (for example the sodium D line, if that were the seed), and the amount of seed introduced could be adjusted to achieve the desired moderate absorption.

CONCLUSIONS

We have obtained good diagnostic results in sooting flames using emission/absorption tomography. For the first time, a self-consistent examination of sooting flame radiation has been performed, with a comparison of measured radiative loss and theoretical predictions based on measured temperature and soot concentration. Although the highly annular, optically thin flames investigated were not optimal for a demonstration of temperature tomography, results were obtained which agreed fairly well with CARS and Kurlbaum. The technique is experimentally simple, and only uncomplicated, fast data reduction procedures are required. It might lend itself to the use of detector arrays for very fast 2D and 3D data acquisition and reduction. The main drawback of the approach is that measurements cannot be made in regions of weak absorption; extreme optical thickness would also present a problem. The accuracy of the temperature measurements, probably now at the several percent level, might be improved by using a laser to make the crucial absorption coefficient measurement. Seeding to achieve a more uniform, moderate level of absorption should also improve the accuracy. Inversions of synthetic projections give reason to believe that the technique can be extended to moderately optically thick axisymmetric media. In summary, there appears to be considerable promise to temperature measurement using tomographic deconvolution of line-of-sight radiation, and the technique is deserving of further attention and refinement.

ACKNOWLEDGEMENTS

The authors would like to thank Joseph Sangiovanni and Robert Santoro for their help in clarifying certain ideas developed in this paper. They would also like to thank Kathi Wicks for her help in preparing the manuscript.

This research was sponsored in part by the Air Force Office of Scientific Research (AFOSR) and the Environics Division of the Air Force Engineering

Services Center (AFESC, Tyndall AFB, FL) under AFOSR contract F49620-86-C-0054. The authors wish to thank Julian M. Tishkoff (AFOSR) and Paul Kerch (AFESC) for their interest in and support of this research.

REFERENCES

1. Porter R.W.: Numerical Solution for Local Emission Coefficients in Axisymmetric Self-Absorbed Sources. SIAM Review 6 (1964) 228.

2. Kuhn G., Tankin R.S.: Spectroscopic Measurements to Determine Temperature and Carbon Particle Size in an Absorbing Propane Diffusion Flame. J.Q.S.R.T. 8 (1968) 1281.

3. Semerjian H.G., Ray S.R., Santoro R.J.: Laser Tomography for Diagnostics in Reacting Flows. AIAA-82-0584 (1982).

4. Emmerman P.J., Goulard R., Santoro R.J., Semerjian H.G.: Multiangular Absorption Diagnostics of a Turbulent Argon-Methane Jet. J. Energy 4 (1980) 70.

5. Uchiyama H., Nakajima M., Yuta S.: Measurement of Flame Temperature Distribution by IR Emission Computed Tomography. Appl. Opt. 24 (1985) 4111.

6. Solomon P.R., *et al.*: FT-IR Emission/Transmission Spectroscopy for In Situ Combustion Diagnostics. in Twenty-First Symposium (International) on Combustion (The Combustion Institute, Pittsburgh, 1986) 1763-1771.

7. Ramachandran G.N., Lakshminarayanan: Three-dimensional Reconstruction from Radiographs and Electron Micrographs: Application of Convolutions instead of Fourier Transforms. Proc. Nat. Acad. Sci. USA 68 (1971) 2236.

8. Shepp L.A., Logan B.F.: The Fourier Reconstruction of a Head Section. IEEE Trans. on Nuc. Sci., NS-21 (1974) 21.

9. Boedeker L.R., Dobbs G.M.: Temperature and Soot Correlations in Sooting, Laminar Diffusion Flames. UTRC (1985) 85-51.

10. Santoro R.J., Semerjian H.G., Dobbins R.A.: Soot Particle Measurements in Diffusion Flames. Combust. Flame, 51 (1983) 203.

11. Santoro R.J., Semerjian H.G.: Soot Formation in Diffusion Flames: Flow Rate, Fuel Species, and Temperature Effects. in Twentieth Symposium (International) on Combustion (The Combustion Institute, Pittsburgh) (1984) 997-1006.

12. Berry M.V., Percival I.C.: Optics of Fractal Clusters Such as Smoke. Optica Acta, 33 (1986) 577.

13. Mountain R.D., Mulholland G.W.: Light Scattering from Simulated Smoke Aggregates. Langmuir, in press.

14. Markstein G.H.: Relationship Between Smoke Point and Radiant Emission from Buoyant Turbulent and Laminar Diffusion Flames. in Twentieth Symposium (International) on Combustion (The Combustion Institute, Pittsburgh) (1984) 1055-1061.

15. Roper F.G.: Soot Escape from Diffusion Flames: A Comparison of Recent Work in this Field. Combust. Sci. Technol. 40 (1984).

16. Bonczyk P.A.: Investigation of Fuel Additive Effects on Sooting Flames. in Final Technical Report under Contract F49620-86-C-0054 for Air Force Office of Scientific Research (July, 1989).

CHAPTER 4

MEASUREMENTS OF RADIATIVE AND CONVECTIVE HEAT TRANSFER

Invited Lecture: **REVIEW OF NON-INTRUSIVE MEASURING PROCEDURES APPLIED TO RADIATIVE GASES**

Hans J. Pfeifer
French-German Research Institute
F 68301 Saint-Louis, France

ABSTRACT

A review will be given on the various optical investigation methods to study radiative flows. Because it is not possible due to the limited space available to describe the physical background of these procedures emphasis is placed on the most recent progress which has been achieved on the basis of more efficient laser sources and computers.

Details will be given on new developments in different kinds of velocity measurements, i.e. particle image velocimetry, laser two-focus, and laser Doppler anemometry as well as on spectroscopic procedures like Raman and CARS spectroscopy.

INTRODUCTION

It is certainly not an easy task to give a review of all the non-intrusive measuring procedures which have been developed so far for application to radiative gases. In particular, if non-optical methods like acoustic sensors in the vicinity of the flow or millimeter waves are taken into account in detail, the size of such a review paper would go far beyond the usual size of a publication of this type.

Therefore, the present paper concentrates mainly on optical methods. But even this approach risks to lead to nothing but a list of existing procedures. This is in particular true if we consider all the classical methods like shadow photography, schlieren photography and the various kinds of interferometry, i.e. Mach-Zehnder, Michelson and differential interferometry. By the simultaneous use of laser radiation and small-band receiving techniques all these methods can be applied to radiative gases even at high luminosity as they could be applied to non-radiative gas flows before lasers have been available. In addition, lasers allow to record photographs in cw-mode as well as in pulsed mode. In the latter case it is possible to record movies with sufficient frame repetition rates to detect high-frequency phenomena.

All this, however, belongs in principle to measuring procedures which have been well-known for a long time and which are subjects of various excellent handbooks.

These methods are very sophisticated. Their potentials and their limits are well evaluated so that they are still widely and successfully applied to all kinds of fluid flows including radiative gas flows. It is obvious that all this can not be detailed here.

In the following, the most modern optical techniques will be described to study radiative gases.

Here, emphasis will be placed on procedures allowing to measure gas temperatures, species concentration, and flow velocities.

If these three quantities are known simultaneously, it is possible to describe a flow completely under the condition that it is not a two-phase or a multi-phase flow and that plasma effects do not play a noticeable role.

The latter two cases are not considered in the present publication.

In addition most of the procedures allow to determine not only mean values of the respective quantity but also the time history of fluctuations. This is mandatory for characterization of turbulent and instationary flows. For each of the techniques it will be outlined to which extent they are capable to meet this requirement.

It is to mention that the present review paper is based on similar papers [1-5] which have been published recently or a few years ago and that it focuses on the progress which has been achieved in the meantime.

We will start with the description of techniques to measure gas velocities.

First, most recent advances in LDA techniques will be outlined, followed by new aspects of two-focus velocimetry and of particle image velocimetry.

After that, active, passive, and semi-passive spectroscopic methods will be considered including spontaneous Raman scattering and, in particular, Coherent anti-Stokes Raman Scattering (CARS).

The paper will conclude with an evaluation of the different methods and a comparison of their capabilities as far as such a comparison is reasonable or possible.

LASER DOPPLER ANEMOMETRY (LDA)

Laser Doppler anemometry may now be considered to be a classical measuring procedure for flow field studies. Its potential is well known and well understood. A number of handbooks describes the principle and applications of this method. Obviously, it cannot be the aim of the present paper to compete with them.

Instead, a review of the most recent developments and progress in this field will be given.

During the last years, laser Doppler anemometry took advantage of the substantial improvement of commercially available material. This ranges from high laser power, much higher than was available a few years ago, to dedicated and very sophisticated optical components and data acquisition systems which were not available until recently, both with respect to hardware and software.

Although for some applications these instruments are quite complex, there is no more need for experts in optics or electronics to operate them.

The three most important advances which can currently be observed are the use of fiber optics, the implementation of laser diodes, and the application of frequency domain data acquisition systems. All of them may be very attractive for applications to radiative flows.

First, fiber optics are considered. Due to the high laser power available with modern argon lasers, even losses of 50% in optical fibers may be tolerated without significant loss in signal quality or data rate.

All the commercially available system produce an excellent beam quality in the probe volume [6, 7].

There are two main advantages of fiber optics with respect to their application to radiative flows. First, most of the optics, including the laser, may be placed far away from the flow facility. In many cases this makes extensive measures against noise and vibration unnecessary, which would be mandatory in the other case. The second advantage is that a traverse system which is necessary to scan a flow field is easy to build.

The second progress consists in laser diode based systems. Although these systems are not yet very much advanced, they offer an extremely promising perspective with respect to applications to radiative flows. First, these systems are extremely insensitive to noise and vibration. Secondly, they are much smaller sized, as compared with gas laser-based systems and thirdly, they allow to reduce the background luminosity emanating from the radiative gases to nearly zero if the diodes are operated in pulsed mode at very high frequencies and if the receiving optics are gated accordingly. So one can expect that in the future small and compact optical LDA sensors will be available to scan radiative flows [8, 9, 10].

The third recent progress in LDA development is the introduction of frequency domain data acquisition systems. Although the idea is not new, the availability of FFT integrated circuits or similar devices allows to recover noisy signals in a much more reliable way than is possible with classical data acquisition systems. Of course, this is of very high interest in the usually hostile environment of radiative flows and one can expect the application of LDA systems to flow situations which have not been -or hardly- accessible so far.

The advantages of this progress have been taken into account in the study of a number of combustion and reactive flows, i.e. [11-16].

LASER-TWO-FOCUS SYSTEM

Laser Two-Focus-Systems have to be considered as a competitive method to measure flow velocities. Their principle is even simpler than that of the fringe type laser Doppler anemometer [17-20].

Their advantages over LDA systems can briefly be summarized in their easier design, their by orders of magnitude higher light intensity in the two spots, which makes them much more suited for backscattering applications and the extremely high precision attainable with respect to the flow angle.

On the other hand, due to the lag of a device comparable to Bragg cells in LDA instruments, the application of two-focus systems is restricted to flows of limited turbulence intensity. These limits have been the subject of several experimental studies. The results disagree, but 30% of turbulence intensity is probably the absolutely highest level which can be handled.

Furthermore, it has to be taken into account that, instead of validating signals of individual particles, two-focus systems validate ensembles of signals, only. This excludes the possibility of getting spectra of velocity fluctuations or similar quantities, at least at the present stage. However, it would be a mistake to underestimate the potential of two-focus velocimeters, in particular with respect to radiative flows. The high concentration of the laser light makes it possible to remove light from the self-luminosity by the use of very narrow optical filters more easily than is possible in LDA arrangements.

So, as long as the mean velocity value along with the turbulence intensity in radiative flows of low turbulence are of interest only, the two-focus system should be preferred to LDA systems. This is particularly true for situations where backscattering arrangements are mandatory.

PARTICLE IMAGE VELOCIMETRY

A very promising method to study combustion flows is Particle Image Velocimetry (PIV). Although this method is not new, it takes more and more advantage of the progress in computer development [21-24].

PIV is in principle very simple: a laser, in most cases a ruby laser [25], emits two pulses with a selectable interval between these pulses. The interval has to be adapted to the actual flow velocity. Optics form light sheets out of these pulses which illuminate particles travelling with the flow.

On photographic plates the resulting double images are recorded and evaluated thereafter.

The advantage of using laser instead of flashlamps is the fact that the narrow band width of laser light allows to apply very narrow interference filters in front of the camera or the photographic plate to remove most of the self-luminosity light of radiative gas flows.

Mainly two different ways to evaluate the recorded images have been developed in the near past.

The first one is adapted to high particle density and is based on an optical Fourier transform. This method is very fast but at the expense of accuracy.

In cases where the particle density is low enough to identify individual pairs of particle spots, image processing techniques are applied more and more successfully. They allow, within acceptable computing times, to obtain a map of two-dimensional flow velocity vectors of the whole observation field.

In addition, since the field is stored in the computer, very effective operations with the data can be performed. For example, the mean velocity value can be subtracted from the individual values to give just the variations or different kinds of false color representation on the computer screen can be monitored which clearly show the areas of different velocities and which give to the observer a quick survey of the total flow field recorded.

On the other hand, it must be noticed that PIV records the flow field at one single instant only. All turbulent quantities, however, are described by time integrals from minus to plus infinity.

Obviously, all existing measuring techniques are only able to give an estimate of this integral, but to give at least an idea of this integral, a number of images have to be recorded in PIV [26]. Further improvements in computer capabilities makes us think that this goal can be reached in a relatively near future.

NON-INTRUSIVE TEMPERATURE MEASUREMENTS

All non-intrusive temperature measurement techniques rely upon the spectroscopic characteristics of the gases to be studied. They may be based on self-luminosity radiation or on excitation by an external light source, either a conventional light source like a light bulb or a flashlamp or by laser light.

In the first case, i.e. conventional sources, progress has been achieved recently by the implementation of image processing systems and of computers with high performance, as will be outlined later. In the second case, i.e. laser-based spectroscopic systems, the availability of lasers with higher power, higher stability and with narrower band width made it possible to build instruments which now give insight into radiative flows, which was not possible in this way before.

Unfortunately, it must be stated that all these measuring procedures are either integrating along the optical path through the gas under study or they are not capable to follow the time history of the temperature fluctuations. There is a certain chance to overcome these problems in the future, as will be outlined later.

COHERENT ANTI-STOKES RAMAN SCATTERING (CARS)

Among the non-intrusive temperature measuring procedures, CARS is the most promising one. It would go far beyond the frame of the present review to detail the physical background and the potential of CARS systems. This has been documented elsewhere [27-39].

CARS allows to determine simultaneously the gas temperature and species concentration. Currently it still suffers from a phenomenon similar to the one occurring in Particle Image Velocimetry. It is able to give information on the above-mentioned flow quantities at one single given instant, or at data rates which, in most cases, can not follow by far the fluctuation histories.

Again, the rapid progress in laser development gives some hope that these problems may be overcome. The promising perspectives for the near future are systems designed to measure the species concentration only, because in this case it is sufficient to record the time history of one or two photodetectors only, whereas to measure both the temperature and the concentration simultaneously would require to record a complete spectrum. This, of course, will be a challenge for the optics as well as for data acquisition and data reduction. However, regarding the rapid development, in particular with respect to metal vapor lasers and to two-dimensional image processing systems, there is a perspective well above zero to develop systems enabling us to measure the time history of both the local temperature and the concentration. Along with velocity measuring procedures which currently are well advanced, this will give us the possibility to describe the flow nearly completely, at least locally.

SPONTANEOUS RAMAN SPECTROSCOPY

Spontaneous Raman Spectroscopy, and, to some extent Rayleigh Scattering [40, 41] may be considered as a competitive method to CARS for gas density measurements and for some studies it has been combined with CARS [42-43]. In contrast to CARS, the optical set-up is much more simple because only one laser source is needed, the wave length of which is nearly arbitrary [44-48]. In addition, the amplitude of the recorded signal is linearly dependent on the number density of the gas molecules under investigation. The linear ratio between the recorded signal amplitude and the concentration, however, may also be regarded as a problem because the Raman scattering cross-section of all gas molecules is extremely low.

Even with the best spectroscopic methods it is difficult to remove stray light from the recorded signals. This stray light may be caused mainly by Mie scattering of residual particles.

So the spectroscopic instruments have to be turned to very narrow optical band widths which, in turn, requires long recording times for a given accuracy if cw laser radiation is applied. In pulsed mode the same statement is true as it has been made for CARS before, i.e. the pulse repetition rate is not high enough to record the time history of concentration variations. Nevertheless, spontaneous

Raman spectroscopy can be considered as a useful tool to measure-time averaged gas concentrations. It should be added that, by means of particular correlation methods, it even gives information on turbulent quantities.

LASER-INDUCED FLUORESCENCE

The availability of turnable lasers in the ultraviolet region with sufficient power makes it possible to build up systems based on laser-induced fluorescence. They are mentioned here because they allow to record both, temperature and concentration, they have a high resolution at low gas concentration, and they are able to react to only one specific species of molecules. However, there are still some problems with calibration to determine absolute values of the concentration and there are some limits with respect to higher pressure. Nevertheless, in the future, laser-induced fluorescence may be considered a competitive candidate to CARS [49-53].

This is particularly true if a direct comparison is made between these two methods [54].

LINE REVERSAL METHOD

Finally, the line reversal method should be mentioned at least [55-58]. Although it is one of the elder spectroscopic procedures to measure temperatures in radiative gases, it is capable to take advantage of modern light sources, detectors, and data acquisition and reduction systems. It relies upon self-luminosity and absorption by small tracer particles embedded in the flow. The particles having the same temperature as the surrounding gas molecules emit a low-level continuous optical spectrum. Out of this spectrum, the color temperature may be derived on the basis of the Wien-Kirchhoff law. Although this method integrates along the optical path through the flow, average quantities may be obtained by means of Abel inversion algorithms on fast computers.

CONCLUSIONS

Although no completely new measuring procedures for the study of radiative flows have been developed in the past few years, substantial progress has been achieved with the existing methods. This is true for all the methods mentioned in the present paper. The basic background for this progress is the availability of much more powerful lasers and electronic equipment like extremely fast A/D converters with large memories and of much more efficient computer hard- and software.

All this together allows the experimenter to get reliable and accurate data from nearly all kinds of radiative flows and on nearly all physical quantities of interest.

However, all those who work in this field are aware that the main problem does not lie in acquiring data but in reducing them and in getting meaningful representation of the result.

So, in the future, more emphasis is expected to be put on the software than on the hardware.

REFERENCES

1. Eckbreth A.C.: Spatially Precise Laser Diagnostics for Combustion. ICIASF'81 Record (1981) 71-89.

2. Goulard R.: Optical Measurements of Thermodynamic Properties in Flow Fields. in AGARD Conference Proceedings, CP 193, paper No. 13 (1976).

3. Witze P.O., Dyer T.M.: Laser Measurement Techniques Applied to Turbulent Combustion in Piston Engines. Experiments in Fluids 4 (1986) 81-92.

4. Attal B., Druet S., Bailly R., Péalat M., Taran J.P.: Techniques RAMAN d'études des écoulements et des flammes par laser. Spectra 2000 7(54) (1979) 41-50.

5. Laurendeau N.M.: Temperature Measurements by Light Scattering Methods. Prog. Energy and Comb. Science 14. Pergamon Press (1988) 147-170.

6. TSI Incorporated, P.O. Box 64394 St. Paul, MN 55164, USA, Brochure "Colorburst" (1990).

7. DANTEK, DK 2740 Skovlunde, Denmark, Brochure Fiber Flow Anemometers (1990).

8. Domnick J., Durst F., Müller R., Naqwi A.: Improved Optical Systems for Velocimetry and Particle Sizing using Semiconductor Lasers and

Detectors in Proceedings Fifth International Symposium on Applications of Laser Techniques to Fluid Mechanics, Lisbon (1990).

9. Dopheide D., Faber M., Reim G., Taux G.: Laser and Avalanche Diodes for Velocity Measurement by Laser Doppler Anemometry. Experiments in Fluids 6 (1988) 289-297.

10. Damp S., Pfeifer H.J.: Anémomètre laser à effet Doppler miniaturisé à diode laser, AAAF. in Proc. 26ème Colloque d'Aérodynamique Appliquée, Toulouse (1989) 23-25 octobre.

11. Hsieh W.H., Chuang C.L., Yang A.S., Cherng D.L., Yang V., Kuo K.K.: Measurement of Flow field in a Simulated Solid-Propellant Ducted Rocket Combustor Using Laser Doppler Velocimetry. in Proceedings AIAA/ASME/SAE/ASEE 25th Joint Propulsion Conference. AIAA-89-2789 (1989).

12. Chigier N.: Velocity Measurements in Inhomogeneous Combustion Systems. Combustion and Flame 78 (1989) 129-151.

13. Labbe J., Jerot A.: Some Problems and Solutions in the Application of Laser Velocimetry to Continuous Combustion. ONERA, 29, avenue de la Division Leclerc, 92322 Chatillon, France, Report T.P. 200 (1987).

14. Gould R.D., Stevenson W.H., Thompson D.: Turbulence Characteristics of an Axisymmetric Reacting Flow. NASA Contractor Report 4110 (1988).

15. Durão D.F.G., Mendes-Lopes J.M.C.: Laser Velocimetry for Combustion. in Proceedings NATO Advanced Study Institute. Kluwer Academic Publishers (1989) 151-177.

16. Cheng R.K., Shepherd I.G., Gökalp I.: A Comparison of the Velocity and Scalar Spectra in Premixed Turbulent Flames. Combustion and Flame 78 (1989) 205-221.

17. Schodl R.: A Laser-Two-Focus (L2F) Velocimeter for Automatic Flow Vector Measurements in the Rotating Components of Turbomachines. Transactions of the ASME, 102 (1980) 412-419.

18. Smart A.E., Wisler D.C., Mayo W.T.: Optical Advances in Laser Transit Anemometry. Transactions of the ASME 103 (1981) 438-444.

19. Kugler P., Langer G.: Laser Anemometry Techniques for Hot Flows. in NASA Technical Memorandum, TT 20077 (1987).

20. Förster W., Schodl R., Beversdorff M.: Design and Experimental Verification of 3-D Velocimeters Based on the Laser-2-Focus Technique. in Proceedings of the 5th Intern. Symp. on Applications of Laser Techniques to Fluid Mechanics, Paper 8.4. Instituto Superior Técnico, Lisbon, Portugal (1990).

21. Reuss D.L., Adrian R.J., Landreth C.C., French D.T., Fansler T.D.: Instantaneous Planar Measurements of Velocity and Large-Scale Vorticity and Strain Rate in an Engine Using Particle Image Velocimetry. SAE Technical Paper Series, 890616 (1989).

22. Reuss D.L., Adrian R.J., Landreth C.C.: Two-Dimensional Velocity Measurements in a Laminar Flame Using Particle Image Velocimetry. Combust. Sci. and Tech. 67 (1989) 73-83.

23. Höcker R., Kompenhans J.: Application of Particle Image Velocimetry to Transonic Flows. in Proceedings of the 5th Intern. Symp. on Applications of Laser Techniques to Fluid Mechanics, Paper 15.1. Instituto Superior Técnico, Lisbon, Portugal (1990).

24. Reuss D.L., Bardsley M.,. Felton P.G, Landreth C.C., Adrian R.J.: Velocity, Vorticity and Strain Rate ahead of a Flame Measured in an Engine Using Particle Image Velocimetry. SAE Technical Paper Series 900053 (1990).

25. Carts A.: Ruby Lasers Shine on Laser Focus World (July 1990) 83-91.

26. Cenedese A., Palmieri G., Romano G.P.: Turbulent Intensity Evaluation with PIV. in Proceedings of the 5th Intern. Symp. on Applications of Laser Techniques to Fluid Mechanics, Paper No. 15.5. Instituto Superior Técnico, Lisbon, Portugal (1990).

27. Eckbreth A.C.: Invited Paper - CARS Applications to Combustion Diagnostics. SPIE, Manufacturing Applications of Lasers, 621 (1986) 116-124.

28. Eichhorn A.: Aufbau einer CARS-Apparatur zur Untersuchung der Durchführbarkeit von Dichte- und Temperaturmessungen mit Hoher Zeitauflösung. French-German Research Institute (ISL), 68301 Saint-Louis France, Report R117/87 (1987).

29. Marie J.J., Cotterau M.J.: Mesure de température par D.R.A.S.C., Application au moteur thermique. Entropie 135 (1987) 64-67.

30. Antcliff R., Jarrett O. Jr: Multispecies Coherent Anti-Stokes Raman Scattering Instruments for Turbulent Combustion. Rev. Sci. Instrum. 58(11) (1987) 2075-2080.

31. Farrow R., Trebino R., Palmer R.: High-Resolution CARS Measurements of Temperature Profiles and Pressure in a Tungsten Lamp. Appl. Optics 26(2) (1987) 331-335.

32. Yueh F.Y., Beiting E.J.: Simultaneous N2, CO, and H2 Multiplex CARS Measurements in Combustion Environments Using a Single Dye Laser. App. Optics 27(15) (1988).

33. Dreier T., Lange B., Wolfrum J., Zahn M.: Determination of Temperature and Concentration of Molecular Nitrogen, Oxygen and Methane with Coherent Anti-Stokes Raman Scattering. Appl. Phys. B45 (1988) 183-190.

34. Boquillon J.P., Péalat M., Bouchardy P., Collin G., Magre P., Taran J.P.: Spatial Averaging and Multiplex Coherent Anti-Stokes Raman Scattering Temperature-Measurement Error. Optics Letters 13(9) (1988) 722-724.

35. Scholten T., Lucassen W., De Mul F., Grewe J.: Compensating Pulse-to-Pulse Fluctuations and Increasing Spectral Reproducibility of Phase-Matched CARS Measurements. Appl. Optics 28(15) (1988) 3225-3232.

36. Goss L.P., Trump D.D.: Combined CARS/LDA Instrument for Simultaneous Temperature and Velocity Measurements. Experiments in Fluids 6 (1988) 189-198.

37. Aldén M., Bengttsson P., Edner H., Kröll S., Nilsson N.: Rotational CARS: A Comparison of Different Techniques with Emphasis on Accuracy in Temperature Determination. Appl. Optics 28(15) (1989) 3206-3219.

38. Kröll S., Aldén M., Bentsson P.E., Löfström C.: An Evaluation of Precision and Systematic Errors in Vibrational CARS Thermometry. Appl. Phys. B49, (1989) 445-453.

39. Kreutner W., Meier W., Plath I.,. Stricker W, Woyde M.: Optical Thermometry in Flames Using CARS Laser Optoelektronik 22(2) (1990).

40. Ötügen M.V., Namer I.: Rayleigh Scattering Temperature Measurements in a Plane Turbulent Air Jet at Moderate Reynolds Numbers. Experiments in Fluids 6 (1988) 461-466.

41. Cladnick P.G., LaRue J.C., Samuelsen G.S.: Simultaneous Optical Measurement of Velocity and Temperature Using Laser Anemometry and Rayleigh Scattering. in Proceedings of the 5th Intern. Symp. on Applications of Laser Techniques to Fluid Mechanics, Paper 8.2. Instituto Superior Técnico, Lisbon, Portugal (1990).

42. Lederman S., Posillico C.: Unified Spontaneous Raman and CARS System.

43. Stricker W., Kreutner W., Just T.: Simultaneous Temperature Measurements with Raman and CARS in Laminar Flames. AGARD CP399. Advance Instrumentation for Aero-Engine Components. paper No. 1 (1986).

44. Leipertz A.: Non-Destructive Probing of Free Jets Using cw-Laser Raman Spectroscopy. Optics and Laser Technology (February 1981) 21-25.

45. Bobin L., Brom G.: Analyse des fluctuations turbulentes de la masse volumique au moyen de la diffusion Raman spontanée. French-German Research Institute (ISL), 68301 St.-Louis, France. Report R103/89 (1989).

46. Bobin L., Brom G.: Mesure par diffusion Raman spontanée de la masse volumique locale dans un écoulement gazeux. French-German Research Institute (ISL), 68301 St.-Louis, France. Report R118/89 (1989).

47. Chabay I., Rosasco G.J., Kashiwagi T.: Species-Specific Raman Spectroscopic Measurements of Concentration Fluctuations in Unsteady Flow. J. Chem. Phys. 70(9) (1979) 4149-4154.

48. Exton R.J., Hillard M.E.: Raman Doppler Velocimetry: A Unified Approach for Measuring Molecular Flow Velocity, Temperature, and Pressure. Appl. Optics 25(1) (1986) 14-21.

49. Salmon J.T., Laurendean N.M.: Concentration Measurements of Atomic Hydrogen in Subatmospheric CH/0/Ar Flat Flames. Combustion and Flame 74 (1988) 221-231.

50. Meier U., Bittner J., Kohre-Höinghaus K., Just T.: Discussion of Two-Photon Laser-Excited Fluorescence as a Method for Quantitative Detection

of Oxygen Atoms in Flames. in Proc. 22nd Symp. on Combustion. The Combustion Inst. (1988) 1887-1896.

51. Eckbreth A.C.: Laser Diagnostics for Combustion Temperature. Abacus Press, Cambridge, Mass, USA (1988) 301-361.

52. Laurendeau N.M., Goldsmith J.E.M.: Comparison of Hydroxyl Concentration Profiles using Five Laser-Induced Fluorescence Methods in a Lean Subatmospheric-Pressure H/0/Ar Flame. Combust. Sci. and Tech. 63 (1989) 139-152.

53. Crosley D.R.: Semiquantitative Laser-Induced Fluorescence in Flames. Combustion and Flame 78 (1989) 153-167.

54. Lawitzki A., Plath I., Stricker W., Bittner J., Meier U., Kohrse-Höinghaus K.: Laser-Induced Fluorescence Determination of Flame Temperatures in Comparison with CARS Measurements. Appl. Phy. B50 (1990) 513-518.

55. Klingenberg G., Mach H.: Investigation of Combustion Phenomena Associated with the Flow of Hot Propellant Gases - I: Spectroscopic Temperature Measurements Inside the Muzzle Flash of a Rifle. Combustion and Flame 27 (1976) 163-176.

56. Mach H., Werner U., Coquerelle P.: Vergleich zweier Verfahren zur spektroskopischen Messung von Flammentemperaturen. French-German Research Institute (ISL), 68301 Saint-Louis, France. Report No. 605/80 (1980).

57. Mach H., Hensel D., Werner U., Masur H.: Durchfürhrbarkeitsstudie zum Bau einer Anlage zur Simulation der Treibgasströmung in einem Waffenrohr. Teil IV. ISL Report R135/86. French-German Research Institute (ISL), 68301 Saint-Louis, France (1986).

58. Eichhorn A., Mach H.: Time Resolved Temperature Measurement in Flows with Cylindrical Symmetry. J. Ballistics, 9(3) (1987) 2311-2334.

A NEW INFRARED EMISSION-ABSORPTION PYROMETER FOR DUST-AIR EXPLOSIONS TEMPERATURE MEASUREMENTS

L. Teixeira de Lemos
Dep. Eng. Industrial
Instituto Politécnico da Guarda
6300 Guarda — Portugal

and

R. Bouriannes
Lab. d'Energétique et de Détonique
E.N.S.M.A.
86034 Poitiers Cedex — France

ABSTRACT

In the present work a new near infrared monochromatic emission-absorption pyrometer, specially designed for dust combustion temperature measurement, is described.

INTRODUCTION

Fine solid combustible particles, when suspended and dispersed in the air, may burn in desired situations (furnaces or boilers) or in undesired conditions (accidental fires or explosions).

In order to improve combustion or to prevent and/or minimize explosion effects, the flame propagation mechanism must be well known.

One of the most important combustion parameters is the flame temperature: in the case of dust flames its knowledge may be an useful contribution to the study of radiative transfer phenomena.

Experimental temperature diagnosis include optical pyrometry.

In the present case a new near infrared monochromatic emission absorption pyrometer, specially designed for dust combustion temperature measurement, is described.

In order to test the performance of the optical pyrometer, the isochoric combustion of starch-air mixtures was studied in a 20 litre spherical chamber with central electric ignition (described in detail elsewhere [1]).

STUDIED MIXTURES

The starch-air suspension (chemical starch particles $(C_6H_{10}O_5)_n$ of about 20 μm in diameter) is generated by elutriation over a fluidized bed.

For this mixture, the stoichiometric composition corresponds approximately to a concentration of 230 g.m^{-3}.

The combustion temperature was determined in the range of equivalence ratios between 0.6 and 3.0.

TEMPERATURE MEASUREMENTS

The temperature measurements were performed with a monochromatic emission-absorption pyrometer (described in detail elsewhere [2]), in which the electronically modulated source signal is easily separated from the flame's signal.

This optical pyrometer was specially designed for measuring solid particles temperature and its wavelength of work is 940 nm.

The optical geometry consists (*see* Figure 1) of a central hot medium growing with time (the burnt products B), surrounded by a cold medium (cold mixture F), the external radius of which (r_0) coincides with the chamber's radius.

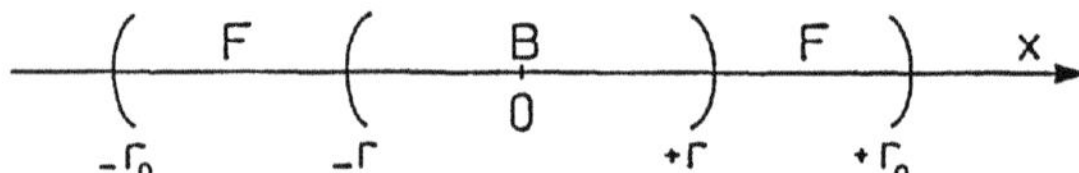

Fig. 1:   Flame's geometry.

<u>Optical method.</u> The pyrometry by generalized emission-absorption is a variation of the reversal technique [3]. This method permits the temperature measurement of flames when the radiation emitted varies rapidly with time.

A detector must collect separately three monochromatic signals:

-   R emitted by the reference source at the brightness temperature TR;
-   B emitted by the flame alone;
-   RB emitted by the reference source through the flame;

The ratio Γ determined by experience, may be written as:

$$\Gamma = \frac{B}{B + R - RB} \qquad (1)$$

In the case of a spherical flame, developing from the central ignition point, and according to the monochromatic radiation intensity equation (*see* [4]), it may be written:

$$R = C.L_o \tag{2}$$

$$B = C.\frac{kab}{keb} . \tau_f (1 - \tau b) . L_B^o \tag{3}$$

$$RB = C.\left[\tau_f^2 . \tau_b . L_o + \frac{kab}{keb} . \tau_f . (1-\tau_b) . L_B^o\right] \tag{4}$$

Where

C $\qquad$ is an equipment factor;

Kab,Keb $\quad$ are respectively the absorption and the extinction coefficient of the burnt products;

τ_f, τ_b $\qquad$ are the transmittances respectively of the unburnt and of the burnt mixture;

L_B^o $\qquad$ is the black body monochromatic radiation intensity at T_B;

L_o $\qquad$ is the incident monochromatic radiation intensity.

The ratio Γ becomes then:

$$\Gamma = \frac{kab}{keb} . \frac{\tau_f . (1-\tau_b) . L_B^o}{(1 - \tau_f . \tau_b) . L_o} \tag{5}$$

For the limiting case, which corresponds to the end of the combustion, the flame touches the wall, $r=r_o$ and $\tau_f=1$. We may write:

$$R = C.L_o \tag{6}$$

$$B = C.\frac{kab}{keb} . (1 - \tau_b^*) . L_B^o \tag{7}$$

$$RB = C.\left[\tau_b^* . L_o + \frac{kab}{keb} . (1-\tau_b^*) . L_B^o\right] \tag{8}$$

$$\Gamma = \frac{kab}{keb} . \frac{L_B^o}{L_o} \tag{9}$$

and $\quad \varepsilon = \alpha = 1 - \tau_b = \dfrac{B + R - RB}{R} \tag{10}$

Wien's law leads to the value of the burnt products temperature T_B:

$$\frac{1}{T_B} = \frac{1}{T_R} - \frac{\lambda}{C_2} \cdot Ln. \frac{kab}{keb} \cdot \Gamma \tag{11}$$

<u>Pyrometer's principle</u>. The determination of the signals B and RB involves the simultaneous measurement of the emission of the burnt gases B and also of the emission of the reference through the burnt gases RB. The optical assembly between the emitting reference source and the detector must have two distinct optical ways.

The conception of the pyrometer MT4 and PIREA of the O.N.E.R.A. [3,5] is based in this principle.

Nevertheless the optical system may have only one optical way if the two signals are obtained one after the other, by modulation of the emitting reference signal. This modulation is often mechanical (rotating disc) [6,7] and the optical assembly becomes simplified.

In order to avoid rotating parts and to achieve an easier treatment of the pyrometer's signal, the signal R is electronically modulated by using a pulsed infra-red diode as reference source.

<u>Optical assembly</u>. The optical system is shown in Figure 2, and consists essentially of three borosilicate lenses of 22.4 mm in diameter. This lens material was chosen because it has a transmittance of 0.93 at the wavelength of work.

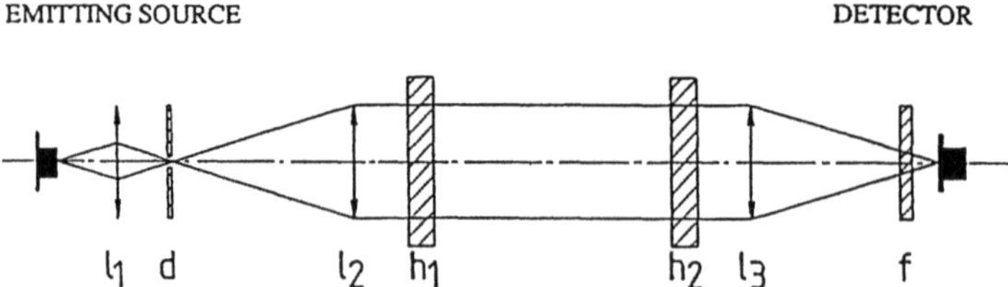

Fig. 2:   Optical assembly.

The L_1 lens (focal distance 20 mm) projects the image of the infrared source to a diaphragm. There, the L_2 lens (focal distance 200 mm) produces a parallel beam. This beam enters the spherical bomb through the borosilicate window h_1, traverses the chamber through its centerline and leaves it through the opposite window h_2.

The L$_3$ lens (focal distance 200 mm) projects the source and the flame (whenever The emitting source is a light emitting diode modulated at a frequency of about 2 kHz. Its brightness temperature is adjustable. In the present work it was fixed at a value of 2000K.

The emitting diode, the detector and the interference filter have their peak response for a wavelength of 940 nm. For this wavelength, the combustion products (at 2100 K) still radiate strongly.

The detector is a photodiode. Prior to making measurements it was calibrated with a carbon rod black body, set in the usual emitting diode's place. The aperture of its radiation cavity is aligned with the optical assembly axis center.

The brighteness temperature of the emitting diode is obtained at each experiment by using the detector's signal and based upon the calibration values previously found.

<u>Electronic assembly</u>. This assembly commands the modulation of the emitting diode and the separation, in the detector, of the continuous component due to the flame B, from the component R'=(RB-B) due to the pulsed emitting diode.

The ratio Γ comes then:

$$\Gamma = B/(R-R') \tag{12}$$

This assembly includes (see Figures 3 and 4):

- A modulation command working at 2 kHz (signal time 3 µs, total time 500 µs);
- A modulation system of the emitting diode;

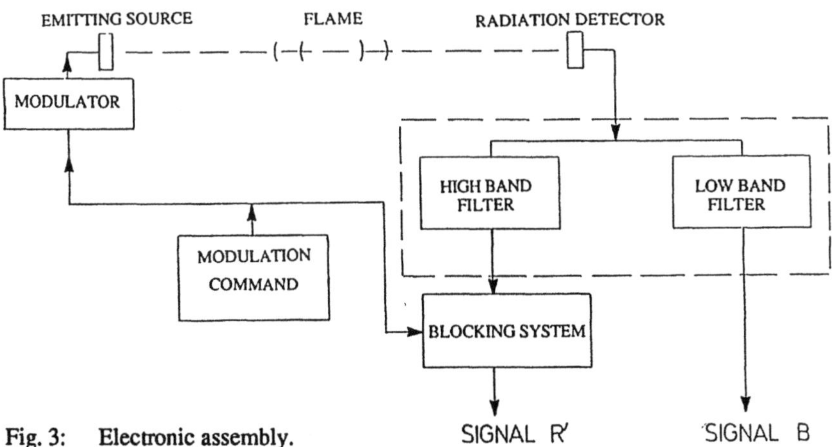

Fig. 3:   Electronic assembly.

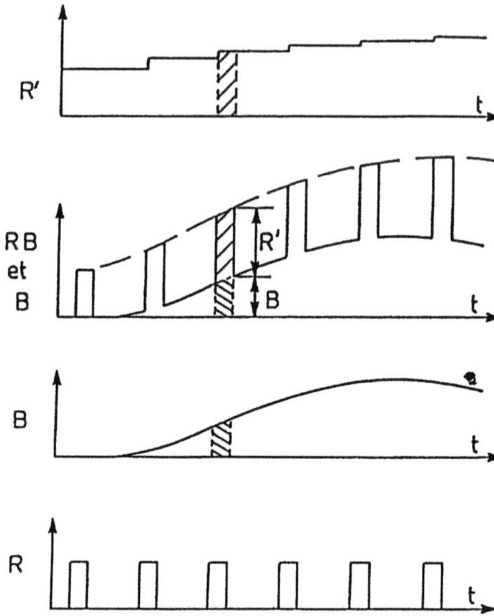

Fig. 4:   Signal separation.

- A separating circuit (a high frequency pass filter and a low frequency pass filter) which allows the separation, in the received signal RB, of the continuous part B from the modulated signal R';
- A blocking and integrating system (also modulated as the emitting diode) to integrate the signal R'.

A signal acquisition system allows the acquisition and the treatment of the three needed signals: R (before the experiment), B and R', which allows the flame monochromatic emissivity and the flame temperature determination.

EXPERIMENTAL RESULTS

For starch-air mixtures the maximal value of the combustion temperature (*see* Figure 5) (2170 K) corresponds to an equivalence ratio of 1.7.

The theoretical temperature value was given by the "QUATUOR" code [8] in the adiabatic isochoric hypothesis and supposing that the starch burns completely in a gaseous phase.

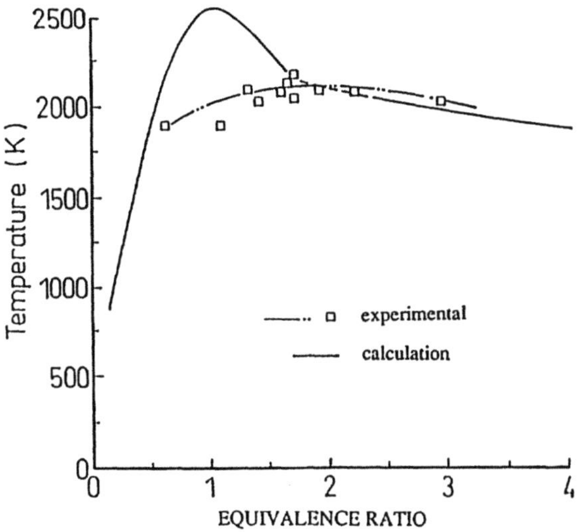

Fig. 5:   Starch-air combustion — Explosion temperatures.

RESULTS AND DISCUSSION

Experimental temperature values are, for lean mixtures, much lower than the theoretical temperatures and for equivalence ratios greater than 1.6 they are very close to the calculated ones.

This situation may be explained by the fact that theoretical calculated temperatures are burnt gas final temperatures while experimental results correspond to mean final solids temperatures and not gas temperatures (gas radiation is negligible at the wavelength of work). For lean mixtures those solids must essentially be starch particles that pyrolyse ahead the flame front at a much lower temperature than the burnt gas temperature. For rich mixtures those solids must be very fine solid burnt particles, whose temperature would be very close to the burnt gas temperature.

Both results are consistent with the observation, after combustion and for equivalence ratios greater than 1.6, of a thick black cloud in suspension inside the chamber, probably containing solid carbon particles.

The measured emissivity (*see* Figure 6) also increases, in an almost linear way with the equivalence ratio, confirming that the solid burnt products are present in growing concentration for rich mixtures.

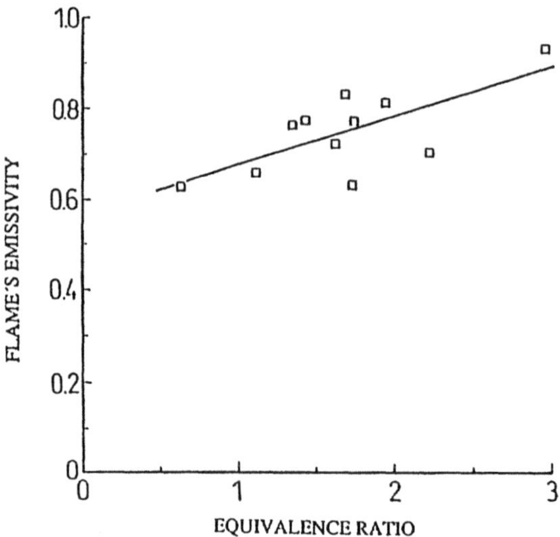

Fig. 6:   Starch-air combustion — Flame emissivity.

For those mixtures, the burnt gas composition also changes, presenting an important concentration of typical pyrolysis products (*see* Table 1) such as: CO, H_2, CH_4 and other low density hydrocarbon gases.

These results would indicate that for rich mixtures, the combustion is not complete.

Table 1:   Burnt composition (% per volume)

Equivalence Ratio	P exp (bar abs)	O_2	Ar	N_2	CO	CO_2	H_2	CH_4	C_nH_n
1.13	7.73	7.78	0.94	78	0.64	12.46	0.11	-	-
1.24	8.63	7.49	0.94	78	0.21	13.30	0.02	-	-
1.32	9.97	5.01	0.94	78	0.15	15.64	0.02	-	-
1.34	-	4.51	0.93	78	0.39	15.85	0.11	-	-
1.84	10.50	2.66	0.93	78	0.40	17.71	0.06	-	-
1.87	12.22	0.57	0.88	73.3	5.55	17.10	2.05	0.34	-
2.97	13.40	0.25	0.60	50.8	23.86	10.85	8.39	3.67	1.5

CONCLUSION

The infrared emitting-absorbing pyrometer described in the present work is an useful tool for measuring particle temperatures in dust-air flames, giving good experimental results. It can be used either in transient flames such as the explosion flames treated in the present case or in stationary flames such as burner flames.

Concerning the starch flames propagation mechanism, previous work has shown that starch particles pyrolyse almost completely prior to flame front arrival [9,10].

Present results let us suppose that, for lean mixtures (equivalence ratios less than 1.6), starch dust particles completely pyrolyse before the flame front arrival, originating a classic gaseous premixed type flame without solid residue formation. On the other hand, for rich mixtures, the presence of the fine solid residue, of the pyrolysis gaseous products and the similarity of the experimental results with the calculated ones (in the hyphotesis of solid carbon formation) indicate that although all the starch is pyrolysed, only a part of it completely burns in the gaseous phase. The rest continues its transformation behind the gaseous flame emitting strong radiation.

REFERENCES

1. Lemos L., Bouriannes R.: Starch dust combustion characteristics in a closed spherical vessel. 12th ICDERS, USA (1989).

2. Lemos L.: Contribution à l'étude de la combustion, en chambre spherique, des mélanges hétérogenes. Cas des mélanges amidon-air. Thèse de Docteur de l'Université de Poitiers, France (1989).

3. Moutet A., Crabol J., Nadaud L.: Températures des gaz et des flammes. Techniques de l'Ingénieur, R 2750 (1974).

4. Lemos L., Bouriannes R: Mesure, par pyrométrie optique, de la température de combustion d'un matérial pulvérulent en suspension dans l'air. Rev. Gen. de Thermique 342,343 (1990) 312-316

5. Charpenel M.: Mesure instantanée par pyrométrie infrarouge des température des gaz de combustion. Application à la turbulence thermique. Rev. Physique Appliquée 14 (1979).

6. Moutet A.: Études de Pyrométrie Pratique. Eyrolles, France (1959)

7. Habib Z.: Étude radiative d'une flamme de charbon pulverisé — mesure de température. Thèse de Docteur de l'Université de Rouen, France (1986).

8. Heuzé O.: Contribution au calcul des characteristiques de detonation de substances explosives gaseuses ou condensées. Thèse de Docteur de l'Université de Poitiers, France (1985).

9. Proust C., Veyssière B.: A new experimental apparatus for studying the propagation of dust-air flames. AIAA Prog. in Astron. and Aeron. 133 (1988).

10. Proust C., Veyssière B.: Fundamental properties of flames propagation in starch dust-air mixtures. Comb. Sci. and Tech. 62 (1988).

INSTANTANEOUS TEMPERATURE PROFILE
MEASUREMENTS, IN A FLAME BY INFRARED
LINE THERMOMETRY TECHNIQUE

B. Bedat, A. Giovannini, S. Pauzin

O.N.E.R.A./C.E.R.T./D.E.R.M.E.S.
BP 4025
31055 Toulouse Cedex

ABSTRACT

Black body radiation from a fibre of βSiC can be used to investigate the temperature profile in a premixed flame. An infrared scanner determines intensity radiation of the fibre, which is related to the fibre temperature by a calibration law. A fast time constant, and excellent spatial resolution of the fibre make the method a very useful tool in examining turbulent or instability flames.

1. INTRODUCTION

Most commonly used techniques for flame temperature measurement are either local and intrusive (thermocouple, cold wire) or non intrusive (Raman, Rayleigh, Induced Fluorescence). These two last techniques lead to 2D images. The first class has been widely used for turbulent flame measurements [1] [2]. Small diameter thermocouple (≤ 50 μm) are more often operating than cold wire as they offer best space resolution. On the other hand, they exhibit quite large response time variable with local velocity and temperature and the rough signal has to be compensated [3] [4].

The second class, based on elastic or non-elastic molecular interactions [5], with an incident laser beam, requires an heavy investment in material, research and development time. More over they necessitate experimental precautions (filtering against dust and soot for Rayleigh diffusion finding a stable particle tracer and operating with a dual laser beam for induced fluorescence). The presented technique named IRLT (InfraRed Line Thermometry) is an improvement of previous works of Vilimpoc and Goss [6]. The radiant energy emitted from a silicon carbide fibre tightened cross wise through the flame is registered in the infrared range and analyzed. The small diameter of the fibre allows a very fast temperature equilibrium with gas.

As we use an Insb detector cooled by liquid nitrogen and operating in the infrared range 2-5 µm we have a good dynamic for the measurement (1000) and we have lowered the minimum temperature level measurable. That paper presents the principle and the IRLT technique performances. Finally applications to a premixed propane air jet flame are shown.

2. EXPERIMENTAL SET-UP

The measured diameter of the fibre (Figure 1) is 15 µm. Its surface is grey and lambertian in the infrared range, with an emissivity factor value of 0.92 [7]. Heat capacity ρc is evaluated at 0.810^6 Jm^{-3} K^{-1}, five times lower than usual metals.

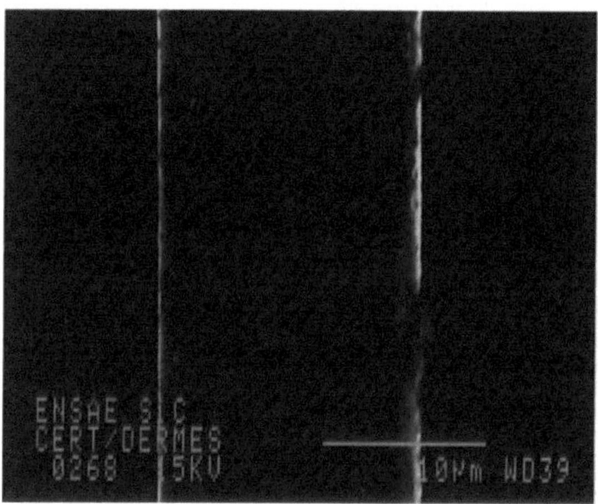

Fig. 1:   Microscopic view of the fibre

The low thermal conductivity $\lambda \simeq 20$ W m^{-1} K^{-1} at the temperature of 1700 K is very interesting to minimize longitudinal conduction of heat in the fibre. The fibre is tightened across the flame flow (Figure 2), its support and the infrared camera are independantly adjustable of each other in height and in rotation since the line swept by the detector during the opto mechanical scanning, achieved by a prism, must be coincident with the image of the fibre. The scanner is a modified AGA 782 SWB camera whose vertical prism has been stopped.

As we apply that radiation detection technique to hydrocarbon flames, we avoid infrared radiation of flame product (CO_2, H_2O, CO) (Figure 3) [8], by interposition of a spectral filter centered at 3.9 µm.

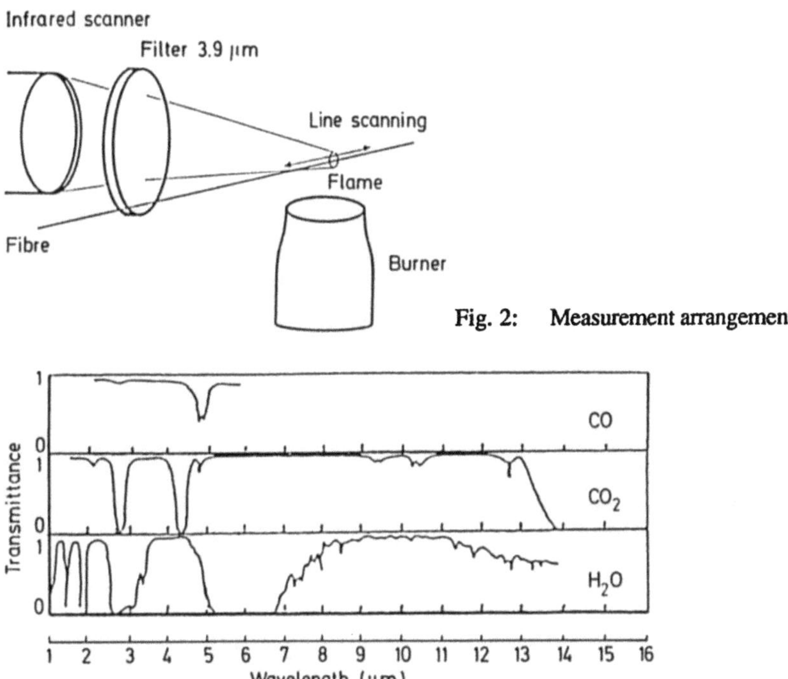

Fig. 2: Measurement arrangement

Fig. 3: Transmittance of flame products

Experiments have been carried out in a non confined premixed propane-air jet flame (Figure 4). The flame is stabilized by a platinium coating at the leading edge of the jet's convergent. The blow-off limit is increased up to 6 m/s for a

Fig. 4: View of the experimental set-up.

stoichiometric upstream mixture and a burner diameter of 40 mm (Re_D=16000). Natural turbulence level is 3%.

3. DATA ACQUISITION AND PROCESSING

The InSb photovoltaic detector has a square shape of side 250 µm. Its maximum specific detectivity is 2.510^{11} cm H $z^{1/2}$ w^{-1} and the response time is smaller than a microsecond. The angle of view is 7° and the field explored varies between 27 mm at a distance of 220 mm, and 120 mm at a distance of 1 m. The horizontal scanning, at a rate of 2500 Hz is achieved by octogonal birefringent rotating prism; but the useful duration is only 280 µs; the others 120 µs being devoted to synchronization tops and interrogation of references in order to correct the signal from internal background. The signal is digitized in real time by the "TIC" card on 256 levels and 140 points per line. A dedicated software named "LINETIC" acquires a sequence of consecutive lines. In the buffer of size 2 Mo, we can register on a Compacq 386-25 computer 14000 lines corresponding to a total time of 5.6 seconds. These lines are stored together by blocks of 140 in order to form pseudo images and active classical primitives of image processing with a software name "CATS". A dedicated software allows us to visualize 448 ms of these data (Figures 11 and 12).

4. TEMPERATURE CALIBRATION

That operation is usually made with an extended black body [9], the sensitive area of the detector receiving an uniform flux. As the image of the fibre is smaller than the detector side, and due to the presence of the spectral filter at 3.9 µm, an accurate calibration, in the same conditions than experiments, is necessary.

We use a filament of tungsten heated by Joule effect and placed in front of a black body. When the filament signal disappears on a line, the apparent temperatures are equal and we obtain a relation between the detector signal (UI) of the filament and the black body. We get an attenuation factor corresponding to the measuring conditions. The factor is multiplied by the constant A of the law corresponding for a black body [9].

$$\Phi_m = \frac{A}{\left(e^{\frac{B}{T}} - 1\right)} \tag{1}$$

The correspondence of the detector signal and the fibre temperature is plotted on Figure 5. The attenuation factor can be determined with an accuracy of 5%.

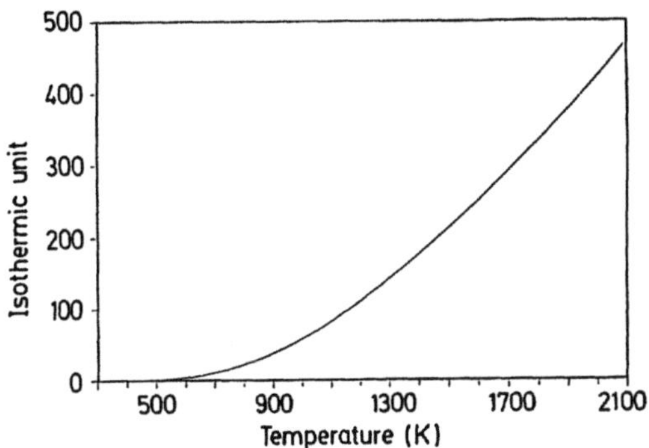

Fig. 5:   Curve of signal unit versus temperature.

That gives an evaluated error of 40 K for a fibre temperature of 1700 K (Figure 6). The temperature difference which corresponds to one discretization level is equal to 40 K at T=600 K, but 5 K only at higher temperature (T=1700 K).

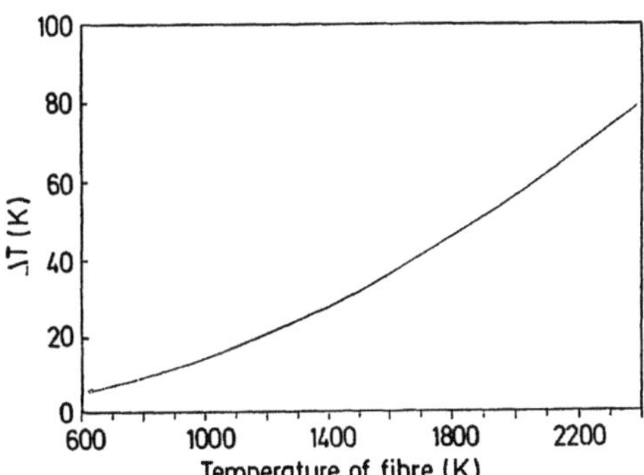

Fig. 6:   Plot of temperature error versus temperature level.

At the present time, the minimum detectable temperature is 600 K (due to the filter attenuation), but we plan to down that value until ambient temperature by changing the filter characteristics and amplifying the output signal.

5.   THERMAL BEHAVIOUR OF THE FIBRE IN COMBUSTION SURROUNDINGS

The heat balance between the fibre (f), the reactants or products (g) and the external surroundings (∞) is written as follows:

$$\rho V c \frac{dT_f}{dt} = hS(T_g - T_f) + S\sigma\varepsilon_f\left(T_\infty^4 - T_f^4\right) + S\sigma\varepsilon_g F_{fg}\left(T_g^4 - T_f^4\right) + \lambda_f A \frac{d^2T_f}{dx^2} \tag{2}$$

The right side terms account successively for exchanges by convection, radiation with surroundings, radiation with flame and finally conduction on the fibre. That basic equation is exploited in order to evaluate characteristics of the operating fibre.

CONVECTIVE RESPONSE TIME

Retaining only the unsteady left term and the convective one, the time constant is given by the expression:

$$\tau = \frac{\rho c d^2}{4 Nu_d \lambda_g} \tag{3}$$

in which ρc is the heat capacity of the fibre, λ_g the gas conductivity and Nu_d the Nusselt number correlated with the Reynolds number by the King law

$$Nu_d = 0.396 + 0.518\, Re_d^{\frac{1}{2}} \tag{4}$$

For flow velocities between 1 and 10 m/s, the time constant ranges from 1.8 ms to 0.8 ms. We recover with a good agreement (Figure 7) these values experimentally by heating the fibre with a 4 Watts argon laser (Figure 8), interrupting suddenly the beam, and monitoring the temperature relaxation of the fibre.

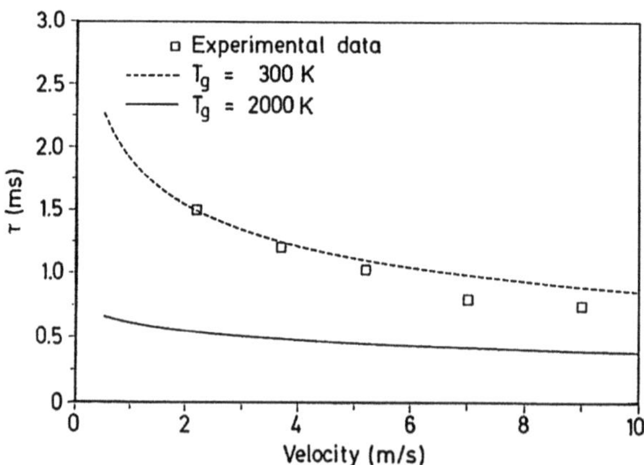

Fig. 7:   Time response of fibre (theoretical curves and q experimental data).

We have to notice that, inside the products, the time constant is smaller than 0.5 ms. Radiation effects are not important because their characteristics time is much larger than τ [10].

Fig. 8:   View of laser calibration set-up.

SPACE RESOLUTION OF THE FIBRE

Taking into account only conduction and convection left terms, eq. (1) becomes:

$$hS(T_g - T_f) = - \lambda_f A \frac{d^2 T_f}{dx^2}$$

(5)

with temperature conditions at x=0 (T=T_0) and x = ∞ (T = T_1). Solution comes as:

$$\frac{T - T_0}{T_1 - T_0} = \exp\left(- \frac{4h}{\lambda_f d} x\right)$$

(6)

with a characteristic length $\delta = \sqrt{\frac{\lambda_f d}{4h}}$.

The Figure 9 shows that at gas temperature of 2000 K and a velocity of 10 m/s that value is equal to δ = 100 μm; the same order of magnitude as the flame thickness and the Kolmogorov scale at a jet turbulent Reynolds number of 10^3. Adding the radiation terms decreases that values of 10%.

TEMPERATURE CORRECTION

This correction is performed, as classical for thermocouple experiments, by taking into account radiation exchanges between the fibre and the surrounding. The real temperature T_g is different from the fibre temperature T_f by the term:

$$T_g - T_f = \frac{\sigma \varepsilon f}{h} (T_\infty^4 - T_f^4)$$

(7)

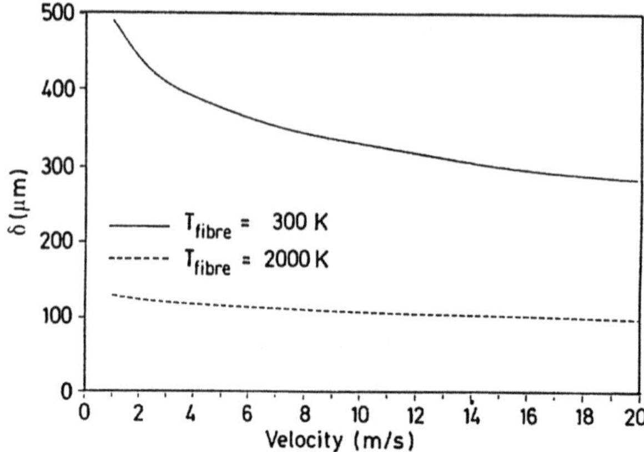

Fig. 9: Theoretical longitudinal thermal diffusion for different velocities and temperatures.

This correction requires an iterative process because of variation of physical properties of gas with temperature. As shown by Figure 10, depending on the velocity, for a temperature level $T_f = 1700$ K, we must add 120 K or 170 K.

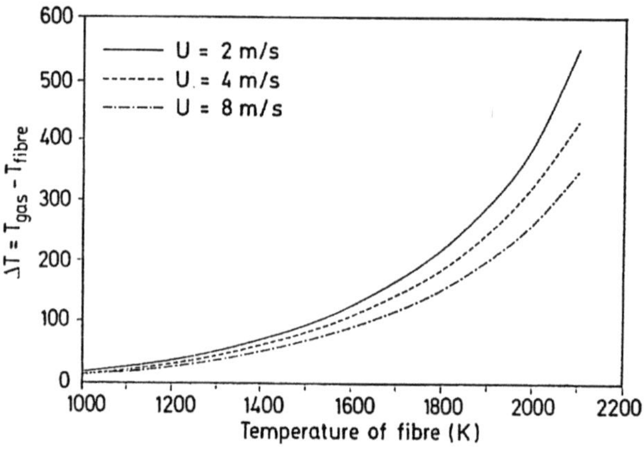

Fig. 10: Correction curve function of velocity and temperature.

6. EXAMPLE OF RESULTS

That technique has been validated on the flame set up described on section 2. Close to the burned lip, varying the equivalence ratio Φ between 0.6 and 1.5, we measure the maximum flame mean temperature. On Figure 11, the dashed curve corresponds to the fibre, the dotted-line is obtained after radiation correction and can be compared with unbroken line plotted using JANAF data and STANJAN [11] code equilibrium calculations. For poor mixtures, agreement is quiet good, the difference corresponds certainly to flame losses

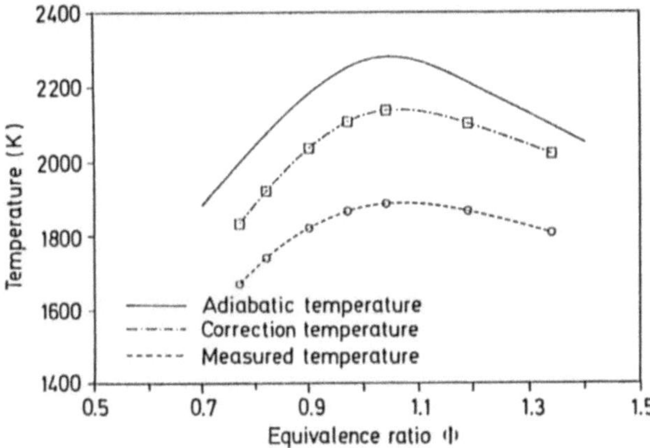

Fig. 11: Experimental and corrected temperature level for a range of equivalence ratio, compared with adiabatic prediction.

and presence of water vapor in the air. On the rich side, as the flame soon radiates, accurate correction is necessary, but not yet being done.

Figures 12 and 13 represent a sequence of thermolines in an x, t mode. On Figure 12, the whole field of the flame is covered (120 mm). The two interfaces between fresh mixture and products (flame front) and between products and

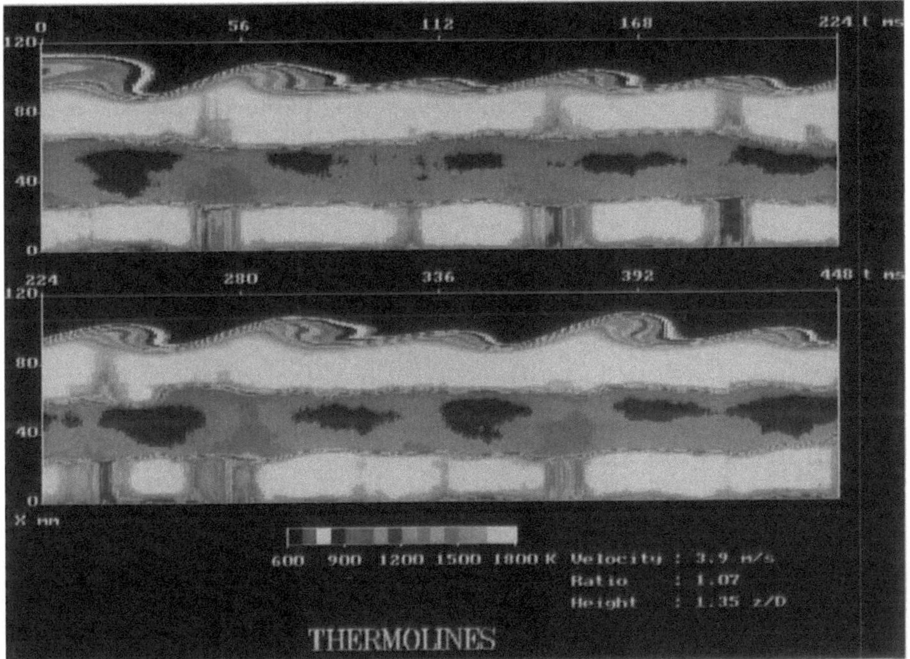

Fig. 12: Thermolines of the whole field of the flame U=3.9 ms-1, ϕ = 1.07, Z/D = 1.35.

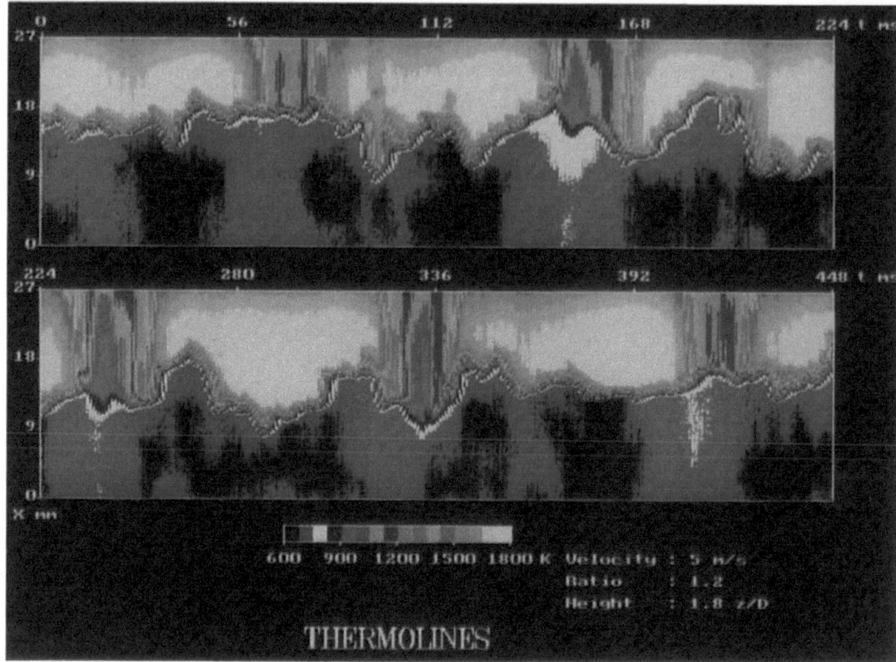

Fig. 13: Thermolines of the flame front U=5 ms⁻¹, ϕ = 1.2, Z/D = 1.80.

still ambient (air/jet) are clearly visible. The cold jet velocity is 3.9 m/s, equivalence ratio Φ = 1.07 and the fibre is tightened on a diameter at an elevation z=1.35 D. Space resolution corresponding to the detector side is poor (3.6 mm), that graph shows only interfaces instabilities. So, if we reduce the field (27 mm), as in Figure 13, we have a better space resolution (800 µm). The

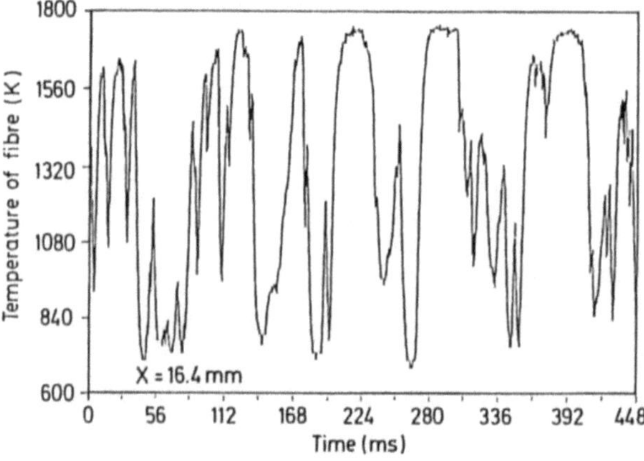

Fig. 14: Example of temporal temperature.

experiments conditions are U=5 m/s, ϕ=1.2 and z=1.8 d. At that distance one can see that the flame is wrinkled.

At a fixed point on a line, at the mean flame position, and close to the flame tip, the Figure 14 exhibits temperature history. We notice the fluctuations due to the hot or cold spots following the flame movement. Treatment has been done in order to derive temperature and flame position statistics, but before, more work is necessary in order to correct signal from space resolution, and down the detectable temperature up to ambient.

7. CONCLUSION

The IRLT technique is able to perform instantaneous measurements of 1D temperature profiles at a rate of 2500 Hz inside a flame. Interesting heat proof and physical properties lead to a response time smaller than 1 ms and a good ability to keep large temperature gradients. The uncertainty is evaluated at 50 K for a temperature level $T_f = 1700$ K, but radiation corrections are necessary and more work is now in progress in order to lower the detectable temperature.

The results give that technique very helpful and a practical tool to obtain statistics on temperature profiles, flame front position, probability density function and to provide input data for B.M.L., PEUL, SLIC models. Frequency analysis coupled with acoustics measurements on combustion noise sources will allow us to study the interaction between acoustics and combustion.

ACKNOWLEDGEMENT

A part of the financial support of this work is from the French Ministry of Defence, Direction des Etudes et Recherches Techniques, under the contract n° 89002.25.

REFERENCES

1. Yoshida A., Günther R.: Experimental Investigation of Thermal Structure of Turbulent Premixed Flames. Combustion and Flame 38 (1980) 249-258.

2. Yoshida A.: An Experimental Study of Wrinkled Flame. In Eighteen Symposium on Combustion (1981) 931-938, The Combustion Institute.

3. Yule A.J., Taylor D.S., Chigier N.A.: On Line Digital Compensation and Processing of Thermocouple on Signals for Temperature Measurements in Turbulent Flames. AIAA Paper (1978) 30.

4. Ballantyne A., Moss J.B.: Fine Wire Thermocouple Measurements of Fluctuating Temperature. Combustion Science and Technology, 17 (1977) 63-72.

5. Eckbreth A.C.: Laser Diagnostics for Combustion Temperature and Species. Ed. Gupta A.K. and Lilley D.G. Abacus Press (1988).

6. Vilimpoc V., Goss L.P.: Sic-Based Thin-Filament Pyrometry: Theory and Thermal Properties. In Twentieth-Second Symposium on Combustion (1988) 1907-1914, The Combustion Institute.

7. Touloukian Y.S.: Thermal Properties of High Temperature Solid Materials, Vol. 5 — Nonoxydes and their Solutions and Mixture Including Miscillaneous Ceramic Materials Properties (1967) 118-140. MacMillan.

8. Yoshida A.: Convective Heat Transfer and Film Cooling in Turbomachinery. V.K.I. Lecture Series (1986) 06. Rhode Saint Genese, Belgium.

9. Gaussorgues G.: La Thermographic Infra-Rouge. Edition Technique et Documentation Lavoisier (1984).

10. Sbaibi A., Lecordier J.C., Paranthoen P.: Réponse en Fréquence d'un Couple Thermoélectrique dans un Environnement Purement Radiatif ou Convectif. Entropie, 135 (1987) 49-53.

11. Kuo K.K.: Principles of Combustion. John Wiley and Sons (1986).

SIMULTANEOUS MEASUREMENT OF TEMPERATURE AND DENSITY OF BURNT GASES BY AN INFRARED RADIATION COMPUTED TOMOGRAPHY

S. Shimizu, S. Sakai and K. Wakai

Mechanical Engineering
Gifu University
1-1 Yanagido, Gifu
Japan, 501-11

ABSTRACT

An iterative method which can compensate the influence of absorption of radiation was introduced into an infrared computed tomography for simultaneous measurement of both temperature and density of self-radiating gases. The characteristics of the method are investigated by simulations and experiments. A feasibility to develop this method into the measurement of transient phenomena is suggested.

INTRODUCTION

For diagnostics of reacting gases at high temperatures, nonintrusive methods have a great potential as they disturb the gases very little. Optical methods which are intrinsically nonintrusive are exerting their excellent abilities in the field of research of combustion [1]. The technique of computed tomography has been applied to optical measurement of gas flows where local states are indispensable factors to clarify the mechanism of flows.

Since the attenuation of optical radiation follows Lambert-Beer's Law, density of gases can be measured by an optical absorption tomography [2] as far as no radiation from the gas is involved. Deflection of light beam is used for the tomographic reconstruction of density distributions [3]. Holographic interferometry also is employed to measure an optical path length through mixing gases and the spacial distribution of gas concentration is determined by a tomographic method [4].

The intensity of radiation from a gas is expressed by a product of two factors, that is black body radiation for the temperature and emissivity of the gas.

Therefore, an accurate value of emissivity must be obtained to measure the temperature of the gas from its radiation. The emissivity is a function of so many factors, such as pressure, temperature, wavelength and geometric size of the gas, that optical pyrometers are contrived to get rid of the influences of these factors to improve the accuracy of measurement. Some of the methods take a measure to reduce the influence of emissivity by taking ratio of the intensities of radiation at two wavelengths [1].

Ray et al. [5] used multiangular absorption spectra observed at two wavelengths to reconstruct the distributions of both temperature and density of a gas by taking the ratio of the two spectra. Density is obtained from the absorption coefficients. No radiation originated in the gas is taken into account in the measured amount of absorption, because the effect of radiation by the gas itself makes it difficult to operate the algorithm of tomography.

In this report, a technique which modified the Kurlbaum [6] and Schmidt [7] methods is used to obtain, by an infrared radiation computed tomography, a two-dimensional distribution of absorption coefficient of a gas even under the condition of strong radiation by the gas itself. Temperature of the gas, then, will be calculated from the local radiative power reconstructed by a tomographic method which can reduce the effect of self-absorption by a method of iteration.

Since acquisition of projections takes time, feasibility of high speed data acquisition for an instantaneous tomography was studied.

PRINCIPLE OF ITERATIVE METHOD

APPARATUS

Outline of the opto-electronic system shown in Figure 1 is similar to the one used in reference [8]. Radiation from a black body passes through the gas whose temperature is to be measured before it reaches a detector. A spectral band of radiation which arrives a detector is selected by a filter placed in front of a detector. When one of 1.8 or 2.7 μm bands of H_2O or 4.3 μm band of CO_2 is used the band width $\Delta\lambda$ is selected in the range of 0.14 ~ 0.48 μm to obtain the optimum signal to noise ratio (S/N) of measured values.

A rotary chopper which operates with a period of 0.2 s is placed just in front of a black body to modulate the signal (D) of detector by intermittently casting the black body radiation onto the gas. The signal which is used to reconstruct the states of gas is compiled into a floppy disc for the following steps of data processing. When the feasibility of instantaneous recording of projection data is checked, sodium's D-line emanated from NaCl which is seeded into the fuel-air

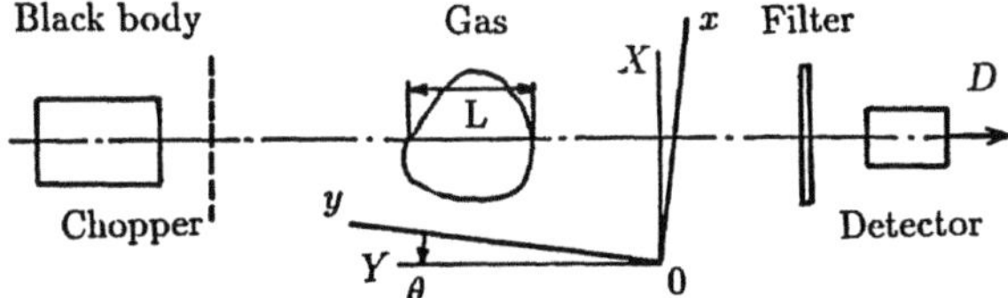

Fig. 1:   Outline of opto-electronic system

mixture is used because array detector for infrared ray is not commonly available at present.

SIGNALS

A signal $D_{1X\theta}$ generated by the black body is used as a reference of measurement. Expression of $D_{1X\theta}$ is

$$D_{1X\theta} = K_{1X\theta}\, I_o , \tag{1}$$

where Io is radiative intensity of the black body. $K_{1X\theta}$, a conversion factor, may be a variable along the X axis of X-Y coordinates which make angle θ against x-y coordinates fixed to the space.

The radiation $D_{2X\theta}$ of a gas can be measured when the chopper is closed to interrupt radiation from the black body, that is

$$D_{2X\theta} = K_{2X\theta}\, I_{GX} = K_{2X\theta} \int_{-\infty}^{\infty} (1-\varepsilon_Y)\, a_{xy}\, I_{xy}\, dY, \tag{2}$$

where I_{GX} is a radiation of gas volume along the line of pass which is parallel to Y axis at X. a_{xy} is a local effective absorption coefficient of the gas at point x,y. The "effective" means an average of the spectral absorption coefficient including spectral characteristics of the opto-electronic components in the range of wavelength $\Delta\lambda$ specified by the components. I_{xy} is a mean black body radiative intensity of the gas at x,y in the same spectral range $\Delta\lambda$. ε_Y is an emissivity of a gas layer from Y=0 to Y, that is

$$\varepsilon_Y = 1 - \exp\left(- \int_0^Y a_{xy}\, dY\right). \tag{3}$$

$K_{2X\theta}$ is a conversion factor which may be a variable of X and θ. This factor is converted to K_{20}, a value obtained by calibration at the center of the field of measurement for one particular θ, by multiplying $K_{20}/K_{2X\theta}$ to $D_{2X\theta}$, that is

$$D_{2X\theta 0} = \left(K_{20}/K_{2X\theta}\right) D_{2X\theta}$$

$$= K_{20} \int_{-\infty}^{\infty} a_{xy}\, I_{xy}\, dY - K_{20} \int_{-\infty}^{\infty} \varepsilon_Y\, a_{xy}\, I_{xy}\, dY. \tag{4}$$

When the black body is uncovered by the chopper the signal will be $D_{2X\theta}$,

$$D_{2X\theta} = (1 - \alpha_{LX\theta})\, D_{1X\theta} + D_{2X\theta}\,, \tag{5}$$

where $\alpha_{LX\theta}$ is an absorptivity of the gas layer L (Figure 1) along the pass of radiation which is parallel to Y axis. That is

$$\alpha_{Lx\theta} = 1 - \exp\left(-\int_{-\infty}^{\infty} a_{xy}\, dY\right). \tag{6}$$

LOCAL EFFECTIVE ABSORPTION COEFFICIENT

From the measured signals $D_{1X\theta}$, $D_{2X\theta}$ and $D_{2X\theta}$, a function $P_{aX\theta}$ will be introduced as follows,

$$P_{aX\theta} \equiv \ln\left[\frac{D_{1X\theta}}{D_{3X\theta} - D_{2X\theta}}\right] = \int_{-\infty}^{\infty} a_{xy}\, dY\,. \tag{7}$$

$P_{aX\theta}$ could be used as a datum of convolution back projection method [9] to obtain the local effective absorption coefficient a_{xy}. Shepp and Logan's filter function [10] is used for the computation.

LOCAL TEMPERATURE BY ITERATIVE METHOD

Eq. (4) is not appropriate for tomographic calculation because the second term has ε_Y. Therefore, this term will be eliminated to start the first step of iteration. $D_{2X\theta0}$ is equated to $D_{2X\theta01}$, that is

$$D_{2X\theta0} = D_{2X\theta01} \equiv K_{20}\int_{-\infty}^{\infty} a_{xy_1} I_{xy_1}\, dY. \tag{8}$$

The effect of neglecting ε_Y will modify the real radiation intensity I_{xy} of the local gas in eq. (4) to I_{xy1}. This term is named the first approximation of I_{xy} and will be calculated by the same algorithm used for a_{xy}. Introduction of I_{xy1} into the second term of eq. (4) will give another $D_{2X\theta02}$ for the second iteration by using ε_Y calculated from eq. (3).

$$D_{2X\theta02} = D_{2X\theta0} + K_{20}\int_{-\infty}^{\infty} \varepsilon_Y\, a_{xy} I_{xy_1}\, dY \equiv K_{20}\int_{-\infty}^{\infty} a_{xy} I_{xy2}\, dY. \tag{9}$$

This equation gives the second approximation of I_{xy}, that is I_{xy2}. Such an iterative calculation is repeated i times until the least square of relative values of difference between $D_{2X\theta0i}$ and $D_{2X\theta0(i-1)}$, which is named Δr_i converges to a value smaller than 0.1. When Δr_i satisfied above condition, the iteration is ceased to obtain the local temperature T_{xy} by applying I_{xyi} to Planck's equation for radiation.

LOCAL DENSITY ρ_{xy} OF MEDIUM GAS

It is assumed that the medium gas follows Lambert-Beer's law, and the local effective absorption coefficient a_{xy} used in eq. (2) has been defined as follows.

$$a_{xy} = \frac{1}{\Delta\lambda} \int_{\Delta\lambda} t_\lambda \cdot A_{\lambda xy} d\lambda, \tag{10}$$

where t_λ is the spectral characteristics of the opto-electronic components, such as filter, detector, lens (not shown in Figure 1) and so forth. $A_{\lambda xy}$ is a real spectral absorption coefficient of the gas at point x,y. When water vapour or carbon dioxide is used as the medium gas the spectral band in near infrared range is expressed by the Random Band Model [11]. The spectral absorption coefficient $A_{\lambda xy}$ will be equated as follows from the function of the curve of growth,

$$A_{\lambda xy} = \frac{k_{\lambda xy} \dfrac{\rho_{xy}}{\rho_s}}{\sqrt{1 + \dfrac{k_{\lambda xy}}{4 \cdot q_{\lambda xy}} \dfrac{\rho_{xy}}{\rho_s}}} . \tag{11}$$

Where $k_{\lambda xy}$ is the absorption coefficient [11] under the condition of standard temperature T_s and pressure P_s, that is at STP, and for the wavelength λ. ρ_s is the gas density at STP and $q_{\lambda xy}$ is the fine structure term of the spectrum [11].

Introducing eq. (11) into eq. (10) and use the mean values of t_λ, $k_{\lambda xy}$ and $q_{\lambda xy}$ in the range of $\Delta\lambda$, that is, $\bar{t}_\lambda$, $\bar{k}_{\lambda xy}$ and $\bar{q}_{\lambda xy}$ respectively, then, the local effective absorption coefficient of eq. (10) will be,

$$a_{xy} \cong \frac{\bar{t}_\lambda \cdot \bar{k}_{\lambda xy} \dfrac{\rho_{xy}}{\rho_s}}{\sqrt{1 + \dfrac{\bar{k}_{\lambda xy}}{4 \cdot \bar{q}_{\lambda xy}} \dfrac{\rho_{xy}}{\rho_s}}} . \tag{12}$$

Solution of this equation on ρ_{xy} is,

$$\rho_{xy} = \frac{-\xi \pm \sqrt{\xi^2 + 4\eta}}{2\eta} \tag{13}$$

where, $\xi = -\bar{k}_{\lambda xy}/(4 \cdot \rho_s \bar{q}_{\lambda xy})$, $\eta = \left\{ \bar{t}_\lambda \cdot \bar{k}_{\lambda xy}/(\rho_s a_{xy}) \right\}^2$.

HIGH SPEED DATA ACQUISITION

The projection data explained in the above chapters are measured by ordinary translate and rotate scanning which necessitates longer time for data acquisition. The use of array-detectors in place of an ordinary detector can reduce the time a great deal. Four sets of opto-electronic systems, similar to Figure 1, with array detectors sensitive to the D-line of sodium are tried to take simultaneously the projection data at four projection angles. Radiation from respective black bodies will be collimated in each set by a cylindrical lens, which is set right next to the chopper, to a parallel beam so as to make the beam passes through the gas with equal view angle at any position of X. Then the beam is cast to respective array detectors after passing through an iris placed at the focus of the second lens placed behind the gas, so that each pixel of the array receives one of fine parallel beams which passed a point X on X axis. This way of data acquisition may greatly reduce the time of scanning.

CHARACTERISTICS OF THE METHOD

Some sample volumes of gases are prepared to investigate by simulation the characteristics of the method of iteration presented in the preceding chapter. Numbers of samplings and projections are named M and N respectively.

INFLUENCE OF ABSORPTION COEFFICIENT

The simplest sample is a circular gas volume of 50 mm in diameter with uniform temperature of 2000 K. Effective absorption coefficients of $4 \times 10^{-4} \sim 2 \times 10^{-2}$ mm^{-1} are used to see the effect of self-absorption of radiation. 2.7 μm band of H_2O vapour at STP has an absorption coefficient of the order of 10^{-4} mm^{-1}, while the one for 4.3 μm band of CO_2 is in the order of 10^{-2} mm^{-1}. The example of reconstructed temperatures for i=1 are shown in Figure 2. As a increases, that is the transmissivity τ_d along a diameter decreases, the temperature in the central part becomes lower. When a is 10^{-2} mm^{-1} the temperature for i=1 at the center is about 200 K smaller than that of the edge (Figure 3).

EFFECT OF ITERATIVE COMPENSATION

Effectiveness of iteration to improve the reconstructed distribution of temperature along a diameter is shown in Figure 4. A circular gas volume with a=7.13×10^{-3} mm^{-1} which gives τ_d=0.7 along the diameter of 50 mm is used.

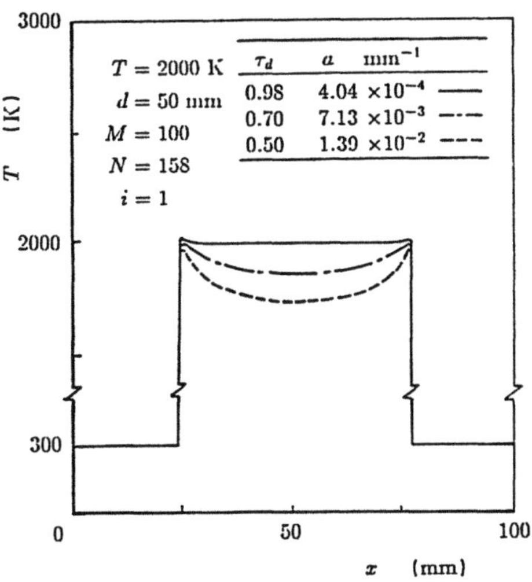

Fig. 2: Effective absorption coefficients and the reconstructed temperature distributions along a diameter of a gas volume, y=50 mm.

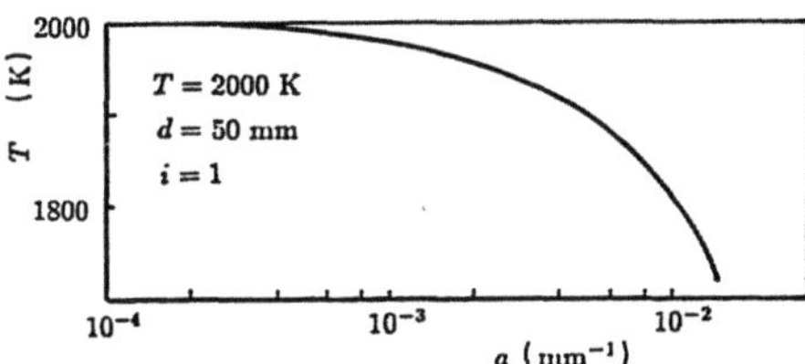

Fig. 3: Effective absorption coefficients and the temperature at the center of a circular gas volume.

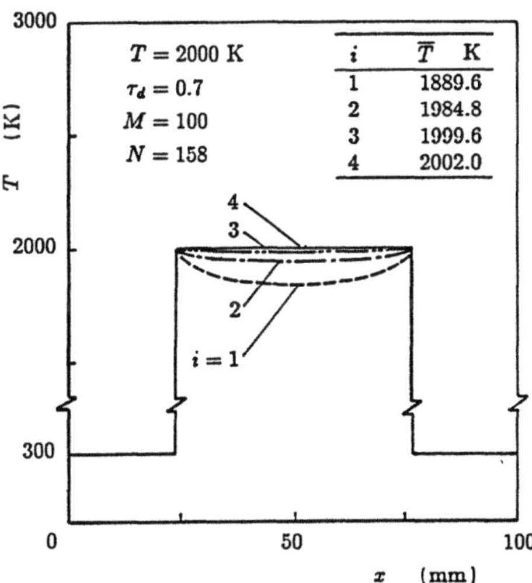

Fig. 4: Demonstration of the effect of iteration for a circular gas volume, y=50 mm.

316

The temperature distribution approaches to the original one (2000 K) with iterations. The average of reconstructed temperatures $\bar{T}$ of the gas volume converges to the original one within 2 K in four iterations. Convergence of Δr_i also is fast (Figure 5) and four iterations are enough in this case. Number of samplings M is related to the minimum size to be reconstructed by tomography. M is set at 100 to see the effect of N on iteration. The results for the same sample gas for Figure 4 are listed in Table 1. As N increases ΔT_{RMS}, root mean square of temperature difference ($T_{xy} - \bar{T}$) for i=4 in the range of d=50 mm, decreases, but no change is seen over 158 which is the optimum value for M=100. As the number of projections increases, the time required for data acquisition becomes so long that it will be impractical. Some kind of trade-off, therefore, should be looked for to facilitate the measurement.

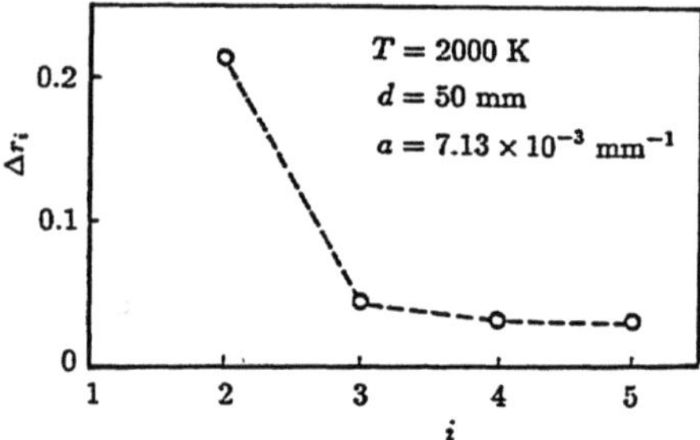

Fig. 5:   Convergence of Δr_i with i.

SPACIAL RESOLUTION

Limit of the smallest size Δx of an object for which an accurate measurement is possible is checked by decreasing Δx of gas with 2000 K and a of 7.13×10^{-3} mm^{-1}. The span of scanning X_L is set to 100 mm. The reconstructed temperature for i=2 increases very slightly, less than 4 K (Figure 6), with $\Delta x/X_L$, because the effect of self-absorption of small optical depth is very little. Interpolation of x,y from the grid of X-Y coordinates gives rise to some amount of errors which smears the values of a at the border of gas volume. Since temperature is calculated by an equation for the i-th iteration similar to the form of eq. (9), which has $D_{2x\theta 0i}$ and a_{xy} on opposite side of equation, the influence of interpolation on temperature will be cancelled during the process of reconstruction. The effective absorption coefficient a_{xy} is influenced by the interpolation, therefore a for $\Delta x/X_L = 0.02$ is only 65% of the original value, while it increases up to 95% and 98% for $\Delta x/X_L = 0.04$ and 0.06 respectively. The spacial

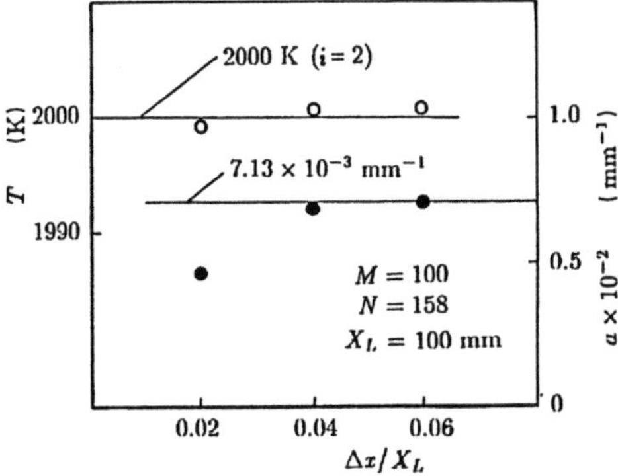

Fig. 6: Size of gas volume Δx and the reconstructed values.

resolution of this method will be 4% of X_L (the domain of measurement) in view of a, though that for temperature will be about 2%.

NOISE IN PROJECTION DATA

The discussion so far was performed for projection data with no noise. To check the algorithm under the more realistic conditions, the effect of a certain amount of random noise upon the final results must be investigated. $D_2X_{\theta 0}$ signals for the circular volume of gas used for Figure 4 and Table 1 is modulated by a random noise of standard deviation δ of 5.6% of the peak value of $D_2X_{\theta 0}$ which corresponds to S/N ratio of 17.9. The projection of $D_2X_{\theta 0}$ signal displayed on the coordinates of projection N and sampling point M is shown in Figure 7. Figure 8 is the temperature distribution along a diameter of the gas. The average temperature $\overline{\overline{T}}$ is very close, within 8 K, to that of Table 1 for i=2 but ΔT_{RMS} is much greater than the one in the same table. The noise of projection data should be suppressed as low level as possible.

Table 1: Effect of N on the mean of reconstructed temperature, M=100.

i	N	$\overline{\overline{T}}$		K	
		30	52	158	312
1		1889.6	1889.6	1889.6	1889.6
2		1985.9	1984.6	1984.8	1984.8
3		2001.2	1999.4	1999.6	1999.6
4		2003.9	2001.8	2002.0	2002.0
ΔT_{RMS} (i=4)		5.87	1.54	0.96	0.96

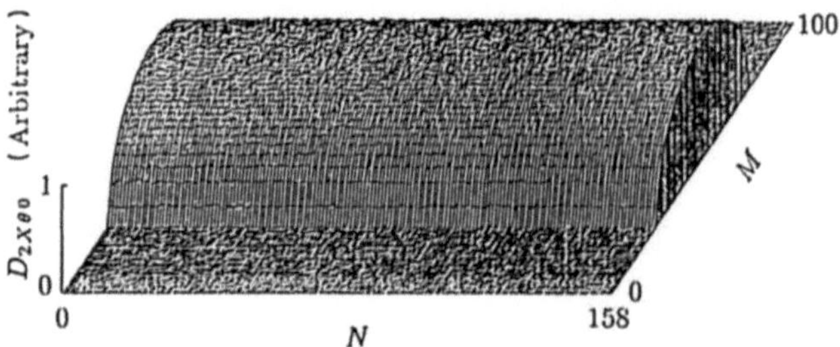

Fig. 7:   Projection data with random noise of δ=5.6%.

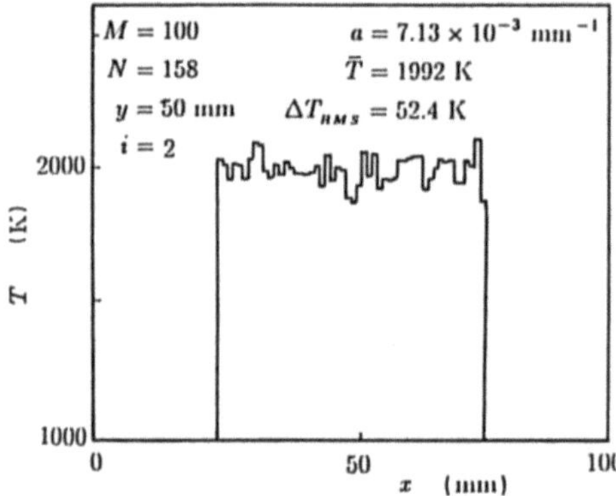

Fig. 8:   Reconstructed temperature distributions with noise in $D_{2x\theta 0}$.

ASYMMETRIC GAS

A pair of two circular gases of 25 mm in diameter are used as an example of asymmetric objects. The temperatures of the two gases are T_1=2000 K and T_2=1500 K, while the absorption coefficients are a_1=7.13x10^{-3} mm^{-1} (τ_{d1}=0.837) and a_2=9.51x10^{-3} mm^{-1} (τ_{d2}=0.788) respectively. The reconstructed distributions along the centers of the two gases are shown in Figure 9. The average temperatures approach the originals within 2 K with i=3.

CPU TIME

The time required for the calculation depends on many conditions and ranges from 20 s to 1000 s with a computer FACOM M760/6 (8MIPS).

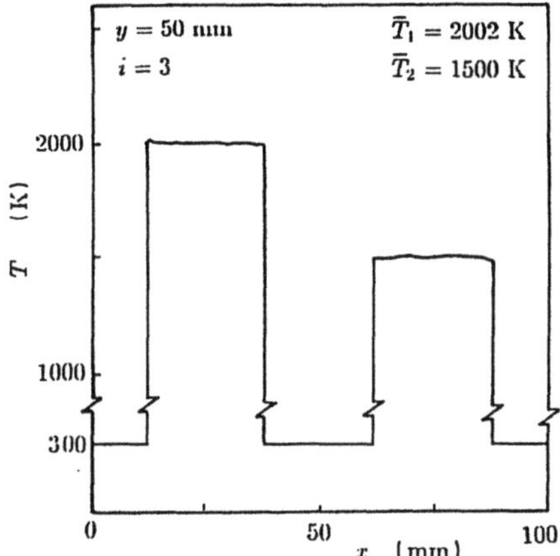

Fig. 9:   Reconstructed distributions of temperature along the line passing through the centers of two circular gases with 25 mm in diameter.

EXPERIMENTAL RESULTS AND DISCUSSIONS

The span of scanning along X axis is 80 mm and the reconstructed results within x=60 mm will be presented. The cross sections of radiative beams are 1.5x4 mm for 1.8 μm, 1.1x3 mm for 2.7 μm and 1.1x1.6 mm for 4.3 μm. 60 projections are used for flat flames while they are 30 for ring and diffusion flames to reduce the time of data acquisition. Number of sampling is 100 for all the experiments. The measurement for a jet flame was carried out under different conditions.

RESULTS WITH 4.3 μm BAND

A stoichiometric premixed flat flame of C_3H_8 with 30 mm in diameter was used to investigate the performances of the iterative method. 4.3 μm band of CO_2 was selected so that the density distribution of carbon dioxide in the flame can be measured. The course of calculation by iteration of temperature distribution at 4.5(=z) mm downstream of the burner outlet is shown in Table 2. The mean temperature $\bar{T}$ converges to 2076 K with four iterations, i=4. Δr_i attains the smallest value 0.0623 at i=3, while ΔT_{RMS}, the root mean square of temperature deviations from the mean value $\bar{T}$ over the range of x=20.8 ~ 39.2 mm increases gradually with i. Figure 10 is a graphical display of the results. (a) is a distribution of the reconstructed temperature. When the radiative intensity is small the S/N ratio of signals are deteriorated to the level which gives rise to reconstructed temperatures with very low accuracy. So the temperatures at the

Table 2: Reconstructed temperature in the central part (x=20.8 ~ 39.2 mm, y=30 mm) of the flame for each step of iteration i with 4.3 μm band, z=4.5 mm, $\bar{T}$: mean temperature, T_{max}: maximum in the range of x, T_{min}: minimum in the range of x, ΔT_{RMS}: root mean square of temperature deviations from $\bar{T}$ in the range of x.

i		1	2	3	4
T_{max}	K	1886	2055	2099	2109
T_{min}	K	1806	1992	2025	2032
ΔT_{RMS}	K	18.1	19.0	20.3	20.8
$\bar{T}$	K	1842	2029	2068	2076
Δr_i		—	0.2688	0.0623	0.0664

points where the absorption coefficients are smaller than a certain value (e.g., 15% of the peak value) were eliminated in the course of calculation. (b) is the temperature distribution along x axis at y=30 mm. The data measured by a Pt-Rh thermocouple, indicated by a broken line, yields a mean temperature $\bar{T}_{TC}$ which is 53 K lower than that of the reconstructed one in the range of x used in Table 2. The difference of 53 K might be caused by the measuring error of thermocouple and the inconsistency of measuring spaces of the two methods, but exact reasons are not clear yet. (c) is the distribution of effective absorption coefficient a of CO_2. The value, about 1.2×10^{-2} mm^{-1}, in the central part of the burner is at the same level of the coefficient presented in reference [11]. The exact value of the gas will be obtained by compensating the effect of opto-electronic characteristics t_λ of the instruments. (d) is the density distribution of carbon dioxide ρ_{CO_2} along a diameter of the burner. The values of the central part coincide well with those (marked •) measured by gas detector tubes (GASTEC CO. LTD) and also with the value of 2.6×10^{-5} g/cm^3 which is calculated for the equilibrium condition of the burnt gas. Results of the tomographic method above the rims of the burner are higher than those of the detector tubes. It will due to the difference of spacial resolution of the two methods and also to the induction of surrounding air which will dilute the burnt gas to lower the results of detector tubes by dilution. The accuracy of density measured by tomography depends on the exactness of eq. (13) and on that of the parameters used [11]. The overall error of the band model is estimated to be below 20% [11]. So the error of density obtained by this method will be on the same level of a or worse.

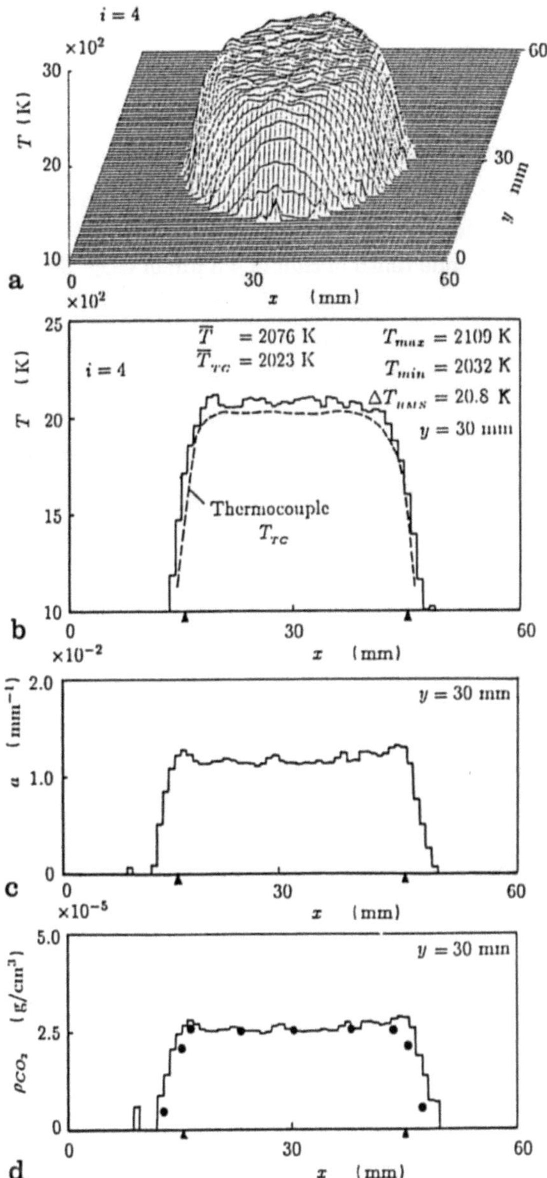

Fig. 10: Reconstructed results of a stoichiometric C_3H_8-air flat flame with 4.3 μm band
▲: burner rims, •: CO_2 density by gas detector tube.

RESULTS WITH 2.7 µm BAND

Figure 11 is the temperature distribution measured with 2.7 µm band along a diameter of the flat flame used for 4.3 µm. Difference between the mean temperature of the constructed one $\overline{T}$ and that of a thermocouple ($\overline{T}$-$\overline{T}_{TC}$) is 38 K in the same part of the gas as Table 2 though the difference is greater outside of the burner rim. The absorption coefficient of 2.7 µm band which is composed of two spectra of H_2O and CO_2 is less than one tenth of that of 4.3 µm of CO_2. So the convergence of Δr_i, as well as $\overline{T}$, is better (Table 3) than the case of 4.3 µm band. The effective absorption coefficient of this band can be estimated by using the composition for the equilibrium state of C_3H_8-air at $\overline{T}$. The density, therefore, depends on the results of equilibrium calculation when the spectra of CO_2 and H_2O are not separated. Temperature distribution is reconstructed with higher accuracy compared with that of 4.3 µm, that is smaller ΔT_{RMS} and $\overline{T}$-$\overline{T}_{TC}$, because the S/N ratio of 2.7 µm band is about 1.7 times better than that of 4.3 µm

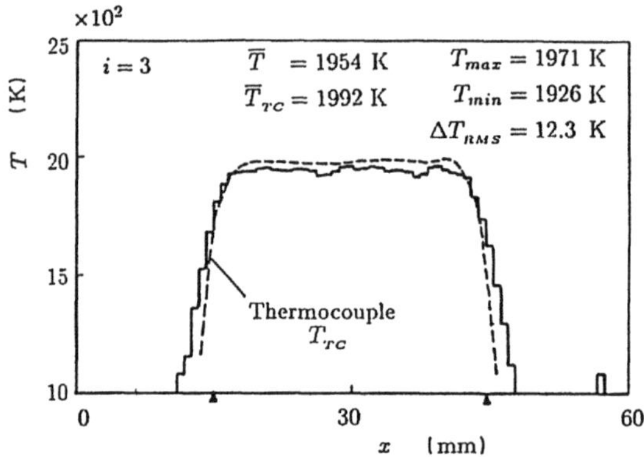

Fig. 11: Reconstructed results of a stoichiometric flat flame with 2.7 µm band, y=30 mm, ▲: burner rims.

Table 3: Effect of iteration i on reconstructed temperatures of gas of Figure 11, x=20.8 ~ 39.2 mm, $\overline{T}$, T_{max}, T_{min}, ΔT_{RMS} are the same as Table 2.

i		1	2	3
T_{max}	K	1966	1971	1971
T_{min}	K	1920	1926	1926
ΔT_{RMS}	K	12	12	12
$\overline{T}$	K	1984	1954	1954
Δr_i		—	0.0831	0.0810

with the opto-electronic system used in this experiment. 2.7 µm band is suitable for the temperature measurement while the measured density can show only that of H_2O-CO_2 mixture as far as the opto-electronic elements of the present apparatus are used.

RESULTS WITH 1.8 µm BAND

The same circular flat flame for the preceding examples was measured with 1.8 µm band of H_2O. Strength of the band is about one ninth of 2.7 µm band and the S/N ratio of projection data is small. Results are shown in Figure 12. The temperature distribution along a diameter (a) has greater ΔT_{RMS} of 60.1 K compared with 20.8 K of 4.3 µm band and 12.3 K of 2.7 µm band. The results of absorption coefficient were roughly equal to the mean value of 8.8×10^{-5} mm^{-1} of the 1.8 µm band calculated from the data of reference [11]. Density of water vapour ρ_{H_2O} (b) also is noisy though average of the flat part is close to the data (marked •) obtained by gas detector tubes. They are not so much different from

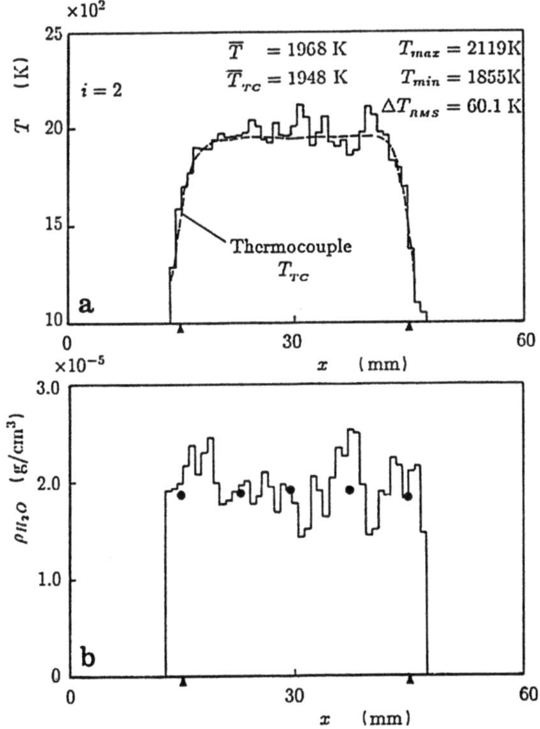

Fig. 12: Reconstructed results with 1.8 µm band, y=30 mm; ▲: burner rims, •: H_2O density by gas detector tube.

1.7×10^{-5} g/cm^3, that is the value of equilibrium state. Results outside of the periphery of the burnt gas were neglected in (b). This band can be used for the estimation of distributions of both temperature and density of water vapour, but the results are too noisy to investigate accurate and minute distributions.

RING FLAME

A thin metal-disc with 15 mm in diameter was placed at the center of a flat flame burner to form a ring shaped temperature field so that a low temperature region will be made in the center. The reconstructed results measured at 6.5 mm downstream of the burner with 4.3 µm band is shown in Figure 13. The tomographic values of temperature throughout the diameter are higher than those of the thermocouple T_{TC} (a). Densities of CO_2 coincide well with the values of gas detector tubes indicated by black circles in Figure 13 (b). Since the temperature above the disc is lower than that over the burner opening the density at the central region is higher than that of the opening while that for a flat flame is uniform throughout the diameter (Figure 10d). Figure 14 is the results for the same ring flame measured with 1.8 µm band. The reconstructed temperature (a) is noisy, though the smoothed values coincide fairly well with

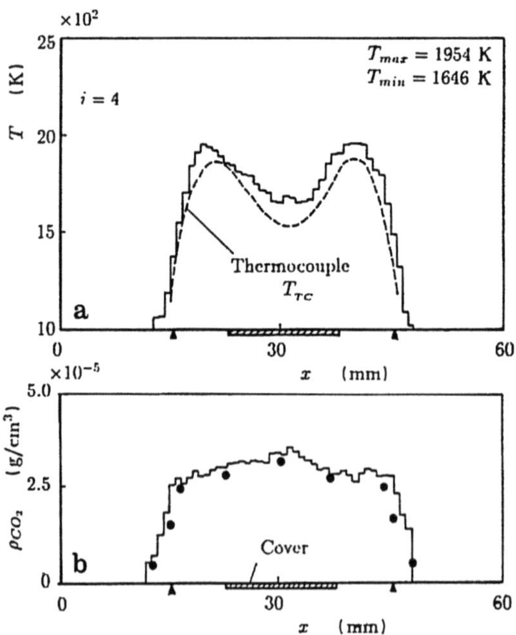

Fig. 13: Reconstructed results of a ring flame with 4.3 µm band, y=30 mm, ▲: burner rims, •: CO_2 density by gas detector tube.

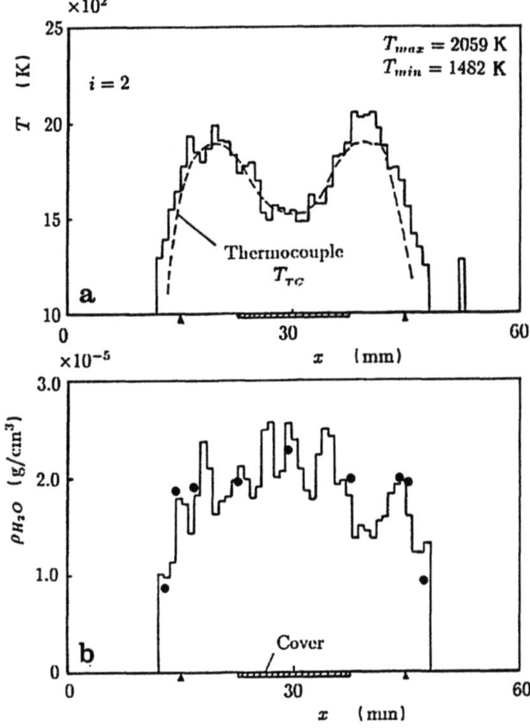

Fig. 14: Reconstructed results of a ring flame with 1.8 µm band, y=30 mm,
▲: burner rims, •: H₂O density by gas detector tube.

that of a thermocouple. The distribution of water vapour density also is noisy
but it has the same tendency as that in Figure 13 b) which implies that the
lower temperature under the same pressure brings higher density.

ASYMMETRIC DISTRIBUTIONS

A semicircular premixed flame was used as an example of asymmetric objects.
The results measured with 2.7 µm band is shown in Figure 15. A semicircular
form of the temperature distribution is clearly displayed in (a), while the one
compared with a thermocouple reading (b) shows similar trends to the
examples presented so far. A pair of two circular premixed flat flames of 15
mm in diameter which was similar to that of Figure 9 was measured with 2.7
µm band and it was checked that the results were the same as the case of
Figure 15.

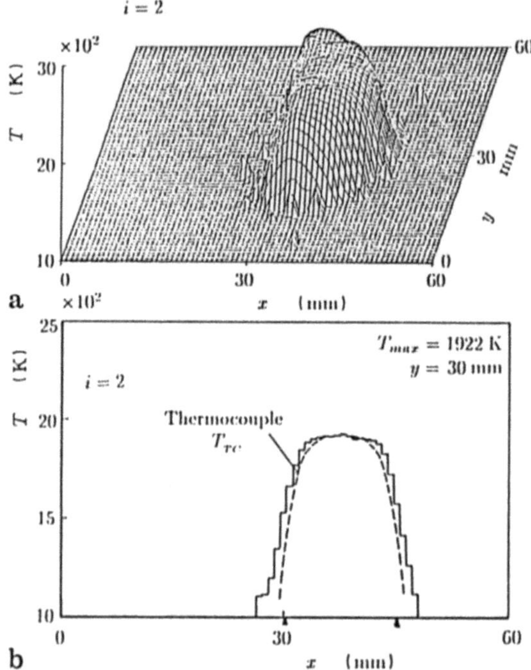

Fig. 15:   Semicircular premixed flat flame measured with 2.7 µm band, ▲: burner rims.

DIFFUSION FLAME

A flame of Bunsen burner which was surrounded by a ring of sideflow to stabilize the flame was employed as an example of diffusion flames. 8% of CO_2 was added to dry air of the sideflow to increase CO_2 density in the flow. Figure 16 is the data measured at 4 mm (=z) downstream from the burner outlet. A diametric cross section (b) of the reconstructed temperature distribution (a) coincides fairly well with that of a thermocouple (dashed line) along the periphery of the gas. The temperature inside the peaks was not measured because the peaks were beyond the thermal durability of the thermocouple. The reconstructed temperature at the center seems much higher than expected, because the flame must maintain the core of cool unburnt mixtures at this section, z=4 mm downstream from the rim. More precise investigations are needed to find exact reasons. CO_2 density coincides well with that of the gas detector tubes, Figure 16(d). Two peaks coincide with the outer peaks of Figure 16(c) which are brought about by the CO_2 mixed into the sideflow.

Figure 17 is the same flame measured at z=25 mm. Reconstructed temperature coincides well with the results of a thermocouple (a). CO_2 density also agrees with the results of gas detector tubes (b).

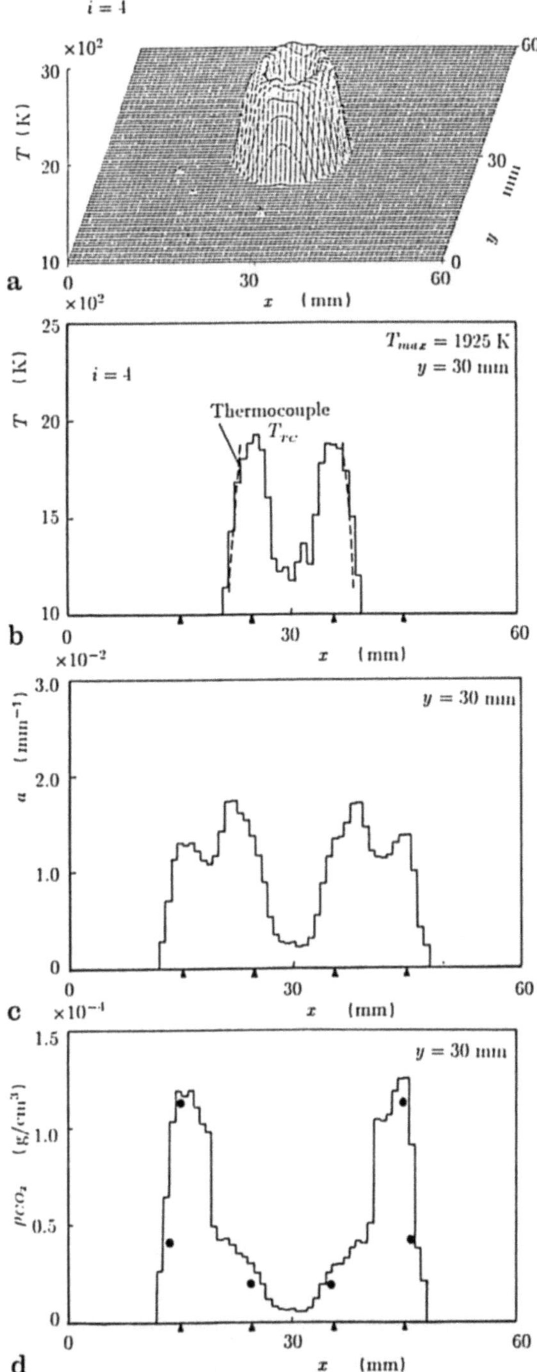

Fig. 16: Results of a Bunsen flame of C_3H_8-air (1.2 stoichiometry) with sideflow with 8% of CO_2, z=4 mm, 4.3 μm band, ▲: burner rims, •: CO_2 density by gas detector tube.

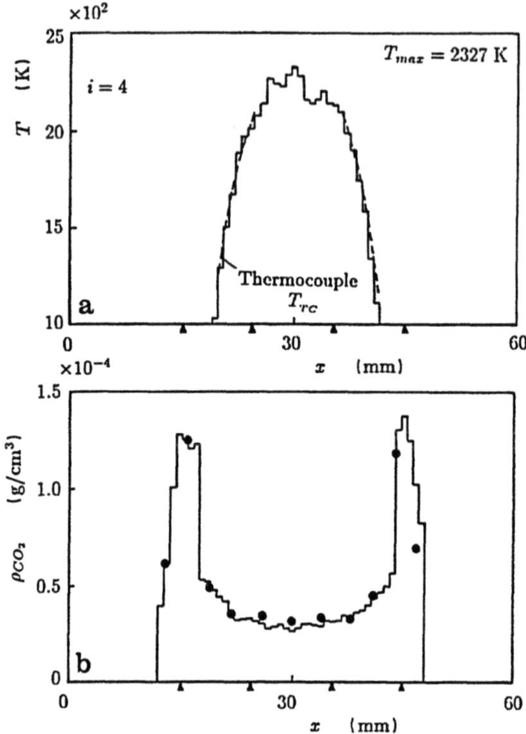

Fig. 17: Results of a Bunsen flame, z=25 mm, 4.3 μm band, y=30 mm, ▲: burner rims,
•: CO_2 density by gas detector tube.

TEMPORAL RESOLUTION

All the measurements presented in the preceding sections were performed
with projections of N=30 and 60 and samplings of M=100. The data acquisition
into a floppy disc took from 50 to 90 minutes. The CPU time is about 2 minutes
for 3 iterations by the same computer used for the simulations. The time for
data acquisition is the biggest obstacle to investigate transient phenomena. A
slab of parallel light beam with a cross section of 4.4x35 mm, in place of a fine
beam of radiation in Figure 1, was cast from a black body through a gas and
detected by a Plasma-Coupled Device array (PCD array, HAMAMATSU
S2313-35Q) of 35 pixels. Each pixel works just like the detector in Figure 1, so
that 35 sampling data can be acquired in one shot. Four sets of these opto-
electronic systems, crossing at each center of the light path, are arranged
radially with angles of 45 degrees. An array for infrared ray is not easy to
acquire at present, so the one for a visible light, 589 nm (D-line) of sodium
spectrum, is used as a preliminary experiment in this report. The fuel-air

mixture was seeded with NaCl. By replacing the opto-electronic elements the system will be altered to be sensitive to infrared active components, such as H_2O, CO_2 and so forth. $D_{1X\theta}$ signal for the black body can be measured before a transient phenomena will occur. Since measurement of one set of $D_{2X\theta}$ and $D_{3X\theta}$ signals takes 1.2 ms for the PCD array used, the measured temperature is a temporal average for this period of time. A slow jet of burnt gas of C_3H_8-air mixture was measured at 10 mm downstream of an orifice with 12.5 mm in diameter which is   placed on one of end-plates of $\phi30x42$ mm drum type chamber.

Figure 18 is one of the data obtained at the time 0.9 ms after the tip of a jet passed the orifice. Numbers of projections and samplings are expanded from 4 and 35 to 40 and 70, respectively, by interpolation. The jet at 10 mm downstream of the orifice might retain a core of flow where the variation of temperature will be not so great. The distribution of the reconstructed temperature under these conditions is shown in (a). The distribution of local radiative intensity E_{xy} has different form (b) because E_{xy} is a function of both temperature and density of the radiative medium.

Nonuniformities among arrays and pixels were compensated but the exactness and stabilities of these elements are important to improve the accuracy of measurement.

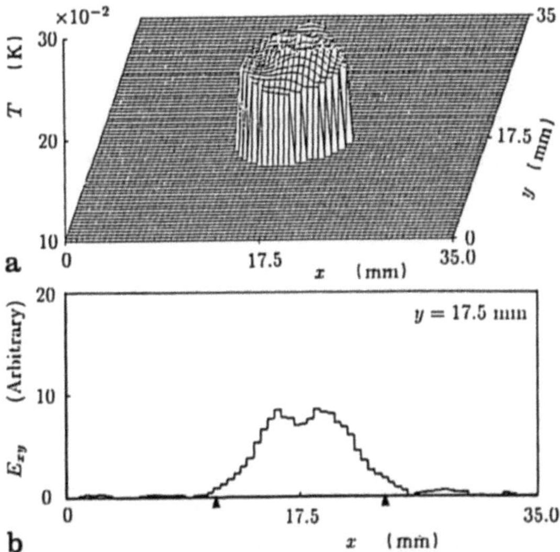

Fig. 18:   Reconstructed data of a flame jet from an orifice of 12.5 mm, stoichiometric C_3H_8-air mixture, z=10 mm, 0.9 ms after ejection, M=70, N=40, ▲: edge of orifice.

The temporal resolution is greatly dependent on the chopper speed and the performances of the PCD array as far as the translate and rotate scanning is employed. Some other methods of data acquisition are needed to improve the temporal resolution.

CONCLUSIONS

A method of iteration was introduced to improve the accuracy of computed tomography which uses infrared radiation to measure simultaneously the distributions of both temperature and density inside a self-radiating gas. The characteristics of the method were checked by both simulations and experiments. Following are the conclusions;

1. Influence of self-absorption of burnt gases causes big error to the temperature reconstructed by an ordinary computed tomography when the transmissivity of a gas volume is less than about 0.9.

2. An iterative method was introduced to compensate the influence of self-absorption. The number of iterations required is three to four for the burnt gases of ordinary burners used in laboratories.

3. The kind of gas components whose density distribution is measured can be selected by choosing a spectral band pertinent to the components. Accuracy of the reconstructed density depends on exactness of absorption coefficient of the band.

4. By employing array detectors (such as PCD), the method of iteration might be used for diagnostics of transient phenomena of high temperature gases. Feasibility of the method was shown with the D-line of sodium vapour.

ACKNOWLEDGEMENT

Authors are grateful to the students of Gifu University for their assistances in performing the experiments.

REFERENCES

1. McCay T.D., Roux J.A. (ed.): Combustion diagnostics by nonintrusive methods. Progress in Astronautics and Aeronautics, 92, AIAA, Inc. New York (1984).

2. Faris G.W., Byer R.L.: Quantitative optical tomographic imaging of a supersonic jet. Optics Letters, 11 (1986) 413.

3. Faris G.W., Byer R.L.: Beam-deflection optical tomography. Optics Letters, 12 (1987) 72.

4. Snyder R., Hesselink L.: Measurement of mixing fluid flows with optical tomography. Optics Letters, 13 (1988) 87.

5. Ray S.R., Semerjian H.G.: Laser tomography for simultaneous concentration and temperature measurement in reacting flows. Progress in Astronautics and Aeronautics, 92, AIAA, Inc. New York (1984) 300.

6. Kurlbaum F.: Über das Reflexionvermögen von Flammen. Physikalishe Zeitschrift, 3 (1902) 332.

7. Schmidt H.: Prüfung der Strahlungsgesetze der Bunsenflamme. Annalen der Physik, 29 (1909) 971.

8. Shimizu S. et al.: Measurement of temperature and density distributions of a jet flame by an infrared radiation CT, 87-Tokyo-International Gas Turbine Conference, 7 (1987) 25.

9. Budinger T.F., Gullberg G.T.: Three dimensional reconstruction in nuclear medicine emission imaging. IEEE Trans, Nucl. Sci, NS-21 (1974) 2.

10. Shepp L.A., Logan B.F.: The Fourier reconstruction of a head section, ibid., 21.

11. Koo I. et al.: Radiative properties of combustion gases. AIAA 14th Thermophysics Conference, Florida (1979) and General Dynamics, Convair: Study on exhaust plume radiation predictions, GD/C-DBE-66-017, Space Science Laboratory (1966).

QUICK MEASURING TECHNIQUE FOR FLAMES (QMF)

P. Stuber, J. Gass, P. Suter

Institute for Energytechnology
Laboratory for Energysystems
Swiss Federal Institute of Technology
ETH Zentrum, 8092 Zürich — Switzerland

ABSTRACT

IR-emission spectra were measured at different positions in an oil flame. They are used to develop an imaging system for temperature zones (named QMF). Methods are discussed on how to obtain temperature information from distinct spectral intensities.

INTRODUCTION

Background. On designing "clean" burners for household furnaces information about temperature inside the flame needs to be obtained. Most of the design companies can not afford expensive equipment for measurement of temperature. Until now personal experience has been the most important instrument in the design of burners. This becomes insufficient when trying to optimize the burners with respect to emission. Temperature charts of the flame at different states would be very useful. Intrusive methods are not desirable. Methods like CARS or LIF are definitely too expensive and too complicated for this industry. A method based on the emitted flame radiation seems reasonable. Spectra from the emitted radiation at different points of a flame are helpful in developing such a method.

New Aspects. The emitted radiation from a flame depends on the temperature and the emission coefficient ε. The coefficient ε is the ratio between the measured emission and the emission of a blackbody at the same temperature. The coefficient ε is mainly a function of the number of radiating molecules. Common low cost methods assume a uniform distribution of the emission coefficient over the flame. The method discussed in this article, QMF, takes into consideration the local variation of the emission coefficient. This is done by measuring the intensity in different wavelength regions.

The QMF will be a tool for designing burners; the result will be rather of qualitative than of quantitative nature.

PRINCIPLE OF "QMF" (Quick measuring Technique for Flames)

IR Camera with Image Processing. The flame is imaged with an infrared camera in three (or more) different wavelength regions using special filters. The images are stored in digital form. For each pixel a temperature is calculated as a function of the three spectral intensities. How this function looks will be discussed later.

The calculated temperatures shall be presented with colours.

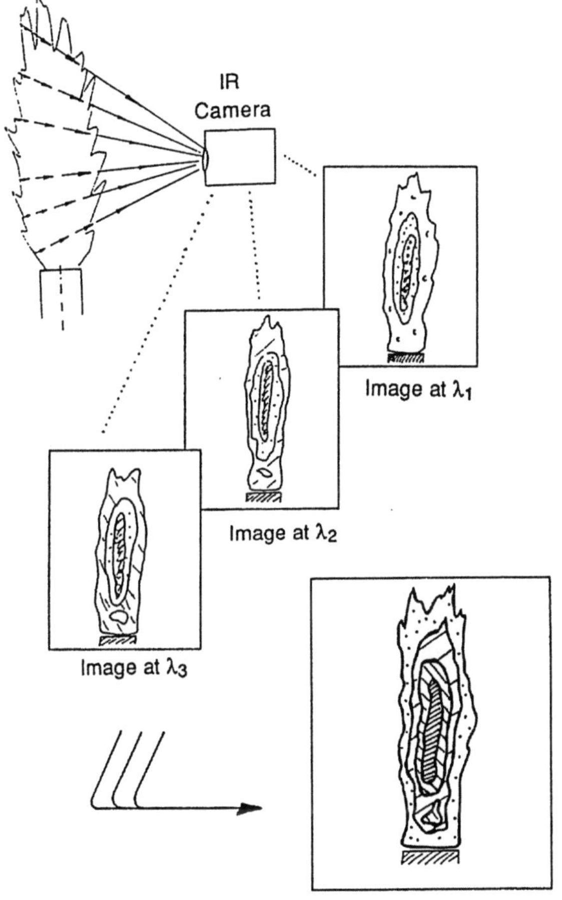

Fig. 1:   Principle of QMF

Two Dimensions. For the time being QMF will yield average two-dimensional temperature information. The value at one point of the calculated temperature chart represents an average radiation temperature along the corresponding line of sight through the flame. For many purposes this two-dimensional information will be sufficient.

Assuming rotation symmetry a three-dimensional temperature field can be calculated with an Abel Transformation. General flames can be treated with Computer Tomography.

CONCLUSION FROM RADIATION INTENSITY ABOUT TEMPERATURE

Intensity and Emission Coefficient. We look at one single line through the flame (Figure 2). The measured intensity in this direction is the product of ε and the intensity that would be emitted by a blackbody at the same temperature and at the same solid angle. ε is called the emission coefficient.

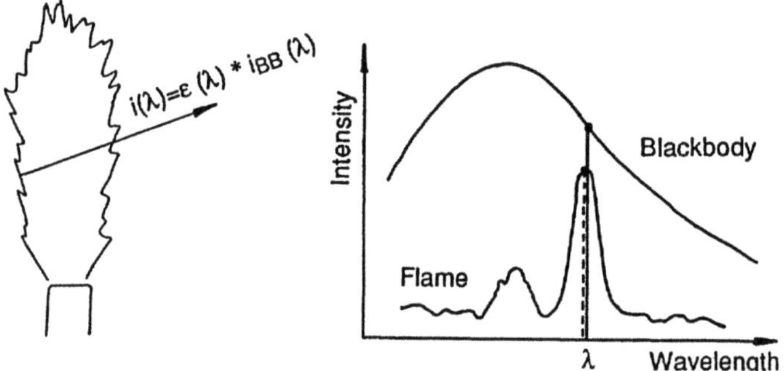

Fig. 2:   Average temperature along line. Flame spectrum in relation to blackbody spectrum.

With the Planck radiation law the temperature can be calculated from the measured intensity and the emission coefficient:

$$T(i(\lambda), \varepsilon(\lambda)) = \frac{c_2}{\lambda * \ln\left(\dfrac{2c_1}{\lambda^5 * \dfrac{i(\lambda)}{\varepsilon(\lambda)}} + 1\right)} \tag{1}$$

c_1 $= 5.9552 * 10^{-17}\ Wm^2$

c_2 $= 1.4388\ 10^{-2}\ mK$

λ = wavelength in m

$i(\lambda)$ = measured intensity in $W/m^3.sr$

$\varepsilon(\lambda)$ = emission coefficient

T = temperature in K

For a given material ε may vary with the wavelength λ; in the case of a flame this variation is very strong. The emission coefficient at a fixed wavelength λ, $\varepsilon(\lambda)$, depends on the point in the flame. $\varepsilon(\lambda)$ becomes stronger as the number of radiating parts on the considered line through the flame increases. Flame radiation is mainly caused by the gases CO_2 and H_2O that radiate in bands; on this is superposed the radiation from soot which shows a continuous spectrum. Due to local variation of these species in the flame the emission coefficient $\varepsilon(l)$ may vary by a factor of 10.

Setting $\varepsilon(\lambda)$ constant at all points in the flame leads to the following problem: a zone of uniform temperature but of varying concentration will be misinterpreted as a zone of varying temperature.

If this error can not be accepted, the value $\varepsilon(\lambda)$ has to be determined for each line through the flame. This may be done by sending known radiation through the flame and measuring the local absorption. Kirchhoff's law states that the spectral absorption coefficient is equal to the spectral emission coefficient: $\alpha(\lambda) = \varepsilon(\lambda)$. A more convenient method is described in the following section.

Ratio Method. For determining the temperature of a solid body the use of a ratio pyrometer is quite popular. Instead of measuring one spectral intensity $i(\lambda)$ together with the emission coefficient $\varepsilon(\lambda)$, the intensity is measured at two wavelengths.

The temperature is given as the ratio of the two intensity values. For $c_2/\lambda T \gg 1$ the Planck formula is simplified to:

$$
T\left(\frac{i(\lambda_1)}{i(\lambda_2)}, \frac{\varepsilon(\lambda_2)}{\varepsilon(\lambda_1)}\right) = \frac{c_2 * \left(\frac{1}{\lambda_2} - \frac{1}{\lambda_1}\right)}{\ln\left(\frac{i(\lambda_1)}{i(\lambda_2)} * \frac{\varepsilon(\lambda_2)}{\varepsilon(\lambda_1)} * \left(\frac{\lambda_1}{\lambda_2}\right)^5\right)}
\tag{2}
$$

The formula shows that this method is practicable under the condition of a constant ratio $\varepsilon(\lambda_2)/\varepsilon(\lambda_1)$. Preliminary investigations show that this ratio varies only a little with the position in the flame. Examples will be described later in this paper.

The advantage of the ratio method over the method described in the last section is that different solid angles under which the flame zones are seen have no influence.

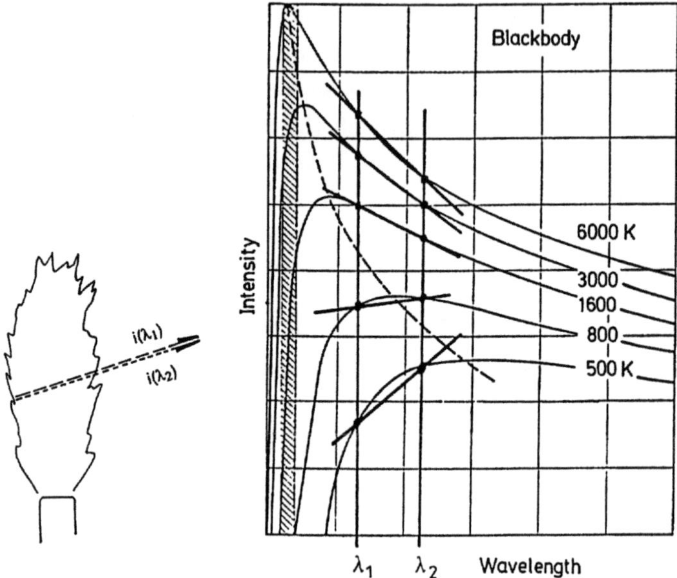

Fig. 3: Ratio method for calculation of average temperature along line. Ratio method for blackbody.

IR-RADIATION OF FLAME

The radiation emitted from a flame in the infrared region comes mainly from the gases CO_2 and H_2O and from the soot particles. These components can be supposed to be in thermodynamic equilibrium. Radiation originating from species not in equilibrium can not be used for calculation of temperature using Planck's law. An example of such radiation is chemoluminescence. Gases radiate in bands whereas soot shows a continuous spectrum.

Radiation from H_2O. The radiation bands are relatively broad and are spread over a wide wavelength region from 1 µm to 10 µm. The bands at 1.9 µm and 2.7 µm are important. For a given pathlength the values $\varepsilon(\lambda)$ increase with increasing temperature, although the density decreases. The increase in molecular emission overcomes the decrease in density. The band wings become stronger with increasing temperature. [1].

Radiation from CO_2. Carbon dioxide shows only two significant bands at 2.7 µm and at 4.3 µm; however these are very strong. In the region of 4.3 µm CO_2 is almost opaque even at short pathlengths. In the same way as H_2O the band wings become stronger with increasing temperature. [1]

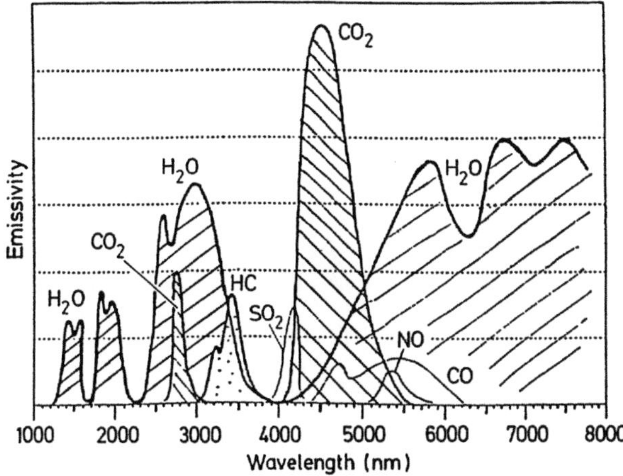

Fig. 4: Radiating gases of a flame. Only the band strength of one and the same gas may be compared.

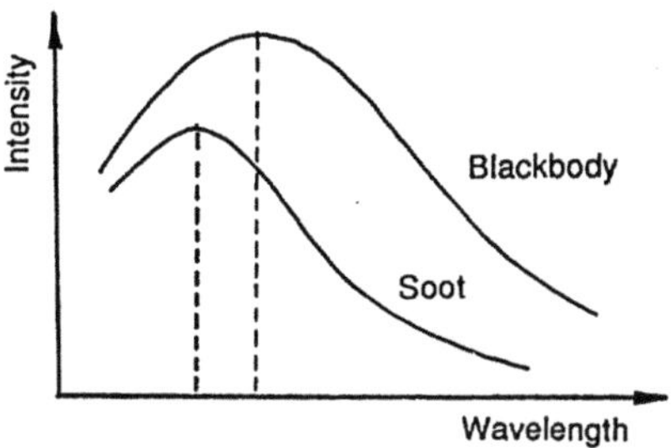

Fig. 5: IR-radiation spectrum of soot compared with spectrum of blackbody at the same temperature.

Radiation from Soot. The radiation spectrum of soot is continuous. The intensity varies strongly with the soot concentration and with the particle size. Towards higher wavelengths the emission coefficient of soot decreases, which results in a radiation maximum at shorter wavelengths compared with that of the blackbody.

Radiation from other Species. Besides the radiation from CO_2, H_2O and soot there is some minor radiation from HC (unburnt fuel), OH, CO, NO and SO_2. Their importance depends on the kind of flame.

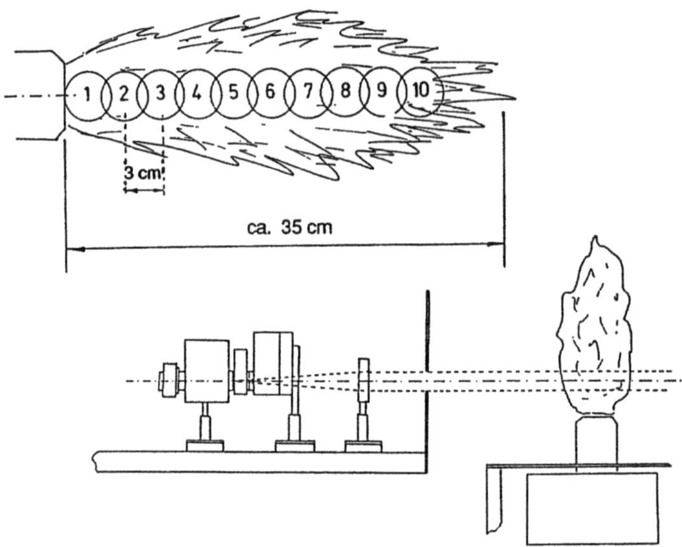

Fig. 6: Positions on the flame, where spectra were measured. Measuring equipment: lens, monochromator with pyroelectric detector.

MEASURED IR-EMISSION SPECTRA OF AN OIL FLAME

For ten positions in a free burning oil flame the IR-emission spectra were measured in the wavelength region 1.2 µm to 8 µm. The background radiation can be neglected.

Description of the Spectra (Figure 7). The two peaks from CO_2 (2.7 and 4.3µm) rise clearly above the general spectrum. At the flame root the continuous spectrum of soot covers the bands of H_2O between 1.3 µm and 2.3 µm. Towards the flame tip these bands appear clearly in the spectrum because the concentration of soot decreases. The peak at 3.5 µm seen in the spectrum at position 1 is interesting. It originates from the radiation of the unburnt fuel. The peak disappears with decreasing concentration of fuel.

Spectral Intensities along Flame Axis (Figure 8). The two upper curves show the intensity of the CO_2 peaks (2.7 and 4.3 µm). The ratio between the two intensities varies only a little (ratios are seen as differences in the log mode). At 2.35 µm and 4 µm the flame radiation originates mainly from soot. The corresponding curves fall more strongly towards the flame tip compared with the CO_2 curves. This means that the soot concentration relative to the CO_2 concentration is decreasing.

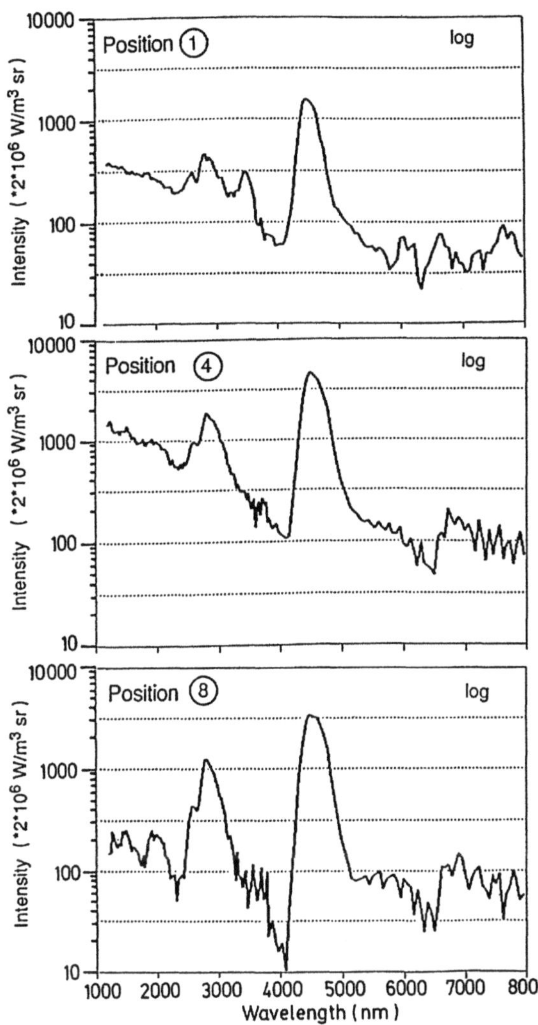

Fig. 7:   IR-emission spectra for three different positions on (through) the flame.

TEMPERATURES FROM MEASURED SPECTRAL INTENSITIES

Three methods of calculating temperatures from spectral intensities are tested for the observed oil flame. The values of the spectral intensities were taken from the spectra presented in the last section.

Temperature from Soot Radiation (Figure 9). The ratio method as described earlier was applied for wavelength regions where mainly soot is radiating. These regions are at 1.2 µm, 2.35 µm and 4.0 µm. The formula 2 was applied for

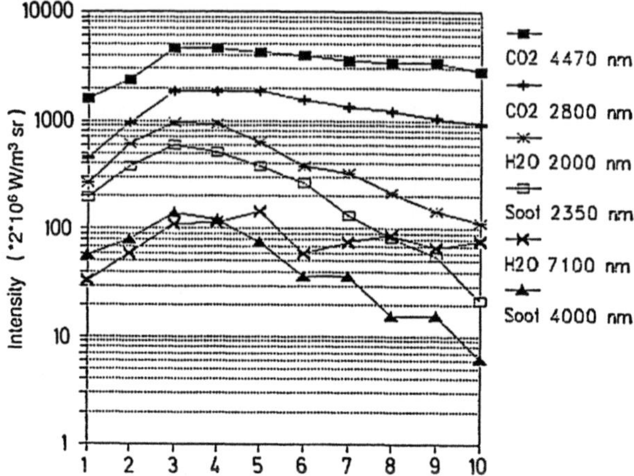

Fig. 8:   Spectral intensities along flame axis. (flame radiation, not partial radiation of components)

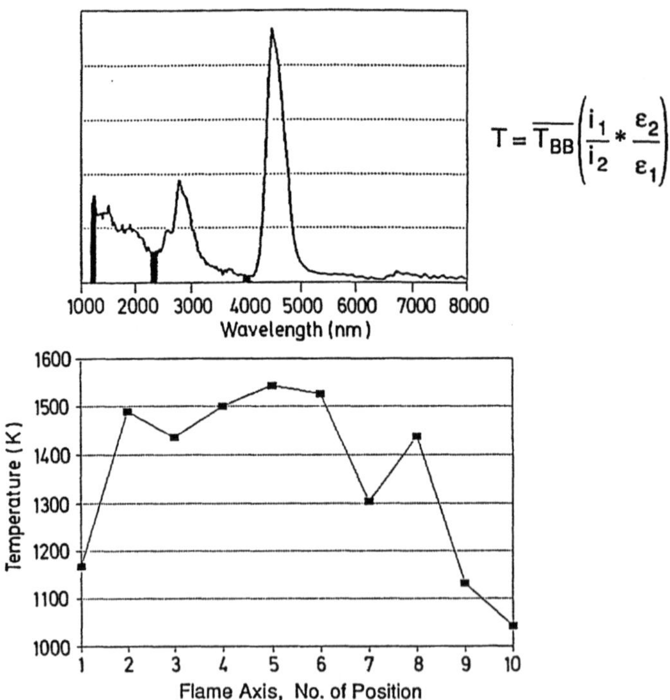

Fig. 9:   Temperature along flame axis. Ratio method for "soot intensities".

all three pairs of wavelengths and yielded three temperatures. The average of these three values is plotted for all ten positions on the flame axis.

The ratio $\varepsilon(\lambda_2)/\varepsilon(\lambda_1)$ was assumed to be constant at all positions. More investigation concerning this assumption remains to be done.

A temperature dip seems to occur at position 3 and 7. The graphs of the spectral intensities do not show these depressions. Obviously the decreasing temperature is compensated by an increasing number of radiating molecules.

Temperature from CO$_2$ Bands (Figure 10). In the same manner as above the ratio method (formula 2) was applied for the two wavelengths 2.7 µm and 4.4 µm, where the peaks of the two CO$_2$ bands are. The ratio $\varepsilon(\lambda_2)/\varepsilon(\lambda_1)$ was set to the value 7 for the whole flame. This may be done in a first approximation.

At 4 µm 95% of the radiation originates from CO$_2$. At 2.7 µm the proportion of radiation coming from species other than CO$_2$ may not be neglected. In practice this radiation from other species, mainly soot and H$_2$O, can be estimated by a measurement at 2.35 µm. The uppermost curve in Figure 10 was calculated

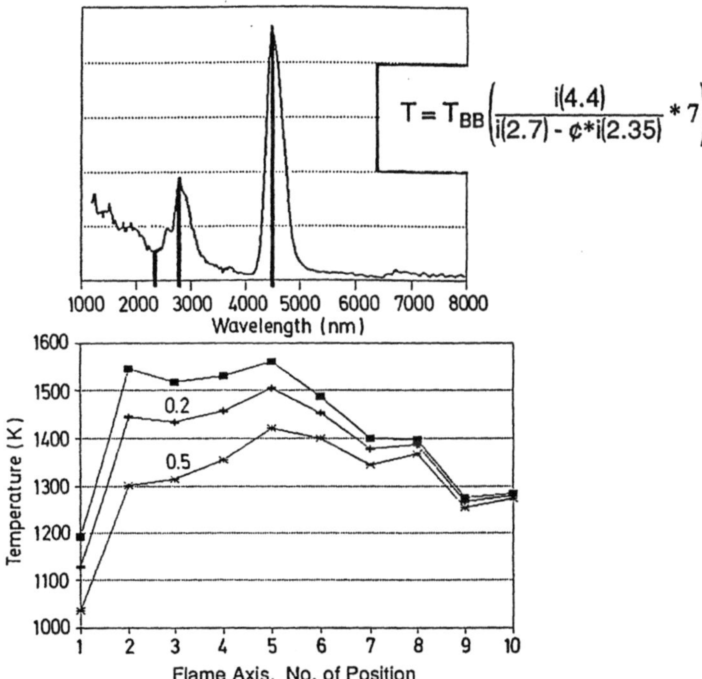

Fig. 10: Temperature from ratio of CO$_2$ bands. Correction of the 2.7 µm intensity.

without correction of the 2.7 µm intensity. For the other two curves the intensity at 2.7 µm was reduced by a fraction (0.2 respectively 0.5) of the intensity at 2.35 µm. The curves show that this correction has a great importance. In order to find the right correction method some other experiments must be done.

"Temperature" from 4.3 µm CO_2 band (Figure 11). Theory predicts that the intensity at the wings of a gas emission band becomes stronger at higher temperatures. In order to see whether this phenomenon can be observed in the measured spectra, the intensity at 4.8 µm was divided by the intensity at 4.4 µm for each position on the flame axis. This ratio is presented in Figure 11.

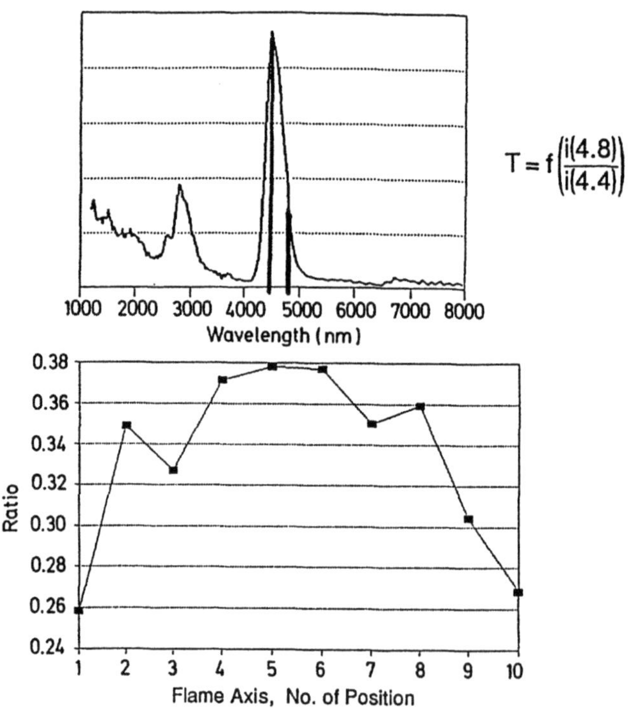

Fig. 11: "Temperature" from 4.3 µm CO_2 band. Ratio between intensities at 4.8 µm and 4.4 µm.

The similarities of the graph of this ratio to the curves in Figures 9 and 10 are very promising. It seems that the effect of change in band shape with varying temperature can in fact be measured even with simple instrumentation.

CONCLUSION

Measurements of intensity of emitted flame radiation at different wavelengths can provide temperature information. Methods as presented above are not suitable to calculate precise temperatures. A main reason for this is the difficulty of "extracting" the radiation of a single component (e.g. CO_2) from the total flame radiation. There is no spectral region (IR) in which the flame radiation originates only from one component. In this regard the region 4 µm is preferable.

Methods as described are useful to show temperature zones in a qualitative sense.

REFERENCES

1. Ludwig M.: Handbook of IR-Radiation from Combustion Gases. NASA SP 3080 (1973).

2. Kang A.L.: Vergleich des mit einer Fernsehkamera gewonnenen Strahlungsintensitätsbildes mit dem aus Temperatur- und Konzentrationsmessungen errechneten Intensitätsbild. Stuttgart (1982).

3. Härtel W.: Entwurf eines Flammenpyrometers auf der Basis der Fernsehtechnik, VGB Krafterkstechnik 53, Helft 6 (1973); Brennstoff-Wärme-Kraft 26, 1 (1974).

4. Mindermann K.-H.: Neue Perspektiven in der Flammenüberwachungs- und -bewertungstechnik, Gas Wärme International. 37 (1988)).

5. Hornbeck G.A.: Optical Methods of Temperature Measurement, Applied Optics, 52 (1966).

6. Tank V., IR-Temperaturmessung mit selbständiger Berücksichtigung des Emissionsgrades. DFVLR (1988).

THERMOGRAPHIC DETECTION OF BOND DEFECTS WITHIN MODELS OF SOLID PROPELLANT MOTORS

Helmut Schneider

Fraunhofer-Institut für Chemische Technologie
Joseph-von-Fraunhofer-Strasse
D-7507 Pfinztal-Berghausen
Federal Republic of Germany

ABSTRACT

Within solid propellant rocket motors there sometimes exist separations or bond defects between case and liner (insulation made of elastic polymer) or between liner and propellant. This paper demonstrates that separations or bond defects can be detected by thermography. Till now the smallest defects that could be detected had a diameter of 20 mm and a separation of 0.05 mm.

INTRODUCTION

Solid propellant rocket motors usually consist of a case (metal) which is coated inside with the liner (elastic polymer). The propellant is casted in this system. The motors may be exposed to mechanical stress and extreme temperature variations during storage and transport. As a consequence motorcase and liner or liner and propellant may be separated at some places. In the first case the motor case may be subjected to high temperature combustion gases during operation and may be destroyed by excessive localized heating. In the second case an increase of combustion chamber pressure can also cause the destruction of the rocket motor. Therefore the investigation of rocket motors with respect to such separations is important.

One method for the detection of these defects is thermography: if the object of investigation is heated from one side it will be penetrated by a heat flow. If there is a flaw, a separation or a bond defect between the different materials of the object, the heat flow will be distorted with the result that the surface temperature above the defect is somewhat higher than in neighbouring areas. Therefore the detection of defects within the object means to determine the distribution of the surface temperature. Since the temperature difference at the surface due to defects inside is very small, a very sensitive detector is necessary. In thermographic measurements the temperature distribution of the surface of the object is determined by the measurement of the IR - radiation. The principle of the method was applied in this field in USA in the sixties already [1,2].

EXPERIMENTAL STUDIES

There are different methods of producing useful temperature patterns. In the transient heating method the motor case is irradiated by means of an infrared source with linear dimensions (rod-shaped) which is oriented parallel to the motor axis (Figure 1). The IR - source was installed within a water cooled shielding in such a way that only a defined section of the case was irradiated and the surrounding was not heated by the IR - source. The motor rotated around its axis at uniform speed; speeds were adjustable between 0.3 and 300 revolutions/minute. To get a uniform irradiation of the case during one revolution, it was necessary that the distance between case and IR - source was kept constant. Since the investigated tubes didn't have completely circular cross sections, the second set of wheels was mounted at the shielding.

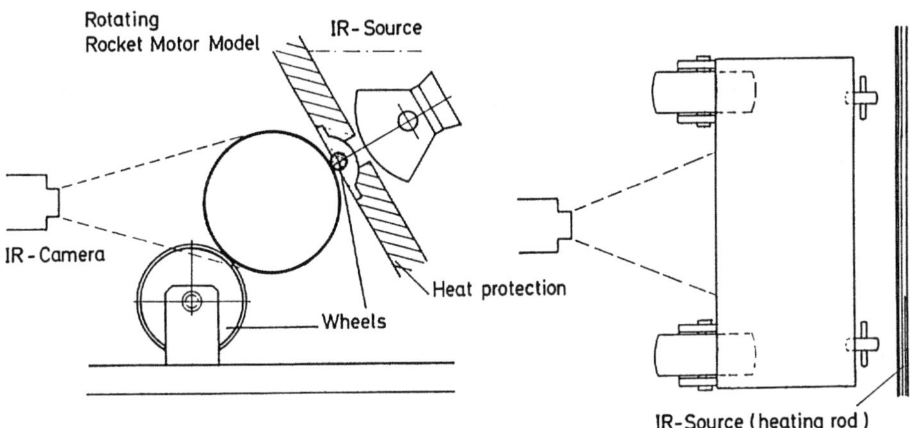

Fig. 1: Test set-up; side view and top view

An IR - camera was looking on the surface of the case from the opposite side of the IR - source. So after a constant and predetermined time delay after irradiation the thermal emission from the case was measured by the camera. The camera has a resolution of 0.1 degree and is sensitive in the region between 2 and 5 microns. The pictures were recorded at a frequency of 25 Hz and could be seen on a monitor. They were stored on Video-Tape and could be transferred to a microcomputer and stored on disc. It is possible to assign to different temperature intervals different colours, so that a thermogram in colours can be plotted on a colour printer.

Since the recorded radiation intensity is not only a function of surface temperature but also of emissivity a uniform surface on the object was desirable. Therefore the case was coated with liquid black lacquer which could easily be removed like a skin after the measurement.

In Figures 2 and 3 the dimensions of the test object are shown. The separation between case and liner and between liner and dummy-propellant consisted of

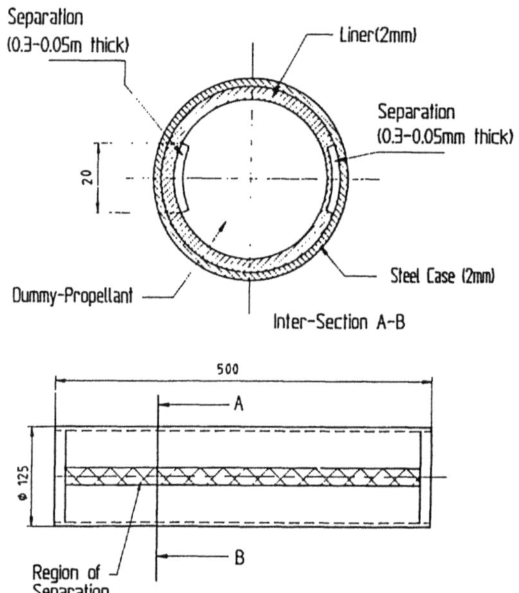

Fig. 2: Strip-shaped built-in separations between case and liner and between liner and propellant

steel 'feeler' stock stripes of various thickness which were arranged parallel to the case axis. After the curing of the liner and the propellant respectively the stripes were pulled from between the case and insulation and from between the insulation and propellant. This left separated regions of known dimensions.

Another kind of defects - so-called bond defects - were also built-in in another model (Figure 3). Circular areas with diameters of 50 mm and 30 mm were coated with substances that prevent the materials (case/liner and liner/propellant) from being bonded. The thickness of the coating was less than 0.05 mm.

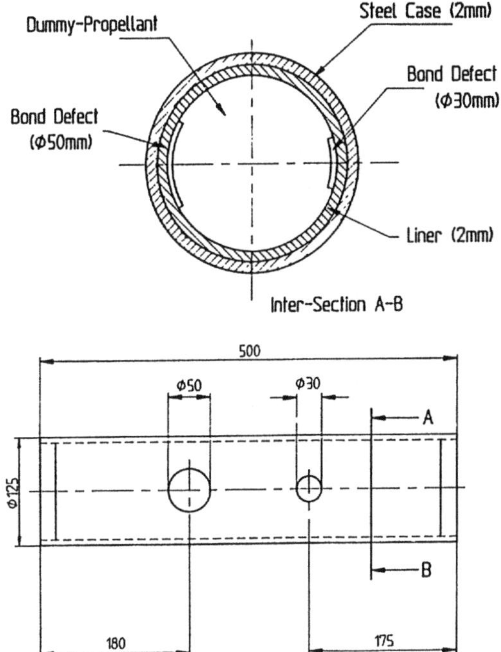

Fig. 3: Circular shaped bond defects between case and liner and between liner and propellant

RESULTS

Figure 4a shows a thermogram of a case without any defects. The picture was taken from the middle section of the case and corresponds to a region with dimension 8 cm x 10 cm. There is a temperature gradient in the horizontal as well as in the vertical direction of the thermogram. The cause for the vertical gradient (Figure 4b) is the cooling of the case. The case rotated from 'top' to 'bottom'; this means that the lower region of the case has more cooled since the time of irradiation than the upper one.

Figure 5a shows a thermogram of the case surface during the passage of a separation located between liner and propellant ('thickness' 0.05 mm, 20 mm wide). Figure 5b shows the corresponding vertical temperature distribution. The gradient of this distribution is clearly stronger than in the case without any separation.

For the detection of the so-called bond defects a sequence of thermograms was recorded. Figure 6 shows these thermograms which were taken from the rotating case with a frequency of 1 picture/sec. The rotational speed of the case

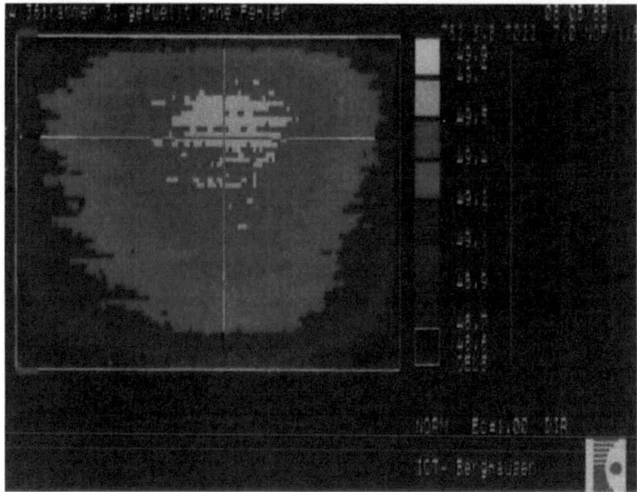

Fig. 4a: Thermogram of a case without separations or bond defects

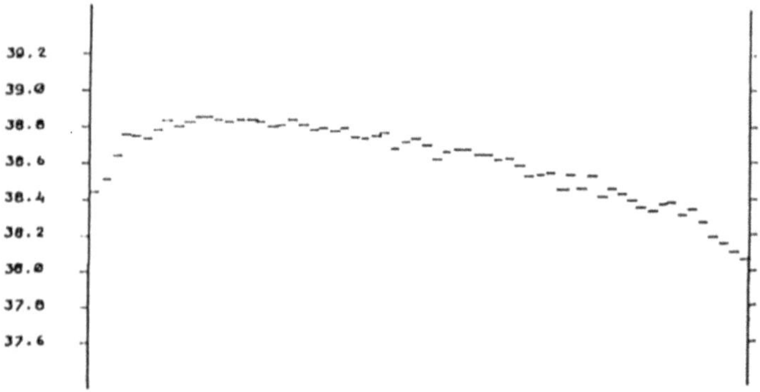

Fig. 4b: Temperature profile of the thermogram of Figure 4a in vertical direction

was 4 revol/min. The case contained the bond defects shown in Figure 3. At first the large bond defect (50 mm diameter) is visible which gradually disappears in the lower left-hand side of the picture; then on the upper right-hand side of the picture the smaller bond defect appears. Finally in the last two pictures the first bond defect appears again.

CONCLUSIONS

Within models of solid propellant motors separations between case and linear as well as between liner and propellant could be detected down to a 'thickness' of 0.05 mm (20 mm wide) by thermographic inspection.

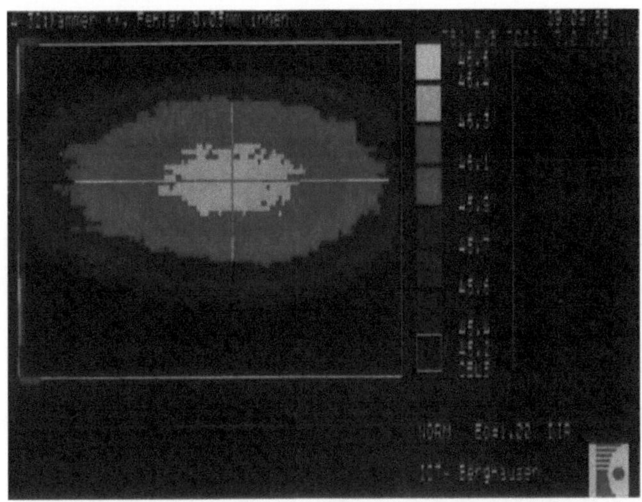

Fig. 5a: Thermogram of a case with separation between liner and propellant (0.05 mm thick, *see* Figure 2).

Fig. 5b: Temperature profile of the thermogram of Figure 5a in vertical direction

The same holds for bond defects with circular dimensions down to a diameter of 30 mm.

In future work we shall apply the test method to still smaller defects and finally the method will be applied to real rocket motors.

REFERENCES

1.   Gericke O.R., Vogel P.E.J.: Infrared Bond Defect Detection System Materials Evaluation, Feb. (1964).

2.   St. Clair J.C.: An Infrared Method of Rocket Motor Inspection, Materials Evaluation, Aug (1966).

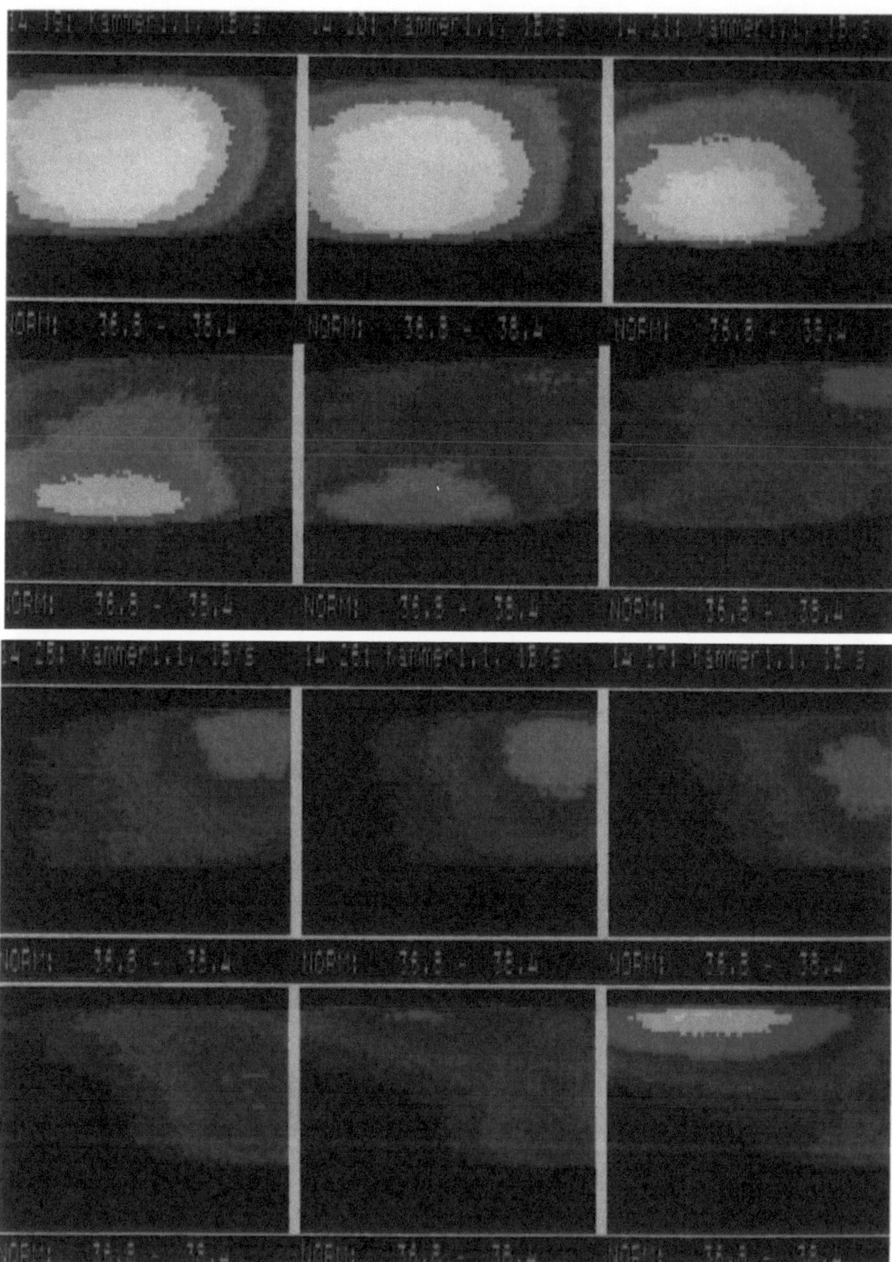

Fig. 6   Thermograms of a rotating case with two circular bond defects between liner and propellant (Figure 3); 1 picture/sec; 4 revol./min

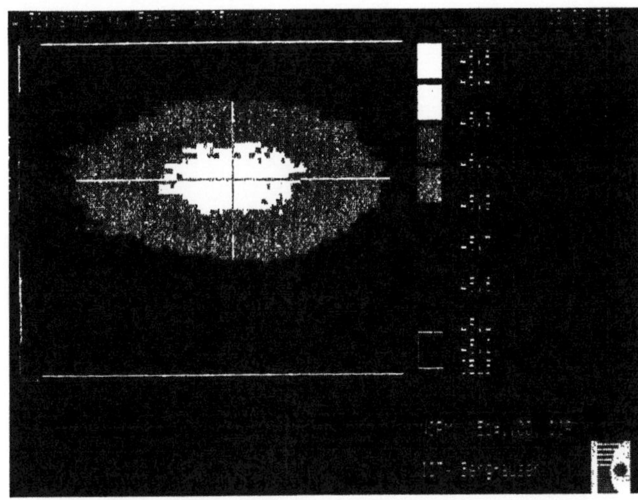

Fig. 5a: Thermogram of a case with separation between liner and propellant (0.05 mm thick, *see* Figure 2).

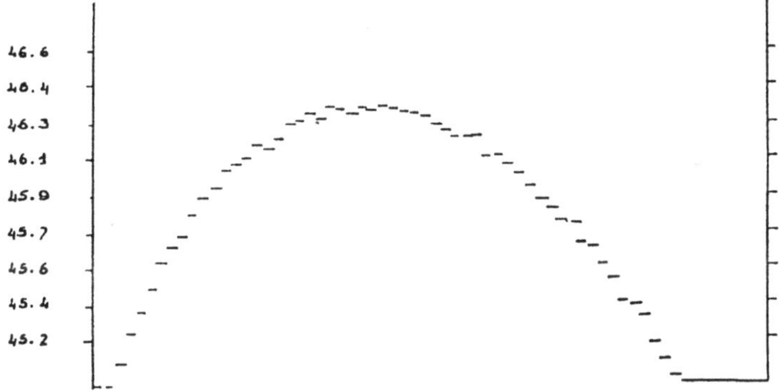

Fig. 5b: Temperature profile of the thermogram of Figure 5a in vertical direction

The same holds for bond defects with circular dimensions down to a diameter of 30 mm.

In future work we shall apply the test method to still smaller defects and finally the method will be applied to real rocket motors.

REFERENCES

1. Gericke O.R., Vogel P.E.J.: Infrared Bond Defect Detection System Materials Evaluation, Feb. (1964).

2. St. Clair J.C.: An Infrared Method of Rocket Motor Inspection, Materials Evaluation, Aug (1966).

HEAT TRANSFER IN INDUSTRIAL FURNACES

K. van Ommen

Shell Laboratorium, Amsterdam Shell Research B.V,
P.O. Box 3003
10033 AA Amsterdam - The Netherlands

The invited lecture is largely based on a part of a recently published paper by Ooms et al [1].

In many areas of the process industry radiative heat transfer plays an important role. The designs of engines, furnaces, burners, waste heat boilers and gasifiers all require careful modelling of radiation, the analysis is complicated because the so-called equation of transfer has to be solved for the intensity of radiation. Solving this equation is quite cumbersome, especially when radiative heat transfer is combined with other modes of heat transfer such as convection and conduction. Here we discuss the interaction between radiation and turbulent conduction in the 'plug' flow part of a gasification reactor. We shall use the Galerkin method, introduced by [2], to solve the one-dimensional equation of transfer for a plane-parallel slab. This method requires little computing time because the equation of transfer is transformed into a set of linear equations whose coefficients can be calculated analytically.

To keep the analysis as simple as possible we make a few more approximations. First of all, we account for the turbulence through a constant turbulent conductivity which we choose to be representative of the 'plug' flow part of the gasification reactor. Temperature fluctuations in the equation of transfer are ignored. Of course, this might result in serious discrepancies because the emissive power of a volume element grows with the fourth power of the temperature. The flow itself is described by a constant mean velocity along the walls. Nevertheless, even with these admittedly severe approximations, our learning model is well suited to shapen our intuition with respect to the interaction between turbulence and radiation. We expect this interaction to be significant because the typical extinction coefficient in a coal or residual fuel gasification reactor is large. Therefore, the mean free path of photons is small, so that radiation is determined by the reactor conditions.

The x axis is chosen parallel to the slab walls; the z axis perpendicular to the walls (see Fig. 1).

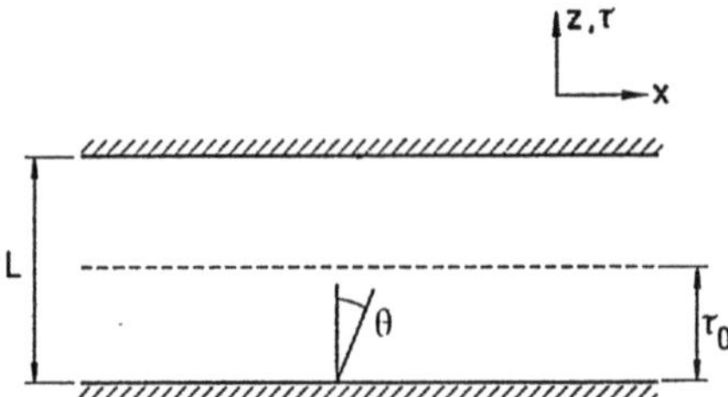

Fig. 1:   Plane-parallel slab.

The pole axis needed to define the polar and azimuthal angles, θ and ϕ, is equal to the z axis. To keep the problem tractable we make the following simplifications. The absorption and scattering coefficients are assumed to be independent of the space coordinate. Furthermore, the scattering is assumed to be isotropic. Because we consider the one-dimensional equation of transfer we in fact assume that gradients of the intensity in the x direction are negligible compared to those in the z direction. Finally, the intensity does not depend on the azimuthal angle ϕ. With these assumptions the equation of transfer and the enthalpy equation read:

$$\mu \frac{\partial}{\partial \tau} I(x,\tau,\mu) + I(x,\tau,\mu) = (1-\omega)\frac{n^2 \bar{\sigma} T^4(x,\tau)}{\pi} + \frac{\omega}{2}\int_{-1}^{1} I(x,\tau,\mu')d\mu' \tag{1}$$

$$\rho\, c_p u \frac{\partial T}{\partial x} = \lambda_t \frac{\partial^2 T}{\partial z^2} - \frac{\partial Q}{\partial z} \tag{2}$$

Here we define:

$\mu \quad = \cos\theta,$
$\omega \quad = k_s/(k_a + k_s),$
$\tau \quad = (k_a + k_s)z$
$\tau_0 \quad = (k_a + k_s)L$

$$Q = 2\pi \int_{-1}^{1} I(x,\tau,\mu')\,\mu'\,d\mu' \tag{3}$$

The density, specific heat at constant pressure, mean velocity along the walls, turbulent conductivity and refractive index are denoted by ρ, c_p, u, λ_t and n, respectively. $\bar{\sigma}$ is Boltzmann's constant. We assume that the opaque walls reflect radiation diffusively. Then the boundary conditions for the equation of transfer equal:

$$I^+(x,0,\mu) = \varepsilon_1 \frac{n^2 \bar{\sigma} T_1^4}{\pi} + 2(1-\varepsilon_1)\int_0^1 I^-(x,0,-\mu')\,\mu'\,d\mu'$$

$$I^-(x,\tau_0,\mu) = \varepsilon_2 \frac{n^{2-}\overline{\sigma}\,T_2^4}{\pi} + 2(1-\varepsilon_2)\int_0^1 I^+(x,\tau_0,\mu')\,\mu'\,d\mu' \tag{4}$$

The boundary equation for the enthalpy equation is obtained as follows. We assume that the boundary layer is so thin that radiation does not influence to convective heat transfer. Next we set the turbulent heat transfer at the edge of the layer equal to the convective heat transfer obtained from Newton's cooling law. This yields a boundary condition of the third kind:

$$\lambda_t \frac{\partial T}{\partial z} = h_w (T - T_{wall}) \tag{5}$$

Consequently, the solution for the temperature will exhibit a discontinuity at the wall. This is, of course, an artefact which can be removed by a proper calculation of the temperature inside the boundary layer. Finally, a start condition is needed when the velocity u is different from 0: for this, it is assumed that the temperature profile at the inlet x=0 is known and equals a constant T_0.

Eqs. (1) and (2), together with the boundary conditions and (4), are solved iteratively. We first consider the equation of transfer and assume that the temperature field is known from a previous iteration. Using formal solutions of the transfer equation and its boundary conditions we transform the radiation equation into an integral equation for the incident radiation $G(x,\tau)$:

$$G(x,\tau) = Y(x,\tau) + \omega \int_0^{\tau_0} K(\tau,\tau')G(x,\tau')\,d\tau' \tag{6}$$

with

$$G(x,\tau) = 2\pi \int_{-1}^1 I(x,\tau,\mu')\,d\mu' \tag{7}$$

$$K(\tau,\tau') = \frac{1}{2}\,E_1(|\tau-\tau'|)$$

$$+ (1-\varepsilon_1)\,\delta\,E_2(\tau)\big[E_2(\tau') + \alpha_2 E_2(\tau_0-\tau')\big]$$

$$+ (1-\varepsilon_2)\,\delta\,E_2(\tau_0-\tau)\big[E_2(\tau_0-\tau') + \alpha_1 E_2(\tau')\big]$$

$$Y(x,\tau) = \int_0^{\tau_0} K(\tau,\tau')S(x,\tau')\,d\tau'$$

$$+ \frac{1}{2}\,A_1\Big[\big(1+\alpha_1\alpha_2\delta\big)\,E_2(\tau) + \alpha_2\delta\,E_2(\tau_0-\tau)\Big]$$

$$+ \frac{1}{2}\,A_2\Big[\big(1+\alpha_1\alpha_2\delta\big)\,E_2(\tau_0-\tau) + \alpha_1\delta\,E_2(\tau)\Big]$$

$$S(x,\tau) = (1-\omega)\frac{n^{2-}\bar{\sigma}\,T^4(x,\tau)}{\pi}$$

$$E_n(x) = \int_0^1 e^{-x/\mu}\mu^{n-2}\,d\mu$$

$$A_i = \varepsilon_i\frac{n^2\bar{\sigma}\,T_i^4}{\pi}\quad (i = 1,2)$$

$$\alpha_i = 2(1 - \varepsilon_i)\,E_3(\tau_o)\quad (i = 1,2)$$

$$\delta = \frac{1}{\left(1 - \alpha_1\alpha_2\right)}$$

Observe that any solution of eq. (7) causes the following functional of G to have an extreme value:

$$J[G] = \omega\int_0^{\tau_o}\int_0^{\tau_o} K(\tau,\tau')G(x,\tau)G(x,\tau')\,d\tau d\tau'$$

$$+ 2\int_0^{\tau_o} Y(x,\tau)G(x,\tau)\,d\tau - \int_0^{\tau_o} G^2(x,\tau)\,d\tau$$

To use this property we expand the unknown solution $G(x,\tau)$ as a polynomial series in τ with unknown coefficients c_n and require that all the derivatives of the functional J with respect to the coefficients c_n equal 0. This yields a set of linear equations for c_n which can be solved in a straightforward manner. If also $S(x,\tau)$ can be expressed as a polynomial series in τ, all coefficients in the linear equations for c_n can be calculated analytically. Of course, this will speed up the solution method. The enthalpy equation is solved by the use of special eigenfunctions, as described by [3].

Results were obtained by means of a computer program developed to perform the calculations outlined above. In view of the general nature of this paper we discuss one result. We consider a stagnant slab (i.e. the fluid velocity is 0), consisting of a wall at z=0 and a semi-infinite medium. Thus the problem is one-dimensional and no start condition is necessary. Despite the fact that the mean velocity is taken as zero, we still assume trhat the turbulent conductivity and the convective heat transfer coefficient are different from zero. Their values are specified as input parameters. The temperature of the wall equals 1700 K, its emissivity 0.5. The convective heat transfer coefficient h_w equals 300 W/m2K. The temperature of the medium far from the wall, T_∞, equals 2000 K. Its turbulent conductivity, absorption and scattering coefficient equal 200 W/mK, 5 m-1 and 5 m-1, respectively. Since our current analysis does not allow for a semi-infinite medium we have to bound the medium by an imaginary wall. This wall is assumed to be black, to have a temperature of 2000 K and to be situated at a distance of 0.5 m from the physical wall. Consequently, the optical thickness of the slab equals 5. In Fig. 2 the radiative heat flux is depicted as a function of z. We have also included the net radiative heat flux in the case $\lambda_t=0$. Then radiative equilibrium yields a constant radiative heat flux. Notice that the turbulent conductivity increases the radiative heat flux near the physical wall. This enhancement is now being studied in more detail.

It is clear that in our analysis many simplifications have been made. For instance, the Galerkian method cannot at present be generalized to solve the three-dimensional equation of transfer, the turbulence conductivity is constant, the geometry is still far from that of a gasification reactor, temperature fluctuations are ignored, etc. Hence, much remains to be done.

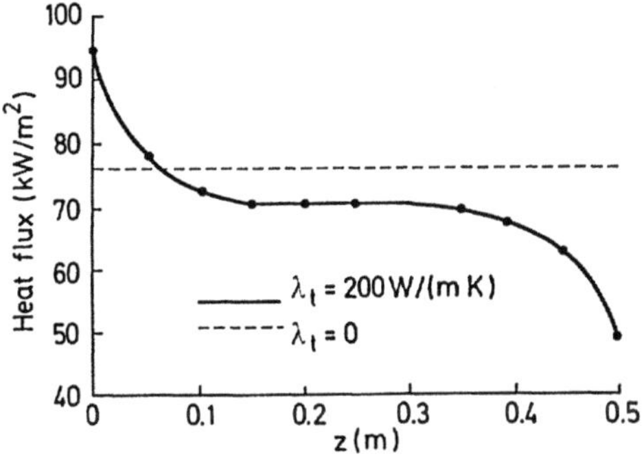

Fig. 2: Radiative heat flux without $\lambda_t=0$ and with $\lambda_t\neq0$ turbulence.

REFERENCES

1. Ooms G. Daverveldt P.H.W.M., Roekaerts D., de Smedt R.J.M.M.: Weisenborn A.J., Felderhof B.U.: Turbulence and the Process Industry. Edited by H.H. Fernholz and H.E. Fielder, Springer Verlag, Berlin and Heidelberg (1989) 484-504.

2. Ozisik M.N., Yener Y.: J. Heat Transfer (1982) 351.

3. Ozisik M.N.: Boundary value problems of heat conduction. Int. Textbook Comp. (1968).

MODELLING RADIATIVE HEAT TRANSFER IN PULVERISED COAL-FIRED FURNACES

F. Liu[*] and J. Swithenbank
Department of Mechanical and Process Engineering
Chemical Engineering and Fuel Technology
University of Sheffield
Sheffield S1 3JD England

[*]*Present address:*
Department of Fuel and Energy
University of Leeds
Leeds LS2 QJT England

ABSTRACT

The P_1-approximation and δ-Eddington phase function approximation are employed to model the highly forward scattering of radiation in pulverised coal-fired furnaces. The effect of scattering on radiative heat transfer is clearly shown by the present version of the P_1-approximation. The accuracy of the P_1-approximation can be improved significantly by using an optimised boundary condition.

INTRODUCTION

Since thermal radiation is the predominant mode of heat transfer at temperatures found in combustion systems, any realistic modelling of a combustion process must take into account radiative heat transfer. Unfortunately, radiative heat transfer is governed by a complex integrodifferential equation which is extremely difficult to solve even for one-dimensional problems. Exact solutions to the radiative transfer equation exist only for some limited situations [1]. Approximate solution methods are of interest for the purpose of modelling radiative heat transfer in practical combustion systems.

Comprehensive up-to-date reviews of the art of modelling radiative heat transfer in combustion systems are available in the literature [1,2] where it is shown that the Zone method [3] and the Monte Carlo method [4] yield numerically exact solutions but are too computationally expensive to be incorporated into a general prediction procedure. The discrete transfer method [5] is another approach to obtain the numerically exact solution of radiative transfer equation with less computational effort. However, it is difficult for this method to handle scattering. The flux methods usually provide high computational efficiency, but the lack of coupling between directional fluxes makes the multiflux method unreliable [1]. On the other hand, the discrete-ordinates approximation cannot handle highly forward scattering with ease which is often the case in a coal-fired furnace due to diffraction by the large particles. The spherical harmonic approximation, unlike the other methods, offers the capability of handling highly forward scattering and its governing equations can be easily solved by the finite difference technique [6,7].

Recently, considerable research attention has been paid to developing radiation models which are compatible with finite difference prediction procedures and are able to take into account the effects of anisotropic scattering in multidimensional absorbing, emitting, and scattering media. This is in response to the requirement for a suitable radiation model which can be used in a general combusting flow prediction procedure to predict radiative heat transfer in pulverised coal-fired furnaces. The spherical harmonic approximation is attractive for use in this context mainly because of its ability to handle anisotropic scattering. To date, the effect of scattering due to the presence of particles on radiative transfer has not been clearly established. Different opinions exist in the literature. The numerical results of Menguc and Viskanta [6,7] obtained by the P_3-approximation indicate that anisotropic scattering has a dramatic effect on radiative heat transfer. Later work of Truelove [8], however, indicates that the results of Menguc and Viskanta are unreliable. It is of great value to elucidate this effect in order to make realistic assumptions in future work.

In this paper, a radiation model based on the first order spherical harmonic expansion of the radiation intensity, called the P_1-approximation, is developed in the three-dimensional cartesian coordinate system and the two-dimensional axisymmetrical cylindrical system. The anisotropic scattering of fly-ash particles is modelled by the δ-Eddington phase function approximation [9,10]. By investigating the arbitrariness of the boundary condition required by the P_1-approximation, a general form of boundary condition for the P_1-approximation is formulated which contains an arbitrary constant. The optimum boundary condition is determined numerically by studying the sensitivity of the P_1-approximation prediction to the value of the constant. Applications of the P_1-approximation using the optimum boundary condition show that its accuracy can be improved significantly. The effect of scattering on radiative heat transfer is also studied using the present version of the P_1-approximation.

FORMULATION

The basis for a quantitative study of the transfer of radiative energy in an absorbing, emitting, and scattering medium is the radiative transfer equation. For engineering problems, the medium is usually assumed to be grey and in local thermodynamic equilibrium. Then the radiative transfer equation takes the form [3]

$$\frac{dI}{ds} = -k_a I - k_s I + k_a I_b + k_s \frac{1}{4\pi} \int_{\Omega=4\pi} \Phi(\Psi) I(\Omega') d\Omega' \tag{1}$$

where I is the radiation intensity, s is measured along the direction of propagation of radiation, Ω. k_a and k_s represent respectively absorption and scattering coefficients. $\Phi(\Psi)$ is the scattering phase function, Ψ is the scattering angle. An element of solid angle in polar coordinates is calculated as

$$d\Omega = \sin\theta d\theta d\phi \tag{2}$$

By expanding the radiation intensity I in terms of the first order spherical harmonics, I can be written as [11]

$$I(\vec{r},\theta,\phi) = \frac{1}{4\pi}\left[I_0 + 3(I_1\xi + I_2\eta + I_3\mu)\right] \tag{3}$$

where $\vec{r}$ is the spatial position vector. The closure conditions for the P_1-approximation are given by

$$I_{ij} = \frac{1}{3} I_0 \delta_{ij} \tag{4}$$

where I_0, I_i, I_{ij} denote the zeroth order moment, first order moments, and second order moments of the intensity respectively. ξ, η, and μ are the direction cosines and can be expressed as

$$\xi = \sin\theta\cos\phi, \quad \eta = \sin\theta\sin\phi, \quad \mu = \cos\theta \tag{5}$$

To formulate the governing equations of the P_1-approximation, an appropriate form of the scattering phase function is required to evaluate the in-scattering term which is the integral term on the right hand side of eq. (1). It has been shown that the δ-Eddington approximation can be used to model a highly peaked scattering phase function in a simple yet accurate fashion [9,12]. The δ-Eddington approximation takes the form

$$\Phi(\Psi) = 2f\delta(1 - \cos\psi) + (1 - f)(1 + 3g'\cos\psi) \tag{6}$$

The symmetry factor of a scattering phase function is an important parameter which is defined as

$$g = \frac{1}{4\pi} \int_{\Omega=4\pi} \Phi(\Psi) \cos \psi \, d\Omega \tag{7}$$

The symmetry factor represents the amount of radiation scattered in the forward direction. For example, $g=1$, 0, -1 correspond to complete forward scattering, isotropic scattering, and complete backward scattering, respectively. Both experimental measurement [13] and theoretical calculation [14] indicate that the symmetry factor of fly-ash particles has a value between 0.7 and 0.8. Substituting eq. (6) into eq. (7) yields the symmetry factor of the δ-Eddington approximation

$$g = f + g' - fg' \tag{8}$$

By applying eq. (3) and eq. (6) to eq. (1) and after some standard derivation procedures [6], the governing equations of the P_1-approximation in a three-dimensional cartesian coordinate system can be written as

$$I_1 = -\frac{1}{3k'_e} \frac{\partial I_o}{\partial x} \tag{9}$$

$$I_2 = -\frac{1}{3k'_e} \frac{\partial I_o}{\partial y} \tag{10}$$

$$I_3 = -\frac{1}{3k'_e} \frac{\partial I_o}{\partial z} \tag{11}$$

$$\frac{\partial}{\partial x}\left(\frac{1}{3k'_e}\frac{\partial I_o}{\partial x}\right) + \frac{\partial}{\partial y}\left(\frac{1}{3k'_e}\frac{\partial I_o}{\partial y}\right) + \frac{\partial}{\partial z}\left(\frac{1}{3k'_e}\frac{\partial I_o}{\partial z}\right) = k_a\left(I_o - 4\pi I_b\right) \tag{12}$$

where k'_e is termed the effective extinction coefficient and defined as

$$k'_e = k_a + (1-g)\,k_s \tag{13}$$

It is worth noting that only the symmetry factor g is present in the governing equations as the result of using eq. (8). Therefore, it is not necessary to determine the individual value of the phase parameters f and g' if the symmetry factor is known. The quantity $(1-g).k_s$ may be called the effective scattering coefficient. A conclusion can be drawn directly from these equations that for complete forward scattering, where g is equal to unity, the effect of scattering vanishes. This is true simply because in this case the increase in intensity due to in-scattering is completely compensated by its decrease due to out-scattering, *see* eq. (1). In addition, an anisotropic scattering problem can be scaled to an isotropic scattering problem through the relation

$$k_s|_i = (1-g)\, k_s|_a$$
(14)

This scaling law is the same as that obtained by Lee and Buckius [15] for one-dimensional problems. If the volumetric heat generation rate S of the medium is specified, eq. (12) takes the form

$$\frac{\partial}{\partial x}\left(\frac{1}{3k'_e}\frac{\partial I_o}{\partial x}\right) + \frac{\partial}{\partial y}\left(\frac{1}{3k'_e}\frac{\partial I_o}{\partial y}\right) + \frac{\partial}{\partial z}\left(\frac{1}{3k'_e}\frac{\partial I_o}{\partial z}\right) = S$$
(15)

In a two-dimensional axisymmetric coordinate system, the governing equations of the P_1-approximation are expressed as

$$I_r = -\frac{1}{3k'_e}\frac{\partial I_o}{\partial r}$$
(16)

$$I_z = -\frac{1}{3k'_e}\frac{\partial I_o}{\partial z}$$
(17)

$$\frac{1}{r}\frac{\partial}{\partial r}\left(\frac{1}{3k'_e}r\frac{\partial I_o}{\partial r}\right) + \frac{\partial}{\partial z}\left(\frac{1}{3k'_e}\frac{\partial I_o}{\partial z}\right) = k_a\left(I_o - 4\pi I_b\right)$$
(18)

BOUNDARY CONDITIONS

To make the governing equations of the P_1-approximation applicable, the boundary conditions must be worked out at each wall surface. For most combustion chambers and furnaces, the surfaces of these enclosures can be assumed to be diffusively reflecting and emitting. The exact boundary condition, obtained by considering the intensity leaving the surface, is written as

$$I_w = \varepsilon_w I_b(T_w) + (1 - \varepsilon_w)\frac{1}{\pi}\int_{\Omega=2\pi} l_i\, I d\Omega$$
(19)

where I_w is the intensity leaving the wall surface, T_w is the wall temperature, ε_w is the wall emissivity, and l_i is the approximate direction cosine. In applying any of the P_N-approximations, the correct boundary condition, eq. (19), cannot be satisfied exactly since the intensity is represented by a truncated spherical harmonics series. Two different schemes of boundary condition approximation have been developed to be applied to the spherical harmonics method in the work related to neutron transport theory, named Mark's and Marshak's boundary conditions [11, 16], respectively. It has been pointed out that Marshak's boundary condition gives better results for low order approximations. Marshak's boundary condition is obtained by taking the integral of the intensity over the appropriate hemispherical space such that

$$\int_{\Omega=2\pi} I\, f(\Omega)d\Omega = \int_{\Omega=2\pi} I_w\, f(\Omega)d\Omega \tag{20}$$

where $f(\Omega)$ is an arbitrary function of direction. It is clear that Marshak's boundary condition involves a certain arbitrariness due to the introduction of $f(\Omega)$. All the work in the literature related to the P_1-approximation employs a boundary condition obtained by replacing the arbitrary function $f(\Omega)$ by an appropriate direction cosine such as [6,7,17,18]

$$f(\Omega) \propto l_i \tag{21}$$

It has been found, however, that the P_1-approximation usually overpredicts radiative heat flux distributions at boundary walls when using this boundary condition convention. Because of this, it is desirable to seek a better boundary condition. A more general form of boundary condition can be obtained by selecting $f(\Omega)$ from among l_i^n rather than l_i and it is written as

$$\int_{\Omega=2\pi} I\, l_i^n\, d\Omega = \int_{\Omega=2\pi} I_w\, l_i^n\, d\Omega \tag{22}$$

where n is an arbitrary positive integer or zero. Substituting eq. (3) and eq. (19) into eq. (22) yields the more general boundary condition for the P_1-approximation

$$I_0 \pm \frac{3k + 2(1 - \varepsilon_w)}{\varepsilon_w}\, I_i = 4\pi\, I_b(T_w) \tag{23}$$

where $\pm$ corresponds to the surfaces at the negative and positive directions, respectively. k is a constant given as

$$k = \frac{n+1}{n+2} \tag{24}$$

The value of n is now open to question. It will be determined numerically in the next section by investigating the sensitivity of the P_1-approximation prediction to the value of n. Note that n=1 corresponds to the widely used boundary condition for the P_1-approximation.

RESULTS AND DISCUSSION

The elliptic partial differential equation of the governing equation of the P_1-approximation has been solved numerically by employing the finite difference technique. The iterative scheme used in this work is the successive over relaxation method.

The first case studied is radiative heat transfer in an infinitely long square cavity. All four wall surfaces are cold and black. The non-scattering medium has a uniform temperature and absorption coefficient. Exact solutions for surface heat transfer rates are available due to Lockwood and Shah [5]. The P_1-approximation predictions are compared with these exact solutions in Figures 1-3. The numerical results of the P_1-approximation are obtained by using a

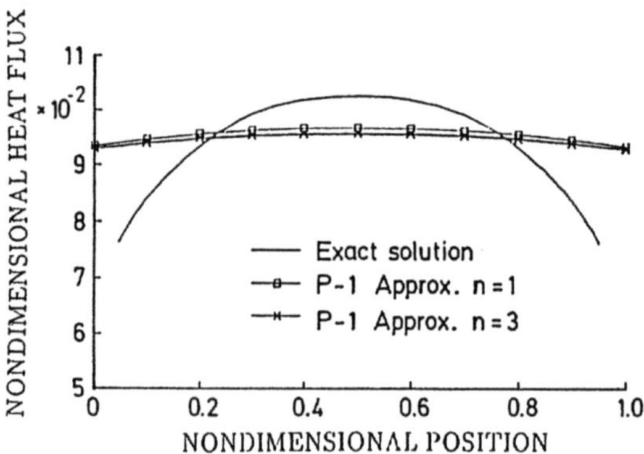

Fig. 1: Influence of boundary condition on the nondimensional surface heat transfer rates: $k_aL=0.1$.

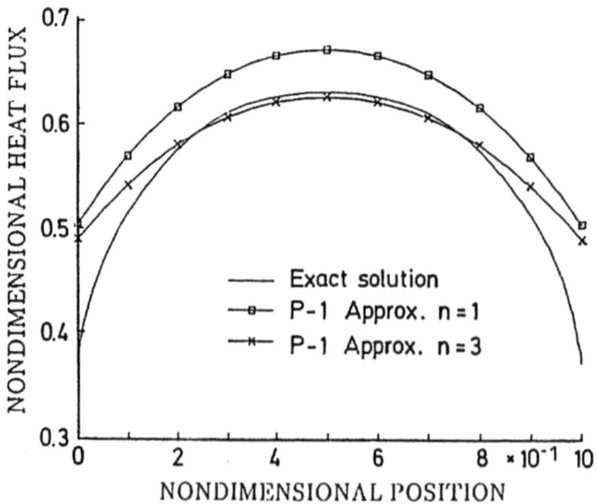

Fig. 2: Influence of boundary condition on the nondimensional surface heat transfer rates: $k_aL=1.0$.

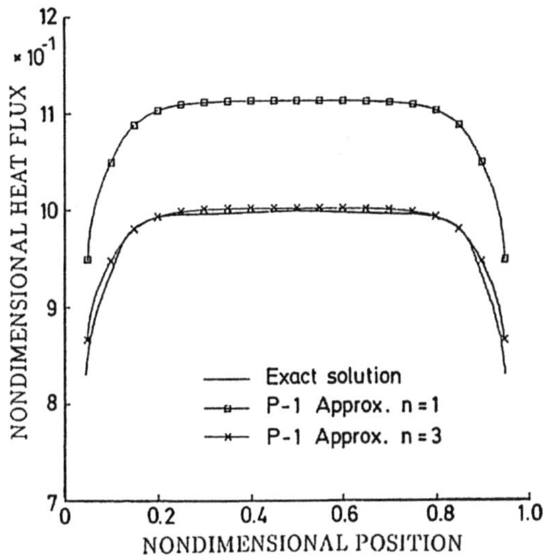

Fig. 3: Influence of boundary condition on the nondimensional surface heat transfer rates: $k_aL=10$.

20x20 uniform grid scheme. The CPU time required is 1, 5, and 13 s when the optical dimension of the cavity is 10, 1.0, and 0.1, respectively. Except using n=1 which represents the conventional boundary condition, different values of n, such as 0, 2, 3 and 4 have been used to calculate the P_1-approximation results. For this problem, it is found that n=3 corresponds to the optimum boundary condition. Therefore, the P_1-approximation predictions of surface heat transfer rates using the boundary conditions of n=1 and n=3 are depicted in the figures. It is clear that the P_1-approximation using the conventional boundary condition of n=1 overpredicts the surface heat flux, especially for the large optical dimension $k_aL=10$ where the P_1-approximation yields unrealistic values of heat flux greater than unity. However, if the optimum boundary condition of n=3 is employed, the accuracy of the P_1-approximation is improved significantly except for the optically thin case, $k_aL=0.1$, where it is believed that the differential approximation loses its validity. In the optically thin case, Figure 1, the boundary condition or the value of n has only a slight influence on the prediction. For the intermediate optical dimension, Figure 2, the P_1-approximation using the boundary condition of n=3 predicts accurate heat flux in the region removed from the corner; however, it still overpredicts the heat flux in the corner region. Based on these findings, it may be concluded that the boundary condition of n=3 is superior to the conventional condition of n=1. However, it is worth noting that the predictions of the P_1-approximation for small and intermediate optical dimensions cannot be further improved, due to the nature of the differential approximation, by selecting different values of n.

The second case considered has the same geometry as the first one but has different thermal conditions. In this case, the bottom wall is at a dimensionless temperature of unity. The other three walls are cold. All the walls are black. The medium is subject to a uniform heat generation of zero, i.e. the medium is at radiative equilibrium. The optical dimension of the cavity is unity. For this problem, numerical results using the zone method have been presented by Ratzel and Howell [18]. The results of the P_1-approximation using different boundary conditions are compared with the zone calculations in Figures 4-5. Numerical results of the P_1-approximation are obtained by a 10x10 uniform grid scheme. Figure 4 shows the nondimensional centerline emissive power distribution. The P_1-approximation underestimates the emissive power near the hot surface and overestimates the emissive power near the cold surface. Note that the P_1-approximation using the boundary condition of n=3 yields a slightly worse centerline emissive power distribution that using the conventional boundary condition. Figure 5 shows the nondimensional heat transfer rate for the hot surface. It is clearly shown that the boundary condition of n=3 is much better than the conventional boundary condition. Although the heat transfer rate for the hot surface can be further improved by using a greater n, it will give rise to further deterioration of the centerline emissive power distribution.

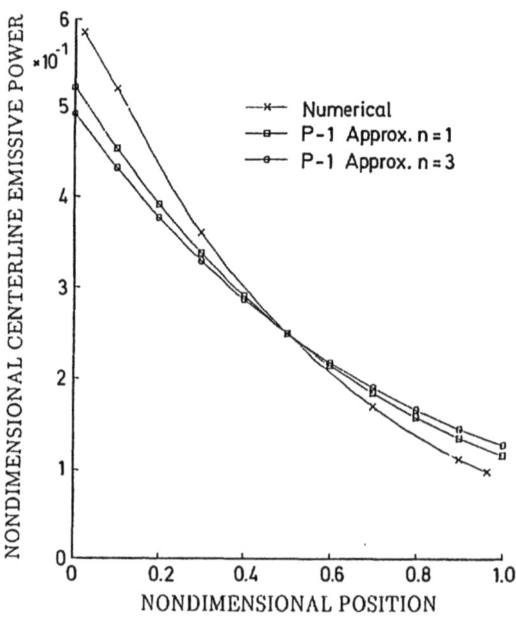

Fig. 4: Influence of boundary condition on the nondimensional centerline emissive power distribution: $k_aL=1.0$.

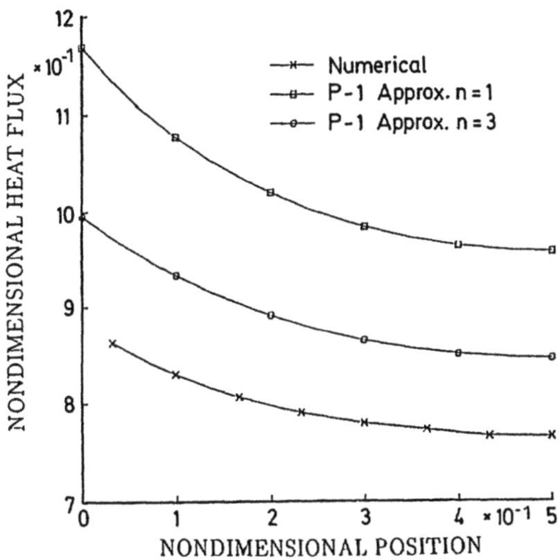

Fig. 5: Influence of boundary condition on the nondimensional hot surface heat transfer rates: $k_aL=1.0$.

In the third example, the P_1-approximation is applied to an idealised furnace containing a scattering medium. Table 1 summaries the physical parameters of the furnace. This problem has been studied by Menguc and Viskanta [6] using the P_3-approximation and by Truelove [8] using the zone method and the discrete-ordinates approximation. The results of Menguc and Viskanta show a dramatic effect of single scattering albedo on the surface heat flux distributions. However, Truelove pointed out that their results are unreliable and should be viewed with suspicion. The numerical results of Truelove show that the effect of scattering on the radiative heat transfer is insignificant in this furnace. The results obtained by the P_1-approximation are here compared with those reported by Truelove. Eq. (15) suggests that the zeroth moment and thus heat flux distributions are unchanged for isotropic scattering compared with non-scattering under the condition of constant extinction coefficient. For isotropic scattering, it is easy to show, from eq. (15), that the medium emissive power E_b is increased by $\omega S/4k_e(1-\omega)$, where ω is the single scattering albedo.

Table 1: Physical parameters of the idealised furnace

Medium	$S = 5.0 \text{ kW/m}^3$
	$x_0 = 2m,\ y_0 = 2m,\ z_0 = 4m$
Boundaries	$z=0,\ T=1200 \text{ K},\ \varepsilon_w=0.85$
	$z=z_0,\ T=400 \text{ K},\ \varepsilon_w=0.7$
	others, $T = 900 \text{ K},\ \varepsilon_w=0.7$

Figure 6 shows the effects of the scattering phase function on the heat flux profiles at the hot and the cold end walls. For isotropic scattering (g=0), the heat flux profiles are the same as for non-scattering as indicated above. For forward scattering (g=0.333), the radiative flux increases at both the hot and the cold walls. Although the P_1-approximation overpredicts the heat flux at both end walls, it correctly predicts the effects of scattering on the surface heat flux. Figure 7 displays the effect of the scattering phase function on the temperature distributions in the medium. For isotropic scattering, the emissive power of the medium is increased uniformly throughout the furnace. For forward scattering, the temperature in the medium tends to be more uniform. This is consistent with the increased heat transfer rates at both end walls shown in Figure 6. The P_1-approximation underpredicts the medium temperature near the hot end wall and at the center of the furnace; while it overpredicts the medium temperature near the cold end wall. Nevertheless, the relative importance of scattering predicted by the P_1-approximation is consistent with that obtained by Truelove. Furthermore, in highly forward scattering media where the symmetry factor g is greater than 0.333, solutions by the discrete-ordinates method are very difficult and expensive to obtain. However, it provides no extra difficulty or computational effort for the P_1-approximation.

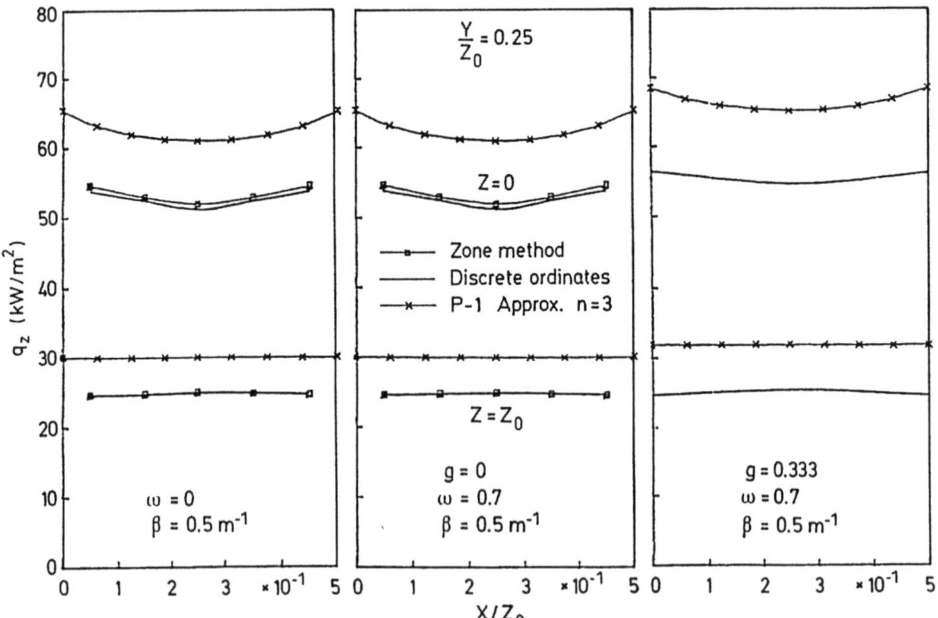

Fig. 6: Effect of scattering on the heat flux profiles at the two end walls.

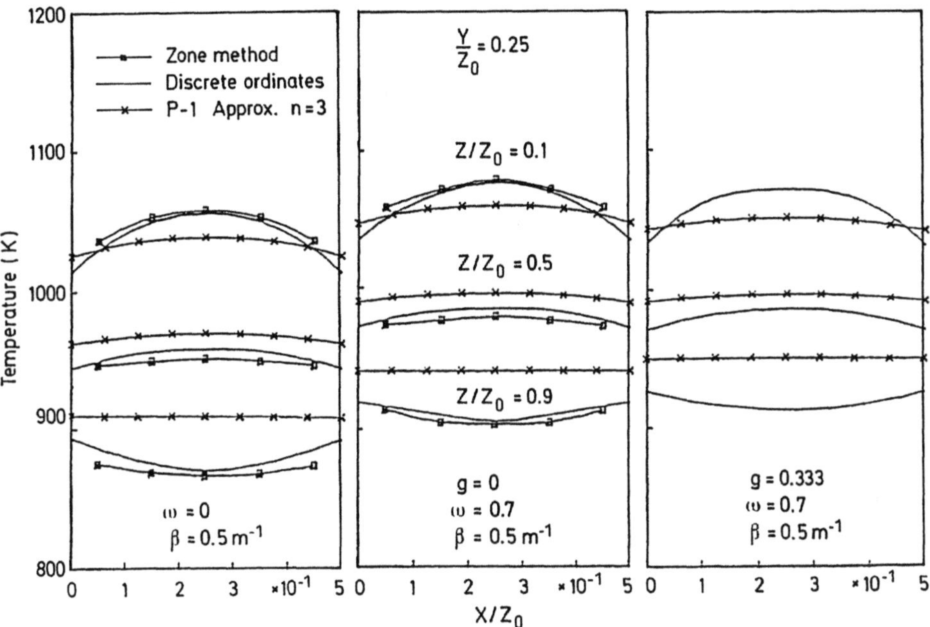

Fig. 7: Effect of scattering on the medium temperature distributions.

The fourth case studied investigates the effect of highly forward scattering on the radiative heat transfer in a three-dimensional furnace for a given medium temperature distribution. The geometrical and physical parameters of the furnace have been described in detail by Selcuk [19]. Since the calculation of anisotropic scattering is very complicated, it is desirable to establish what error can be introduced if the scattering is either neglected or is assumed to be isotropic. Note that in this case the effect of scattering is considered under the condition of constant absorption coefficient rather than constant extinction coefficient. In pulverised coal flames, the highly forward scattering fly-ash particles have a symmetry factor of about 0.7 to 0.8 [13,14]. The effect of the symmetry factor on the source term is given in Figure 8. The physical parameters are also given in the caption. It can be seen that the highly forward scattering has an insignificant effect on the source term distributions and in particular, its effect becomes negligible in the furnace corner. By increasing the symmetry factor from 0 (isotropic scattering) to 1 (non-scattering or complete forward scattering), the source term near the furnace centerline increases by only 10%. Figure 9 shows the effect of the symmetry factor on the side wall heat flux. It can be seen that the effect of symmetry factor near the center of the side wall is negligible, however, it becomes noticeable near the corner of the side wall. It should be pointed out that the results for highly

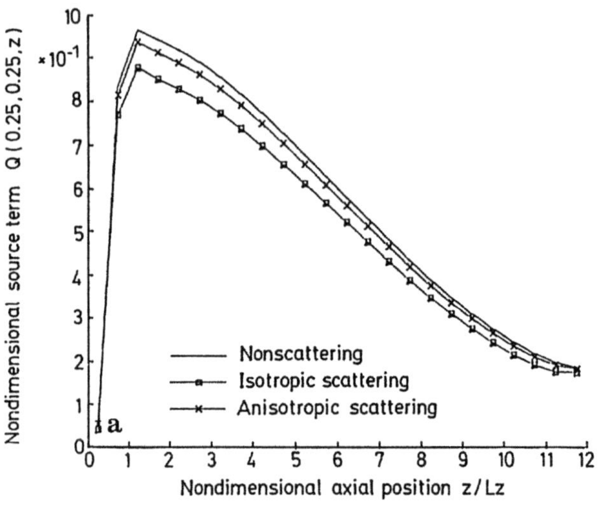

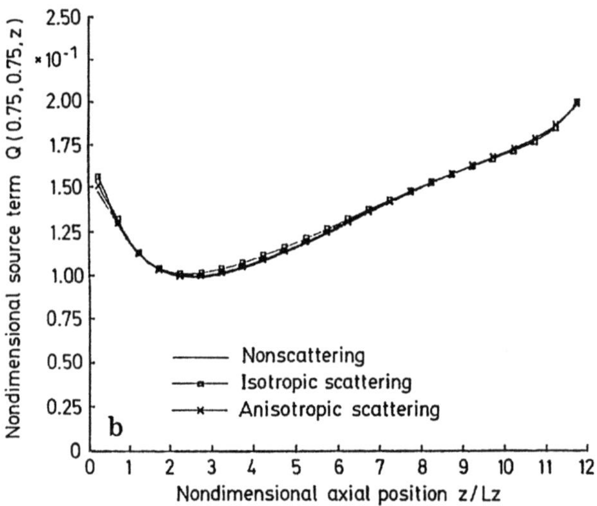

Fig. 8: Effect of the symmetry factor on the source term distribution: optical dimension=0.5, ε_w=1.0, ω=0.5, for anisotropic scattering g=0.7

forward scattering lie between the results of non-scattering and isotropic scattering and they are closer to those of non-scattering. Therefore, the conclusion can be drawn that it is more realistic to neglect the effect of anisotropic scattering than to assume it to be isotropic if the two extreme assumptions have to be made.

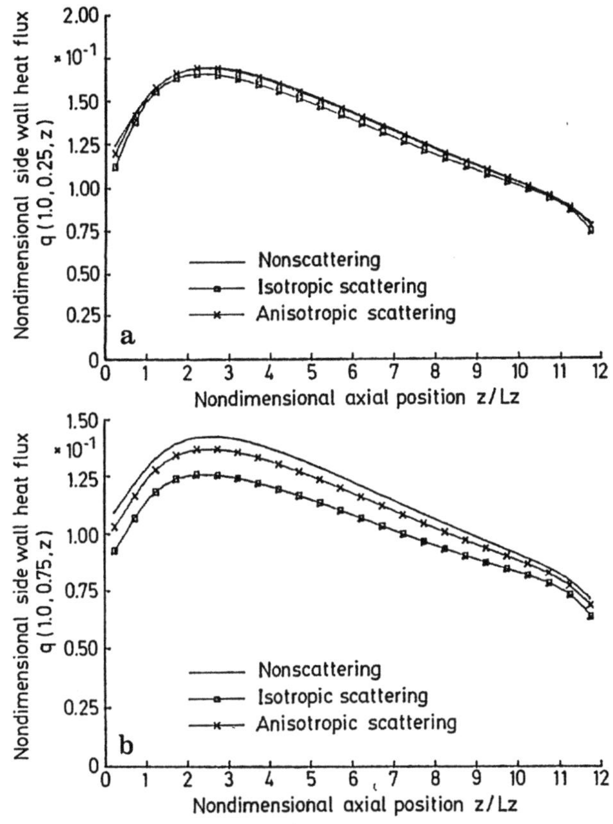

Fig. 9: Effect of the symmetry factor on the side wall heat flux distribution: optical dimension=0.5, ε_w=1.0, ω=0.5, for anisotropic scattering g=0.7

CONCLUSION

The accuracy of the P_1-approximation has been compared with more reliable results of radiative heat transfer in multidimensional absorbing, emitting, and scattering media. It is found that the accuracy of the P_1-approximation can be improved significantly by using the optimum boundary condition of n=3, especially for surface heat transfer rates. The P_1-approximation can also correctly predict the relative importance of scattering. There is no difficult in handling anisotropic scattering by the P_1-approximation method and it does not require any extra computing time. In the radiation model derived in this work, the symmetry factor of the scattering phase function plays a key role in modelling the effect of scattering. When the two extreme assumptions have to be made in order to simplify the calculation, it is more realistic to neglect highly forward scattering than to assume it to be isotropic. The P_1-

approximation, when using the optimum boundary condition, offers the advantages of simplicity, high computational efficiency, and relatively good accuracy.

REFERENCES

1. Viskanta R., Menguc, M.P.: Radiation Heat Transfer in Combustion Systems. Progress in Energy and Combustion Science 13 (1987) 97.

2. Howell J.R.: Thermal Radiation in Participating Media: The Past, the Present, and Some Possible Futures. J. Heat Transfer 110 (1988) 1220-1229.

3. Hottel H.C.; Sarofim A.F.: Radiative Transfer. MacGraw-Hill, New York (1967).

4. Siegel R., Howell J.R.: Thermal Radiation Heat Transfer. 2nd ed., Hemisphere, Washington, D.C. (1981).

5. Lockwood F.C., Shah N.G.: A New Radiation Solution Method for Incorporation in General Combustion Prediction Procedures. 18th Symp. (Int) on Combustion. The Combustion Institute, Pittsburgh, (1981) 1405-1414.

6. Menguc M.P., Viskanta R.: Radiative Transfer in Three-Dimensional Rectangular Geometries Containing Inhomogeneous, Anisotropic Scattering Media. J. Quant. Spectrosc. Radiat. Transfer. 33 (1985) 533-549.

7. Menguc M.P., Viskanta R.: Radiative Transfer in Axisymmetric, Finite Cylindrical Enclosures. J. Heat Transfer 108 (1986) 271-276.

8. Truelove J.S.: Three-Dimensional Radiation in Absorbing-Emitting-Scattering Media Using the Discrete-Ordinates Approximation. J. Quant. Spectrosc. Radiat. Transfer 39 (1988) 27-31.

9. Joseph J.H., Wiscombe W.J., Weinman J.A.: The delta-Eddington Approximation for Radiative Flux Transfer. J. Atmospheric Science 33 (1976) 2453-2459.

10. Crosbie A.L., Davison G.M.: Dirac-Delta Function Approximation to the Scattering Phase Function. J. Quant. Spectrosc. Radiat. Transfer. 33 (1985) 391-409.

11. Case K.M., Zweifel P.F.: Linear Transport Theory. Addison-Wesley, Reading, Massachusetts (1967).

12. McKellar B., Box M.A.: The Scaling Group of the Radiative Transfer Equation. J. Astronspheric Sciences 38 (1981) 1063-1068.

13. Boothroyd S.A., Jones A.R., Nicholson K.W., Wood W.: Light Scattering by Fly Ash and the Applicatbility of Mie Theory. Combustion and Flame 69 (1987) 235-241.

14. Goodwin D.G., Mitchner M.: Flyash Radiative Properties and Effects on Radiative Heat Transfer in Coal-Fired Systems. Int. J. Heat Mass Transfer 32 (1989) 627-638.

15. Lee H., Buckius R.O.: Scaling Anisotropic Scattering in Radiation Heat Transfer for a Planar Medium. J. Heat Transfer 104 (1982) 69-75.

16. Davison B.: Neutron Transport Theory. Clarendon, Oxford (1958).

17. Higenyi J., Bayazitoglu Y.: Differential Approximation of Radiative Heat Transfer in a Gray Medium. J. Heat Transfer 102 (1980) 719-723.

18. Ratzel III A.C., Howell J.R.: Two-Dimensional Radiation in Absorbing-Emitting Media Using the P-N Approximation. J. Heat Transfer. 105 (1983) 333-340.

19. Selcuk N.: Exact Solutions for Radiative Heat Transfer in Box-Shaped Furnaces. J. Heat Transfer 107 (1985) 648-655.

MATHEMATICAL MODELLING OF HEAT TRANSFER IN AN INDUSTRIAL GLASS FURNACE

M.G. Carvalho and M. Nogueira
Mechanical Engineering Department
Instituto Superior Técnico/Technical University of Lisbon
Av. Rovisco Pais
1096 Lisboa Codex
Portugal

ABSTRACT

In the present work a mathematical model of the fluid flow and heat transfer phenomena occuring in industrial glass furnaces is presented. The model is divided in two main submodels: the combustion chamber and glass tank parts. The submodel for the combustion chamber incorporates physical modelling for the turbulent diffusion flame, soot formation and consumption and thermal radiation. The discrete transfer procedure is used to solve radiative transfer. The glass melting tank modelling considers the buoyancy driven laminar flow, the radiative transfer in the glass and the air bubbling effect. The time-averaged conservation equations set was solved by finite volume technique. The method was applied to an end-port furnace for the prediction of the fluid flow and temperature field inside both the combustion chamber and the glass melting tank, as well as distributions of soot and fuel concentration inside the combustion chamber.

NOMENCLATURE

A   Area (m^2)

Cp   Specific heat (J/kgK)

E   Black body emissive power (W/m^2)

f   Mixture fraction

g   Mixture fraction fluctuations

I   Radiation intensity (W/m^2)

k   Conductivity (W/mK)

K   Absorption coefficient (m^{-1})

p   Partial pressure (Pa)

q   Energy flux (W/m^2)

Q   Total energy flux (W)

s   Distance (m)

S   Stoichiometric ratio

S_ϕ Source term

T   Temperature (K)

u   Velocity (m/s)

Greek Letters:

β Thermal expansion coefficient

ϕ Scalar

Γ_ϕ Diffusion coefficient (kg/ms)

Ω Solid angle

ρ Specific weight (kg/m^3)

μ Viscosity (kg/ms)

Subscripts:

i   x, y or z component

t   turbulent

o   reference values

w   Wall

Superscripts:

—   time averaged

'   turbulence due to fluctuations

1. INTRODUCTION

The resources limitations growth, on one hand, and the pollution problem awareness in the other are turning heat transfer research attention towards the importance of improving performance of industrial process furnaces, such as those used in glass, cement and steel assemblies manufacture and power plants.

That commitment cannot be achieved without improving control strategies, personal training, furnace design, particularly combustion equipment design.

Traditional zero-dimensional methods based on empirical knowledge, largely ignoring spatial variations are unable to assist that purpose. The physical understanding of the phenomena occuring in the furnace/boiler environment based on three-dimensional mathematical modelling is becoming an essential requirement to assist operating conditions changes and further design modifications.

A three-dimensional simulation of a glass furnace combustion chamber in which the flow field and heat transfer were predicted, was presented at a first time by Gosman et al. [1].

Lockwood and Shah [2] presented an accurate and efficient technique for handling thermal radiation, the "discrete transfer method". This method combines ease of use, economy and flexibility of application which is of particular importance in geometrically intricate applications.

The discrete transfer assumptions were used in the first simulations of a full scale industrial glass furnace (Gosman et al. [1] and Carvalho [3]).

In the work of Carvalho [3] a geometry with a horseshoe flow shape and combustion of fuel oil were considered. The same end-port regenerative glass furnace was considered in the study of Carvalho and Lockwood [4] which brought into analysis the batch and glass tank flows assuming a two-dimensional batch flow.

Ungan and Viskanta [5] presented a simulation of the transport processes inside a glass melting tank, solving CO_2 bubbles concentration, air bubbling and electrical boosting effect. The radiative transport was handled setting an effective diffusion coefficient. Carvalho et al [6] presented a complete three-dimensional modelling of a cross fired glass furnace, comprising two submodels for combustion chamber and glass tank.

In spite of recent progress in this area, the need of widening the flexibility of modelling techniques emphasizes the new applications appearance necessity.

2. THE PHYSICAL-MATHEMATICAL MODELLING

The heat transfer, fluid flow and combustion within the combustion chamber and glass tank of the furnace are predicted by solving the governing partial differential equations set in its steady-state time-averaged form.

Transport of momentum, mass, energy, chemical species and soot concentration, turbulent quantities is solved for the three-dimensional space taking a general conservation equation which can be stated as:

$$\text{div}\left(\rho\, \vec{u}\, \phi\right) = \text{div}\left(\Gamma_\phi\, \text{grad}\, \phi\right) + S_\phi \tag{1}$$

Where ϕ represents the transported scalar, Γ_ϕ the diffusion coefficient and S_ϕ denotes the source term. The quantities which ϕ, Γ_ϕ and S_ϕ taken to each differential equation are referred in table 1.

Table 1: Transport Equation for Scalar ϕ.

EQUATION	ϕ	Γ_ϕ	S_ϕ
CONTINUITY	1	0	0
MOMENTUM (Gas Flow)	$\bar{u}_i$	$\mu_t + \mu$	$-\dfrac{\partial \bar{p}\,\delta_{ij}}{\partial x_j} + \dfrac{\partial}{\partial x_i}\left(\mu_{eff}\dfrac{\partial \bar{u}_j}{\partial x_i} + \dfrac{2}{3}\dfrac{\partial \bar{u}_k}{\partial x_k}\delta_{ij}\right)$
TURBULENT KINETIC ENERGY	k	$\Gamma_{k,t}$	$\mu_t \dfrac{\partial \bar{u}_i}{\partial x_j}\left(\dfrac{\partial \bar{u}_i}{\partial x_j} + \dfrac{\partial \bar{u}_j}{\partial x_i}\right) - \bar{\rho}\,\varepsilon$
k DISSIPATION RATE	ε	$\Gamma_{\varepsilon,t}$	$C_1\dfrac{\varepsilon}{k}\mu_t\dfrac{\partial \bar{u}_i}{\partial x_j}\left(\dfrac{\partial \bar{u}_i}{\partial x_j} + \dfrac{\partial \bar{u}_j}{\partial x_i}\right) - C_2\bar{\rho}\dfrac{\varepsilon^2}{k}$
ENTHALPY (Gas Flow)	h	$\Gamma_{h,t} + \Gamma_h$	$\dot{S}_r$
MIXTURE FRACTION	$\bar{f}$	$\Gamma_{g,t} + \Gamma_f$	0
FLUCTUATIONS OF f	g	$\Gamma_{s,t} + \Gamma_g$	$C_{g1}\mu_t\left(\dfrac{\partial \bar{f}}{\partial x_j}\dfrac{\partial \bar{f}}{\partial x_j}\right) + \dfrac{C_{g2}}{k}\bar{\rho}\,\varepsilon\,g$
SOOT	$\bar{m}_s$	Γ_s	$P_{s-} - P_{s+}$
MOMENTUM (Molten Glass Flow)	U_i	μ	$-\dfrac{\partial p\,\delta_{ij}}{\partial x_j} + \dfrac{\partial}{\partial x_i}\left(\mu\dfrac{\partial U_j}{\partial x_i} + \dfrac{2}{3}\dfrac{\partial U_k}{\partial x_k}\delta_{ij}\right) +$ $+ g_i(\rho - \rho_0) + \dot{S}_{U,B}$
ENERGY (Molten Glass Flow)	h	$Cp.k_{eff}$	$\dot{S}_r + \dot{S}_{h,B}$

2.1 COMBUSTION CHAMBER MODELLING

<u>Turbulence Model</u>. The "Two-equation" model, [7] in which equations for the kinetic energy of turbulence, k, and its dissipation rate, ε, are solved, was considered to be appropriate. (see table 1).

A "turbulent" viscosity, μ_t, that may be related to k and ε by dimensional arguments is defined as:

$$\mu_t = C_\mu \bar{\rho}\, k^2/\varepsilon \tag{2}$$

Where C_μ is a model constant. The turbulent exchange coefficient, $\Gamma_{\phi,t}$' (see Table 1) for any variable, ϕ, may be expressed by:

$$\Gamma_{\phi,t} = \mu_t / Pr_{\phi,t} \tag{3}$$

Where $Pr_{\phi,t}$ is a turbulent Prandtl number of unity order. Turbulent transport correlations involving ϕ' are determined from the Boussinesq approximation:

$$\overline{\rho\, u'_j\, \phi'} = \Gamma_{\phi,t} \overline{\partial\phi/\partial x_j} \tag{4}$$

The model constants appearing in these equations were assigned to the following values, taken unchanged from Launder and Spalding [7]: $C_\mu = 0.9$, $C_1 = 1.44$, $C_2 = 1.92$, $Pr_{k,t} = 1.0$, $Pr_{\varepsilon,t} = 1.3$. These model constants are well established in many previous furnace applications.

<u>Combustion Model</u>. The combustion model is based on the ideal fast single step reaction between the vaporised fuel and the oxidant. Equal effective turbulent mass diffusion coefficients for the fuel and oxidant are also assumed [8] and [9]. As a consequence of these assumptions the flame thermodynamic state becomes related to a single passive scalar by:

$$\psi \equiv s\, m_{fu} - m_{ox} \tag{5}$$

Where s is the stoichiometric oxygen required by mass, and m_{fu} and m_{ox} are the fuel and oxidant mass fractions.

The mixture fraction, f, is related to this quantity by:

$$f \equiv (\psi - \psi_o)/(\psi_1 - \psi_o) \tag{6}$$

Where the subscripts 1 and 0 designate the fuel-and oxidant-bearing streams. Furthermore, the authors have assumed that the chemical kinetic rate is fast with respect to the turbulent transport rate; then fuel and oxidant cannot coexist, so $m_{fu} = 0$ for $m_{ox} \geq 0$ and $m_{ox} = 0$ for $m_{fu} \geq 0$.

A transport equation for the mixture fraction, f, is solved allowing to predict the mass concentration of the chemical species inside the combustion chamber (see Table 1).

The fluctuating nature of the turbulent reaction may be accommodated through a modelled equation for the variance of the f fluctuations.

A statistical approach to describe the thermal nature of that fluctuations is followed.

The time-averaged value of any property ϕ solely depending on f can be determined from

$$\phi = \int_0^1 \phi\,(f)\,P\,(f)\,df \tag{7}$$

In the present work we have assumed the "Clipped normal" probability density function [10] which is characterized by just two parameters, and the mean square of the fluctuations $g \equiv (f - \overline{f})^2$. The g transport equation is solved (see table 1), adopting C_{g1} and C_{g2} as adjustable parameters.

<u>Soot Model</u>. The largest concentrations of soot are present just in a limited region, in consequence, the accurate prediction of local soot concentration is not a prerequisite for good calculation of radiative transfer. This is indeed fortunate since the mechanisms of soot formation are far from being well established even in the simplest laboratory flames.

A simple global expression similar to that used by Khan and Greeves [11] was chosen to characterize soot production:

$$P_{s+} = C_f p_{fu}\, \phi^{n}\, \exp\left(- C_e / RT\right) \tag{8}$$

where C_f is a function which depends on an easily definable fuel property such as the C/H ratio. p_{fu} represents the vaporized fuel partial pressure, ϕ the equivalence ratio and T the local gas temperature. The adopted values for the model constants were the following: $C_f = 0.01$ Kg/Nms (from Abbas et al. [12]), $C_e = 167.5$ kJ/mol and n = 3 (from Khan and Greeves [11]). A straightforward method of estimating the soot oxidation rate has been proposed by Magnussen and Hjertager, [13]. Their expression for the soot consumption rate is:

$$P_{s-} = A\,\overline{m}_s\,\overline{\rho}\,(\varepsilon / k) \tag{9}$$

where A is a model constant which is assumed to be A = 4. That relation will not be satisfactory in regions where the reaction rate is limited by oxygen deficiency, in which case Magnussen and Hjertager [13] propose:

$$P_{s-} = A \left(\frac{\overline{m}_{ox}}{\overline{m}_s S_s + \overline{m}_{fu} S_{fu}} \right) \overline{m}_s \left(\frac{\varepsilon}{k} \right) \overline{\rho} \tag{10}$$

where S_s and S_{fu} are the soot and fuel stoichiometric ratios. The alternative that gives the smallest reaction rate P_{s-} is to be used.

A transport equation for the soot mass concentration was solved accounting the source term as $S_s = P_{s+} - P_{s-}$ (see Table 1).

<u>Radiation Model</u>. The "discrete transfer" radiation prediction procedure of Lockwood and Shah [2] has been utilized in this study.

The "discrete transfer" method is founded on a direct solution of the radiation transfer equation for a direction which runs:

$$\frac{dI}{ds} = (K_g + K_s)\left(\frac{E}{\pi} - I\right) \tag{11}$$

where I is the radiation intensity in a direction Ω, s is the distance in that direction, E is the black body emissive power, and K_g and K_s are respectively the gas and soot absorption coefficients.

That relation was applied along a chosen Ω from known conditions at a point Q say (either guessed or pertaining to those of the previous iteration) on one wall to a point of impingement, P say, of the direction, Ω, on the opposite wall.

If the hemisphere above P is discretized into sub-angles, $d\Omega$, within which the intensity is considered to be uniform; then the energy arriving at P is the sum of the energy flux arriving from all this directions.

The authors have solved the integrated form of the eqn (11) within the discretization $d\Omega$; of the whole solid angle, Ω, about selected directions, taking finite distance increments ds, and where $K = K_g + K_s$

$$I_{n+1} = \frac{E}{\pi}\left(1 - \exp(-Kds) + I_n \exp(-Kds)\right) \tag{12}$$

The net heat gain loss within a small control volume is:

$$S_r = (I_{n+1} - I_n)\,\Omega\,d\Omega\,dA \tag{13}$$

where the locations n and (n+1) correspond to the "entry" and "exit" of a direction into and from a control volume, dA is the cell area projected in the plane normal to Ω, and ds represents the distance between n and (n+1) locations.

The relation (12) is applied along the chosen Ω from known conditions at a point Q say (either guessed or pertaining to those of the previous iteration) on one wall to a point of impingement, P say, of the direction, Ω, on an opposite wall.

The hemisphere above P, was discretised into sub-angles, $d\Omega$, within which the intensity is considered to be uniform; the energy flux arriving at P is:

$$q_{+,p} = \int_p I_p \Omega\,d\Omega = \sum_{\text{all } r} I_{p,r}\Omega_r d\Omega_r \tag{14}$$

where $\Omega = \cos\theta$ with θ representing the polar angle measured from the normal to the impingement surface.

The wall boundary condition is:

$$q_{-,p} = \left(1 - \varepsilon_w\right) q_{+,p} + \varepsilon_w E_w \tag{15}$$

where $q_{-,p}$ is the energy leaving the wall at P, ε_w is the wall emissivity, and $E_w = \sigma T^4_w$ is the wall emissive power. The value of I_0 at point Q, the initial value required for the application of the recurrence relation, eqn (12), is $q_{-,p}/\pi$. The net radiation heat flux is:

$$q_p = q_{+,p} - q_{-,p} \tag{16}$$

At the glass melt surface, a prescribed temperature was imposed. This temperature was taken from the glass tank calculations described below. At the remaining surfaces constant flux condition was used, except for the inlet and outlet openings, where imposed temperature condition was considered.

The radiation source, S_r, is appended to the energy transport equation (see table 1).

The gas plus soot absorption coefficient, K, is calculated from the "two gray plus a clear gas" model due to Truelove [14].

2.2 GLASS-MELT TANK MODELLING

<u>Radiation Model</u>. The proposed modelling is based on the fact that in diathermanous materials absorption and emission of radiation are bulk phenomena. In the bulk of the diathermanous medium, the radiation mechanism is comparable to thermal conduction through the concept of radiative diffusity. This concept, however, is not applicable near boundaries, for which a semi-empirical representation is proposed.

The diathermancy of a glass is determined by the wave-length dependence of the absorption coefficient. The nature of the present application (due to the complexity of the problem and lack of relevant data) demands that the wave length application should be avoided. A gray material effective diffusity is a property experimentally accessible as:

$$\Gamma_{eff} = - Q / (dT/ds) \tag{17}$$

where Q is the total transferred heat between two layers, distanced by ds, with a temperature difference dT.

For a gray material the radiative diffusity may be given by:

$$\Gamma_{rad} = \frac{16n^2 \sigma T^3}{3K} \tag{18}$$

Data [15], which account for different spectral coefficients, allow the following representation of an effective conductivity for a gray glass medium:

$$k_{eff} = 2.555\,T^3 + 18.96 \tag{19}$$

The used thermal conductivity was k=2.0906. The knowledge of that value, and equations.(18) and (19) allow to compute a gray absorption coefficient as

$$K_{eff} = -0.01848\,T + 62.58 \tag{20}$$

The most important limitation to the application of the effective diffusion concept is its application near the boundaries.

In opaque materials, in which heat only flows by conduction, temperature distributions are continuous throughout the boundaries. In the case of diathermanous materials, this does not append. Thus, in an ideal glass, assumed to have zero conductivity, the temperature varies continuously to a value To at the surface. However, the solid opaque material bounding the glass will be at a different temperature, yielding a temperature discontinuity in the boundary surface.

This temperature discontinuity may be explained by the radiative exchanges between the glass and the surface. The phenomena may be interpreted as a transfer between adjacent layers of glass. The first layer of glass would itself become hotter and therefore, radiate more energy. The layer that absorbed it would also become hotter and radiate, and thus, the process goes on up to the extinction of the influence of the radiative flux at the boundary. The flux absorbed in the transfer between two such layers may be given as:

$$W_n = I_n \left(1 - \exp(-Kds)\right) dA \tag{21}$$

At the boundary, the intensity I_n would be equal to the exchanged flux between the glass and the bounding medium. The above referred discontinuity does not allow, in the present modelling, the use of imposed temperature boundary condition. The interaction between the glass and the gas is therefore represented in this model by an imposed flux at the surface of the glass and melting batch.

Due to the large ratio between the size of the area of the glass melted surface boundary surface and the beam penetration length, a unidimensional transfer assumption was followed for the description of the influence of the boundary fluxes. The optical thickness of the considered kind of glass allows that simplification.

<u>Air Bubbling Model</u>. The air bubbling effect was handled using a lagrangian formulation to predict the bubble rising flow accounting for the heat and momentum transfer with the glass melt. A spherical shape is assumed for the rising air bubbles. The initial bubble radius is determined from experimental observation in the furnace of the actual number of bubbles reaching the free surface for a given air flow.

<u>Thermodynamic Relations</u>. The glass density was calculated by the following standard relation

$$\rho = \rho_o\left[1 - \beta\left(T - T_o\right)\right] \tag{22}$$

using $\beta=5\times10^{-5}$ as thermal expansion coefficient. The kinematic viscosity was considered as represented by the Fulcher-Vogel-Tamman equation:

$$\mu = \rho \, \exp\left(- 6.21 + \frac{4570.98}{T - 525.6}\right) \tag{23}$$

The included constants were extracted from [5].

2.3 COUPLING ALGORITHM

The combustion chamber and the molten glass flow were studied by a cyclical iterative way, by separated calculations, matched by the relation between the heat flux from the flame to the glass surface and its temperature. Using the heat flux to the glass, calculated by the chamber model, as a boundary condition, the flow and the temperature distribution of the molten glass were predicted by the tank model. A similar procedure is described in [16].

2.4 NUMERICAL SOLUTION

The finite difference method used to solve the equations entails subdividing the combustion chamber into a number of finite volumes or "cells".

The convection terms were discretized by the hybrid central/upwind method [17]. The velocities and pressure are calculated by a variant of the SIMPLE algorithm described in [18]. The solution of the individual equations sets was obtained by a form of Gauss-Seidel line-by-line iteration.

3.   RESULTS AND DISCUSSION

The furnace which is of the end-port regenerative kind, is shown in Figure 1 .

Combustion Results. Figure 2 presents a predicted gas temperature field represented by a row of vertical planes, one of which crosses the middle burner.

The largest temperature gradients appear in the interface between air and fuel stream denoting the presence of combustion reaction. Near the glass melt surface the temperature distribution denotes the cooling effect of the batch melting process and presents an hot spot below the flame tip.

Figure 3 presents the soot concentration distribution. The highest concentration value can be found by the flame tip, as expected. A strong soot oxidation, supported by the high temperatures inside the chamber, determines the appearance of strong decreasing gradients after the position of highest concentration values.

Figure 4 shows the distribution of fuel mass fraction in a row of vertical planes.

The main flow recirculation from the inlet ports to the outlet imposes a displacement of the time-averaged flame shape towards the nearest side wall. This effect, which certainly will yield a wall degradation increase in that region, may be avoided introducing changes in the furnace design. This task might be assisted by mathematical modelling.

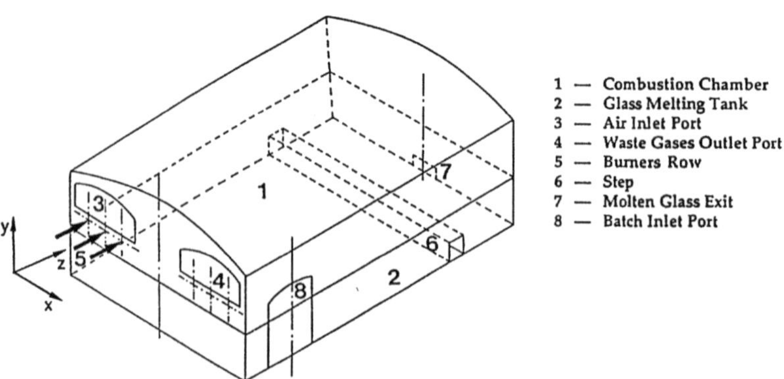

Fig. 1:   Diagram of the furnace.

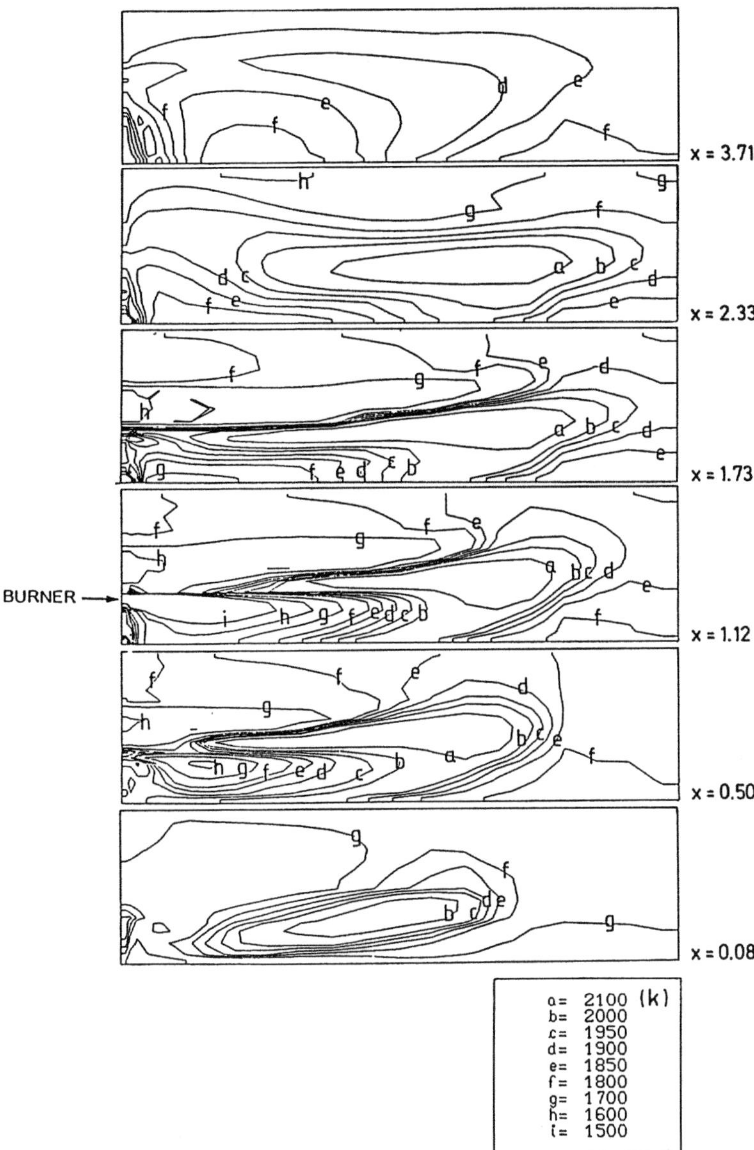

Fig. 2:   Predicted temperature field.

The inlet air and fuel relative location imposes the flame shape displacement towards the glass surface, the effect which is explained by a secondary recirculation below the burners row (see Figure 5), producing a reductive atmosphere just above the batch melting region, even for oxigen-rich flames. This effect cannot be neglected in the melting of some kinds of glass compositions.

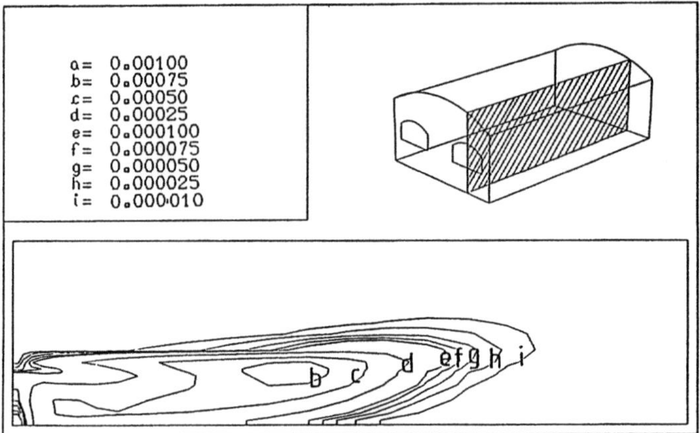

Fig. 3: Predicted soot concentration.

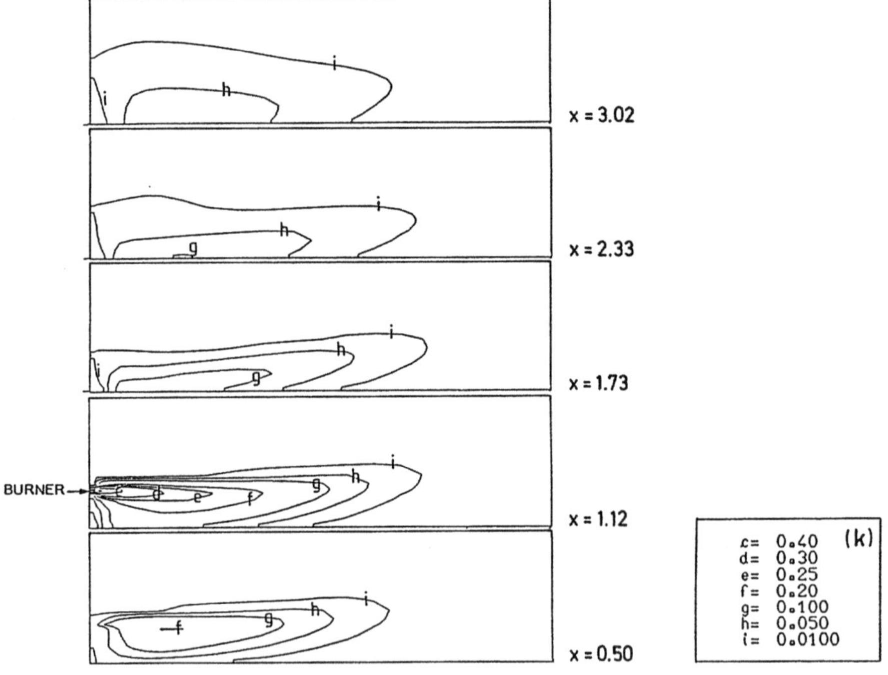

Fig. 4: Predicted field of the fuel mass fraction.

A control strategy able to avoid this effect, without damaging the thermal efficiency, may be studied using mathematical modelling skills.

The flow is characterised by a main recirculating motion as shown in the Figure 5 which presents the trajectories of unweighted particles imaginary seeded in the

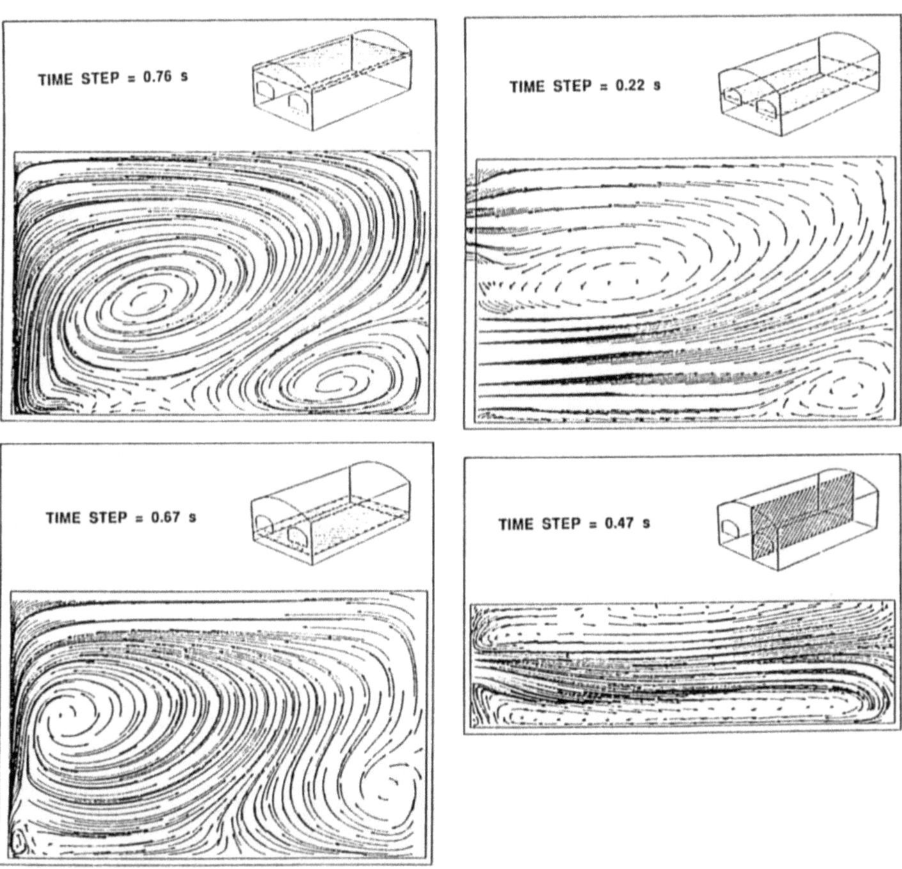

Fig. 5: Predicted gas flow in the chamber.

flow. The produced streamlines show the "horse shoe" shape of the predicted main recirculation typical of this kind of furnace. The time step referred is the time during which the trajectories were followed.

<u>Glass Tank Results</u>. This results were obtained assuming an imposed distribution for the unmelted materials, which is characterised in Figure 6.

Free convection effect, which predominantly drives the flow when no air bubbling is being used, can be observed in the pattern shown in Figure 7, in which a vertical plane in the middle of the tank and an horizontal plane near the free surface are represented. These representation uses the unweighted particles trajectory technique above referred. The imaginary particles were seeded in the melting region. A minimum residence time of eight hours was obtained. An interesting feature of the flow is its ability to produce a backward motion at the

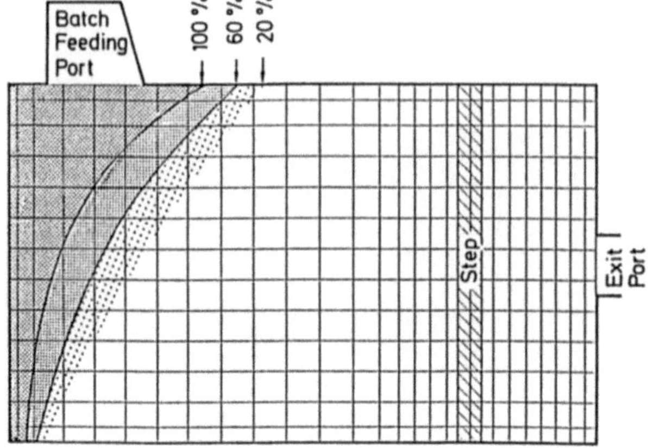

Fig. 6: Unmelted materials imposed pattern at the molten glass surface.

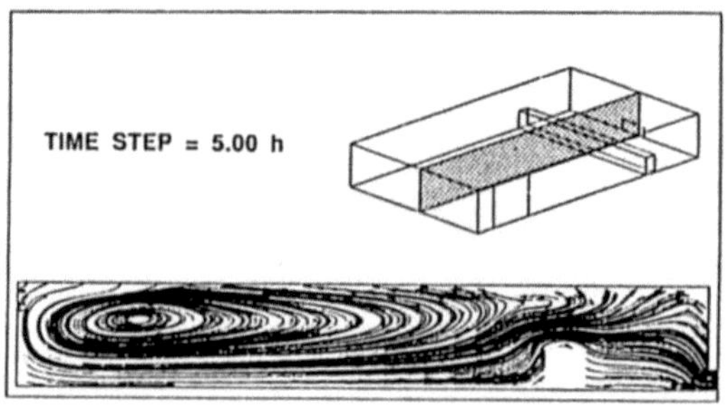

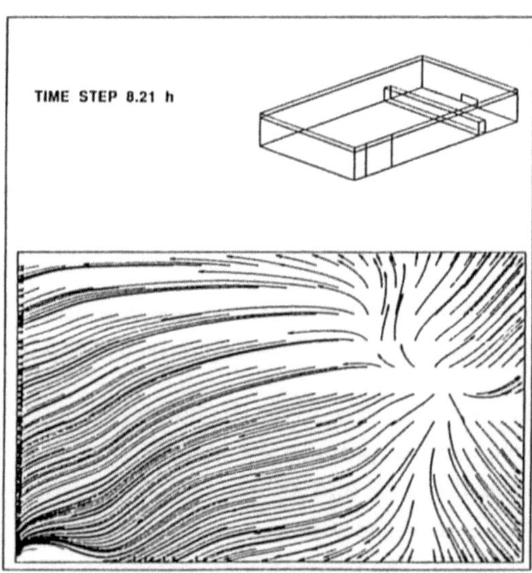

Fig. 7: Predicted fluid flow in the tank
(without air bubbling).

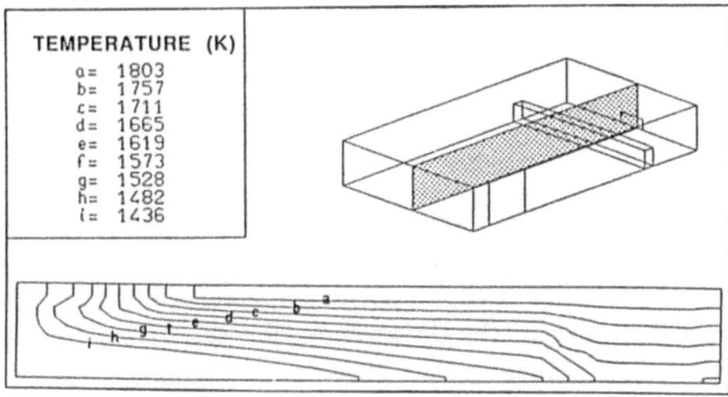

Fig. 8: Predicted temperature field in the tank
(without air bubbling).

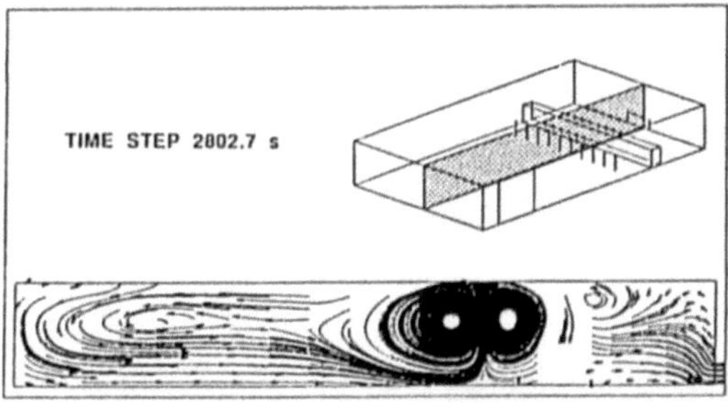

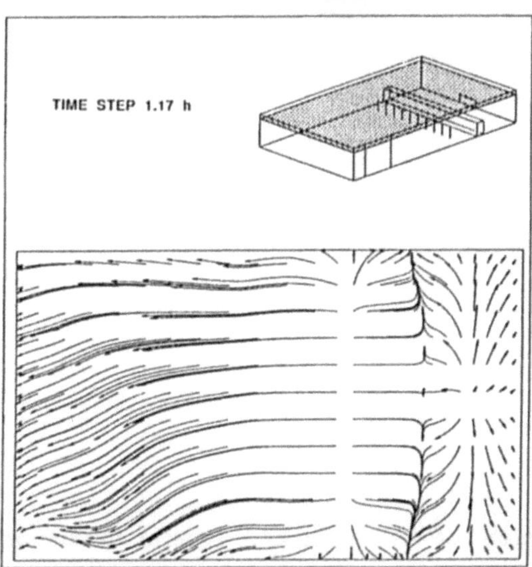

Fig. 9: Predicted fluid flow in the tank
(with air bubbling).

free surface, keeping the batch materials well conditioned in the melting region, as may be observed in the shown horizontal plane. High velocity values can be observed in the near outlet region and above the step, where severe corrosion effect may be expected.

The temperature field in a plane near the middle of the tank, in which the cold regions below the batch zone are well apparent, can be seen in Figure 8. For no air bubbling conditions, a stratified temperature field is produced as shown in that figure, effect which does not help the glass refining process. Air bubbling is used however to produce a barrier in the molten glass flow, preventing the motion of unmelted batch particles towards the refining zone of the furnace. Therefore a completely different flow pattern occurs for air-bubbling operating conditions as can be seen in Figure 9, where the main flow is represented by a vertical plane in the middle of the furnace and by an horizontal plane near the free surface. The effect of the bubbling produces two strong recirculations which actually ensure the barrier effect above mentioned. These recirculations also produce a mixing in the central region of the tank yielding a homogenizing effect which is well characterised in the temperature field presented in Figure 10.

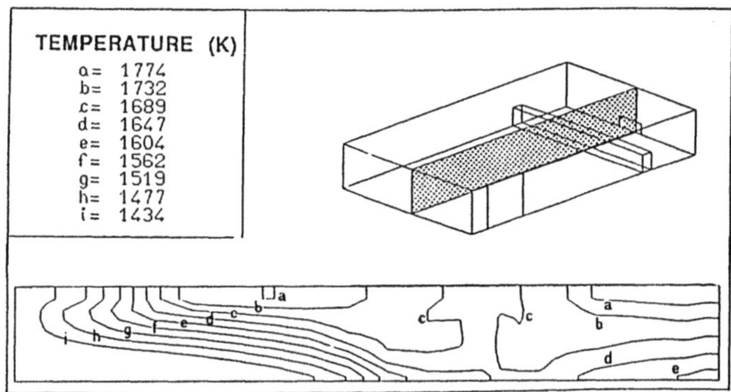

Fig. 10:   Predicted temperature field
           (with air bubbling).

4.   CONCLUDING REMARKS

This paper describes the application of an useful and general prediction procedure to real industrial furnaces. Predictions were obtained by the three-dimensional solution of the fluid flow and heat transfer following a physical-mathematical modelling method described in this paper.

The authors suggest that the processes occuring in the combustion chambers can be predicted with an acceptable engineering accuracy by mathematical modelling techniques. The modelling of relevant physical processes in industrial furnaces can supply an important basis for improvements in control strategies, furnace design and training.

REFERENCES

1. Gosman A.D., Lockwood F.C., Megahed I.E.A., Shah N.G.: The prediction of the Flow, Reaction and Heat Tranfer in the Combustion Chamber of a Glass Furnace. AIAA 18th Aerospace Sciences Meeting, January 14-46, Pasadena, CA, U.S.A. (1980).

2. Lockwood F.C., Shah N.G.: A New Radiation Solution Method for Incorporation in General Combustion Prediction Procedures. 18th Symp. (Int.) on Combustion, The Combustion Institute (1980).

3. Carvalho M.G.: Computer Simulation of a Glass Furnace. Ph.D. Thesis, London University (1983).

4. Carvalho M.G., Lockwood F.C.: Mathematical Simulation of an End-Port Regenerative Glass Furnace. Proc. Inst. Mech. Engrs., 199 (C2) (1985) 113-120.

5. Ungan A., Viskanta R.: Three-Dimensional Numerical Modelling of Circulation and Heat Transfer in a Glass Melting Tank. Glastech. Ber. 60 (1987) 71-78 and115-124.

6. Carvalho M.G., Oliveira P., Semião V. A.: Three-Dimensional Modelling of an Industrial Glass Furnace. Journal of the Institute of Energy, Sep (1988).

7. Launder B.E., Spalding D.B.: Mathematical Models of Turbulence. Academic Press, New York (1972).

8. Pun W.M., Spalding D.B.: A Procedure for Predicting the Velocity and Temperature Distributions in a Confined Steady, Turbulent, Gaseous Diffusion Flame. in Proc. Int. Astronautical Federation Meeting, Belgrade (1967).

9. Bilger R.W.: Turbulent Flows with Non-Premixed Reactants, Turbulent Reacting Flows. in Topics in Applied Physics, Edited by Libby P.A. and Williams F.A.. Springer-Verlag (1980).

10. Lockwood F.C., Naguib A.S.: The Prediction of the Fluctuations in the Properties of Free, Round Jet, Turbulent Diffusion Flame. Combustion and Flame 24(1) (1975) 109.

11. Khan I.M., Greeves G.: A Method for Calculating the Formation and Combustion of Soot in Diesel Engines, Heat Transfer in Flames. Edited by Afgan and Beer, (1974) 391-402.

12. Abbas A.S., Koussa S.S., Lockwood F.C.: The Prediction of a Variety of Heavy Oil Flames. in Proc. ASME Winter Combustion Annual Meeting, Spec. Sess. on Two-Phase Combustion Liquid Fuels, Washignton, Nov. (1981).

13. Magnussen B.F., Hjertager B.H.: On Mathematical Modelling of Turbulent Combustion with Special Emphasis on Soot Formation and Combustion. 16th Symp. (Int.) on Combustion, Combustion Institute (1976).

14. Truelove J.S.: Mathematical Modelling of Radiant Heat Tranfer in Furnaces. Heat Tranfer & Fluid Flow Service, Chemical Engineering Division, Aere Harwell Rept., N°. HL76/3448/KE, Sept. (1974).

15. Genzel L.: Zur Berechnung Der Strahlungsleitfähickeit der gläser (Calculation of Radiation Conductivity of Glasses), Glastech. Ber., 26(3) (1953) 69-71.

16. Spalding D.B.: A Novel Finite Difference Formulation for Differential Expressions Involving both First and Second Derivatives. Int. J. Num. Methods Eng. 4 (1972) 557.

17. Carreto L.S., Gosman A.D., Patankar S.V., Spalding D.B.: Two Calculation Procedures for Steady, Three-Dimensional Flows with Recirculation. in Proc. of 3rd Int. Conf. on Numerical Methods in Fluid Dynamics. Springer-Verlag, New York (1972) 60.

MATHEMATICAL MODELLING OF THERMAL NOχ EMISSIONS IN COMBUSTION CHAMBERS

K.A. Pericleous[1], I.W. Clark[2] and N.C. Markatos[3]

[1] Centre for Numerical Modelling and Process Analysis
   Thames Polytechnic, London, UK
[2] Flomerics Inc., USA
[3] National Technical University, Athens, Greece

ABSTRACT

A mathematical model of combustion in typical burners is presented. The model differs from other similar ones presented by other investigators, in that it features a two-step approach, which allows the calculation of the main exothermic reactions of fuel in air, and of those responsible for the formation of pollutants, specifically nitrogen oxides (NO_x) to be performed in tandem. The model uses CFD techniques to compute the flowfield in the burner and transport equations are solved for all relevant reacting species. To ensure industrial relevance all factors affecting combustion performance and heat transfer have been included, i.e. three-dimensionality, an exact representation of geometry, the air swirler, turbulence and radiation.

In the first step, the calculation deals with the combustion of methane in air for which a global, single step reaction is assumed. The limiting effect of turbulence on reaction rate is taken into account using an eddy-break-up model. Radiation between the burner walls, the flame and intervening gas is accounted for using a flux model.

In the second step, the product stream resulting from the first step calculation is analysed to determine the concentrations of O, OH and H radicals, subsequently used in a Zeldovich type scheme to form oxides of nitrogen. The convective and diffusive transport terms of the first step are used to solve conservation equations for N and NO.

The paper illustrates the technique in an application to an axial recirculating burner.

NOMENCLATURE

C_R Empirical expression in combustion model

F   Species concentration

k   Turbulent kinetic energy per unit mass, J/kg

M   Mole fraction

m   Mass fraction

R   Ideal gas constant

R   Reaction rate per unit volume, kgs/m^3

s   Stoichiometric oxidant to fuel ratio

S_φ Source term in finite difference equations

T	Temperature, K
T_{act}	Activation temperature, K
$\vec{V}$	Velocity vector

Greek Symbols

Γ_φ	Turbulent exchange coefficient for variable φ
ε	Turbulence energy dissipation rate per unit mass, W/kg
μ	Dynamic viscosity, Ns/m^2
ρ	Density, kg/m^3
φ	General dependant variable

Subscripts

fu	Fuel
MO	Nitrogen oxides
N_2	Nitrogen
OX	Oxygen
ox	Oxidant
pr	Product

INTRODUCTION

Combustion products are recognised as a threat to the natural environment. If we exclude the main products of combustion, CO_2 and H_2O which threaten to alter the climate of our planet, other minor constituents act as pollutants. The most obvious air pollutant is smoke, being visible to the naked eye. Other pollutants of importance are, carbon monoxide (CO), unburned hydrocarbons, sulfur oxides (SO_x) and finally nitrogen oxides (NO_x). Virtually all of these pollutants have a harmful effect on the ecosystem [1]. Emotive phrases have evolved over the years which describe these effects, as in the "greenhouse effect", "acid rain" and the "ozone layer hole".

Nitrogen dioxide (NO_2) constitutes one of the most dangerous air pollutants in modern cities, because in addition to its harmful effects on the human respiratory tract, it partakes in the cycle of photochemical reactions that result in the creation of smog, or photochemical fog. This yellow-brown fog characterises pollution in cities such as Los Angeles and Athens and is due mainly to the presence of NO_2 and secondary produced aerosols of nitric particles.

Nitrogen oxide is formed by direct oxidation of NO in the atmosphere. In the stratosphere, where ozone supplies the oxygen, a chain reaction takes place which contributes to the depletion of the ozone layer. The quantitative influence of this chain mechanism is uncertain. However, this gap in knowledge and the fact that holes in the ozone layer have appeared over both poles, has led to a worldwide tightening of legislation governing the sources of NO_x emission which include aircraft jet engines, modern lean-burn internal combustion engines, coal- and gas-fired power stations and industrial furnaces.

The European Community in directive No. 85/203 has set a limit for NO_x concentrations to 200 ng/m^3. This directive has been in force since 1/1/1987; however, in regions where the said limit is exceeded, a transition period has been allowed for the gradual introduction of necessary measures, up to 1/1/1994. In the example of Athens, emergency measures are planned in two stages of severity when critical limits of 500 ng/m^3 and 700 ng/m^3 are reached respectively.

From monitoring stations set up by ITEPITA in 1984, it has been found that the whole of the Athens basin is faced with a severe problem of NOx pollution. Concentrations exceed the prescribed limits in the city centres of Athens and Piraeus, where maximum hourly values of 634 ug/m^3 and 574 ug/m^3 have been recorded respectively. Measurements over successive years reveal a particularly worrying tendency for these values to increase. The Athens problem is mainly due to motor car pollution and is therefore particularly acute in road junctions during rush hour.

Away from city centres main sources of NO_x pollution are due to coal/gas fired power stations. Typically, 40% of all NO_x in the UK is due to power stations. Particularly stiff national controls are already in operation concerning these in certain EC countries, specifically Germany and Netherlands. Typically, Dutch legislation requires limits of 200 ppm NO_x for liquid fuel and 120 ppm NO_x for gas fuel burners [2]. In the near future these figures are expected to be halved.

Under this climate of tightening legislation new, improved methods of design are called for to replace burners/combustors, and improved methods of operation needed to prolong the use of existing units.

The term NO_x usually implies two major oxides, nitrogen oxide (NO) and nitrogen dioxide (NO_2). In combustion, NO is the dominant of the two components, NO_2 being mainly derived from NO. It is the prediction of NO we are concerned with in this paper, loosely referred to as NO_x. The following processes lead to NO production:

- From reaction of N_2 with oxygen, "thermal NO";
- from N existing in the fuel, "fuel NO"; and,
- from reactions of fuel derived radicals with N_2 ultimately leading to NO, the "prompt NO".

It is the first of these three mechanisms we are concerned with here, since it is the one that can be most affected by burner design.

MEANS OF NO_x REDUCTION

The main factor controlling thermal NO_x is temperature. In fact, NO_x formation is insignificant at temperatures below 1800K. It has been observed [3] that

$$NO_x \ \alpha \ \exp(0.009T) \tag{1}$$

where T=reaction temperature in Kelvin. All other factors can be viewed as means of affecting temperature. Therefore, the burner designer needs to avoid hot spots and also keep the time available for NO_x formation to a minimum. Both these factors are affected by an increase in the flow of air into the combustion zone. However, care needs to be exercised if CO and unburned hydrocarbon levels are to be kept to a minimum. Practical means of control include:

- Low excess air combustion since O_2 concentration is reduced.
- Exhaust, or flue gas recirculation; effective if the recirculation gas is cooled before being injected into the combustion zone;
- Water injection into the primary combustion zone of a burner, as used in industrial turbines; utilises the latent heat of water droplets to reduce temperature.
- Staged combustion, which introduces primary, secondary or even tertiary combustion zones and hence allows tighter control on temperature limits. In order to be effective, both air and fuel delivery needs to be staged. Design complexities result.

In order to be able to evaluate the influence of these factors, a mathematical model which represents correctly the physico-chemical aspects of the problem is needed. The object of this paper is to report on the development of such a model and present sample calculations.

EXISTING PREDICTION METHODS

To be successful and relevant, any theoretical model attempting to predict NO_x formation must accommodate the complex flow behaviour occuring in a burner and in addition the chemical kinetics characterising the main reactions. A satisfactory emissions model needs to achieve an even balance between accuracy, utility, ease of use, economy of operation, and capacity for further improvement. A considerable amount of modelling work has been done in recent years in the field of gas turbine combustors and furnaces, by Scheffer [4], Quan [5], Mizutomi [6] and Khalil [7], who utilise the solution of a transport equation for NO. This contains a source term based on a single step global Arrhenius expression, based on temperature and mole fractions of N_2 and O_2. The rate constants are derived from experiments, i.e.,

$$\overline{R}_{NO} = 8.39 \times 10^{16}\, (\overline{T})^{\frac{1}{2}}\, \overline{M}_{N2}\, \overline{M}_{OX}^{\frac{1}{2}}\, \exp\,(-134900/R\overline{T}) \tag{2}$$

where, the overbar denotes time averaged quantities. Comparison with data showed a varied degree of success, with agreement being mostly qualitative in nature. A modification of eq. (2) which takes into account temperature fluctuations was adopted by Hutchinson [8] who reported particularly good agreement with experiments in furnace calculations. It might be pointed out that the lack of good agreement in predicting NO_x may be due to inaccuracies in the prediction of the temperature field resulting from combustion, rather than deficiencies in eq. (2).

The high cost and complexity of some of the above models, combined with their rather modest success, has led to the development of semi-empirical models, which place more emphasis on the physical aspects of the problem with chemical kinetics given a relatively minor role. The work of Hung [9] is typical in this respect.

The present work which first appeared in [10] suggests an alternative two-step reaction of methane which separates the main exothermic terms from those responsible for NO_x production.

DESCRIPTION OF METHOD USED

In simulating a real burner accurately one has to contend with complex flow fields which often include swirl and regions of flow reversal and are highly turbulent, and also tackle complex chemical kinetics which govern the combustion and subsequent formation of trace chemicals.

It is clear that the combustion and NO_x reactions in a burner are inseparable, and the latter is a direct consequence of the former. However, it has been observed in experiments [11] that the main oxidising reaction between fuel and air is not affected by the formation of trace species which lead to NO_x. Hence, although the NO_x reactions tend to be endothermic, the radicals involved occur in minute concentrations and they do not affect the exothermic reaction occuring in the bulk of the burner. On the other hand, as pointed out earlier, the temperatures generated by the combustion are the determining factor in NO_x formation. With these observations in mind, one can treat the combustion and NO_x formation steps as separate.

CONSERVATION EQUATIONS SOLVED

Flow and reaction in a burner is defined in terms of a set of elliptic partial differential equations that express the conservation of mass, momentum, energy and other fluid variables in three-dimensional recirculating flow. With the primitive variable φ representing velocity, chemical species, temperature and turbulence variables the equations can be written in a generalised form:

$$\frac{\partial}{\partial t}(\rho\varphi) + \operatorname{div}(\rho\vec{v}\varphi - \Gamma_\varphi \operatorname{grad}\varphi) = S_\varphi$$

(3)

where ρ, v, Γ_φ and S_φ are density, velocity vector, turbulent exchange coefficient and source rate per unit volume respectively. The sources and exchange coefficients for velocities and turbulence have been discussed in detail elsewhere [12, 13] together with the effects of buoyancy. The k-ε model of turbulence is used to calculate the effective viscosity [13] in all regions of the flow, except where strong swirl exists. In these regions a mixing length model is used instead [14].

The numerical analogue of eqs. (3) was solved using CFD code PHOENICS [15] on a cylindrical polar finite difference grid.

COMBUSTION MODEL

Calculations were performed using a single step combustion model which could be either diffusion controlled or kinetically controlled. In both cases the equation for a mixture fraction [16] is solved; in the first case the mass fraction of fuel at any point is determined from it, while in the second, an additional conservation equation is solved for the mass fraction of fuel. The mass fractions of products and oxidant are then derived from these quantities [17].

The disappearance rate of fuel (CH4) was calculated in the kinetically controlled model as being the minimum of that given by an Arrhenius expression and eddy-break-up expression [18].

$$R_{fu} = - C_R \rho \frac{\varepsilon}{k} \min\left[m_{fu}, \frac{m_{ox}}{s}, \frac{m_{pr}}{(1+s)}\right] \tag{4}$$

where, C_R either takes a constant value of 4 or is calculated locally as a function of k and ε, i.e.

$$C_R = 23.6 \left[\frac{\mu}{\rho}\frac{\varepsilon}{k^2}\right]^{\frac{1}{4}}. \tag{5}$$

The effect of radiation was included using a flux model [7, 19] which solves ordinary differential equations for radiation fluxes along each coordinate direction.

The ideal gas law was used to calculate density variations and the specific heat was allowed to vary as a function of temperature and composition. The combustion calculation yielded field values of all the dependent variables, which were stored on disc for use in the NO_x prediction.

NO_x CALCULATION

The chemistry involved in the formation of NO_x will now be described. The model employed incorporates one of the simplest and most widely used mechanisms for calculations involving nitric oxide formation, namely the Zeldovich mechanism:

$$N_2 + 0 \Leftrightarrow NO + N, \tag{6}$$

$$O_2 + N \Leftrightarrow NO + O, \tag{7}$$

$$OH + N \Leftrightarrow NO + H. \tag{8}$$

As mentioned earlier, the basic assumption in this work is that reactions (6)-(8) can be considered as decoupled from the main fuel-burning reaction. In addition, the species appearing in these reactions are assumed to form part of the product stream of the single combustion reaction which in the first step of the calculation was assumed to be a single entity.

The species O, O_2, OH, H, H_2, CO and CO_2 are assumed to exist at chemical equilibrium. Their concentration is determined from a minimisation of Gibbs free energy subject to the following constraints: (a) mass balance of the elements present in the system, (b) specified enthalpy and, (c) specified pressure. The polynomial equations for species concentrations are solved in a single pass by calling the code CREK [20] in the user attachment of PHOENICS. The mass fractions of N and NO are determined by solving conservation equations (eq. (3)) which have as convection and diffusion fluxes the ones determined in the combustion step. It is assumed that the concentrations of the other species remain invariant, with the justification that they are produced by reactions which are considerably faster than those involving nitrogen. If this proves to be a severe restriction then the method can be extended to include further "slow" reactions.

The source term for the two conservation equations comes from a summation of the reaction rates for all reactions in which each species participates; i.e. the Zeldovich reactions. The forward and backward reactions are given by Arrhenius expressions of the form:

$$R = K \rho^2 F_1 F_2 \tag{9}$$

where F_1 and F_2 are the concentrations of participating species and K the reaction rate coefficient, given by

$$K = 10^B T^N \exp(-T_{act}/T) \tag{10}$$

with $10^B T^N$ being the pre-exponential factor, and T_{act} the activation temperature. The values of these quantities are given in Table 1.

Table 1: Constants used for the forward and reverse reactions (eqs. (9), (10)).

REACTION	FORWARD			REVERSE		
	10B	T_{act}	N	10B	T_{act}	N
$N_2 + 0 \Leftrightarrow NO + N$	7.6×10^{10}	3.8×10^4	O	1.6×10^{10}	0.0	0
$O_2 + N \Leftrightarrow NO + O$	6.4×10^6	3.15×10^3	1	1.5×10^6	1.95×10^4	1
$OH + N \Leftrightarrow NO + H$	6.3×10^8	0.0	0.5	2.5×10^9	2.45×10^4	0.5

The temperature used in the reaction rate coefficient (10), is that of the product stream, not the mixture value. This was obtained by interpolation. Also note that the oxygen appearing in the table is that produced by back-reaction in the products stream.

METHOD APPLICATION

The model described above was derived to predict the NO_x emissions of a typical burner used in large steam generators [10].

It is widely recognised [21], [22] that significant NO_x emission reduction can be achieved through burner design. However, better understanding of the flame structure produced by a given burner is necessary in order to identify and apply successful design modifications.

Such modifications involve fuel injection distribution and pressure, air flow characteristics, and swirl stabilisation. Furnace geometry also needs to be investigated since it conditions the heat transfer rate and the combustion products flow pattern [23].

As a precursor to this calculation the model was tested in a one-dimensional situation so that the influence of stoichiometry, temperature and residence time on NO_x production could be investigated. The results were given in reference [10] and will not be repeated here. For this simple situation, the model predicts:

(a) NO_x concentration increases with temperature, with maximum values occurring downstream (on the lean side) of the maximum temperature position.

(b) NO_x production increases with residence time, given a temperature distribution along the duct.

(c) Generally higher values of NO_x are observed on the lean side of stoichiometry at a temperature of 2000 K, as more molecular oxygen O, becomes available for the reactions.

A TYPICAL BURNER

The burner under investigation is the Todd "Dynaswirl" axial air flow type, which uses a blade swirler to impart a tangential velocity component to a portion (15%) of the total airflow. The latter principle is commonly used to stabilise high intensity industrial flames [24].

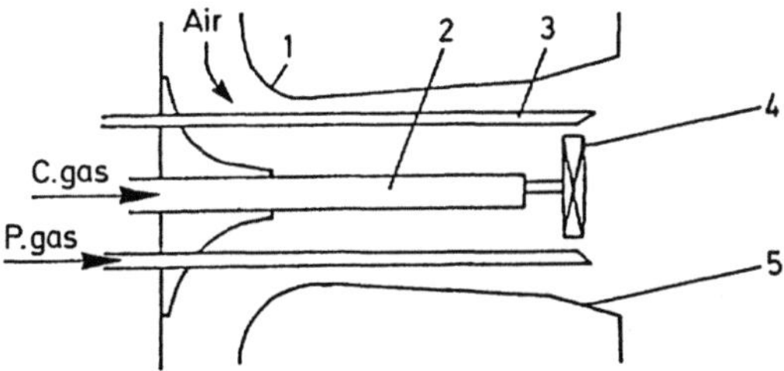

Fig. 1: Typical burner investigated

As shown in Figure 6, the major burner components include a venturi air register (1), a central fuel injector (2), six peripheral fuel injectors (3), a swirler (4), and an exit quarl (5).

The characteristics of the burner being simulated were:

- Maximum heat output: 100 MMBTU/hr
- Combustion air temperature: 500 degrees F
- Burner air side pressure drop: 10" WG
- Fuel: Natural gas (100% CH4)

A grid was defined using a polar system of coordinates, to simulate a cylindrical combustion chamber 9ft in diameter by 20ft long.

At the upstream end of this combustion chamber, inlet boundary conditions were set, thus allowing for the burner outflow to be specified. A constant pressure boundary was assumed downstream, at the exit.

Parametric studies covering the influence of furnace size, combustion air temperature, fuel injection mode, burner pressure drop, burner heat input, and number of burners on thermal NO_x emission tendencies were described in reference [10]. A sample of the results is presented here, for an axisymmetric control fuel injection design and a 3D peripheral injection equivalent.

CENTRAL FUEL INJECTION MODE

Results are presented in Figures 2 to 4 where the combustion air temperature is increased from 70 degrees F. to 300 degrees F. and finally 500 degrees F. The

PARAMETRIC STUDY 1

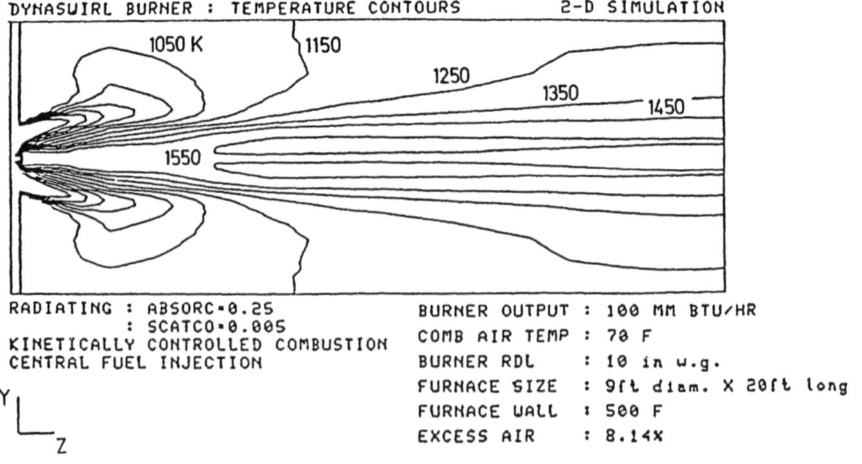

Fig. 2

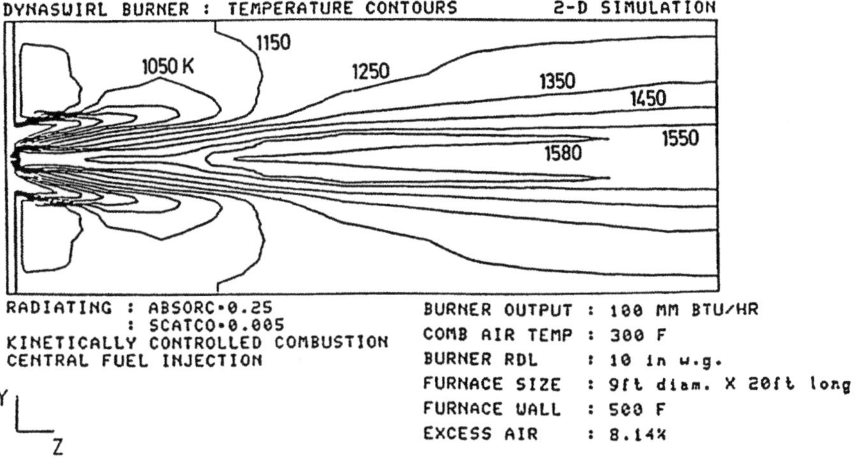

Fig. 3

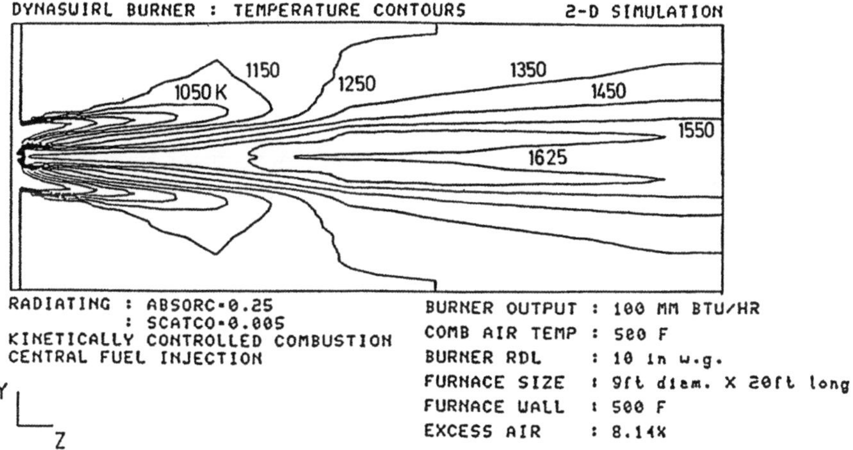

Fig. 4

flame temperature contours reveal the evidence that flame length is decreasing as combustion air temperature increases. NO_x concentration contours in the flame are shown in Figure 5 to 7 and display a significant increase when air temperature is increased. A synopsis of these results is presented in Table 2.

Another point worth noting concerns the flame aerodynamics in the furnace as depicted in Figures 8 to 10. As the air temperature is increased, the extent of

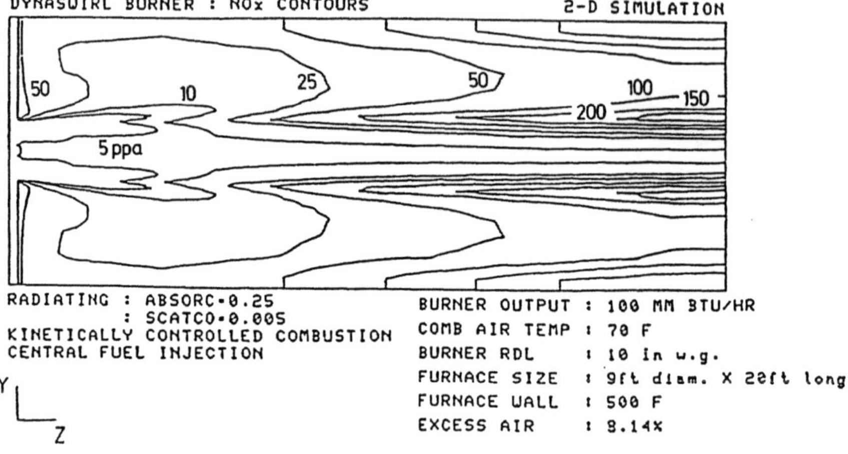

Fig. 5

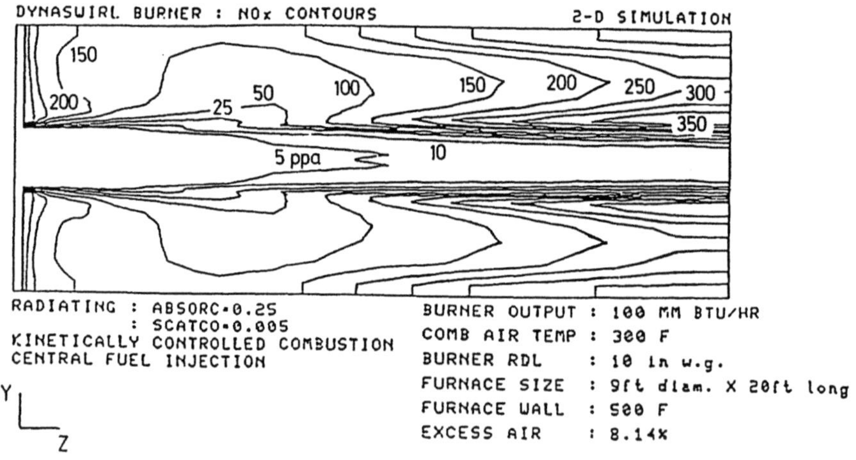

Fig. 6

both internal and external recirculation zones is enhanced significantly. Internal recirculation is generated by the swirler and is responsible for the stabilisation of the flame, thus suggesting a better flame stability as air temperature is increased.

Although these are well known practical facts, the simulations were able to reflect their implications on the flame structure.

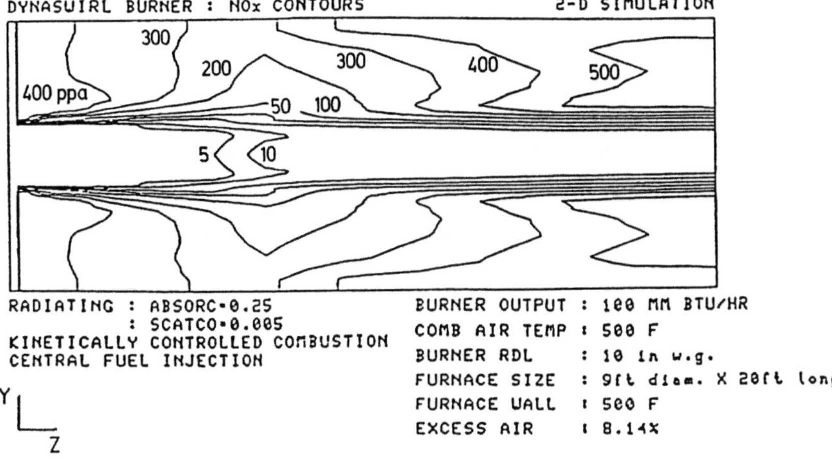

Fig. 7

Table 2: Summary of NOx parametric study 1

Combustion air temp (F)	70	300	500
Primary air ratio (%)	15.5	13.3	12.03
Adiabatic flame temp (K)	2162	2226	2283
Maximum field temp (K)	1557	1591	1650
Maximum NOx conc (PPM)	214.4	396.6	587.2
Exit NOx conc (PPM)	148.2	285.1	482.9

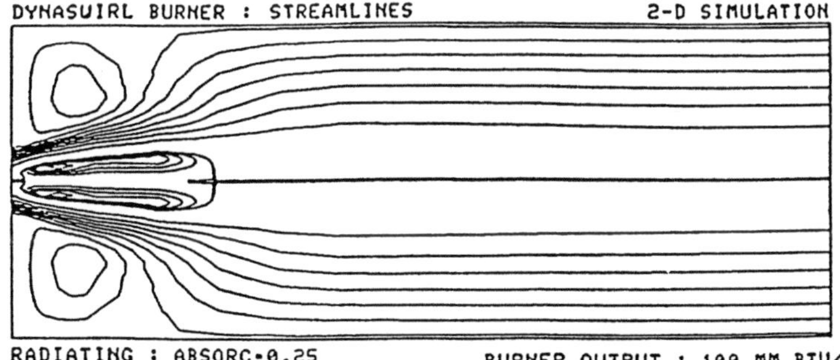

Fig. 8

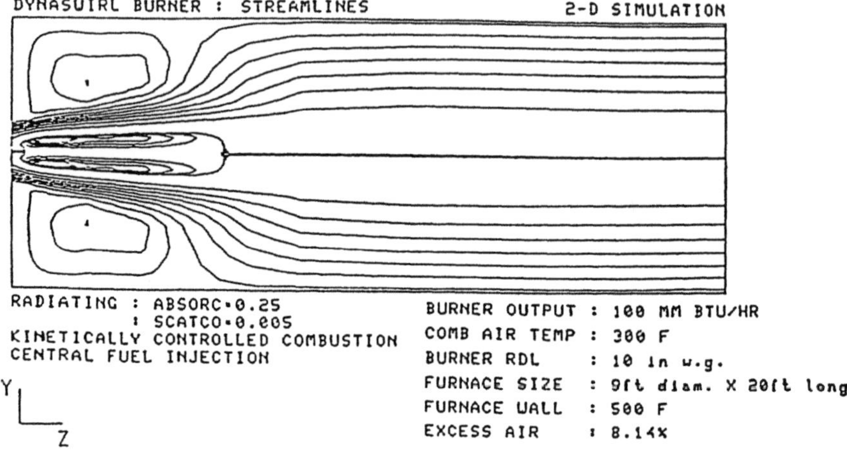

Fig. 9

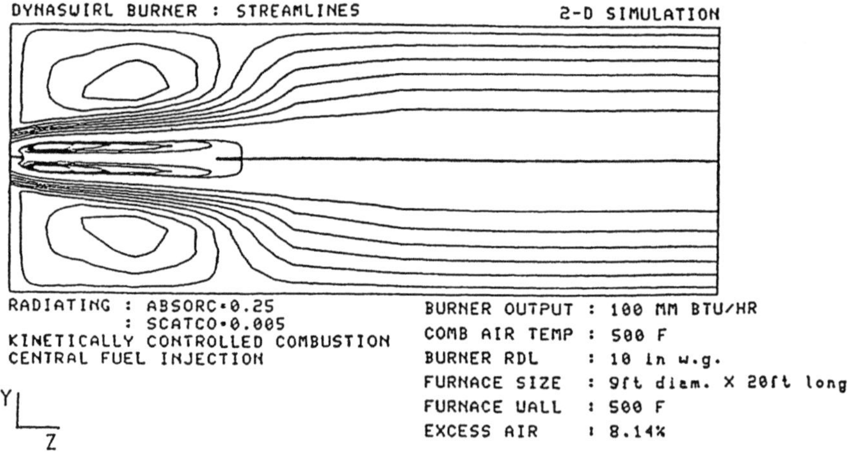

Fig. 10

PERIPHERAL FUEL INJECTION MODE

In this case, the burner uses six axial "spuds" located on a radius immediately outside the swirler to inject the fuel. The spud geometry allows the gas to be injected with a variable tangential and radial velocity component, while the axial component is maintained constant. This design along with its multitude of variations is used in practice when shorter flames are desired to satisfy furnace depth limitations. Amongst the design variations, the annular injector or "gas ring" falls into this category.

As shown in Figure 11, the flame structure is significantly altered. The fuel concentration is much lower (near stoichiometry) in the stabilisation zone than with the central injector mode. This will have major consequences on temperature and NO_x concentrations. In addition, the fuel depletion rate is faster and results, as expected, in a shorter flame.

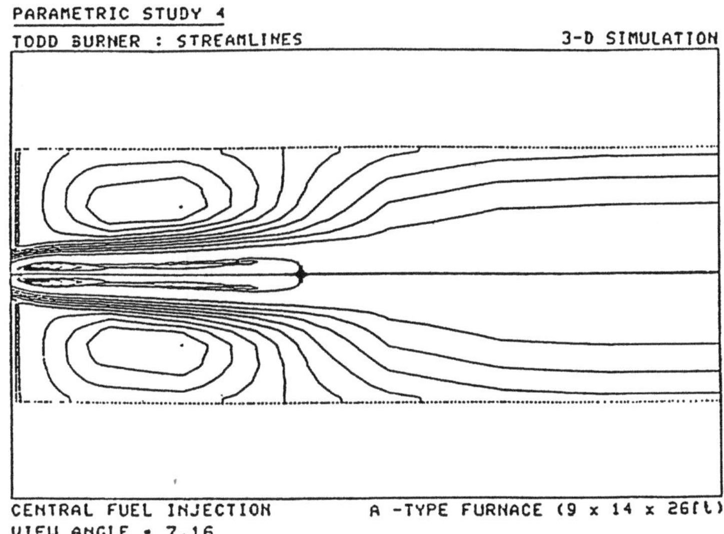

Fig. 11

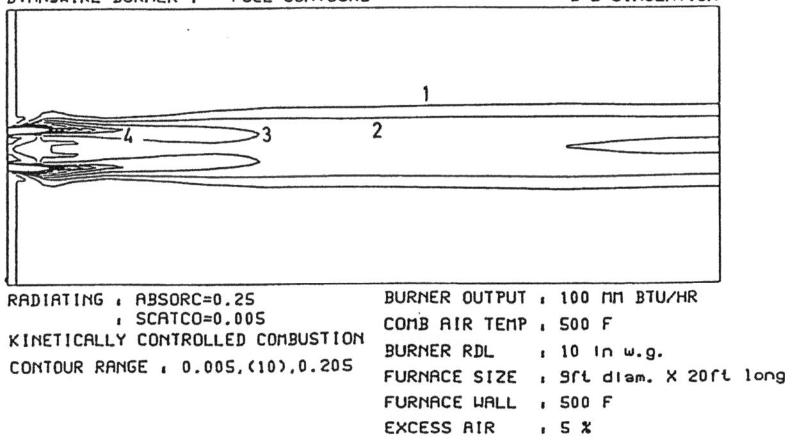

Fig. 12

Temperature contours (Figure 12) show higher maxima in the stabilisation zone and in the flame periphery. This actually explains the fact that this flame is usually more stable at high firing rates. It also indicates that thermal NO_x emissions will be higher than with the central fuel injection mode. This was also experienced during field tests where an approximate 30% NO_x emission increase was observed.

Although this design is easier to deal with for flame stability and length considerations, low NO_x burner designers would be advised to treat it with caution.

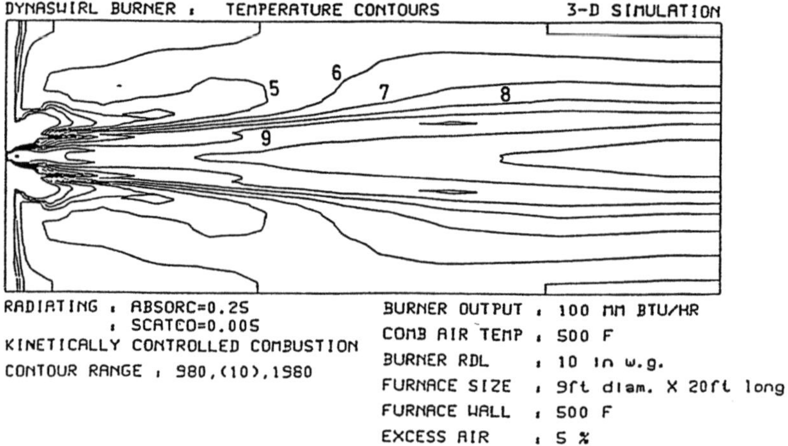

Fig. 13

DISCUSSION OF RESULTS

The partial results of the study indicate that the design of low NO_x combustion equipment must take into consideration the combustion chamber geometry. It also demonstrates that NO_x emission is intimately related to the flame structure. The flame structure was seen to be governed mainly by the fuel injection mode and the aerodynamics prevailing inside the furnace.

The results shown here suggest that the central fuel injection mode where a fuel rich core is created in the flame front is well adapted for low NO_x operation. However, some stability problems may arise when fuel concentration in the internal recirculation becomes too high to sustain combustion. This phenomenon was experienced in the field and was solved by the introduction of a small quantity of air flowing through the co-axial inner conduit of the gas injector. This model derived design alteration has proved to be very effective in the field in improving the flame stability, while leaving NO_x emissions unchanged.

CONCLUSIONS

The present mathematical model is a powerful design tool to assist the burner designer in predicting NO_x emission trends. It allows comparative evaluations to be successfully performed on a series of burner design variations.

However, more experimental work in conjunction with the model is required in order to predict absolute values of NO_x emissions on specific applications. Particularly, a fuel-NO_x model and an oil droplet combustion model must be incorporated to widen the application range of the present work.

Nonetheless, the results partially presented here were sufficient, with the help of some empirical burner design guide rules, to accomplish major improvements in the NO_x emission performance of recent burner installations. For instance, the modelling results have suggested the effectiveness of a selective flue gas recirculation concept through the burner. Such a flue gas recirculation system was designed accordingly and implemented in a refinery early in July 1987. Impressive NO_x reduction levels (80%) were achieved with a relatively small amount of the gas being recirculated (less than 16% of total). This is already a significant contribution to better burner design.

Empirical burner design methods have long reached their limits. But in the light of mathematical modelling work, new audacious burner designs can be attempted to meet current and future environmental constraints.

REFERENCES

1. Sawyer R.F.: Atmospheric Pollution by Aircraft Engines and Fuels. AGARD Advisory report 40 (1972).

2. Process Engineering, 41 (March 1987).

3. Lefebvre A.H.: Gas Turbine Combustion, MacGraw Hill (1983) 481

4. Shefer R.W., Sawyer R.F.: Lean Premixed Recirculating Flow Combustion for Control of Oxides of Nitrogen. Proc. 16th Int. Symposium on Combustion, The Comb. Inst. (1976) 119-133.

5. Quan V., Kiegel J.R., Peters R.L.: Predicted Effects of Fluid Dynamic Parameters on Nitric Oxide Formation in Turbulent Jet Diffusion Flames. Combustion and Flame, 25 (1975) 67.

6. Mizutani Y., Katsuki M.: Emissions from Gas Turbine Combustors. Bulletin of JSME 19 (1976) 1360.

7. Khalil E.E.: Numerical Computation of Combustion Generated Pollutants. Proc. Italian Flames Days (Lar Rivista du Combustibili) (1981).

8. Hutchinson P., Khalil E.E., Whitelaw J.H.: Measurement and Calculation of Furnace Flow Properties. AIAA Journal of Energy 1 (1977) 212.

9. Hung W.: An Experimentally Verified NOx Emission Model for Gas Turbine Combustors. ASME 75-CT-71 (1975).

10. Pericleous K., Clark I.W., Brais N.: The Modelling of Thermal NOx Emissions in Combustion and its Applications to Burner Design. 2nd Int. PHOENICS User Conf., London (Nov. 1987).

11. Zeldovich Y.B., Sadovinkov P.Y., Frank-Kamenctskii D.A.: Oxidation of Nitrogen in Combustion. Academy of Sciences of USSR, Moscov (1974).

12. Markatos N.C., Malin M.R., Cox G.: Mathematical Modelling of Buoyancy-Induced Smoke Flow in Enclosures. Int. J. Heat Mass Transfer 25 (1982) 63-75.

13. Rodi W.: Turbulence Models and their Applications in Hydraulics — a State of the Art Review. SFB 80/T/127, Univ. of Karlsruhe (1978).

14. Pericleous K., Rhodes N.: The Hydrocyclone Classifier — A Numerical Approach. Int. Jour Mineral Processing 17 (1985).

15. Rosten H.I., Spalding D.B.: PHOENICS Reference Manual. CHAM TR/200, CHAM Limited (1987).

16. Spalding D.B.: GENMIX: A General Computer Program for Two-Dimensional Parabolic Phenomena. Pergamon Press (1977).

17. Markatos N.C., Pericleous K.A.: An Investigation of Three-Dimensional Fires in Enclosures. ASME-MTD 25 (1983) 115-124.

18. Magnussen B.F., Hjertager B.H.: On Mathematical Modelling of Turbulent Combustion with Special Emphasis on Soot Formation and Combustion. proc. 16th Int. Symp. on Combustion. The Comb. Inst. (1976) 719-729.

19. Hamaker H.C.: Radiation and Heat Conduction in Light Scattering Material, Philips Res. Rep. 2 (1947) 103-111.

20. Gordon S., McBride B.: CREK: A Computer Program for Calculations of Complex Chemical Equilibrium Compositions. NASA SP-273 (1971).

21. Siegmund C.W., Turner D.W.: NOx Emissions from Industrial Boilers: Potential Control Methods. Journal of Engineering for Power (Jan. 1974) 1.

22. Lisauskas R.A., Snodgrass R.J. et al.: Experimental Investigation of Retrofit Low NOx Combustion Systems. Joint Symp. on Stationery Combustion NOx Control. Boston (May 6-9 1985).

23. DeSoete G.: Etude Parametrique des Effets de la Stratification de la Flamme Sur les Emissions d'Oxides d'Azote. Revue de l'Institut Français de Petrole 32(3) (1977) 427.

24. Gupta A.K., Liley D.G., Syred N.: Swirl Flows. Abacus Press (1984).

HEAT TRANSFER PREDICTIONS IN AIRCRAFT ENGINES COMBUSTORS AND REHEATS

M. Desaulty and S. Meunier
Combustion Department
SNECMA
77550 Moissy Cramayel

ABSTRACT

To cope with the important thermal fluxes which are met in aeronautical combustors, efficient cooling technologies are to be used. In order to calculate the thermal behavior, several methods are available at SNECMA.

As aerodynamics is very complex a satisfactory prediction of the wall temperatures is obtained by wall thermal calculations using boundary conditions coming from 2D or 3D Navier-Stokes calculation of the aerothermal field.

INTRODUCTION

The evolution of aircraft engine cycle leads to higher and higher nominal pressure in the combustor and outlet temperatures.

To provide the combustor walls with required life time, efficient cooling technologies have to be used (such as film cooling, multihole cooling).

In order to optimize the distribution of cooling air along the combustor liners a set of computer methods is used at SNECMA. It includes codes which use empirical correlations and 3D calculations of the aerothermochemical field associated with 2D or 3D calculations of the thermal transfer in the walls.

Two examples of this methodology are described below. They deal with the prediction of the wall temperatures of the main combustor and afterburner liners.

1. PREDICTIONS OF THE MAIN COMBUSTOR WALL TEMPERATURES

1.1. GENERAL POINTS

As the flight domain of an aircraft is very wide, thermodynamical conditions in front of the main combustor vary over a large range.

For instance the inlet temperature varies between 300K and 800K, the pressure between 0.3 bar and 40 bar. In order to avoid the overheating of the turbine, the temperatures at the exit of the combustor should not be too high. The mean inlet temperatures for turbines which are state of the art today, are about 1850K. In these conditions the nominal fuel/air ratio must be well under the stoichiometric value. So the air coming out of the compressor must be progressively introduced in the combustor to have satisfactory combustion stability and altitude relight. Hence two important zones are defined (Figure 1).

- primary zone: its fuel/air ratio is about stoichiometry and the temperature around 2400K at full power regime;

- dilution zone: its function is to obtain an acceptable temperature level, and profile for the turbine.

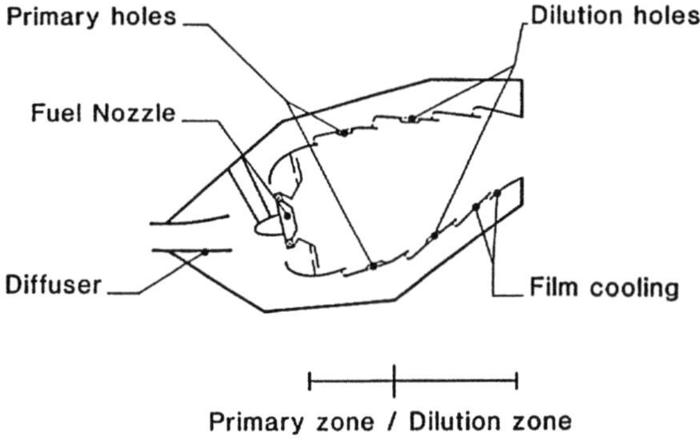

Fig. 1: Main combustor geometry

All these constraints lead to very complex three dimensional aerodynamics.

Because of very high thermal fluxes an effective and air saving cooling technology has to be implemented to prevent disturbance of the overall combustor operation.

1.2. METHODS USING EMPIRICAL CORRELATIONS

The thermal life of the combustor liner is assessed through an external air convective cooling (i.e. a cooling by the air flowing in the inner and outer passage), and by film cooling or multihole cooling.

Hence the thermal field in the liners depends thoroughly on the mixing between the air coming, for example, out of the films and the hot gases produced by combustion.

In the performance design step of an engine, empirical correlations of the measured efficiency are used. They link a parameter called the adiabatic efficiency, η, to a Reynolds number based on a characteristic length scale of the film geometry and on a reduced parameter M equal to the ratio between the film and hot gases momentum.

The film temperature can be obtained by solving the 3 fundamental one dimensional conservation equations (mass, momentum and energy).

The local temperature in the immediate vicinity of the wall is obtained by the relation

$$\eta(x) = \frac{Th - T(x)}{Th - Tc}$$

where

> Th is the hot gases temperature
> Tc is the cooling air temperature
> x the distance along the wall

In a second step to improve the results an experimental characterization of the adiabatic efficiency can be made on elementary tests rig or be obtained by a two dimensional calculation of the mixing between the film and the hot gases.

We are going to describe the latter methodology below.

1.3. NAVIER-STOKES FINITE ELEMENT 2D CALCULATION OF THE FILM

A two dimensional finite element Navier-Stokes code NADIA developed in cooperation with Ecole Centrale de Lyon have been routinely used at SNECMA for 5 years to predict the aerodynamic field around combustors [1], [2].

Its main characteristics are:

- triangular meshing P1 ISO P2
- Galerkin weak formulation
- the coupling between velocity and pressure is made by a UZAWA's algorithm
- semi-implicit solving
- k-ε turbulence model

Figure 2 shows the gross meshing used to make the calculation of a film cooling geometry.

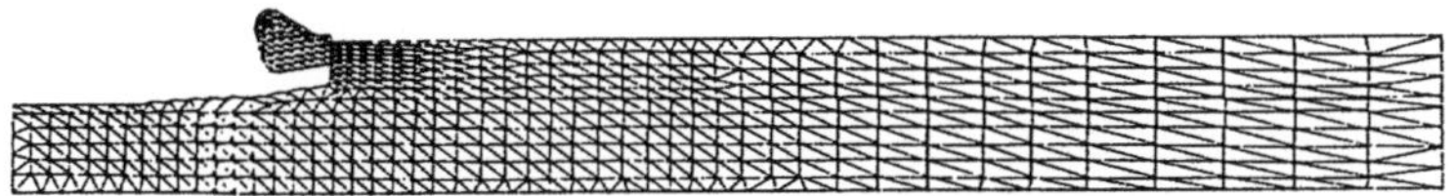

Fig. 2:   Gross meshing of the calculation domain

1.3.1. One dimensional boundary conditions

In a first step boundary conditions (velocities, temperatures) at the entrance of the calculation domain have been obtained by solving the one dimensional conservation equations.

A detail of the fine meshing and of the velocity vectors around the film is shown in Figure 3. The thermal field is given in Figure 4.

The comparison between the calculated values of the adiabatic efficiency coefficient and the measurements made on elementary test rig shows a good agreement Figure 5.

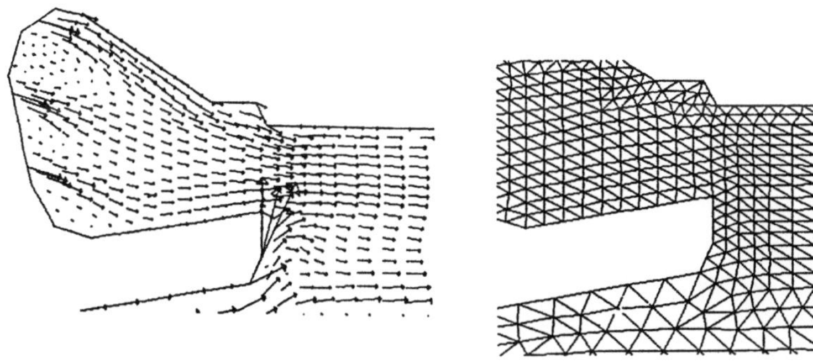

Fig. 3:   Detail of the calculation domain.

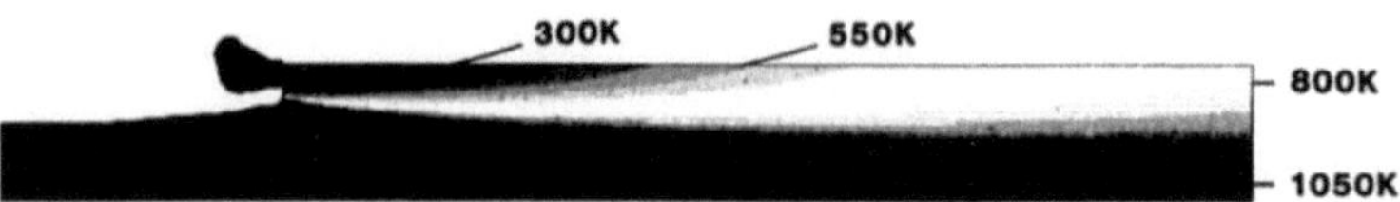

Fig. 4:   Isotemperature lines around the film.

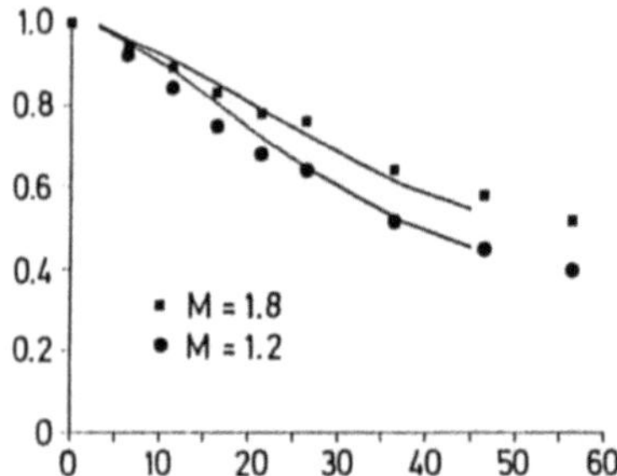

Fig. 5: Comparison between measured and calculated adiabatic efficiency

The adiabatic efficiency coefficient is then used in a code which solves the energy equation (with conduction in the wall, radiation given by an empirical law [3], and convection) and predicts the wall temperatures and hence the operating life.

1.3.2. Coupling with 3D calculations

When further information is needed (especially when local hot spots are expected) the boundary conditions can be obtained by a 3D Navier-Stokes calculation of the entire combustor with the ECRIN code [4]. An example of the meshing and of the results is shown in Figures 6 and 7.

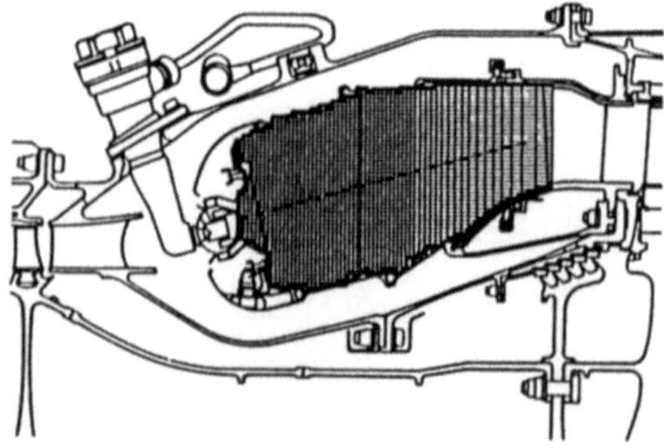

Fig. 6.:   Combustor calculation — geometry and mesh.

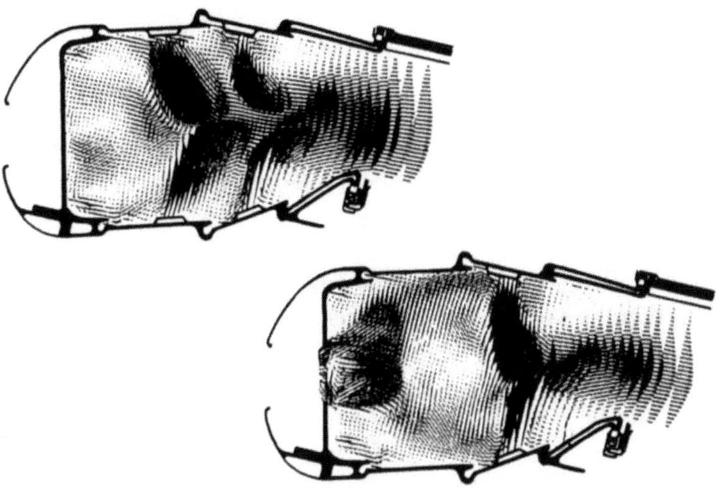

Fig. 7: Combustor chamber 3D calculation — velocity vectors.

The main features of this software are

- cartesian coordinates
- finite volume approach
- structurated mesh
- k-ε model
- E.B.U. Arrhenius combustion model
- SIMPLE algorithm

2. PREDICTIONS OF THE AFTERBURNER LINER TEMPERATURE

2.1. GENERAL POINTS

The main combustor fuel/air ratio is smaller than the stoichiometric value. Moreover on by-pass engines the outer flow (fan flow) is not concerned by combustion. So an important oxygen mass fraction remains in the plane situated approximately at the exit of the turbine.

On military and supersonic transport aircraft (CONCORDE) engines this can be profitably used to obtain an extra thrust in some areas of the flight domain.

The second combustor, called afterburner has also to cope with the fact that the thermodynamical conditions vary over a wide range. The temperatures vary, for example, between about 1000K when the afterburner is off and more than

2000K when it is at full power. Moreover in the latter case the gradients of the temperature and mass fractions are very important in the flame front regions.

The mixing between the two flows and the thermal pressure loss give a complex aerodynamic field. Liners are used to protect the afterburner walls. The relatively cold air from the fan flow is partly introduced between the walls and the liner to provide air for multihole cooling.

The technology used at SNECMA on the latest engines consists of corrugated perforated liners [5] (figures 8 and 9). Diaphragms, the flowing section of which is calculated taking into account the relative dilatation of the different metals and the combustion pressure losses, optimize the cooling air flow. Hence there

Fig. 8:   Afterburner — View from the exit nozzle.

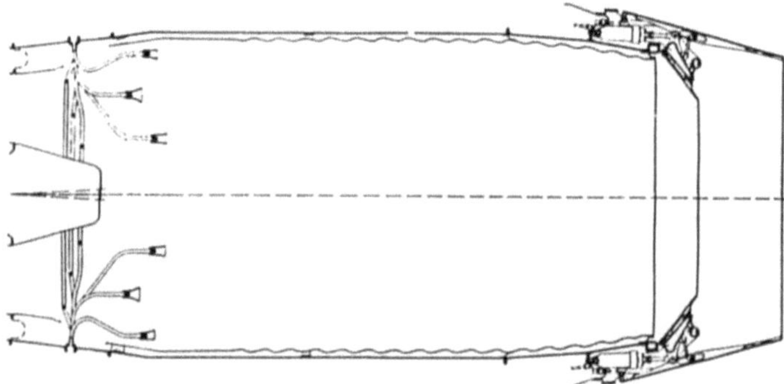

Fig. 9: Afterburner geometry

is a very close interaction between the thermal field in the walls (liner and duct) which causes dilatation and so affects multihole cooling, and the three-dimensional aerothermal flow inside the afterburner.

As for the main burner, different methods are used, according to the required level of accuracy.

2.2. SIZING OF THE THERMAL LINERS

When the thermodynamic cycle of a new engine has been chosen, the mean value of each thermodynamic variable is know in each flow. A one dimensional formulation of the conservation equations of mass, momentum and energy is used [5].

The boundary conditions which are necessary to solve these equations are detailed below.

2.2.1. Pressure boundary condition on the inner face of the thermal liner

In the preliminary design step the total pressure losses due to obstacles can be given by empirical correlations, the velocities and temperatures needed to convert total pressures in static pressures are also obtained by correlations.

The value of the static pressure at the downstream end of the liner is used to determine iteratively the flow between liner and duct.

2.2.2. Temperature boundary conditions

a) Convective temperature

The temperatures in the burner near the wall (noted Th) may be obtained from calculations taking into account either a law of evolution of the combustion efficiency along the burner, or a law of mixing between primary and secondary flow (Figure 10).

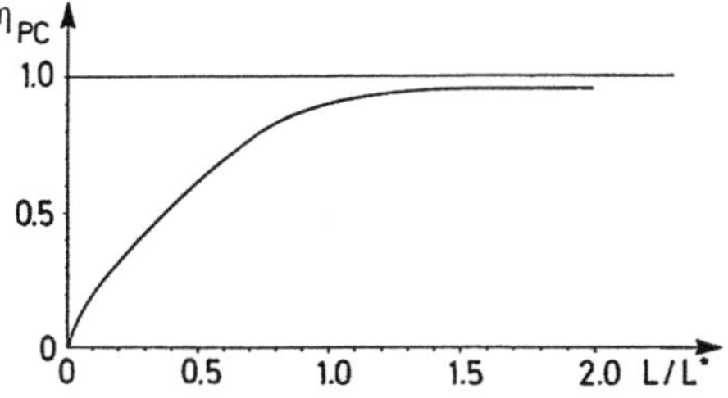

Fig. 10: Evolution of combustion efficiency along the afterburner.

The local temperature, in the immediate vicinity of the wall is, as in the case of the main burner, calculated with an adiabatic efficiency coefficient obtained from measurements on elementary test rig configurations, or from Navier-Stokes calculations.

b) Radiation temperature

Thermal radiative transfer may be evaluated with global laws (such as the Reeves law [3]). They use a temperature that can be the mean hot gases temperature.

2.2.3. Velocity boundary conditions

The local velocities (which are used in the Reynolds number in convective laws) are also obtained by empirical correlations.

2.3. PERFORMANCE EVALUATION

After each component of the engine has been designed, more elaborate methods are used to evaluate its performance.

For the liners conductive transfer and dilatation are relatively well known, as far as the materials properties are known.

The accuracy of thermal convective heat transfer calculations is directly related to the accuracy of the values of the local air temperatures and velocities. The radiative flux coming from emissive hot gases is much more difficult to calculate.

Hence in the performance evaluation step an improvement of the prediction is obtained essentially by a better knowledge of the boundary conditions and by the use of more detailed radiative models.

2.3.1. Aerodynamic field prediction

The complete 3D calculation of the afterburner aerodynamic field is made with the ECRIN code. The cooling along the liner is simulated by source terms. The local value of the massflow is determined by a first, one dimensional, aerodynamic calculation.

A 3D calculation using a rough meshing (Figure 11) is enough to obtain the local temperatures and velocities (Figures 12 and 13).

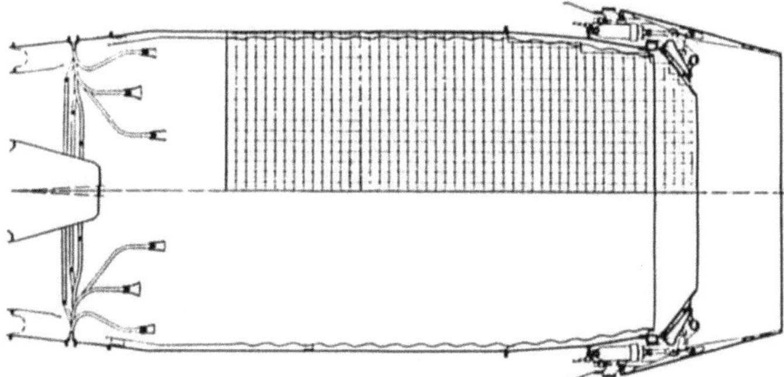

Fig. 11:. Afterburner geometry and mesh.

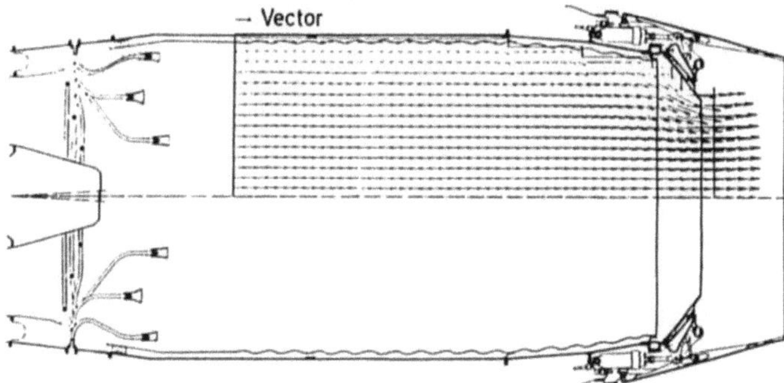

Fig. 12:   Velocity vectors.

Fig. 13:   Afterburner isothermal lines.

2.3.2. Radiation modelling

The mass fractions of emissive gases (H_2O and CO_2 when the afterburner is off — H_2O, CO_2 and CO when operating) are obtained from the 3D Navier-Stokes calculation. An emission absorption calculation is then made along some representative optical paths. The radiation model used is described below:

- Narrow band statistical model NBSM
- Spectral resolution: 5 cm^{-1}
- A Curtiss-Godson approximation is used to transform the heterogeneous mass fraction profile along the optical path into an homogeneous column. The spectral emissivity of such a column is shown in Figure 14.

This model has been assessed by comparison with a line by line calculation [6].

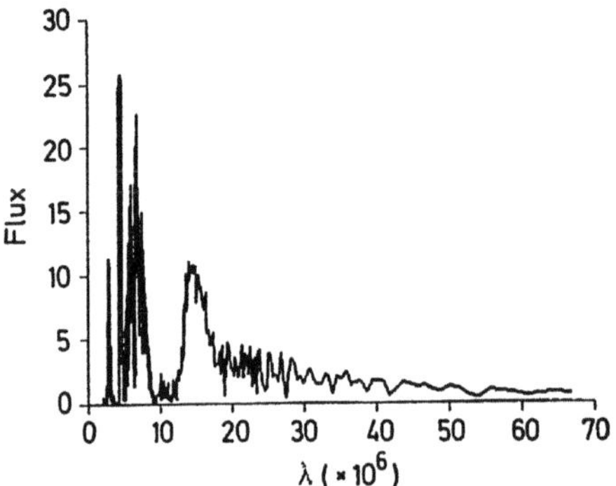

Fig. 14:   Spectral emissivity NBSM with 5 cm^{-1} resolution.

With these hypotheses in a case where the afterburner is off (the total pressure is about 4 bar and the temperature 1000K) the NBSM and the Reeves model give the same order of magnitude for the radiative fluxes (Figure 15).

However the study of specific technologies, such as these improving mixing between primary and secondary flow, can be made only with the NBSM model.

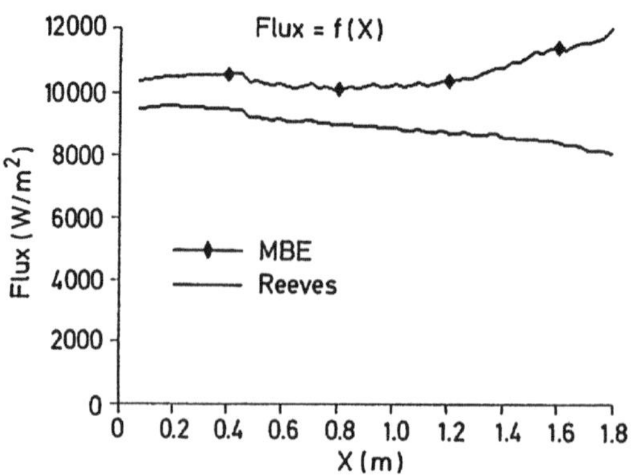

Fig. 15:   Comparison between radiative fluxes obtained with models: Reeves and NBSM.

2.3.3. Comparison with experimental data

In the above thermodynamic conditions (4 bar, 1000 K), the results given by the more complex model (i.e.: NBSM radiation model-local boundary conditions given by ECRIN) are in good agreement with the experimental data.

See Figure 16 comparison between predicted and measured liner temperatures.

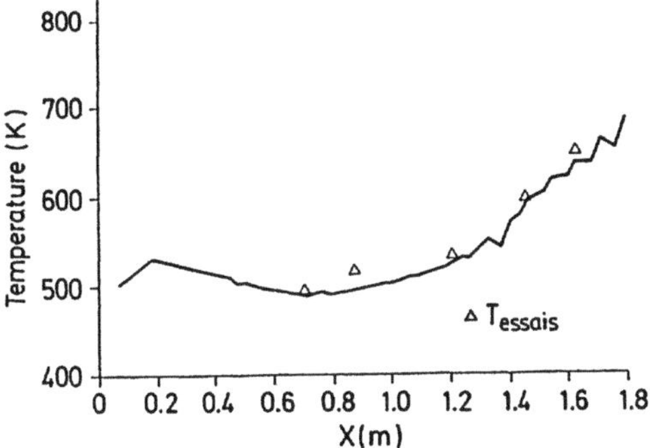

Fig. 16: Afterburner liner temperatures: Comparison between measurements and calculations.

3. CONCLUSION

The analysis of thermal transfers in the complex 3D combustors geometry is performed with different methods according to the status of the project.

As soon as the geometry of the combustor has been designed a close coupling between a code predicting heat transfers in the walls and a Navier-Stokes code calculating the 3D flow inside the combustor is routinely used at SNECMA. However this methodology needs several steps.

In a near future the use of calculation methods with subdomain formulations and a simultaneous prediction of the flow and of the heat transfer in walls with the same code will save much time. The physical phenomenon will be described by the latest models which are under development in laboratories [7], [8].

ACKNOWLEDGEMENTS

The authors would like to express their thanks to Mrs S. Coutor and Messrs M. Cazalens, C. Favreau, J.L. Schultz and W. Toulgoat for their contribution to some of the calculations.

REFERENCES

[1] Brun G., Buffat M., Jeandel D., Schultz J.L., Desaulty M.: Finite element simulation of compressible turbulent flows, Validation and Application to internal aerodynamics in gas turbine engines. in Proc. VII Int. Conf. on FEM in flow problems, Huntsville, USA 1989, edited by Chung and Karr, Huntsville press: (1989) 1592-1597.

[2] Jeandel D., Brun G., Meunier S., Desaulty M.: Numerical Simulation of Diffusor/Combustor Dome Interaction. Eight International Symposium on Air Breathing Engines, Cincinnati (1987), Published by AIAA.

[3] Lefebvre A.H.: Flame radiation in gas turbine combustion chambers. International Journal Heat Mass Transfer, 27 (9) (1984) 1493-1510.

[4] Karadimas G.: Application of computational systems to aircraft engine components development. 9th Isabe, 3-8 September (1989).

[5] Desaulty M., Trouillot P., Coutor S.: Techniques de refroidissement des canaux de réchauffe des turboréacteurs, AGARD C.P. 390 (1985).

[6] Soufiani A., Hartmann J.M., Taine J.: Validity of band model calculations for CO_2 and H_2O applied to radiative properties and conductive-radiative transfer. J.Q.S.R.T., 33, 243-257.

[7] Gilbank P.: Contribution à la modelisation de la combustion turbulente dans le cas d'une stabilisation par accroche-flammes. Thèse soutenue à Paris VI, 9 May (1989).

[8] Veynante D., Lacas F., Candel S.M.: Cohérent flame model in non uniformly premixed turbulent flames. 7th Turbulent Shear Flows, Stanford University, August 21-23 (1989).

INFLUENCE OF THE PRESSURE-RATIO ON RADIATIVE HEAT TRANSFER IN GAS TURBINE COMBUSTORS

R. Koch, S. Wittig, B. Noll
Lehrstuhl und Institut für Thermische Strömungsmaschinen
Universität Karlsruhe
Germany

ABSTRACT

Radiative absorption and emission of combustion gases is strongly increased by raising total pressure due to the effect of pressure broadening. A new method, the Harmonical Transmission Model (HTM), has been developed to account for this effect in multidimensional radiative heat transfer calculations. A set of differential equations is derived, which is of the same mathematical form as the equation of radiative transfer. The equations have been solved numerically by the P1-approximation.

The Harmonical Transmission Model (HTM) has been employed to study the effect of pressure broadening on radiative heat transfer in a gasturbine combustor. The results show, that the radiation spectra can be influenced considerably by pressure broadening, whereas the spectrally integrated radiative fluxes inside the combustor as well as the radiative heat load of the combustor walls are only slightly influenced, because of predominant soot radiation in gas turbine combustors.

NOMENCLATURE

d	line spacing
$F_{o...\pm n}$	fourier coefficients of monochromatic intensity
$\vec{F}$	intensity vector
$g_{o...\pm n}$	fourier coefficients of monochromatic absorption coefficient
I_λ	monochromatic intensity
$I_{b,\lambda}$	monochromatic intensity of the black body
k_λ	monochromatic absorption coefficient
$\vec{n}$	normal vector of the wall
S	line intensity
$\vec{s}$	direction vector

Greek

γ	line half width
ϕ	dimensionless wavelength parameter
λ	wavelength
ρ	reflectivity of the wall

$\vec{\Omega}+$ direction vector of the outgoing intensity at the wall

$\vec{\Omega}-$ direction vector of the incoming intensity at the wall

INTRODUCTION

Increasing radiative heat transfer in gas turbine combustors with raising pressure ratios is a well known phenomenon. Besides the influence of pressure on reaction kinetics (higher soot formation rate) increasing radiation is mainly caused by two effects: the optical depth in the combustor is increased by higher total pressures, and the absorption and emission behaviour of the combustion gases is changed by the effect of pressure broadening.

For one-dimensional problems, the influence of pressure broadening can be accounted for by use of band models (e.g. Elsasser model, Goody model). When dealing with combustion problems, the multi-dimensional equations of fluid mechanics and radiative transfer have to be solved simultaneously. Preferentially those methods are advantageous, where the radiative transfer equation can be written as a set of differential equations of the space coordinates. Among these methods, at present the most established ones are the Flux method [1], the Moment or Spherical Harmonics method [2], [3], [4] and the Discrete Ordinate method [5], [6]. However, it is impossible to incorporate band models in these numerical methods for multi-dimensional radiative heat transfer computations.

In the following, a completely new numerical method, the Harmonical Transmission Model (HTM), is presented, which enables to account for pressure broadening also in multi-dimensional radiative heat transfer calculations. The Harmonical Transmission model is based on the same physical principles, that are used in the established band models to describe the line structure of a vibration-rotation band. The basic idea of the Harmonical Transmission model is, that the spectral averaging is done directly over the monochromatic radiative transfer-equation in contrast to the band models, where the averaging is done over the monochromatic transmissivity.

ANALYSIS: THE HARMONICAL TRANSMISSION MODEL (HTM)

If radiative transfer in gases is considered, scattering can be neglected and the stationary monochromatic equation of radiative transfer in thermal equilibrium can be written as

$$\vec{s} \cdot (\vec{\nabla} I_\lambda) = - k_\lambda (I_\lambda - I_{b,\lambda}) \tag{1}$$

Because of the linestructure of the vibration-rotation bands of infrared-active gases, the spectral absorption coefficient k_λ is strongly varying with the wavelength λ. In addition, k_λ also depends on the properties of the gas, as well as on the temperature and the partial and total pressure of the gas.

From a quantum mechanical analysis of the absorption process of two-atomic gases [7] it can be shown, that the line structure of a vibration-rotation band consists of nearly equally spaced spectral lines of the same line intensity S and half-width γ. Such a regular structure is also a good approximation for linear, symmetrical three-atomic molecules like Carbondioxide. Therefore, the adequate band model to describe the absorption behavior of such gases is the regular Elsasser model [8].

By the Elsasser band model, the structure of a vibration-rotation band is assumed to consist of regularly spaced lines of the same intensity. The lines show a Lorentz profile and do partially overlap. The absorption coefficient of such an Elsasser band is given by

$$k_{\lambda,\text{Elsasser}}\left(\phi = 2\pi \ \frac{\Delta\lambda}{d}\right) = \frac{S}{d} \cdot \frac{\sinh 2\pi \ \dfrac{\gamma}{d}}{\cosh 2\pi \ \dfrac{\gamma}{d} - \dfrac{1}{2}\left(e^{i\phi} + e^{-i\phi}\right)} \tag{2}$$

The derivation of the Harmonical Transmission model is based on the fact, that the absorption coefficient k_λ of a vibration-rotation band shows a periodic structure over a narrow wavelength interval. Because of the periodicity of the line structure, the absorption coefficient k_λ and the spectral intensity I_λ are expanded into complex Fourier series

$$k_\lambda(\phi) = g_0 + g_1 \cdot e^{i\phi} + g_{-1} \cdot e^{-i\phi} + \ldots \tag{3}$$

$$I_\lambda(\phi) = F_0 + F_1 \cdot e^{i\phi} + F_{-1} \cdot e^{-i\phi} + \ldots \tag{4}$$

where

$$\phi = 2\pi \frac{\Delta\lambda}{d}$$

As the spectral intensity of the black body $I_{b,\lambda}$ is not varying within the narrow wavelength interval $[0 < \Delta\lambda/d < 1]$, the Fourier expansion of $I_{b,\lambda}$ contains only zero-order terms

$$I_{b,\lambda}(\phi) = I_{b,\lambda} \tag{5}$$

The coefficients g_n of the Fourier expansion of the absorption coefficient (eq. 3) are determined from eq. 2 using a least-square fit

$$g_n = \frac{1}{2} \int_{\phi=0}^{2\pi} k_{\lambda,\text{Elsasser}} \cdot e^{-in\phi} \, d\phi \tag{6}$$

The resulting coefficients are

$$g_0 = \frac{S}{d} \tag{7a}$$

$$g_1 = g_{-1} = \frac{S}{d} \cdot e^{-2\pi\frac{\gamma}{d}} \tag{7b}$$

$$g_2 = g_{-2} = \frac{S}{d} \cdot e^{-4\pi\frac{\gamma}{d}} \tag{7c}$$

$$\vdots \quad \vdots$$

$$g_n = g_{-n} = \frac{S}{d} \cdot e^{-2n\pi\frac{\gamma}{d}} \tag{7d}$$

Because $k_{\lambda,\text{Elsasser}}$ is symmetrical to $\phi=\pi$, it is evident that the imaginary part of the Fourier coefficients g_n vanishes. Therefore, the complex expansion of k_λ can be written as a real one

$$k_\lambda = g_0 + 2g_1 \cdot \cos\phi + 2g_2 \cdot \cos 2\phi + \dots \tag{8}$$

In Figure 1 it is shown, how the line structure of an Elsasser band is approximated by the Harmonical Transmission model (eq. 8) of first and second order.

In order to derive the transport equation for the Fourier coefficients F_i of the intensity, the Fourier expansions of the absorption coefficient and the intensity (eq. 3,4) are inserted into the monochromatic equation of radiative transfer (eq. 1) and the equation of radiative transfer is averaged over the period interval $[\phi=0, \phi=2\pi]$. Because of the orthogonality-relation of the Fourier-series, averaging in the sense of a least-square fit is achieved by multiplying the equation of radiative transfer by $e^{ni\phi}$, where $n = \dots-2, -1,0,1,2\dots$ and integrating from $\phi=0$ to $\phi=2\pi$.

$$\int_{\phi=0}^{2\pi} (\text{eq.1}) \cdot e^{ni\phi} \, d\phi \tag{9}$$

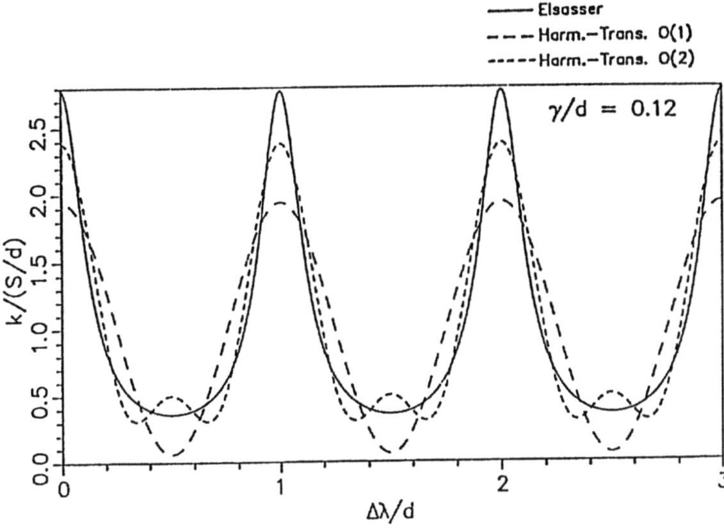

Fig. 1: Line structure as approximated by the Harmonical Transmission model for $\gamma/d = 0.12$.

As a result of that quadrature scheme, a coupled system of transfer-equations for the complex Fourier coefficients of the intensity $F_0, F_1, F_{-1},\dots$ is obtained

$$\vec{s}\cdot\vec{\nabla}\begin{bmatrix} \vdots \\ F_{-2} \\ F_{-1} \\ F_0 \\ F_1 \\ F_2 \\ \vdots \end{bmatrix} = -\begin{bmatrix} \vdots & \vdots & \vdots & \vdots & \vdots & \vdots & \vdots \\ \cdots & g_0 & g_{-1} & g_{-2} & \cdots & \cdots & \cdots \\ \cdots & g_1 & g_0 & g_{-1} & g_{-2} & \cdots & \cdots \\ \cdots & g_2 & g_1 & g_0 & g_{-1} & g_{-2} & \cdots \\ \cdots & \cdots & g_2 & g_1 & g_0 & g_{-1} & \cdots \\ \cdots & \cdots & \cdots & g_2 & g_1 & g_0 & \cdots \\ \vdots & \vdots & \vdots & \vdots & \vdots & \vdots & \vdots \end{bmatrix}$$

$$\cdot\left(\begin{bmatrix} \vdots \\ F_{-2} \\ F_{-1} \\ F_0 \\ F_1 \\ F_2 \\ \vdots \end{bmatrix} - \begin{bmatrix} \vdots \\ 0 \\ 0 \\ I_{b,\lambda} \\ 0 \\ 0 \\ \vdots \end{bmatrix}\right) \tag{10}$$

MULTIDIMENSIONAL RADIATIVE TRANSFER USING THE HARMONICAL TRANSMISSION MODEL

In a first order approximation, the set of differential equations (eq. 10) is reduced to a system of three equations, which can be decoupled by means of matrix transformations (*see* [9] for details). The resulting set of three decoupled differential equations

$$\vec{s} \cdot \left(\vec{\nabla} \begin{bmatrix} F_0 + \frac{1}{\sqrt{2}} \left(F_1 + F_{-1} \right) \\ \sqrt{2} \left(F_1 - F_{-1} \right) \\ F_0 - \frac{1}{\sqrt{2}} \left(F_1 + F_{-1} \right) \end{bmatrix} \right) = - \begin{bmatrix} g_0 + \sqrt{2} \, g_1 & 0 & 0 \\ 0 & g_0 & 0 \\ 0 & 0 & g_0 - \sqrt{2} \, g_1 \end{bmatrix}$$

$$\cdot \left(\begin{bmatrix} F_0 + \frac{1}{\sqrt{2}} \left(F_1 + F_{-1} \right) \\ \sqrt{2} \left(F_1 - F_{-1} \right) \\ F_0 - \frac{1}{\sqrt{2}} \left(F_1 + F_{-1} \right) \end{bmatrix} - \begin{bmatrix} I_{b,\lambda} \\ 0 \\ I_{b,\lambda} \end{bmatrix} \right) \tag{11}$$

must be solved on the multidimensional domain. The coefficients g_0 and g_1 of the matrix are determined by means of eq. (7a-d).

For ideal diffusely reflecting walls, the boundary conditions are

$$F_0(\vec{\Omega}^+) = (1-\rho) I_{b\lambda}(\vec{\Omega}^+) + \frac{\rho}{\pi} \int_{\dot{\Omega}=0}^{2\pi} F_0(\vec{\Omega}^-) \cdot (\vec{n}, \vec{\Omega}^-) \, d\Omega \tag{12}$$

$$\left[F_1(\vec{\Omega}^+) + F_{-1}(\vec{\Omega}^+) \right] = \frac{\rho}{\pi} \int_{\dot{\Omega}=0}^{2\pi} \left[F_1(\vec{\Omega}^-) + F_{-1}(\vec{\Omega}^-) \right] \cdot (\vec{n}, \vec{\Omega}^-) \, d\Omega \tag{13}$$

$$\left[F_1(\vec{\Omega}^+) - F_{-1}(\vec{\Omega}^+) \right] = \frac{\rho}{\pi} \int_{\dot{\Omega}=0}^{2\pi} \left[F_1(\vec{\Omega}^-) - F_{-1}(\vec{\Omega}^-) \right] \cdot (\vec{n}, \vec{\Omega}^-) \, d\Omega \tag{14}$$

The direction-vector Ω^- is pointing to the wall and Ω^+ is pointing away from the wall. The direction of $\vec{n}$ is perpendicular to the wall and pointing away from the wall.

With the approximation made here, it can be shown, that the second equation of the system

$$\vec{s} \cdot \left(\vec{\nabla} \left[\sqrt{2} \left(F_1 - F_{-1} \right) \right] \right) = - g_0 \cdot \left[\sqrt{2} \left(F_1 - F_{-1} \right) \right] \tag{15}$$

together with the boundary condition

$$\left[F_1(\vec{\Omega}^+) - F_{-1}(\vec{\Omega}^+)\right] = \frac{\rho}{\pi} \int_{\Omega'=0}^{2\pi} \left[F_1(\vec{\Omega}') - F_{-1}(\vec{\Omega}')\right] \cdot (\vec{n},\vec{\Omega}') \, d\Omega' \qquad (16)$$

has the trivial solution

$$(F_1 - F_{-1}) = 0 \qquad (17)$$

This fact is easily understandable, when the vanishing sin-terms in the Fourier expansion of the absorption coefficient are considered. Therefore, also no sin-terms in the Fourier expansion of the intensity should be expected, which is formally expressed by $F_1 - F_{-1}$.

Thus it is sufficient to solve the remaining two differential-equations

$$\vec{s} \cdot \left(\vec{\nabla} \begin{bmatrix} F_0 + \frac{1}{\sqrt{2}}(F_1 + F_{-1}) \\ F_0 - \frac{1}{\sqrt{2}}(F_1 + F_{-1}) \end{bmatrix} \right) = \begin{bmatrix} g_0 + \sqrt{2}\,g_1 & 0 \\ 0 & g_0 - \sqrt{2}\,g_1 \end{bmatrix}$$

$$\left(\begin{bmatrix} F_0 + \frac{1}{\sqrt{2}}(F_1 + F_{-1}) \\ F_0 - \frac{1}{\sqrt{2}}(F_1 + F_{-1}) \end{bmatrix} - \begin{bmatrix} I_{b,\lambda} \\ I_{b,\lambda} \end{bmatrix} \right) \qquad (18)$$

together with the boundary conditions

$$F_0(\vec{\Omega}^+) = (1-\rho)\, I_{b,\lambda}(\vec{\Omega}^+) + \frac{\rho}{\pi} \int_{\Omega'=0}^{2\pi} F_0(\vec{\Omega}') \cdot (\vec{n},\vec{\Omega}') \, d\Omega' \qquad (19)$$

$$\left[F_1(\vec{\Omega}^+) + F_{-1}(\vec{\Omega}^+)\right] = \frac{\rho}{\pi} \int_{\Omega'=0}^{2\pi} \left[F_1(\vec{\Omega}') + F_{-1}(\vec{\Omega}')\right] \cdot (\vec{n},\vec{\Omega}') \, d\Omega' \qquad (20)$$

If the substitutions

$$A = F_0 + \frac{1}{\sqrt{2}}(F_1 + F_{-1}) \qquad (21)$$

$$B = F_0 - \frac{1}{\sqrt{2}}(F_1 + F_{-1}) \qquad (22)$$

$$a = g_0 + \sqrt{2}\,g_1 \qquad (23)$$

$$b = g_0 - \sqrt{2}\,g_1 \qquad (24)$$

are introduced, the resulting equations show the same formal structure as the monochromatic equation of radiative transfer in a participating medium (eq. 1).

$$\vec{s} \cdot (\vec{\nabla} A) = -a \, (A - I_{b,\lambda}) \tag{25}$$

$$A(\vec{\Omega}^{+}) = (1-\rho) I_{b\lambda}(\vec{\Omega}^{+}) + \frac{\rho}{\pi} \int_{\Omega^{-}=0}^{2\pi} A(\vec{\Omega}^{-}) \cdot (\vec{n},\vec{\Omega}^{-}) \, d\Omega \tag{26}$$

and

$$\vec{s} \cdot (\vec{\nabla} B) = -b \, (B - I_{b,\lambda}) \tag{27}$$

$$B(\vec{\Omega}^{+}) = (1-\rho) I_{b,\lambda}(\vec{\Omega}^{+}) + \frac{\rho}{\pi} \int_{\Omega^{-}=0}^{2\pi} B(\vec{\Omega}^{-}) \cdot (\vec{n},\vec{\Omega}^{-}) \, d\Omega \tag{28}$$

Here, the main advantage of the Harmonical Transmission model is obvious. As the resulting equations are of the same mathematical form as the monochromatic equation of radiative transfer (eq. 1), any numerical method, which is usually applied to solve the basic equation of radiative transfer (eq. 1), may also be used to solve the equations (25, 27) resulting from the Harmonical Transmission model. The only difference arising from the employment of the Harmonical Transmission model is, that two equations (one for 'absorption coefficient' a and one for b) have to be solved simultaneously.

From the solutions A,B of these two equations, the line structure of the spectral intensity at every local position can be computed from

$$I(\phi) = \frac{1}{2}(A+B) + \frac{1}{\sqrt{2}}(A-B) \cdot \cos(\phi) \tag{29}$$

where

$$\phi = 2\pi \, \frac{\Delta\lambda}{d} \tag{30}$$

APPLICATION OF THE HARMONICAL TRANSMISSION MODEL: COMPUTATION OF RADIATIVE TRANSFER IN A GAS TURBINE COMBUSTOR

The Harmonical Transmission model of first order has been employed to study the radiative heat transfer in a gas turbine combustor with special emphasis on the influence of total pressure. As a new approach to radiative transfer computation, the calculation has been done on a spectral basis, where radiation from soot as well as from combustion gases is considered. The radiative properties of the combustion gases, which are usually expressed by the wavelength dependent parameters S/d, $S^{1/2}/d$ and γ, have been calculated on basis of quantum-mechanical methods, given for example in [10], [11]. In this numerical study only CO and CO_2 are considered, water vapor has been neglected for the sake of simplicity. The radiative property of soot has been computed by means of the Rayleigh approximation [12].

In the overlapping bands of the different combustion gases the absorption coefficient k_{Gas} and line-width-to-spacing ratio γ/d have been calculated using the equivalent line model [13]

$$k_{Gas} = \sum \left(\frac{S}{d}\right)_i p_i \tag{31}$$

$$\frac{\gamma}{d} = \frac{\left[\sum \left(\frac{S^{\frac{1}{2}}}{d}\right)_i \gamma_i^{\frac{1}{2}} p_i^{\frac{1}{2}}\right]^2}{\sum \left(\frac{S}{d}\right)_i p_i} \tag{32}$$

$$g_{1,Gas} = k_{Gas} \cdot e^{-2\pi \frac{\gamma}{d}} \tag{33}$$

The contribution of soot radiation is accounted for by

$$k = k_{Gas} + k_{Soot} \tag{34}$$

$$g_1 = g_{1,Gas} \cdot \frac{k_{Gas}}{k_{Gas} + k_{Soot}} \tag{35}$$

The two differential equations resulting from the Harmonical Transmission model (eq. 25, 27) have been solved numerically, using the well-known P1-approximation [2], [3], [4].

The geometry of the combustor and the gridpoints of the numerical grid are shown in Figure 2. For the numerical study the combustor has been approximated by cylinder with the air inlet located at the left side and the combustor outlet located at the right side.

The temperature and species distribution inside the combustor (Figure 3), which are typical for gas turbine combustors, show the structure of a recirculation zone with lower temperatures on the centerline of the cylinder.

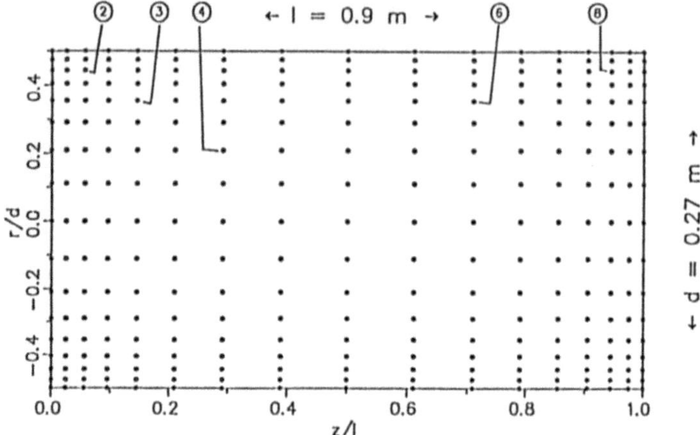

Fig. 2: Geometry of the combustor and numerical grid.

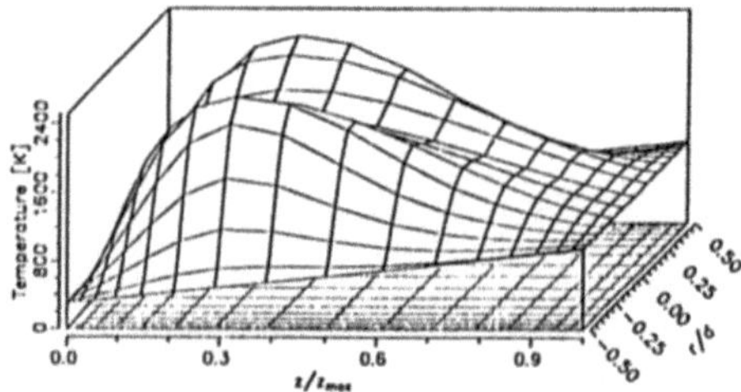

Temperature-Distribution

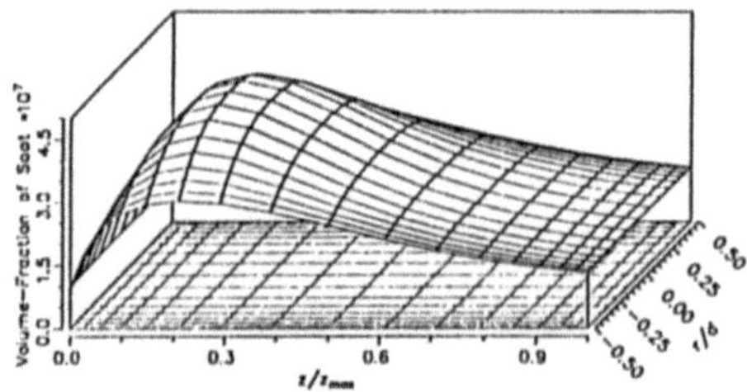

Soot Concentration

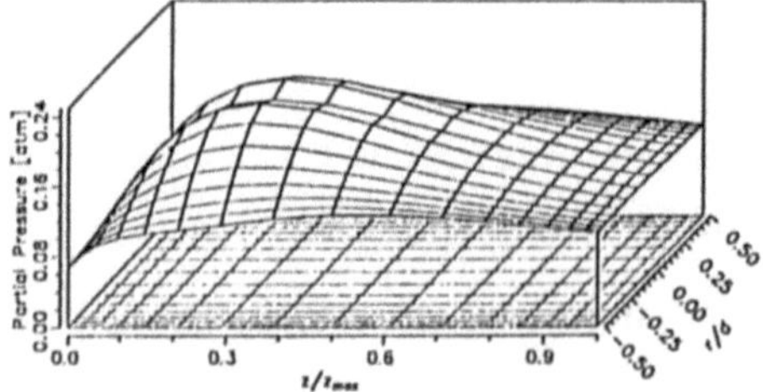

Partial Pressure of Carbondioxide

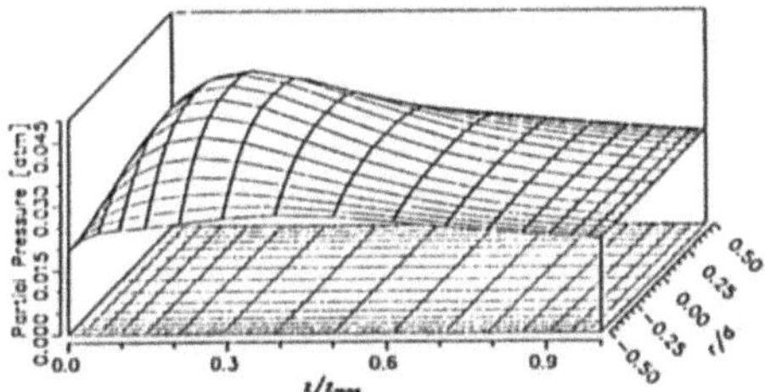

Partial Pressure of Carbonmonoxide

Fig. 3:   Temperature and species distribution inside the combustor.

RESULTS

The principal objective of this numerical study is to evaluate the effect of pressure broadening of the combustion gases on the radiative heat transfer. Therefore, two calculations at different pressures have been performed. The temperature and species distribution inside the combustor have been left unchanged. The only parameter, that was changed, was the total pressure: One computation was performed at a total pressure of 1 bar and the other at a total pressure of 10 bar. The results of these two computations are compared in the following section.

As mentioned above, the computation has been done on a spectral basis, with a discretisation into 50 wavelength intervals. The spectrally integrated radiative flux distribution inside the combustor, as it is presented in Figure 4, was obtained by an integration of the spectral fluxes over the 50 wavelength intervals. In Figure 4 the results of the computation at a total pressure of 1 bar is presented.

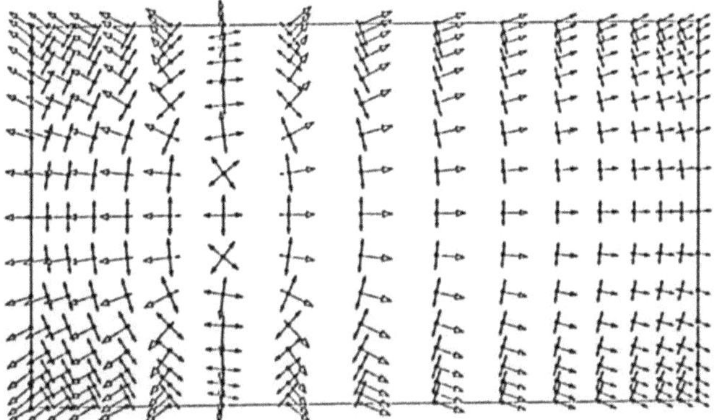

Fig. 4:   Spectrally integrated radiative fluxes (total pressure 1 bar)

436

The radiative flux distribution resulting from the computation at a total pressure of 10 bar is almost the same, showing differences of about 2%. The fact, that pressure broadening of the combustion gases shows only a slight effect on the spectrally integrated radiative fluxes is not surprising, because soot radiation is predominant over gaseous radiation in the present example. However, it can be shown from the radiation spectra, that gaseous radiation is considerably changed by pressure broadening. To demonstrate this, the radiation spectra at the grid points no. 2, 3, 4, 6, 8 (for numbering of grid points *see* Figure 2) are presented in Figure 5. In the wavelength region from 4—5.5 μm, where gaseous radiation of carbondioxide and carbonmonoxide is predominant over soot radiation, significant differences between a total pressure of 10 bar and a total pressure of 1 bar can be detected especially at grid points with low temperature (points no. 2, 3, 8).

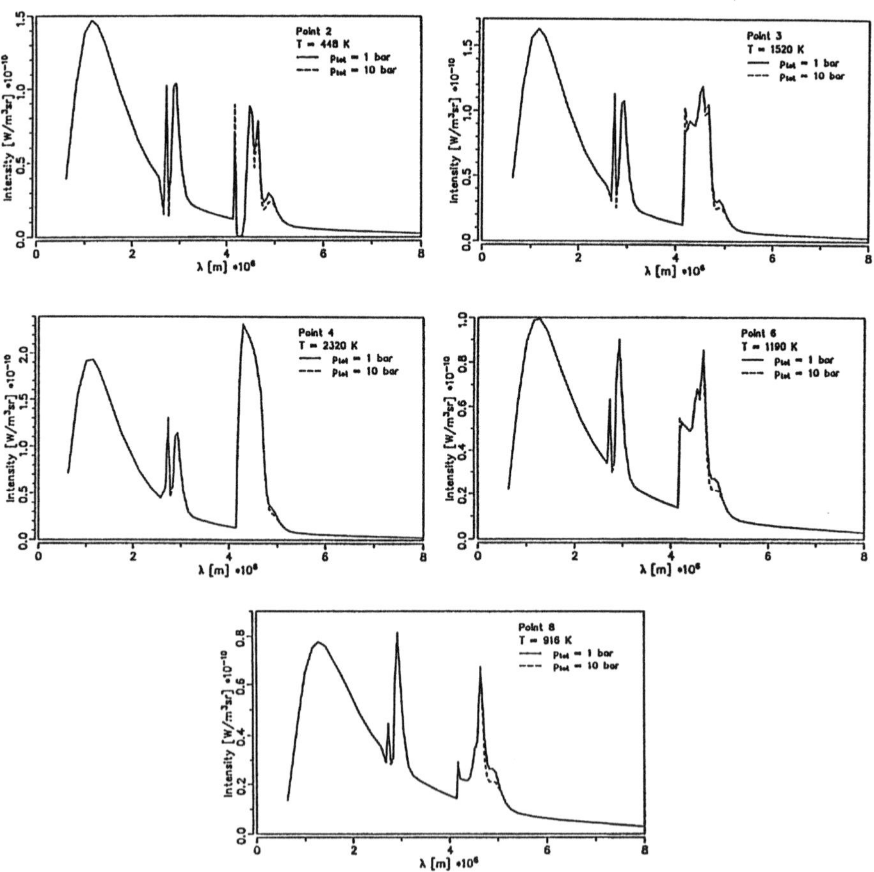

Fig. 5:   Spectral radiation at different grid points

One of the most important parameters for the design of a gas-turbine combustor is the heat load to the combustor wall. The result of the two computations is given in Figure 6. As well as the spectrally integrated radiative fluxes the radiative heat load to the outer combustor wall is only slightly effected by higher total pressure, because of predominant soot radiation.

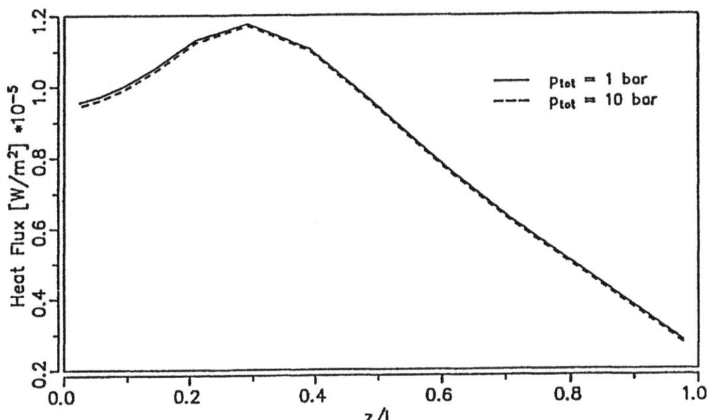

Fig. 6:   Comparison of the radiative heat load to the outer combustor wall

ACKNOWLEDGEMENT

This work was performed under support of TECFLAM and the Sonderforschungsbereich 167 of the Deutsche Forschungsgemeinschaft. The encouragement of Prof. R. Viscanta of Purdue University during the discussions of the work is also gratefully appreciated.

REFERENCES

1.   Sidall R.G., Selcuk N.: Evaluation of a new Six-Flux Model for Radiative Transfer in Rectangular Enclosures. Trans. IChemE 57 (1979) 163-168.

2.   Szu-Cheng S. Ou, Kuo-Nan Liou: Generalisation of the Spherical Harmonic Method to Radiative Transfer in Multidimensional Space. J. Quant. Spectrosc. Radiat. Transfer 28 (1982) 271-288.

3.   Mengüc P.M., Viskanta R.: Radiative Transfer in Three-dimensional Rectangular Enclosures containing inhomogeneous, anisotropically scattering Media. J. Quant. Spectrosc. Radiat. Transfer (1985) 33.

438

4. Ratzel A.C., Howell J.R.: Two-dimensional Radiative Transfer in Absorbing/Emitting Media using the P-N Approximation. J. Heat Transfer 105, (1983) 333-340.

5. Truelove J.S.: Three-dimensional Radiation in Absorbing-Emitting-Scattering Media using the Discrete-Ordinate Approximation in Three-dimensional Rectangular Enclosures. J. Quant. Spectrosc. Radiat. Transfer 39 (1988) 27-31.

6. Fiveland W.A.: Discrete-Ordinate Solutions of the Radiative Transport Equation for Rectangular Enclosures. J. Heat Transfer 106, (1984) 699-706.

7. Penner S.S.: Quantitative Molecular Spectroscopy and Gas Emissivities. Addison-Wesley (1959).

8. Elsasser W.M.: Heat Transfer by Infrared Radiation in the Atmosphere, Phys. Rev. 54 (1938).

9. Koch R., Wittig S., Noll B.: The Harmonical Transmission Model: A New Approach to Multidimensional Radiative Transfer Calculation in Gases Under Consideration of Pressure Broadening. Accepted for publication in Int. Journal of Heat and Mass Transfer (1990).

10. Malkmus W., Thompson A.: Infrared Emissivity of Diatomic Gases for the Anharmonic Vibrating-Rotator-Model. J. Quant. Spectrosc. Radiat. Transfer 2 (1961) 17-39.

11. Malkmus W.: Infrared Emissivity of Carbon Dioxide (4.3 μm - Band). J. Opt. Soc. Am. 53 (1963) 951-961.

12. Kerker M.: The Scattering of Light and other Electromagnetic Radiation. Academic Press (1969).

RADIATION CHARACTERISTICS OF SURFACE COMBUSTION BURNERS

M. Golombok and L.C. Shirvill
Shell Research Ltd., Thornton Research Centre
P.O. Box 1, Chester, CH1 3SH
England

ABSTRACT

The radiation characteristics of two types of surface combustion burner are reported both in terms of the quantity and the spectral distribution of the energy emitted as a function of thermal input. Surface emissivities deduced from the infra-red spectra are accounted for through microscopic examination of the radiating surfaces. For one burner type, the wide distribution of temperature thus revealed is shown to be consistent with the measured emissivity and with the steep temperature gradient known to exist in the fibre layers of the burner surface.

INTRODUCTION

Surface combustion is a gas-burning technique in which a combustible mixture burns within the surface layer of a permeable medium, heating it to incandescence and thereby releasing a proportion of the input energy as thermal radiation. Figure 1 shows the basic configuration of a surface combustion burner. Practical burners may have a flat surface (as shown and used in this study) or a cylindrical geometry. Surface combustion burners offer uniformity of heating, low thermal inertia and low NO_x emissions in a wide range of heating applications in the domestic, commercial and industrial sectors.

In this paper we describe the radiation characteristics of two types of surface combustion burner as a function of thermal input. The burners studied were, the Alzeta Pyrocore Burner [1], made from ceramic fibres, and the Metal Fibre Burner [2] (MFB), made from refractory metal fibres.

Small, sample burners were examined under laboratory conditions and results obtained by infra-red spectroscopy, thermal imaging and long-range visible microscopic techniques are described.

The layered fibrous structure of the MFB proved more amenable to detailed investigation and its radiative behaviour has been studied both isothermally and

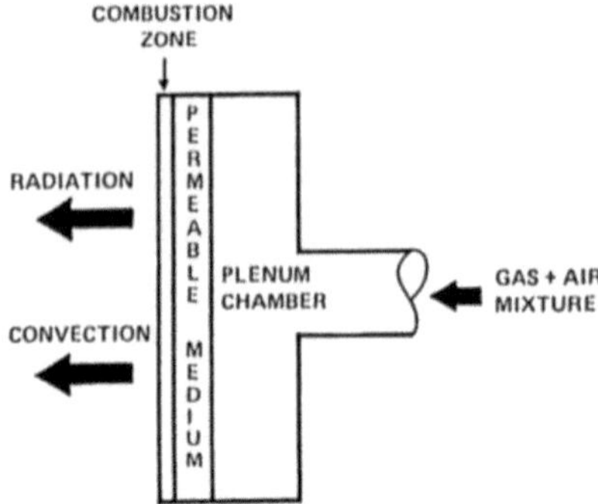

Fig. 1:   Surface combustion burner configuration.

non-isothermally to explain the characteristics observed when it is operated as a surface combustion burner.

BURNER TYPES

<u>Alzeta Pyrocore Burner</u>. The permeable medium of the Alzeta Pyrocore Burner is vacuum-formed from ceramic fibres onto a stainless steel supporting mesh. Figure 2 is a close-up photograph of the surface. The structure consists of

Fig. 2:   Surface structure of pyrocore burner.

interconnected pores surrounded by a web of very fine (1-3 µm) ceramic fibres. The active surface area of the burner was 0.017 m^2.

<u>Metal Fibre Burner</u>. The permeable medium of the MFB consists of fine fibres (22 µm) of a refractory metal (Fecralloy) sintered together to produce a rigid panel. The labyrinth structure of the material formed by the randomly laid fibres is shown in Figure 3, a scanning electron micrograph of the surface. The material used in this study had a porosity of 80% and a thickness of 4 mm. The active surface area of the burner was 0.019 m^2.

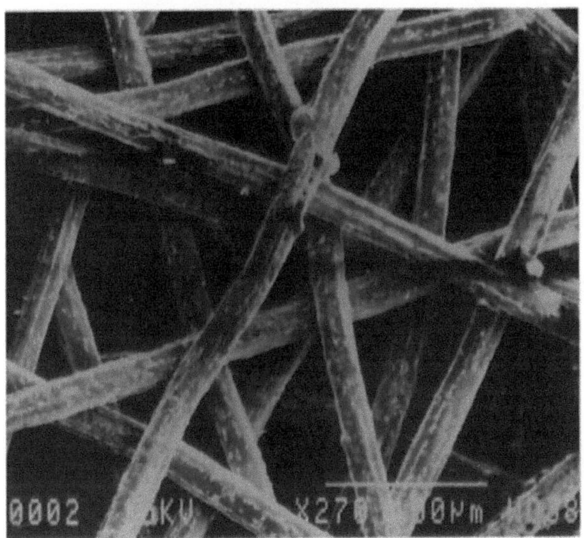

Fig. 3: Fibrous structure of metal fibre burner material viewed under scanning electron microscope.

BURNER PERFORMANCE

The burners were operated on a stoichiometric mixture of natural gas and air at thermal inputs of 250, 400 and, in the case of the MFB, 800 $kW.m^{-2}$ (based on the gross calorific value of the fuel and the superficial area of the burner surface). All the measurements were made with the burners freely radiating to an ambient temperature environment. Figures 4 and 5 show the Pyrocore and the MFB in operation, respectively.

<u>Spectral Analysis</u>. Radiation from surface combustion burners originates from two sources, emission from the heated surface and emission from the hot combustion products leaving the surface. The solid surface produces a continuous

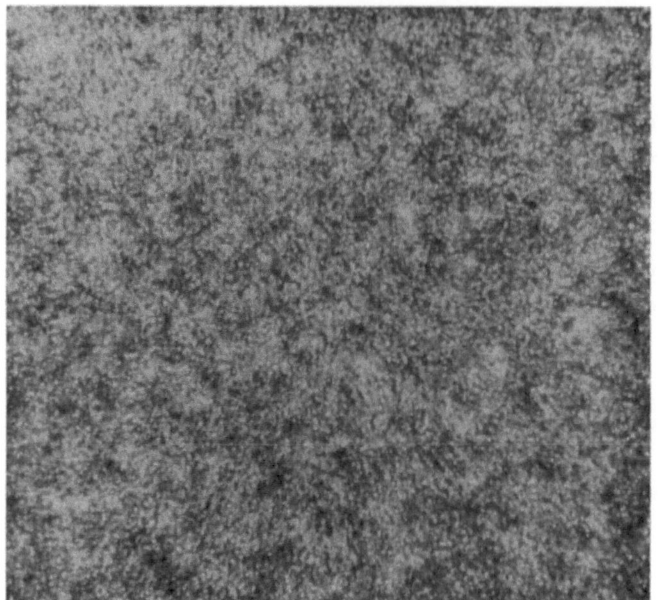

Fig. 4:   Pyrocore burner in operation

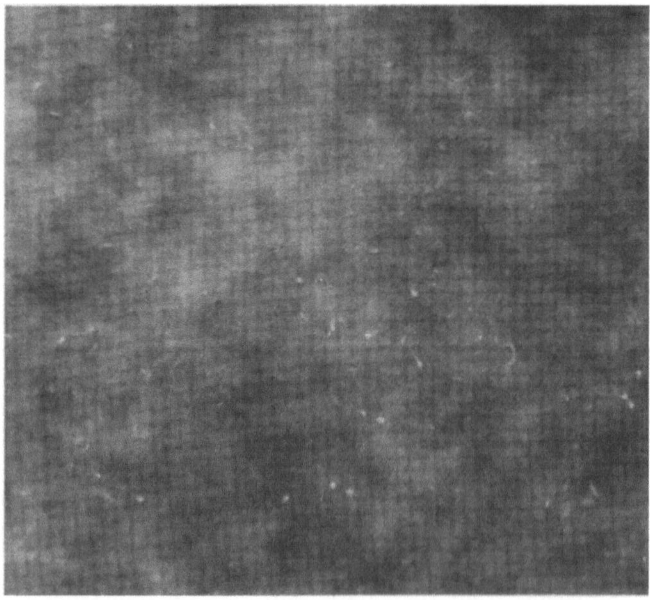

Fig. 5:   Metal fibre burner in operation

emission spectrum whereas the gases emit only in discrete spectral bands, the major emitting species in the combustion products being CO_2 and H_2O.

The combined effect of these two contributions can be seen in Figure 6, which shows the infra-red emission spectra between 1.2 µm and 7.2 µm of the MFB operating at thermal inputs of 250, 400 and 800 $kW.m^{-2}$.

The spectra were measured using a Spex spectrometer containing a 200 lines/mm grating and a Golay cell detector. The system was calibrated against the spectrum obtained from a heated conical black body.

The monochromatic emissive power of the surface, measured by the spectrometer, increased with thermal input at all wavelengths. A Planck grey-body energy distribution was fitted to the underlying continuum attributed to emission from the surface. A grey-body emissivity and temperature for the surface was thus derived. The calculated surface temperature increased with thermal input from 865 °C at 250 $kW.m^{-2}$ to 1045 °C at 800 $kW.m^{-2}$. The calculated normal emissivity decreased with thermal input from 0.55 at 250 $kW.m^{-2}$ to 0.45 at 800 $kW.m^{-2}$.

The energy emitted by the hot combustion products, H_2O and CO_2 between 2.5 and 3.0 µm and CO_2 at 4.4 µm, also increased with thermal input. This is mainly because of the higher temperature of the gases at high thermal input, with a small contribution from increased pathlength through the gas plume leaving the burner.

Figure 7 shows the emission spectra of the Pyrocore burner when operating at thermal inputs of 250 and 400 $kW.m^{-2}$. The emission spectra are similar to those described for the MFB. The calculated surface temperature is 1000 °C at 250 $kW.m^{-2}$ thermal input and 1050 °C at 400 $kW.m^{-2}$. The respective emissivities are 0.30 and 0.31.

<u>Thermal Imaging</u>. Spatial variations in surface temperature were investigated using an AGA-780 Thermovision camera containing an InSb detector cooled by liquid nitrogen. In order to remove the hot gas banded emissions observed above, a band pass filter at 3.9 µm was used.

Figure 8 shows the mean surface temperature of the MFB operating at thermal inputs between 200 and 800 $kW.m^{-2}$, using the average emissivity of 0.5 obtained from the spectral experiments. The surface temperature increases rapidly from 900-1100 °C over the thermal input range 200 to 500 $kW.m^{-2}$, but there is little increase over the remainder of the input.

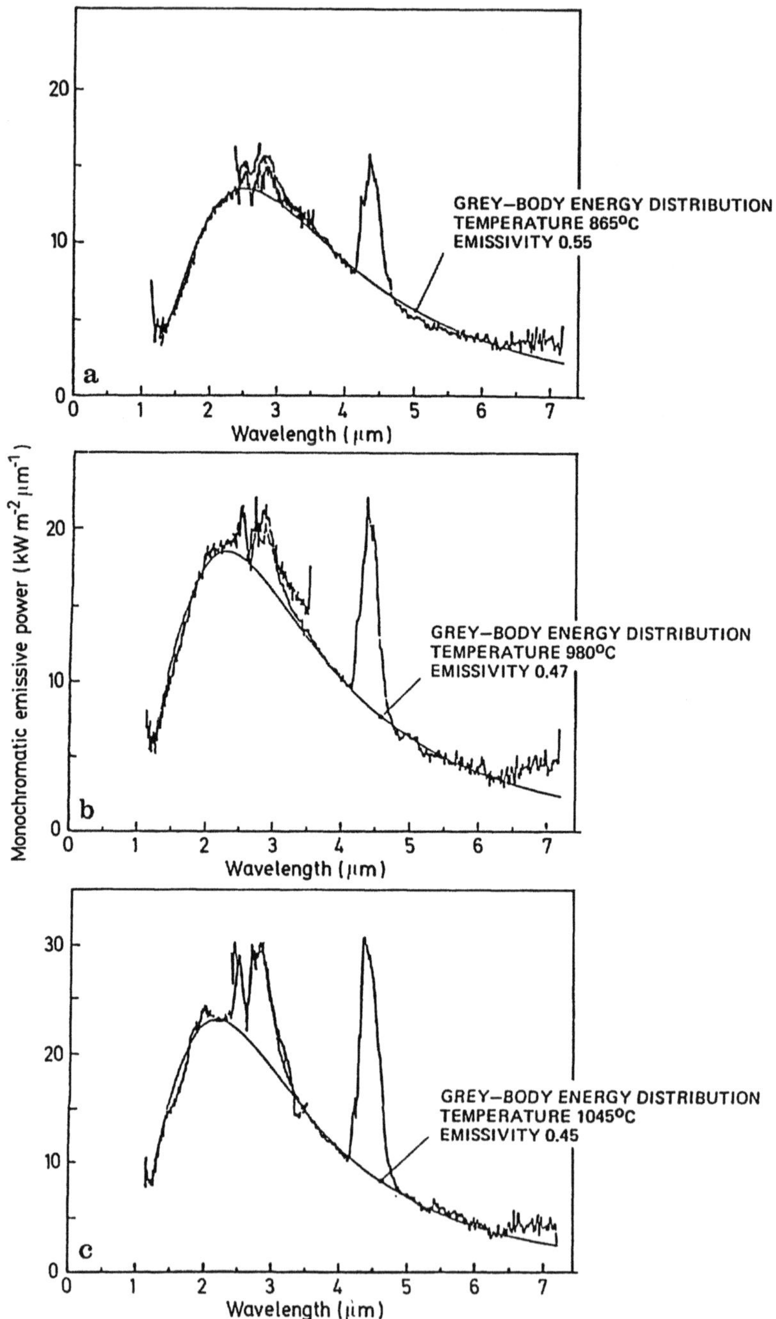

Fig. 6: Emission spectra of the MFB at three thermal inputs (a) 250 kW.m^{-2}, (b) 400 kW.m^{-2}, (c) 800 kW.m^{-2}.

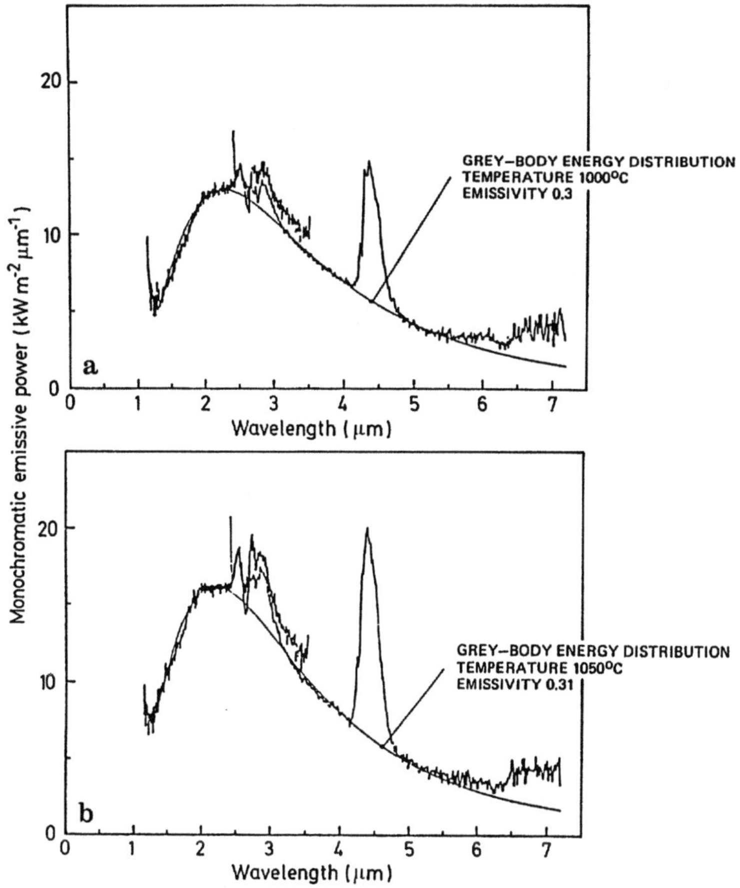

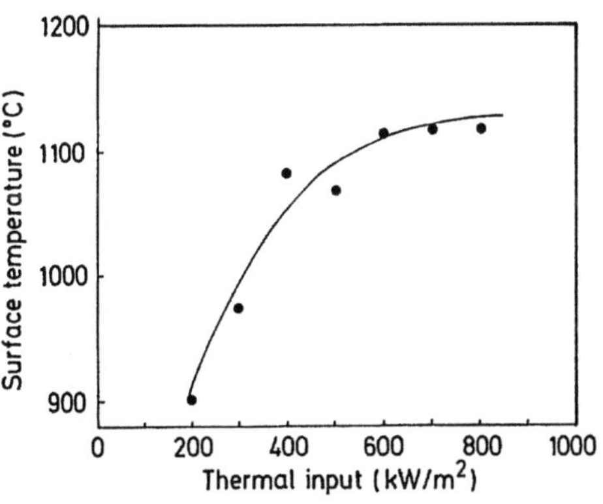

Fig. 7: Emission spectra of the pyrocore burner at two thermal inputs (a) 250 kW.m^{-2}, (b) 400 kW.m^{-2}.

Fig. 8: Mean surface temperature of MFB measured by AGA thermovision with assumed emissivity of 0.5, as a function of thermal input.

Above 800 kW.m^{-2}, patches of free flame occur. At these points the combustion zone is downstream of the burner surface, which remains cool. As the input is increased further these free flame patches occupy greater zones of the surface until they become the only form of burning and there is no surface combustion heating of the burner at all. The onset of a free flame patch above 800 kW.m^{-2} is associated with small variations in burner properties across the surface. Figure 9 shows that the standard temperature deviation across the surface increases linearly with thermal input, nearly doubling in spread over the turn down range. Moreover, we note that over the range 500-800 kW.m^{-2} the surface temperature changes very little, but the spread in surface temperatures continues to rise from 27 °C to 37 °C. We use this as a basis for a model of emissivity decrease due to non-uniform temperature (see below).

One drawback of the thermal camera is that it is only able to resolve gross variations on the order of 1 mm so that each image pixel really represents a spatial average over many fibre diameters. The average pore size of the MFB is only 66 µm so that the fine details cannot really be seen using this system. In the next section, we describe a method for partially overcoming this.

<u>Long-range Microscopy</u>. Examination of the burners in operation using a Questar long-range microscope reveals a hot top layer of fibres with a cooler background as demonstrated by Figure 10 (Pyrocore) and Figure 11 (MFB), despite the very different structures of the burners. The maximum temperature occurs at the

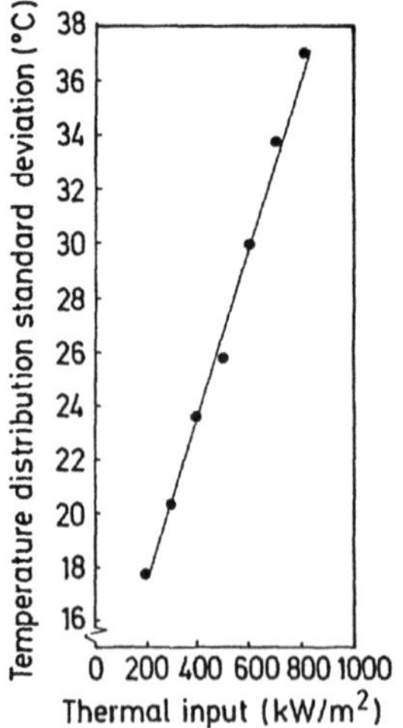

Fig. 9:   Standard deviation of temperature distribution on MFB surface as a function of thermal input.

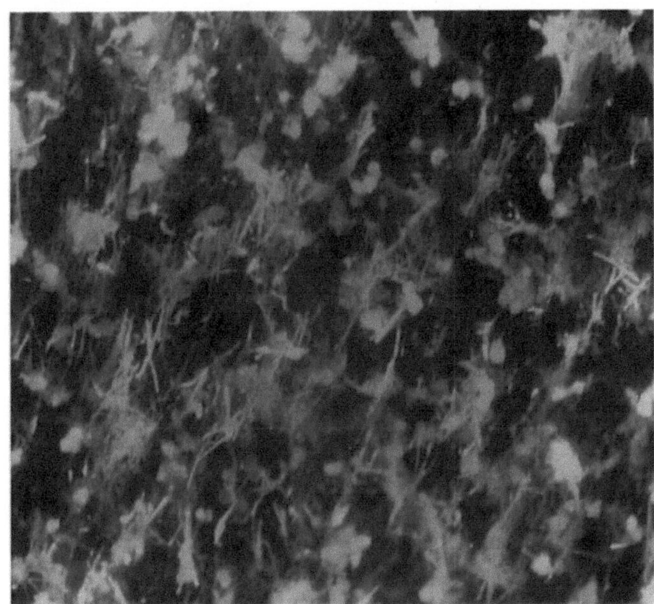

Fig. 10:   Pyrocore burner in operation viewed through long range microscope.

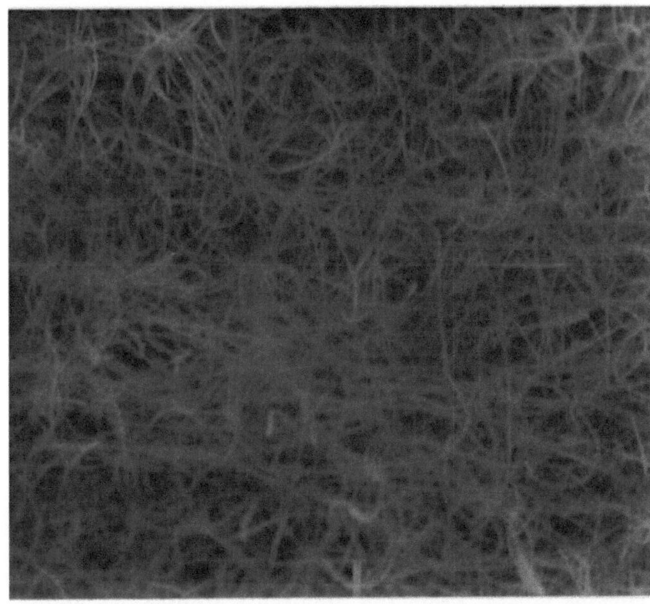

Fig. 11:   Metal fibre burner in operation viewed through long range mircoscope.

downstream surface as the temperature falls rapidly with depth into the material, i.e. there is a positive temperature gradient with respect to gas flow in the surface boundary layer of the burner material. This means that rather than an isothermal radiation temperature, there is a thermal distribution which, in part, determines the net radiant characteristics in surface combustion.

The temperature distribution was analysed using a grey-scale image analysis system. The microscopic image was recorded on video. The image of a black body at a known temperature was also recorded as a calibration reference. For maximum sensitivity in the required range, the video camera exposure parameters were set to give minimum and maximum readings at black-body temperatures of 600 °C and 1000 °C respectively. The black-body images were processed by a Kontron image analyzer and sorted in a 512 x 512 pixel array. Each pixel has an 8-bit grey-level intensity value assigned to it. Thus the relative brightness of the radiation at a point is graded on a scale of 0 to 255 which can be matched against known temperatures. This provides the reference for measurements on the burner. The calibration curve of grey scale intensity against black-body temperature corresponds to the convolution of the spectral response of the camera with the Planck radiation function over the visible regime. The grey-scale analysis is found to be sensitive to levels of illumination corresponding to black-body temperatures of 530 °C to 965 °C, which corresponds to the 680 °C to 1200 °C range of MFB operation, using the emissivities derived above.

When this analysis has been carried out on the MFB operating at 400 and 800 kW.m^{-2} (Figure 12) the grey-scale distributions were evaluated to give mean surface temperatures of 986 °C and 1117 °C and standard deviations of 17 °C and 32 °C respectively, both in excellent agreement with the infra-red camera measurements. However, the real advantage of these observations is the insight into the temperature gradients at the surface during combustion. Unfortunately, it is not possible to obtain the true temperatures of the different layers of radiating fibres because the effective emissivity varies with depth of the fibre layer from the surface. Fibres below the topmost level will reflect radiation from above.

For the top layer, all radiosity originates from the surface emissive power. Spectral measurements have shown that the solid material has an emissivity of 0.6 which is applicable to these top fibres which cannot reflect radiation as they are at the downstream surface. This surface layer constitutes about 25% of all emitting fibres in the surface layers. From the grey-scale studies, we obtain temperatures of 970 °C and 1090 °C at 400 and 800 kW.m^{-2} inputs respectively.

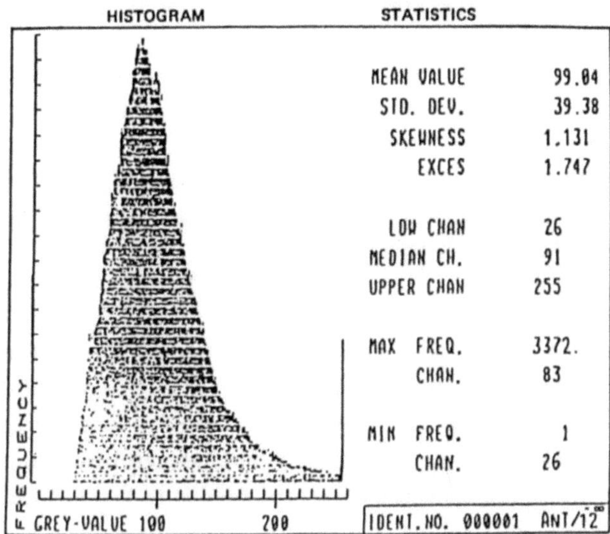

Fig. 12: Grey scale distribution of MFB operating at 800 kW.m^{-2}.

The lower lying levels will have a much higher effective emissivity because they reflect a lot of radiation from the higher levels. They will be effectively "black" at the bottom of the surface boundary layer. By interpreting the grey-scale measurements as pure thermal emission, it is possible to put an estimated lower bound on the temperature gradient in the boundary layer of the material. Given that the lower levels originate only a small amount of radiation and are located up stream of the high temperature reaction zone, they must be at a temperature of below 700 °C, which implies typical mean gradients in the surface boundary layer of around 2 K.μm^{-1}. This is a minimum gradient because the radiation originating from lower levels has a much greater reflected component than that from higher levels. This makes the lower levels appear hotter than their true temperature as the brightness temperature derived experimentally assumes that all radiation arises from emissive power, as discussed above.

It should be emphasized that these measurements are sensitive only to the optical wavelength regime (400-700 nm) and thus do not encompass any significant fraction of the energy; most of which we have shown to be in the range 1-7 μm. The only reason for using visible radiation is the greater resolution obtained. The implications of these experimental observations have important ramifications for calculations of emissivity. This problem of differential reflection with depth is solved in the next section where these experimental results are compared to a theoretical surface boundary layer model.

RADIATIVE PROPERTIES OF MFB MATERIAL

<u>Isothermal Measurements</u>. The emissivity of the fibrous material used in the MFB is clearly dependent on the emissivity of the solid Fecralloy metal from which the burner fibres are made. The oxidized surface of the material when used as a burner leads to a higher emissivity than that for the bare metal. Figure 13 is a scanning electron microghraph of the surface of an oxidized fibre, clearly showing oxide ridges and nodules of the order of 2-3 μm.

To evaluate the geometric effect of the fibres, the spectral profile of a piece of solid oxidized Fecralloy and a sample of 80% porosity burner material were examined using an infra-red spectrometer. The samples were heated in a muffle furnace and compared with a black body at the same temperature in order to obtain values of the emissivity at different wavelengths. The temperature of the solid material was measured using a thermocouple, spot-welded to the surface. The temperature of the fibrous material was obtained by flowing air along the furnace and out through the porous material, (which was convectively heated) past a fine thermocouple located on its surface. The temperature for solid and fibrous material was 592 °C, this may be confirmed from the position of the spectral maxima using the Wien displacement law, as used for the identical measurements on the burner in operation.

Fig. 13:   Scanning electron micrograph of the surface of a fecralloy fibre.

Figure 14 shows the spectral radiosity of the solid and fibrous materials compared to a black body at the same temperatures. The solid emissivity drops from 0.6 to 0.45 over the wavelength range 2-6 µm. The fibrous material emissivity falls from 0.9 to 0.8 over the same spectral range. Thus the two main features are the enhanced emissivity of the fibrous material and the dispersive decrease in emissivity with increasing wavelength.

The greater emissivity of the fibrous material is due to the cavities in the porous structure. Multiple reflections at and near the surface give rise to an increased absorption cross-section. In effect, the porous material has a more black-body cavity character than the flat surface. This increase in emissivity due to fibre configuration in a porous material has been modelled both numerically and analytically [3].

The decrease in emissivity with increasing wavelength for the two samples can be attributed to the surface structure discussed above. The oxide nodules are 2-3 µm

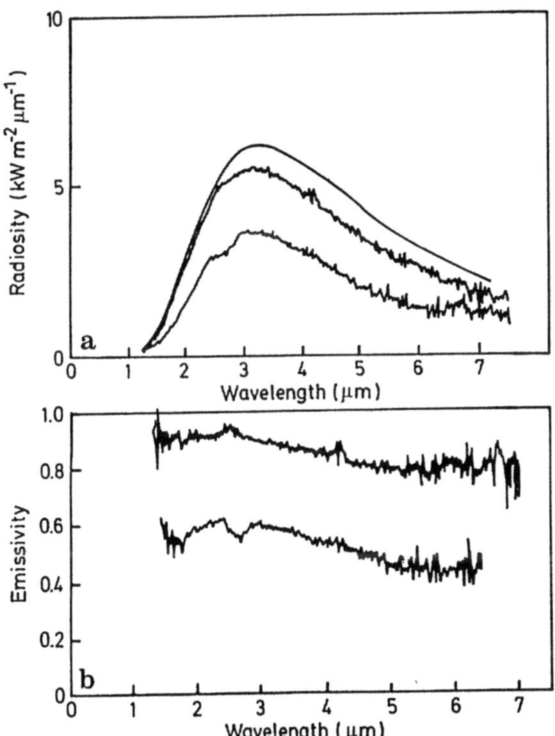

Fig. 14: (A) Spectral radiosities of solid fecralloy (lower curve), fecralloy MFB material (middle curve) and calculated black body profile (top smooth curve) at 592 °C; (B) emissivities of solid fecralloy (lower) and fecralloy MFB material (upper).

in size, implying that at shorter wavelengths light can enter the surface micron size cavities [4]. However, longer wavelengths "see" a flatter surface thus resulting in a lower emissivity for wavelengths longer than the dimension of the surface structure. The surface topography thus determines emissivity dispersion. In this case it is however, sufficiently small to permit the use of a grey body for modelling purposes.

<u>Non-isothermal Behaviour</u>. Previous simulations [5,6] of the radiant characteristics of cones have shown that variations in temperature can change the effective emissivity. The fibrous material contains many distributed cavities and so variations in temperature can be expected to have a pronounced effect on the effective emissivity. We first consider the effect of having an arbitrary temperature distribution on a flat surface. The distribution consists of a large number of areas each radiating at different temperatures and we sum the spectra to calculate the net effective radiation temperature. This idea is then extended to relate surface variations to depth in the emitting non-occluded surface boundary layer of the fibrous material. We have extended our earlier work [3] based on geometric cross-section view factors to take in the non-isothermal case. This is possible by use of an analytic solution computer routine.

We consider a collection of small isolated areas f_i radiating at different temperatures T_i. Because the surface is flat, the view factors for radiation exchange between each small area are zero. All that is required is that some heat generation mechanism, such as surface combustion followed by heat transfer from gaseous products to the burner surface, maintains a steady-state temperature distribution. We approximate the surface by a collection of small isothermal areas, each radiating according to the Planck radiation distribution as determined by the local temperature. The energy distribution is summed over the surface and equated to a net radiosity. The mean surface intensity spectrum is thus obtained by integrating the radiant fluxes over the range of temperature variations and fitting the result to a grey-body distribution to determine the effective emissivity and temperature T_b, the latter being derived from the wavelength maximum as in the experiments described above.

The functional form for the area temperature distribution was taken to be a Gaussian probability density function of the form

$$f(T_i) = 1/\sigma_T \, (2\pi)^{-1/2} \exp{-1/2\{(T-T_m)/\sigma_T\}^2}$$

where σ_T is the spread in the distribution and T_m is the mean first moment temperature as opposed to T_b which is a modal radiation temperature derived from a grey-body distribution.

In the numerical simulation we take the isothermal fibrous emissivity, to be 0.85. From an assigned T and σ_T, $f(T_i)$ is calculated and summed using the Gaussian distribution. The effective burner temperature is obtained from the wavelength maximum of the summed functions, and is used to calculate the total radiation intensity. This is in turn compared with the black-body spectrum to obtain the burner emissivity, which is found to vary only slightly with wavelength.

Figure 15 shows that the parameters required to simulate the emissivity of 0.55 and temperature of 865 °C observed for a thermal input of 250 kW.m^{-2} imply a standard deviation of more than 200 °C which could not be sustained given the relatively high surface thermal conductivity [7]. The spread in temperature decreases the effective emissivity. The values of spread obtained from thermal imaging measurements of the burner in operation are an order of magnitude smaller than the values required to yield a flat surface emissivity corresponding to the observed values of around 0.5.

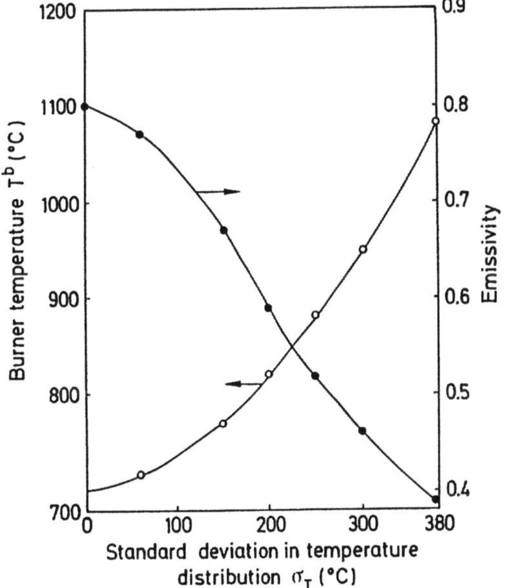

Fig. 15:   MFB temperature and emissivity as a function of normal temperature distribution.

As the lack of resolution in the AGA Thermovision measurements led to a consideration of microscopic structure, so now we look at the emission arising from different depths of material being able to radiate to the environment.

<u>Surface Boundary Layer Model</u>. We define the surface boundary layer, as being those layers of fibres that have a non-occluded view factor to the downstream

environment. Image analysis using the long-range microscope has shown that different levels of fibres within the surface boundary layer of the material emit radiation to the environment. The size of the area that emits to the environment is clearly a function of depth within the boundary layer.

Previously we have described a model for the numerical solution of the isothermal emissivity problem where the non-occluded levels (i.e. those within the surface boundary layer) have geometric cross-sections between layers which are the same as the radiation shape factors. These factors represent heat flows between different levels in the surface boundary layer. The whole system may be interpreted in terms of an electrical resistance network, an analogy which we have developed previously, in considering the isothermal case. Further details may be found in an earlier paper [3]. The resulting heat flux network yields a set of equations which have been solved numerically for the isothermal case. The expression for the fibrous material emissivity is bivariate polynomial in terms of the porosity and solid emissivity.

In the non-isothermal case, the system of equations is no longer closed by the condition for a uniform temperature throughout the surface boundary layer. Each level acts as an independent radiant energy source. The problem was made tractable by using an analytical solution computer routine REDUCE [8] which can solve the equations in terms of the desired variables. These are, as for the

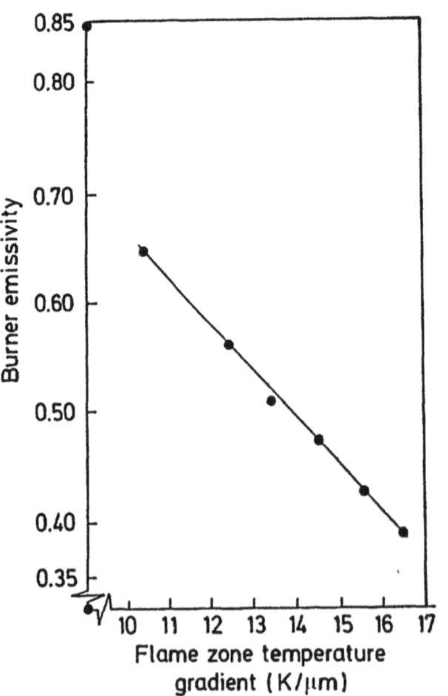

Fig. 16: MFB emissivity as a function of surface temperature gradient.

isothermal case, but with the addition of the temperature distribution. The equations are of the Kirchoff form, setting net heat flux into a cell equal to zero since the system is in a steady state during continuous combustion.

Unlike the isothermal case, there is no single black-body reference temperature for comparison in order to obtain an effective emissivity. This requires a form for the temperature distribution which is known from the image analysis to fall off with depth. An exponential form was assumed with the surface at 1100 °C, as measured in the grey-scale analysis. The result indicates that at a surface gradient of 13.5 K.μm^{-1} (Figure 16), the emissivity of the burner will be around the observed value of 0.5. However, the mean gradient in the surface boundary layer as a whole is around 3 K.μm^{-1} which is in good agreement with the result obtained from image analysis.

CONCLUSIONS

1. The radiation characteristics of two types of surface combustion burner have been measured under freely radiating laboratory conditions. It is recognized that in practical applications the characteristics will be enhanced by confinement and radiation interchange.

2. Infra-red dispersed spectral measurements have shown that the radiation emitted originates from two sources, the burner surface itself and the hot combustion products leaving the surface. Emission from the surface has been quantified in terms of an effective emissivity and temperature.

3. Measurements using an AGA thermal imaging system have shown spatial variations in burner surface temperature but these are insufficient to explain the measured effective emissivity. The spatial resolution of the system was only about 1 mm.

4. Greater resolution long-range microscopic analysis has shown that during surface combustion there is a temperature gradient in the burner which is positive with respect to gas flow. Grey-scale image analysis was used to measure the temperature of the top layer. This is derived using our knowledge of fibre emissivity since the top layer does not reflect any irradiance. The true temperatures of lower levels cannot be measured directly because multiple reflections give rise to a distribution of effective emissivities, but a minimum temperature gradient of 2 K.μm^{-1} in the surface boundary layer may be reasonably assigned.

5. The emissivity of solid Fecralloy and of the fibrous burner material made from it, measured under isothermal conditions, show that the emissivity is enhanced by the fibrous structure.

6. A calculation of the net radiant effect of a flat surface with a collection of non-isothermal individually non-resolvable radiating points is presented. The net effective temperature is determined by the maximum of the superposed grey distributions. A wide distribution of temperatures is required in order to lower the effective emissivity on a flat surface.

7. A computer algebra routine was used to solve the heat flow network for the surface boundary layer model of the fibrous material. By using effective radiant temperatures, the estimated top level surface temperature gradients are $0(10)$ K.μm^{-1} which is reasonable experimentally.

REFERENCES

1. Schreiber R., Krill W., Kesselring J., Vogt R., Kukasiewizc M.: Intl. Gas Res. Conf. (1983)/DO6-83.

2. Shirvill L.C.: Metallic Fibre Surface-Combustion Radiant Gas Burners. in Proc. Intl. Gas Res. Conf. (1986) 837.

3. Golombok M., Shirvill L.C.: Emissivity of layered fibrous materials. Appl. Opt. 27 (1988) 3921-2925.

4. Hutchins M.G., Wright P.J.: Microstructure and Composition of Solar Absorber Coatings. in Coatings for Energy Efficiency and Solar Applications. UK-ISES 38 (1984) 59-70.

5. Shirley J.H., Eberly J.H.: Local Effective Emissivity of Conical Cavities. Appl. Opt. 18 (1979) 3810-3814.

6. Bedford R.E., Ma C.K., Chu Z., Sun Y., Chen S.: Emissivities of diffuse cavities. Appl. Opt. 24(18) (1985) 2971-2980.

7. Golombok M., Shirvill L.C.: Laser flash thermal conductivity studies of porous metal fibre materials. J. Appl. Phys. 63(6) (1988) 1971-1976.

8. Hearn A.C.: REDUCE 3.2 (1985) (RAND CP78).

SIMULATION OF RADIATION HEAT EXCHANGERS

B. Stapper, D. Köneke, P.-M. Weinspach
Universität Dortmund
Fachbereich Chemietechnik
Lehrstuhl für Thermische Verfahrenstechnik
Postfach 500 500 - 4600 Dortmund 50
Germany

ABSTRACT

A model based on the Monte Carlo method is introduced in order to calculate temperature distributions and radiation heat fluxes to cooled surfaces in radiation boilers. The model is verified with data from a coal gasification plant. A parameter study reveals the application of the model to fluidized bed combustion systems.

NOMENCLATURE

A	$[m^2]$	area
B	$[kg/m^3]$	solid loading
$\vec{e}_s$	$[-]$	unit vector
h	$[kJ/kg]$	enthalpy
I	$[W/m^3]$	radiation intensity
K_e	$[1/m]$	extinction coefficient
K_a	$[1/m]$	absorption coefficient
K_s	$[1/m]$	scattering coefficient
k_a	$[-]$	absorption parameter (for particles)
k_s	$[-]$	scattering parameter (for particles)
ℓ, L	$[m]$	length
$p(\Theta^x)$	$[-]$	scattering function
p	$[bar]$	pressure
q	$[W/m^2], [W]$	heat flux
r, R	$[m]$	radius
S_h	$[W/m^3]$	enthalpy of reaction
S_m	$[m^2/kg]$	specific surface
T	$[K]$	temperature
u	$[m/s]$	velocity
V	$[m^3]$	volume
x	$[m]$	length
Z	$[-]$	random number

ε	[-]	emissivity
η,φ	[-]	angles of emission
η_s	[-]	angle of scattering
λ	[W/(m^2 K)]	heat conductivity
ρ	[kg/m^3]	density
σ	[W/(m^{2}K^4)]	Stefan-Boltzmann's constant $(5.67 \cdot 10^{-8}$ W/(m^{2}K^4))
Ω	[-]	solid angle

Indices

b	black
conv	convective
g	gas
i,j	zone indices
in	incoming
out	outgoing
rad	radiative
s,scat	scattering
w	wall
λ	wavelength

INTRODUCTION

Radiation heat transfer is of great importance in industrial processes which produce energy by combustion of fossilized fuels such as coal. The combustion heat is transferred to a cooling agent in a heat exchanger. This medium is preferably water being converted to superheated steam. The heat transfer process is dominated by radiation due to the high temperature level. The model described in this paper tries to determine the amount of radiation in those processes in order to achieve a more detailed description of heat transfer in radiating systems.

The composition of the radiating combustion products has great influence on radiative heat transfer. Those products may be absorbing gases such as water vapour, carbon dioxide, carbon oxide etc., or they may be particles which occur as fly ash or soot. The processes analysed are power plants, which utilize gas or coal, or coal gasification processes. Considering fluidized bed combustion and partial coal gasification, particle radiation takes over a great amount of the heat to be transferred. Therefore detailed knowledge of the radiation process is necessary.

When preparing an energy balance on radiation heat exchangers several approaches are possible. Simple total energy balances may be utilized [1]. If a detailed calculation is necessary so-called zone methods [2,3] have to be used in order to simulate the inner volume of the heat exchanger.

This work introduces the application of the Monte Carlo method. In comparison to other methods it has the advantage of a physical description of radiation phenomena. As the memory and calculation speed of computers have increased significantly during the last years the application of Monte Carlo methods even on complicated systems now is feasible. The model is applied to a radiation boiler of a partial coal gasification plant. A heat exchanger similar to this could be used in fluidized bed combustion processes.

Because of the high solid loadings the main goal of the work is to model the influence of particles on radiation heat transfer. Therefore a study on the variation of process parameters is carried out. The parameters are set to meet fluidized bed combustion conditions. From the results aspects for an improved heat exchanger design could be derived. This is necessary considering the economical and ecological aspects of the use of fossilized fuel resources.

THEORY

The simulation of radiation heat exchangers with zone methods requires an enthalpy balance of single volume elements or zones, respectively. This balance can be written for steady state conditions as follows

$$0 = \frac{\partial}{\partial x_j}\left(\lambda \frac{\partial T}{\partial x_j} + q_{radj}\right) + S_h - \frac{\partial(\rho u_j h)}{\partial x_j}$$
$$\quad\quad\; I \quad\quad II \quad\quad III \quad IV \tag{1}$$

Term I describes heat transfer by conduction,
term II the radiation heat flux,
term III is the energy source by chemical reaction and
term IV the convective heat transfer by mass flow.

As term I is negligible compared to term II and IV and no chemical reaction takes place in the heat exchanger, only terms II and IV are considered. Convective heat transfer is included in the model, but only in a simplified manner, as radiation dominates the transfer process.

The heat flux from a black body radiator is given by Stefan-Boltzmann's law as

$$q = \sigma A\, T_b^4. \tag{2}$$

The radiation behaviour of real surfaces is different. In such cases the radiation heat flux of a surface having the temperature T is ε-times that of a black body radiator. If this emissivity does not depend on the wavelength λ of the radiation, the body is called gray.

In order to achieve an equation that describes radiation heat transfer a volume element is balanced. This balance gives [2, 3, 4]

$$\vec{e}_s \, grad \, I_\lambda = - (K_{a\lambda} + K_{s\lambda}) \, I_\lambda + K_{a\lambda} I_{b\lambda} + \int_0^{4\pi} p_\lambda (\phi^x) \, I_{\lambda,scal} d\Omega.$$

(3)

Eq. (3) introduces the parameters K_a and K_s which are called absorption and scattering coefficients. A solution of (3) is possible if the values of K_a and K_s are known. They depend on the material and have to be determined by experiments or theoretical approaches.

Radiating media are combustion gases (CO_2, H_2O, CO, SO_2, etc.) and particles (fly ash, soot etc.). Absorption coefficients can be calculated from emissivity data obtained by Hottel [2] or using the wide band model [5] with data given by Görner [3]. Absorbing gases produce non-luminous radiation that is concentrated in spectral lines and bands. An emissivity can be obtained by comparing the radiative heat flux of an emitting gas to that of a black body radiator (*see* Figure 1). From emissivity data the absorption coefficient can be calculated as follows

$$K_{a,g} = \lim_{pl \to 0} \left(- \log \frac{1 - \varepsilon_g}{P_g l} \right).$$

(4)

Luminous radiation occurs when particles are involved. Continuum radiation is emitted throughout the visible and infrared spectrum. Data on the emissivity of particles are not available in many cases. Biermann [6] measured emissivities

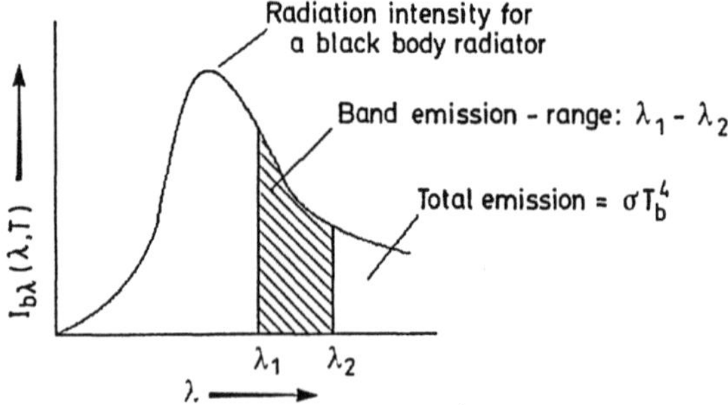

Fig. 1: Intensity distribution for a radiating gas

of different fly ashes from pulverized coal combustors. From this he calculated absorption coefficients using Beer's law. From the results he concluded that there is no temperature dependence of $K_{a,s}$. Recent experimental [4,7] and theoretical [8] investigations show that the absorption and scattering parameters of small particles are temperature dependent. Elsner's [4] results also provide values on the emissivity of quartz and fly ash from a fluidized bed combustor. It is shown that the emissivity is independent of wavelength considering the accuracy of the experiments. According to this particles can be approximated as gray body radiators.

MONTE CARLO METHOD

The application of Hottel's zone method [2] demands a single energy balance on each volume element. This results in a so-called view factor matrix which has to be solved. For complex geometries this procedure is quite difficult. Thus the time consuming factor in the calculation procedure is the numerical solution of the heat transfer balance.

The Monte Carlo application to radiation heat transfer belongs to the zone methods, too. However, the view factor matrix is substituted by the result of a simulation of the mechanism of radiation and the physical properties of the radiating media. In the model the radiation is described as photons being emitted in discrete directions. Combining a certain amount of photons gives a so-called energy bundle. Thus one bundle represents a discrete amount of energy transferred. The method consists of following the path of a bundle from the point of emission until its final absorption. This path is determined by random numbers according to the random nature of radiation. Point of emission, direction and beam length are calculated by appropriate functions. These functions are given by Howell [9].

In order to calculate the temperature distribution in the system the following procedure is employed. Neglecting heat transfer by conduction, an energy balance on each volume element j (*see* eq. (1)) gives

$$q_{rad,out,j} = q_{conv,out,j} - q_{conv,in,j} + q_{rad,in,j}. \tag{5}$$

The heat fluxes by convection are known by superposing a mass flow pattern. The outgoing radiation can be calculated [10] as

$$q_{rad,out,j} = 4\,K_{e,j}\,\sigma\,T_j^4\,dV_j, \tag{6}$$

using the extinction coefficient K_e as described below. The incoming radiation heat flux has to be obtained by calculating the total amount of absorptions in the Monte Carlo simulation.

The calculation procedure starts by dividing the system into discrete volume elements (zones) and postulating a temperature profile in the system. From this and a given mass flow pattern the quantities $q_{conv,out,j} - q_{conv,in,j}$ can be calculated. For each zone this value represents the net amount of radiation to be simulated. Adding all values gives the total amount of radiation to be simulated. By dividing this energy through the total number of bundles one obtains the energy per bundle. This also gives the number of bundles to be emitted from each volume zone. Now the simulation has to be run using the functions described below. The number of absorptions in each volume element gives $q_{rad,in,j}$. Using the mass flow pattern and the calculated incoming radiation heat fluxes, the system of non-linear equations given by (5) can be solved to produce a new temperature profile. This iteration is repeated until all zone temperatures converge. The result is the final temperature distribution and the radiation heat fluxes which can be obtained from the final simulation loop.

Random Functions. The simulation of stochastic functions is based on sets of uncorrelated numbers. These numbers are called random numbers. Usually they are evenly distributed accross the intervall [0;1). Computers cannot provide such sets of numbers. They use algorithms that produce a sequence of so-called pseudo-random numbers.

The point of emission and the direction of an energy bundle in a cylindrical system is given by [11]

$$l = Z \, L, \tag{7}$$

$$r = \sqrt{Z} \, R, \tag{8}$$

$$\cos \eta = \sqrt{Z}, \tag{9}$$

$$\varphi = \pi \, Z. \tag{10}$$

The geometry is illustrated by Figure 2.

Beam Course. After calculating point of emission and direction from eqs. (7-10) the energy bundle is followed until its final absorption. Therefore the mean

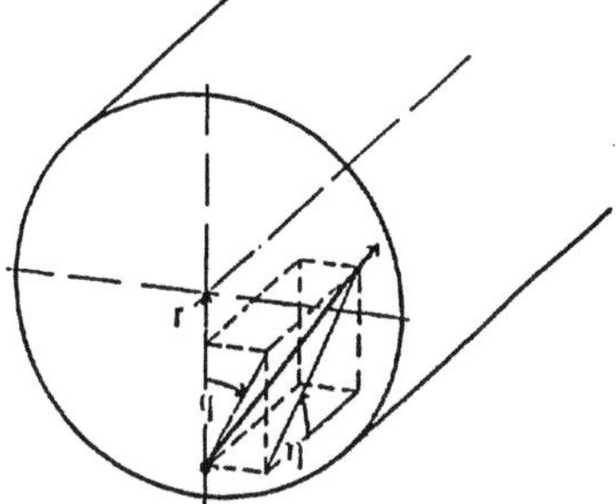

Fig. 2: Geometry of emission

beam length has to be calculated from

$$l_j = \frac{-\left(\ln Z + \sum_i (K_{e,i}\, l_i)\right)}{K_{e,j}},$$

(11)

with

$$K_e = k_a S_m B \sqrt{1 + 2\frac{k_s}{k_a}} + K_{a,g}.$$

(12)

The sum of single absorption coefficients in (11) reflects the temperature dependence of this parameter. As each volume element has a unique temperature, the absorption coefficient of each zone is different.

The energy bundle is now followed until it reaches one of the boundary surfaces or it is absorbed in a volume element. If the bundle is incident on a surface the reflection is determined by comparison of a random number to the emissivity of the wall. If the number's value is less than ε_w the beam leaves the system and the next bundle is released from a volume element. If the beam is reflected the course of the bundle is followed to the next absorption. When leaving the system the energy is added to the heat flux incident on this particular surface element.

If the beam is absorbed before reaching a surface or after reflection from a surface, a decision has to be made whether or not the beam will be reemitted. The net radiative flux from a zone can be negative due to the mass flow pattern. In this case the bundles are absorbed until the total energy absorbed equals the negative net radiation. Then all incoming bundles are reemitted because there can be no accumulation of energy with steady state conditions. The same reemission takes place if the net radiation from a volume element is positive. Each absorption is registered and adds to the amount of incoming radiation heat flux. The beam is then followed until it is absorbed by a surface or another element with a negative net emission.

Scattering. Earlier Monte Carlo applications to radiation heat transfer only analysed gaseous radiation [13] or the particles involved in the process were supposed not to scatter radiation [3]. The model described in this study also considers scattering of incoming radiation. Each time an absorption takes place, a decision is made if the beam is finally absorbed or scattered depending on the scattering albedo

$$\mu = \frac{K_s}{K_{a,g} + K_{a,s} + K_s} < 1, \tag{13}$$

with $K_{a,s} = k_a\, S_m\, B$, $K_{a,g}$ given by (4), and $K_s = K_a - K_{a,s}$.

If the calculated random number is less than μ, scattering occurs. Then the beam is reemitted from the point of absorption without registration, which means a redistribution of radiation. The angle of scattering can be calculated for isotropic or anisotropic scattering. In case of isotropic scattering relations (9-10) are used to determine the new direction. Anisotropic scattering is given by the following equation [14]

$$\Theta(\eta_s) = \frac{8\left(\sin(\eta_s) - \eta_s \cos(\eta_s)\right)}{3\pi}. \tag{14}$$

For a cylindrical geometry the angle η_s has to be transformed correctly.

The influence of scattering is negligible for weak scattering particles and low solid loadings. These conditions occur in coal gasification processes, for example. In other industrial applications scattering causes a redistribution of radiation which leads to a variation of the heat transfer to the surrounding wall.

SIMULATION OF A RADIATION HEAT EXCHANGER

Temperature distribution and heat fluxes to the surrounding wall were calculated for the radiation heat exchanger of a partial coal gasification plant. The heat exchanger is simply a tube with watercooled walls. Therefore it was divided into concentric ring elements and equally spaced axial segments. The radial segmentation was done in order to obtain the same volume for each element. The symmetric segmentation also reduced the dimensions in the calculation from three to two. Figure 3 shows the design of the heat exchanger.

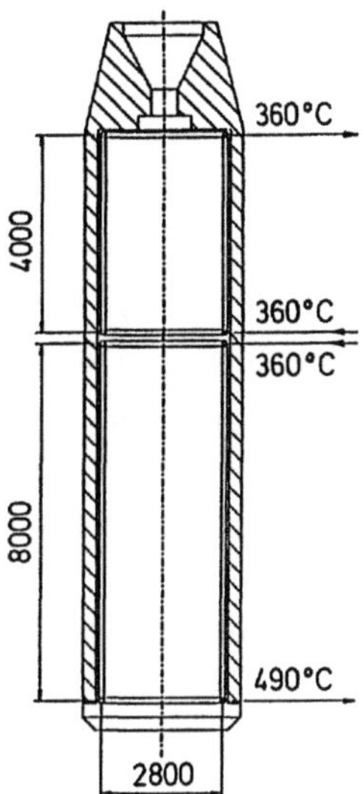

Fig. 3. Radiation boiler (geometry)

Verification. First the heat exchanger was simulated without particles. Experimental data of an operating condition with a gas mixture as radiating medium were available. As convective heat transfer has minor influence in high temperature systems above 1000 °C, this transfer mechanism was modelled in a simplified manner. The axial transfer is given by the mass flow of the gas and its temperature. Inside the tube turbulent flow occurs. Therefore radial convective heat transfer was determined by estimating the turbulence intensity and calculating a radial velocity from this value. The heat transfer coefficient to the surrounding walls was determined by an equation given by Hausen [15]. There were no internal energy sources in the system. As no reactions should take place the gas concentrations were known and the absorption coefficients could be calculated using (4).

The criterion for validity was the known mean outlet temperature of 620 °C. The simulation result was 622 °C. The mean error of both values was larger than the difference between the results. Thus the model is supposed to work properly.

Calculation with Particles. The same heat exchanger was simulated with a gas coming from a partial coal gasification reactor. The gas concentrations were different to those of the gas operation. In addition to this particles were present in the radiating medium. As the solid loading was low, scattering phenomena were neglected.

466

No data on the absorption coefficient of the particles were available. Therefore a simulation was run in order to determine this parameter. As a simplification the absorption coefficient was taken to be independent of temperature. Considering the turbulence and mass flow pattern in the same way as for the first simulation, the outlet temperature was the criterion for the accuracy of the absorption coefficient. The given outlet temperature was 780 °C. With an estimated absorption parameter of k_a = 0.61 for the particle phase an outlet temperature of 778 °C was obtained. The value of 0.61 agrees with experimental data given by Biermann [6]. He analysed particles from pulverized coal combustion plants. The values for k_a varied from 0.2 to 0.41 in the temperature region from 900 to 1100 °C. In the present study temperatures in the radiating system were higher than 1100 °C. Therefore the obtained value for k_a and the temperature distribution were considered to be sufficiently exact.

Figures 4 and 5 show the axial and radial temperature profile.

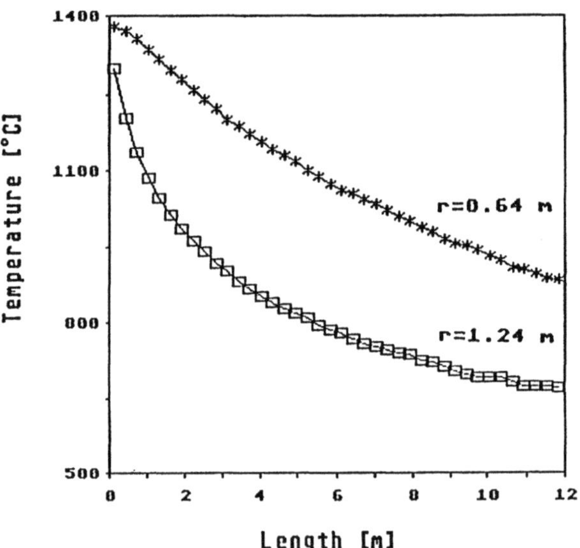

Fig. 4. Simulation result: axial temperature profile

The slope of the temperature gradient for r=1.24 m in Figure 4 shows that the largest amount of heat is transferred in the entrance section of the heat exchanger. This is confirmed by Figure 6 which shows the total heat flux to the surrounding wall. Near the axis the temperature decrease is almost linear.

Figure 5 shows that there is a region with almost no temperature decrease near the axis of the cylinder. Close to the wall the temperature decrease becomes

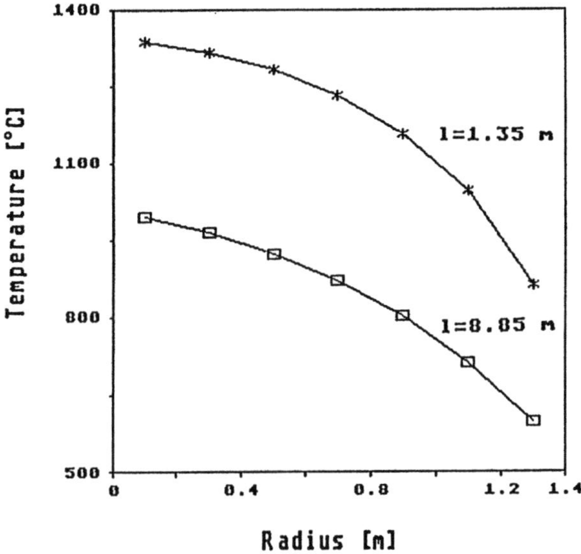

Fig. 5. Simulation result: radial temperature profile

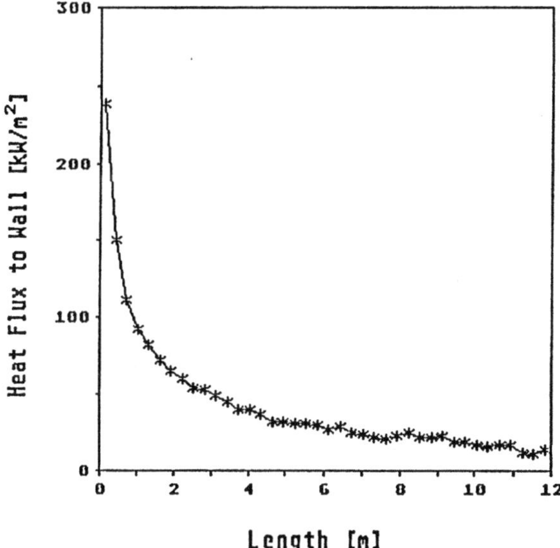

Fig. 6. Heat flux distribution

higher. The shape of the radial temperature profile does not vary towards the end of the tube. The reason for these effects is the high optical thickness of the medium. The use of eq. (5) to determine the optical thickness yields a value of 18 for a layer of the same size as the radius of the tube. Thus radiation heat transfer to the walls is only possible from segments close to the surrounding surfaces. Heat from the center of the cylinder has to be transferred mainly by convection, which results in the temperature graph shown in Figure 5.

A comparison between heat transfer by convection and by radiation shows that the latter amounts to 78% of the total energy transferred. In the end section of the heat exchanger the temperature gradient and temperature level are low. The amount of radiation heat transfer decreases and convection becomes more important. As convective heat transfer is very low in gaseous media this explains the slow reduction of the temperature towards the end of the tube.

In order to verify the preceding simulation and the estimated absorption coefficient the boundary conditions were changed. Results from the computer calculation and experimental data again showed good agreement.

PARAMETER STUDY

The main objective of this model was the simulation of radiation heat transfer with particulate matter as radiating medium. There is a lack of radiation parameters for substances like fly ash or soot which are often encountered in high temperature processes. Biermann [6] published radiation properties for different fly ashes from pulverized coal combustion plants. However, he did not observe any temperature influence. Recently Elsner [4] and Brummel [8] proved that there is a dependence of the absorption and scattering coefficients of small particles on temperature. Brummel applied Mie's theory [16] to calculate these parameters theoretically. Elsner carried out experiments on fly ashes from fluidized bed combustors. His data were used to simulate a radiation boiler placed in the freeboard region of a fluidized bed combustor. The parameters varied were solid loading, material and temperature dependence of the radiation parameters. The influence of scattering was analysed, too. In this paper only solid loading and the scattering influence are discussed. More detailed information is given in [12]. Boundary conditions were a length to diameter ratio of 4:1 and an inlet temperature of 600 °C. The gas concentrations were taken from a typical combustion calculation. As the parameter study should only reveal qualitative effects the length of the tube was set to 6 m in order to reduce calculation time. All other parameters were appropriately set to simulate fluidized bed conditions. Radiation properties were taken from Elsner's work.

Solid Loading. Figures 7 and 8 shows the axial and radial temperature profiles. The dependence on the solid loading is shown clearly. With high solid loading

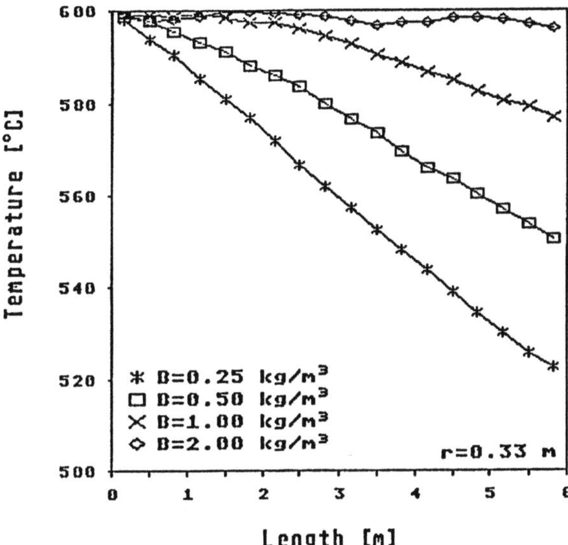

Fig. 7. Influence of solid loading — axial

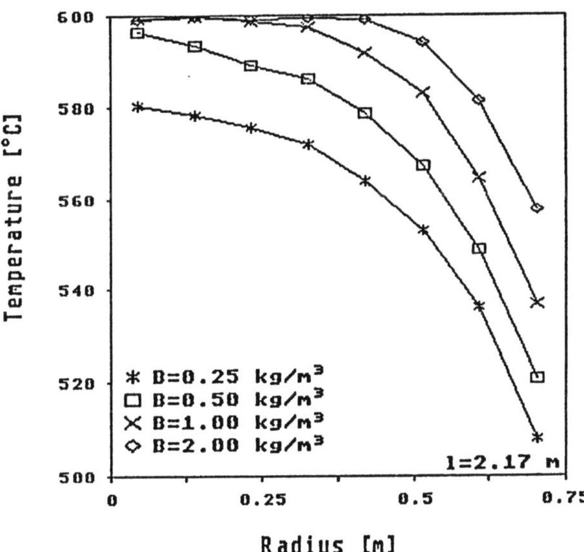

Fig. 8. Influence of solid loading — radial

an isothermal zone is built up near the axis (Figure 8). Figure 7 shows the decreasing heat transfer with increasing solid loading. Table 1 shows the percentage of radiation heat transfer on the total energy transferred. With high solid loadings the percentage of radiation heat transfer decreases though the

470

Table 1: Percentage of radiation heat transfer

Solid loading [kg/m^3]	q_w [MW]	q_{conv} [MW]	q_{rad} [%]
0.25	0.40	0.05	87.5
0.50	0.43	0.06	86.0
1.00	0.47	0.07	85.1
2.00	0.49	0.08	83.7

total heat transferred increases. The reason can be found in the high optical thickness which has been discussed in the chapter before.

Scattering. The effect of scattering is presented by Figures 9 and 10. Results are plotted for a fly ash with temperature dependent properties (*see* Table 2), a specific surface of 45 m^2/kg and a solid loading of 0.5 kg/m^3. Calculations were carried out considering no scattering and isotropic scattering. As the scattering coefficient is very low an effect due to scattering can hardly be seen. The temperature distributions do not differ very much. Table 2 gives results from Elsner's [4] work. He measured very low scattering parameters for fly ashes from fluidized bed combustors. Thus the calculation result can be explained. For industrial applications an effect of scattering may be neglected. Further investigations on this subject have to be carried out.

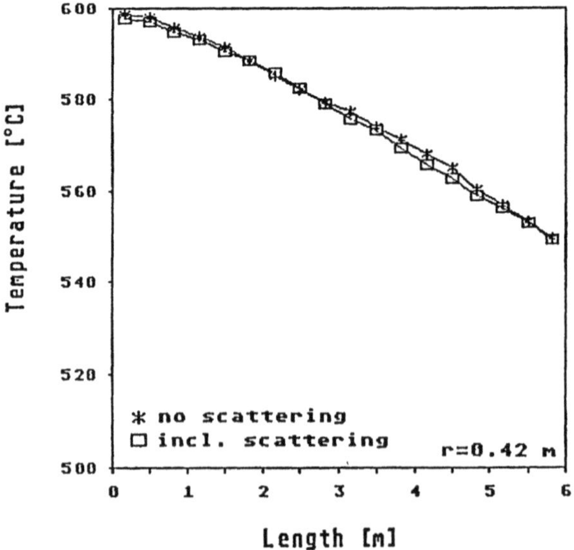

Fig. 9:   Influence of scattering — axial

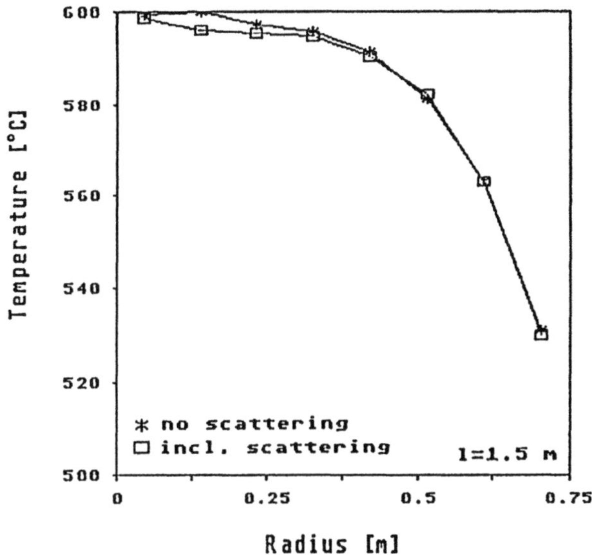

Fig. 10:   Influence of scattering — radial

Table 2. Radiation parameters of fly ash

T [°C]	400	425	450	475	500
k_a [-]	0.189	0.227	0.25	0.312	0.384
k_s [-]	$2.8 \cdot 10^{-4}$	$3.5 \cdot 10^{-4}$	$4 \cdot 10^{-4}$	$5.1 \cdot 10^{-4}$	$6.4 \cdot 10^{-4}$

Effects on the Design of Heat Exchangers. The parameter study provides information on the behaviour of a radiating system. From the effects discussed some directives to construct radiation heat exchangers can be derived. The diameter of the heat exchanger has to be as small as possible. If this is not feasible, additional surfaces have to be introduced in the system in order to minimize the optical thickness and thus to improve radiative heat transfer. Especially under conditions with high solid loadings scattering should not be neglected in order to give detailed information about the process. However, the effect is small and for industrial applications simulations may be carried out without the scattering mechanism.

SUMMARY

This paper describes a possibility to model radiation heat transfer in a radiation heat exchanger. The application of the Monte Carlo method avoids the solution of integro-differential equations. One disadvantage is the high number of calculation steps due to the simulation process, but the complicated physical phenomena can be simulated very easily. In order to give a detailed description of radiative transfer this procedure is an alternative to traditional simplifying calculation methods. It can be useful to design radiation heat exchangers to achieve high heat transfer rates. This increases the efficiency of industrial processes.

The model described considers gas and particle radiation including scattering. Convective heat transfer can be superposed. The model was verified by simulating the radiation boiler of a coal gasification plant. Good agreement between model and experiment was obtained. A parameter study shows the effect of solid loading and scattering due to particles.

In future the model will be combined with burner calculation models. The objective is to obtain a detailed description of combustion processes, especially for fluidized bed combustions.

REFERENCES

1. VDI-Wärmeatlas: Chapter K, 5th edition. VDI-Verlag, Düsseldorf (1988).

2. Hottel H.C.: Heat Transmission. McGraw-Hill, New York (1954).

3. Görner R., Dietz U.: Strahlungsaustauschrechnungen mit der Monte-Carlo Methode. Chem.-Ing.-Tech. 62 (1990) 23-33.

4. Elsner T., Köneke D., Weinspach P.-M.: Thermal Radiation of Gas/Solid Mixtures. Chem. Eng. Technol. 11 (1988) 237-243.

5. Tien C.L., Lee S.C.: Flame Radiation. Prog. Energy Combust. Sci. 8 (1982) 41-59.

6. Biermann P.: Wärmestrahlung staubhaltiger Gase in Dampfkesseln, Ph.D. Thesis, Univ. Stuttgart (1967).

7. Köneke, D.: Kühlung und Erhitzung von Gas-/Feststoff-Strömen in Wärmeaustauschern bei höheren Temperaturen. Ph.D. Thesis, Univ. Dortmund (1983).

8. Brummel H.-G., Kakaras E.: Wärmestrahlungsverhalten von Gas-Feststoffgemischen bei niedrigen, mittleren und hohen Staubbeladungen. Wärme- und Stoffübertragung 25 (1990) 129-140.

9. Howell J.R.: Monte Carlo Applications in Heat Transfer. Advances in Heat Transfer 5 (1968) 2-54.

10. Jakob M.: Heat Transfer, Vol. I, John Wiley & Sons, New York (1949).

11. Howell J.R., Perlmutter M.: Radiant Transfer Through a Gray Gas Between Concentric Cylinders Using Monte Carlo, J. Heat Transfer 86 (1964) 169-179.

12. Hessberg J.: Mathematisches Modell zur Modellierung von feststoffbeladenen Strahlungskühlern, Thesis, Univ. Dortmund (1990).

13. Steward F.R., Cannon P.: The Calculation of Radiative Heat Flux in a Cylindrical Furnace Using the Monte Carlo Method. Int. J. Heat Mass Transfer 14 (1971) 245-262.

14. Taniguchi H., Yang W., Kudo K., Hayasaka H., Fukuchi T., Nakamachi I.: Monte Carlo Method of Radiative Heat Transfer Analysis of General Gas-Particle Enclosures. Int. J. Numerical Methods Eng. 25 (1988) 581-592.

15. Hausen H.: Neue Gleichungen für die Wärmeübertragung bei freier oder erzwungener Strömung. Allgemeine Wärmetechnik 9 (1959) 75-79.

16. Mie G.: Beiträge zur Optik trüber Medien, speziell kolloidaler Metallösungen. Annalen der Physik 25 (1908) 377-445.

HEAT TRANSFER TO PIPES SUBMERGED
IN TURBULENT JET DIFFUSION FLAMES

J.E. Hustad
The Foundation for Scientific and Industrial
Research at the Norwegian Institute of Technology
and
O.K. Sonju
Division of Thermal Energy
The Norwegian Institute of Technology
7034 Trondheim, Norway

ABSTRACT

Experimental studies of total heat transfer to a pipe submerged in impinging jet flames are performed. The experiments show that the heat flux varies considerably depending on the position in the flames. The maximum heat flux occurs at the flame center-line at about the middle of the flame height. The obtained correlations show that the convective heat flux is dominant. The correlations make it possible to determine the heat flux to a pipe submerged in a flame as a function of the flame height and the nozzle exit jet velocity.

NOMENCLATURE

A_p - Probe surface area, m^2

A_{th} - Thermocouple bead area, m^2

c - Exit sonic velocity, m/s

C_p - Specific heat capacity of the probe, kJ/m^2K

d - Nozzle exit diameter, m

D - Average visible flame diameter, m

D_s - Pipe diameter, m

D_{th} - Thermocouple bead diameter, m

h - Heat transfer coefficient, W/m^2K

H - Total flame height, m

H* - Flame radiation height, $H-L_f$, m

L_f - Lift off distance, m

M - Exit Mach number, U_e/c

m_p - Probe weight, kg

Q - Total heat flux, kW/m^2

T_b - Average flame temperature, K

ΔT_p - Temperature increase in the probe, K

T_∞ - Ambient temperature, K

T_{th} - Thermocouple temperature, K

Tu	-	Turbulence intensity, %
Δt	-	Test time, s
U_e	-	Nozzle exit velocity, m/s
x	-	Radial distance from flame centerline, m
y	-	Vertical distance from nozzle exit, m
Nu	-	Nusselt number
Re	-	Reynolds number at the nozzle exit
Re_b	-	Reynolds number of bulk flow
Pr_b	-	Prandtl number of bulk flow
Pr_p	-	Prandtl number at the probe surface
Q_R	-	Radiation heat flux, kW/m^2
Q_c	-	Convective heat flux, KW/m^2
ε_{th}	-	Thermocouple emissivity
λ_f	-	Thermal conductivity of the gas, W/mK

INTRODUCTION

In the process industry accidental fuel-gas leaks can occur. Ignition of the leaking gas can result in structural elements of different sizes and shapes being submerged in high velocity impinging flames or jets causing high heat loadings and damage to the structure. Convective heat rates around cylinders in cross flow have been studied by many authors during the last 40 years. The early work of McAdams [1] treats heat transfer rates at low temperatures. Eckert and Soehngen [2] studied the distribution of the heat transfer coefficient around circular cylinders, as did Giedt [3] but at considerably larger Reynolds numbers. Churchill and Brier [4] and Douglas and Chruchill [5] studied local heat transfer to a cylinder in a nitrogen gas stream at temperatures up to approximately 1000 °C. Also, Kilham [6] studied total heat transfer from flames at 2000 °C to tubes with wall temperatures of 1100 °C to 1500 °C. His tubes were, however, oxide coated to study catalytic surface effects, and in addition, he used a very small flame with a rectangular discharge nozzle. Milson and Chigier [7] studied total heat fluxes from methane flames impinging on cold plates. Both premixed flames and diffusion flames were studied, but at relatively low nozzle exit Reynolds numbers (less than 36000) and with the plate positioned relatively near the nozzle exit where the heat flux at the centerline is rather low because of the cold flame core.

Since the heat flux depends on both the geometry of the structural member and the nozzle exit velocity, it is of great importance to study various geometries for a large range of fuel jet velocities. The experiments reported here are carried out with both methane and propane jet diffusion flames with circular nozzles using exit velocities in the range of 5-200 m/s for the propane flames and 10-125 m/s for the methane flames. Most of the measurements were made at the pipe front stagnation point and at the flame centerline for various vertical position above the nozzle exit. Several traverses across the flames in the horizontal direction have also been made. In addition, some experiments with the probe located at an angle of 90° and 180° from the pipe front stagnation point along the flame centerline were performed. To determine the convective part of the total heat flux, average velocity distribution along the flame centerline was measured with a pitot tube. The paper represents a continuation of the work published previously by authors [8, 9, 10] on flame size and flame radiation measurements for jet diffusion flames.

EXPERIMENTAL APPARATUS

The experimental set-up is shown in Figure 1. The gas storage consists of several gas bottles manifolded together. The gas flow is measured with standard flow meters.

Straight tubes with nozzle diameters of 5, 8.6, 10 and 40 mm were used. Tests with both methane and propane were performed. The flame/gas stream

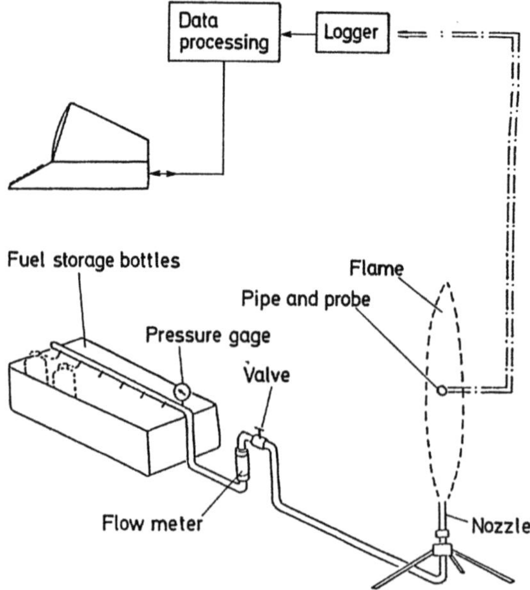

Fig. 1: Experimental set-up.

temperature is measured using a Cr/Al thermocouple located 10 mm from the probe surface. The thermocouple radiation loss is calculated from the equation:

$$\frac{Q}{A_{th}} = \varepsilon_{th} \cdot \sigma \, (T_{th}^4 - T_{\infty}^4) \tag{1}$$

where the thermocouple emissivity is set equal to 0.2. The convective term is given by:

$$\frac{Q}{A_{th}} = h \,\, (T_b - T_{th}) \tag{2}$$

The Nusselt number for the thermocouple bead is given by Ranz and Marshall [11]:

$$Nu = \frac{h \cdot D_{th}}{\lambda_f} = 2 + 0.6 \cdot Re^{0.5} \cdot Pr^{0.33} \tag{3}$$

The viscosity and the conductivity of the gas mixture can be approximated by pure air relations [12]. The heat conduction loss due to the temperature gradient in the vicinity of the thermocouple bead is neglected [13]. The average flame temperature can now be calculated by the equation:

$$T_b = T_{th} + \frac{\varepsilon_{th} \cdot \sigma \left(T_{th}^4 - T_{\infty}^4\right)}{h} \tag{4}$$

The average velocities within the flames are measured by a pitot tube. The total heat flux is measured using a special copper probe/sensor mounted at the pipe surface as shown in Figure 2.

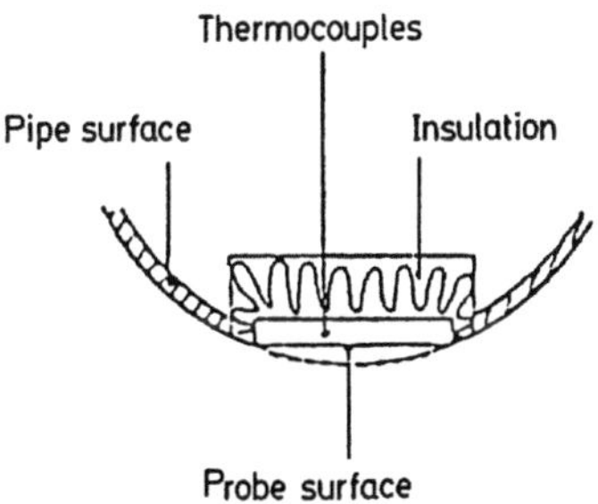

Fig. 2: Location of probe in the pipe

The probe is instrumented with four Cr/Al thermocouples to measure the copper temperature, T_p. The copper probe is 25 mm in diameter and 3 mm thick. The steel pipe is 50 mm in diameter and the probe is insulated from the pipe with

Alundum heat resistance insulation. The probe covers a total angle of 60 degrees, or 30 degrees on either side. The results thus are an average over this region. The heat flux measurements are made during the heating of the probe from 250 °C to about 350 °C. The duration of the measurement period is approximately 6 seconds. The total heat flux is calculated based on the measured accumulated heat in the copper probe as a function of time. The total heat flux consists of both a radiative and a convective part, and it is calculated by using a heat storage method given by the equation:

$$Q = \frac{m_p \cdot C_p \cdot \Delta T_p}{\Delta t \cdot A_p} \ (kW/m^2) \tag{5}$$

Determination of heat fluxes from transient temperature response measurements can be complicated. With our probe construction, however, the heat flux is very nearly one-dimensional in nature. The internal resistance is negligible, and the overall heat transfer process is controlled by the surface resistance. Thus the temperature in the probe is nearly uniform.

Schneider [14] presents many time-temperature charts covering a variety of body shapes for a wide range of Fourier and Biot numbers. The accuracy of the heat storage method was checked using Schneider's result. A time-temperature chart for low Biot number (negligible internal resistance) and high Fourier number was used. The comparisons were made near the nozzle exit where the radiation is small since Schneider did not include radiation in his results. The results of the comparisons were found to be satisfactory, in that a maximum difference of ±5% was observed. The heat transfer coefficient is calculated from the heat flux using the average temperature difference between the probe and the flame. In this way the total average heat flux and heat transfer coefficient to the steel pipe is obtained. The experimental results are discussed in the next section.

RESULTS AND DISCUSSIONS

Figure 3 shows average flame temperatures for a propane flame (34.3 kW) and a methane flame (37 kW) measured along the flame centerline. The data were taken at the same fuel mass flow through the nozzle. The data show a steeper increase from the flame front for the methane flame and also higher flame temperatures than in the propane flame as expected.

Figure 4 shows a typical heat flux profile measured along the centerline of a 150 kW (heat release) methane flame with a nozzle exit diameter of 10 mm and a nozzle exit jet velocity of 50 m/s.

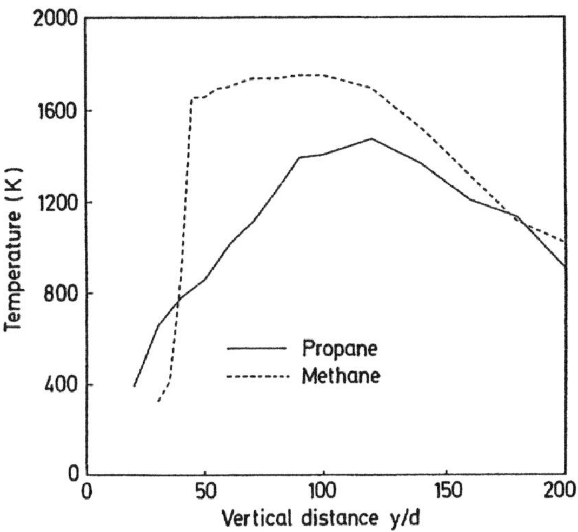

Fig. 3: Average flame temperatures along the flame centerline for a propane (34.3 kW) and a methane (37 kW) flame.

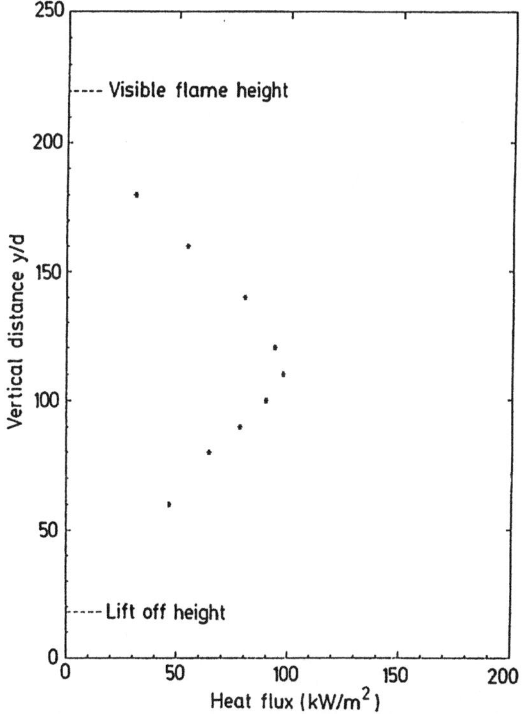

Fig. 4: Heat flux profile along the flame centerline for a methane flame

480

As seen, the heat flux increases with increasing vertical position approximately halfway up the visible flame height and then it decreases again. The data exhibit an almost symmetric profile around the middle of the visible flame height. The maximum centerline total heat flux for this flame was measured to be approximately 100 kW/m^2 and located 110 nozzle diameters above the nozzle. For the same flame, horizontal traverses across the flame diameter at distances of 50 and 90 nozzle diameters above the nozzle exit are shown in Figure 5. The data show a symmetric profile around the flame centerline as one would expect.

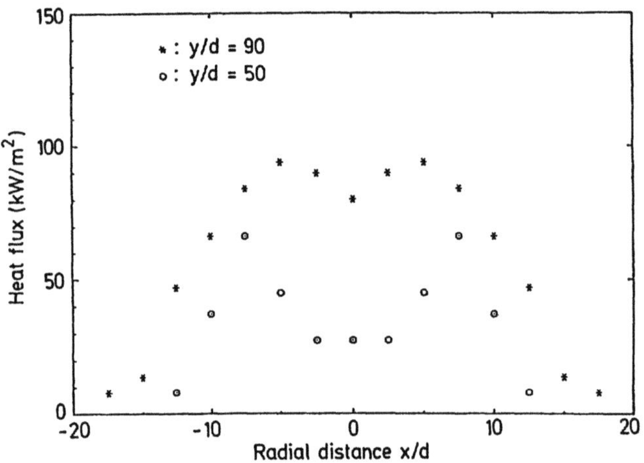

Fig. 5: Heat flux profiles across the flame at a distance of 50 and 90 nozzle diameters above the nozzle exit for a methane flame.

The maximum heat flux is located approximately 8 and 5 nozzle diameters away from the flame centerline for y/d=50 and y/d=90, respectively. This indicates a cold zone in the center of the flame. The difference between the centerline value and the maximum value is larger closer to the nozzle exit because of the lack of combustion near the centerline. The cold zone in the center of diffusion flames (in the lower part of the flame) is caused by entrainment of air giving lower temperatures and high oxygen concentrations [9].

Milson and Chigier [7] obtained heat flux values from a methane diffusion flame impinging on a flat plate. The flow field in this case is quite different from around a cylinder, but it would nevertheless be interesting to compare their results with the present results obtained at the pipe front stagnation point. The plate temperature levels used by Milson and Chigier were different from the present probe surface temperatures, therefore the comparison is made on the basis of the heat transfer coefficients and not the heat flux values. At a nozzle exit Reynolds number of 35300 and at y/d=16, Milson and Chigier obtained a heat flux at the centerline of approximately 4 kW/m^2. Assuming the presently measured flame temperatures, a calculated heat transfer coefficient of 45 W/m^{2}K is obtained from their experiments at the stagnation point. Calculation of the heat transfer coefficient from the present measurements of heat flux at the same location in the flame and the same nozzle exit Reynolds number gives the same result, namely 45 W/m^{2}K. A comparison at nozzle exit Reynolds number of 7700 and at y/d=16 gives a heat transfer coefficient from Milson and Chigier's results of 91 W/m^{2}K, and from the present measurements of 93 W/m^{2}K. The lower heat transfer coefficient at higher Reynolds number is caused by the increase of air entrainment and liftoff height and thus the increase of the height of the cold center core.

In Figure 6 heat flux profiles are shown for a 740 kW propane flame at a nozzle exit velocity of 100 m/s.

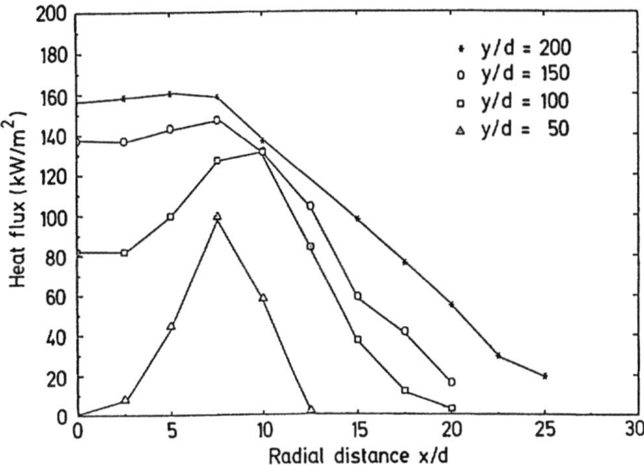

Fig. 6: Heat flux profiles at various distances from nozzle exit for a propane flame.

The profiles obtained are seen to be similar to the results for the methane flame, showing a cold center core up to the middle of the flame height (y/d=200) where the maximum heat flux is measured. The heat flux profile is relatively flat at y/d=200. At locations nearer the nozzle exit, the maximum heat flux occurs off the centerline. For instance at y/d=50, the maximum heat flux occurs at approximately 10 nozzle diameters away from the centerline. As was seen from the methane flame, the horizontal profiles were symmetric around the flame centerline, so here only one side of the flame centerline is shown. The data show larger heat flux values for propane than methane at the same nozzle exit velocities. This is caused by higher bulk velocities for propane flames than methane flames at the same nozzle exit velocity as will be seen from Figure 9.

In Figure 7 the maximum total heat flux values are plotted as a function of the nozzle Mach number both for methane and propane flames. The total heat flux values plotted are measured at the middle of the flame height where they are at their maximum. The figure shows that the total heat flux increases with increasing nozzle exit velocity. The heat flux increases rapidly up to a Mach number of approximately 0.3; the further increase then levels off as the velocity increases and reaches a value for the propane flames of about 200 kW/m2 at a Mach number of 0.8. It can be seen from the figure that the total heat flux for the propane flames increases five times as the exit velocity increases from 5 m/s (M=0.02) to 200 m/s (M=0.8). The same trends are observed for the methane flames, but these flames are only tested up to a Mach number of approximately 0.3. The total heat flux has a radiative and convective component. An estimate of these components has been made, and their sum has been compared with the measured total heat flux as discussed below.

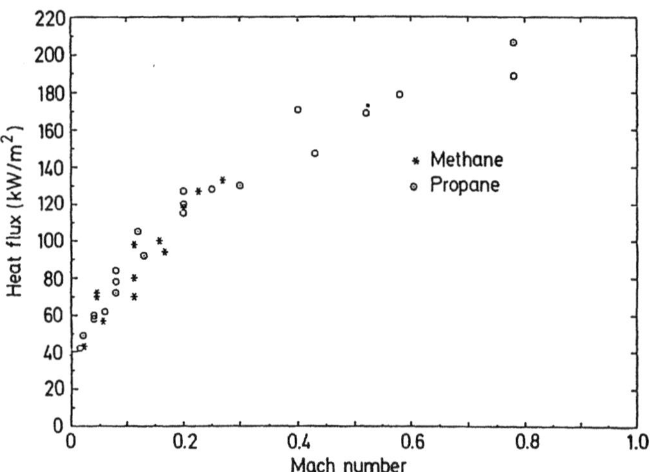

Fig. 7:   Maximum heat flux as a function of exit Mach number for methane and propane flames.

RADIATIVE HEAT FLUX

The authors have proposed a model [9,10] for predicting flame radiation by approximating the flame as a radiating cylinder with a constant surface radiation and then calculating the radiation to targets outside the flame in terms of shape factors. The proposed model is in agreement with the measured distributions. As a first approximation it is here assumed that the average surface radiation flux is approximately constant from the flame surface towards the flame centerline at the middle of the flame height. This can be argued from the fact that the flame is relatively thick compared to the submerged pipe and the measured average temperature profile is relatively flat, the flame is thus assumed to be optically thick. The equation for the average radiation surface flux is [9]:

$$Q_R = \; <\varepsilon\sigma T^4> \; = 10 + 7.8 \cdot H^* \; (kW/m^2) \tag{6}$$

for propane flames. The average flame surface radiation flux as a function of flame radiation height is shown in Figure 8.

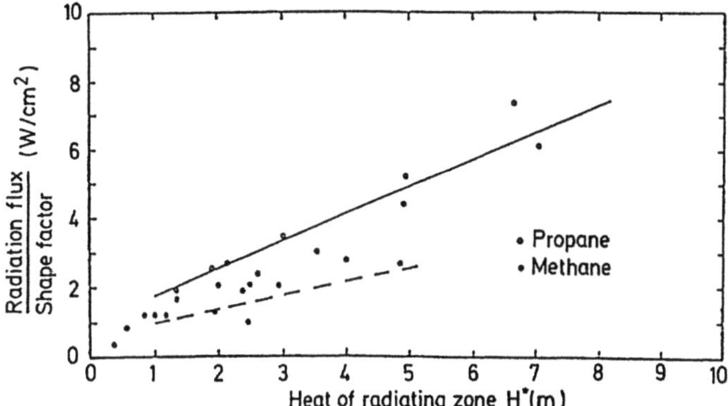

Fig. 8: Average surface flame radiation as a function of flame radiation height.

CONVECTIVE HEAT FLUX

The convective heat flux is given by the difference between the total heat flux and the radiative heat flux:

$$Q_c = Q-Q_R = h \; (T_b - T_p) = \frac{Nu(T_b-T_p)\lambda_f}{D_s} \tag{7}$$

To estimate the convective heat transfer, the average velocity at various locations within the flame is measured and typical profiles for a methane flame with a nozzle exit velocity of 50 m/s and two propane flames with nozzle exit velocities of 50 m/s and 100 m/s are shown in Figure 9.

The figure shows that the velocity along the flame centerline decreases rapidly for all flames. The figure shows further that the velocity in propane flames are higher than in methane flames at the same exit velocity which is caused by the higher momentum (density) of propane flames. From the measured average

484

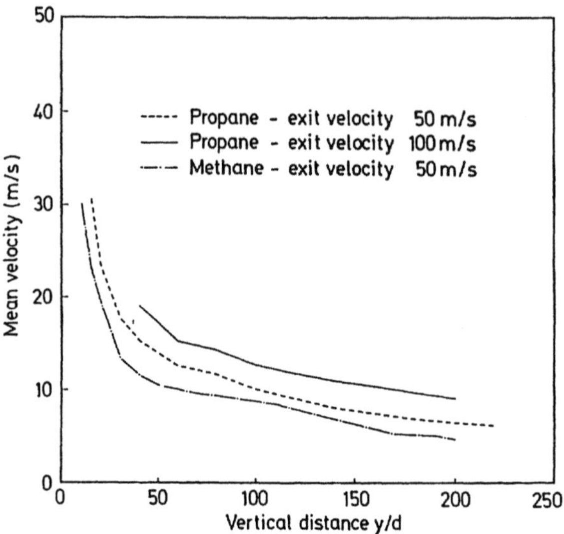

Fig. 9: Average velocities along the flame centerline as a function of distance from nozzle exit for a methane flame and two propane flames.

velocities and temperature the convective heat transfer along the flame centerline can be calculated. The results are shown in Figure 10 for the front stagnation point Nusselt number and show good agreement with the experimental results of Krall and Eckert [15] and Soehngen [2].

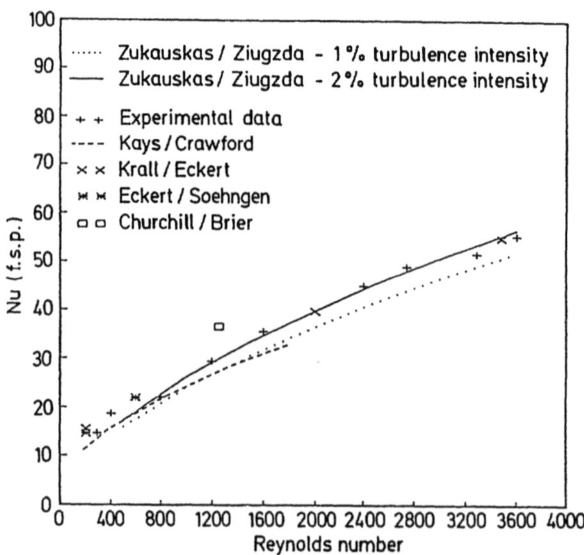

Fig. 10: Front stagnation point Nusselt number as a function of Reynolds number

The results of Churchill and Brier [4] give a somewhat higher value. Zukauskas and Ziugzda [16] proposed the following equation for calculating the Nusselt number to a cylinder in cross flow at the front stagnation point:

$$Nu = 0.41 \cdot Re_b^{0.6} \cdot Pr_b^{0.35} \cdot Tu^{0.15} \left(\frac{Pr_b}{Pr_p}\right)^{0.25}$$

(8)

A comparison of the theoretical and experimental values of the Nusselt number as a function of the Reynolds number is also shown in Figure 10, and agreement is evident assuming a freestream turbulence intensity of 2%. The result from the equation of Kays and Crawford [17] assuming laminar flow, constant freestream velocity and constant wall temperature is shown in Figure 10 to give lower estimation of the front stagnation point Nusselt number than our results. Their equations compare well with Zukauskas and Ziugzda [16] at a turbulence intensity of 1%. The results of Sarma and Sukhatme [18] in the Reynolds number range from 1200 to 3600 also give good agreement with our results. The results further show that the convective heat transfer in a diffusion flame is approximately the same as in other convective heat transfer situations. A comparison between the estimated and measured values of the total heat flux can now be made. The theoretical value of the total heat flux is calculated as the sum of the convective and radiative heat fluxes. The comparison is shown in Figure 11 as a function of exit Mach number for propane flames. In Figure 11 the total

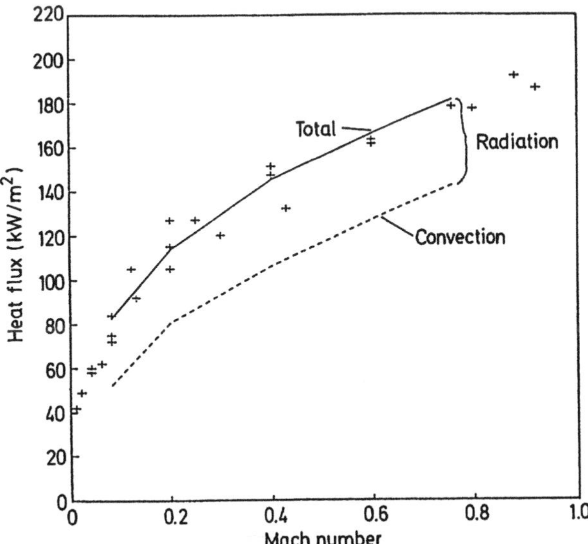

Fig. 11: Radiative and convective part of the total maximum heat flux as a function of exit Mach number in a propane flame.

maximum centerline heat flux (at the middle of the flame height) is compared to the calculated values and shows definite agreement considering the simplifications and assumptions made (especially in the radiation calculations).

The local total heat fluxes around the pipe surface have also been measured. In Figure 12 measurements with the probe turned 90° from the front stagnation point are compared with the value at the front stagnation point as a function of radial distance from the flame centerline. The results are shown for a propane flame having a nozzle discharge of 10 mm and at a nozzle exit velocity of 100 m/s. The measurements show that the total heat flux at 90° from the stagnation point, but still in the flame centerline, is approximately 40% of the total heat flux at the stagnation point. The difference between these two heat fluxes decreases from the flame centerline towards the outer surface of the flame.

Figure 13 shows the variations of total heat flux as a function of the angle from the front stagnation point at nozzle exit jet velocities of 100 and 180 m/s. The general trend shows a decrease in total heat flux up to an angle of 90° to 135°, but then the heat flux increases again. The maximum total heat flux is observed at the front stagnation point. Our results show the same general trends as obtained by Eckert and Soehngen [2] and Krall and Eckert [15] for convective heat transfer around a cylinder. Their results are also confirmed by detailed numerical calculation of convective heat transfer to cylinders in cross flow [19]. However, at considerably higher Reynolds numbers when the boundary layer becomes turbulent, the maximum local heat flux seems to occur behind the flow separation point of the cylinder [20].

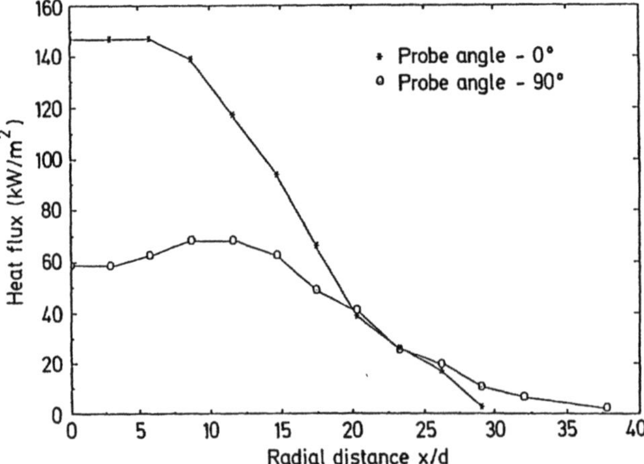

Fig. 12: Heat flux values at an angle of 90° from stagnation point compared to the values at the stagnation point as a function of radial distances in a propane flame.

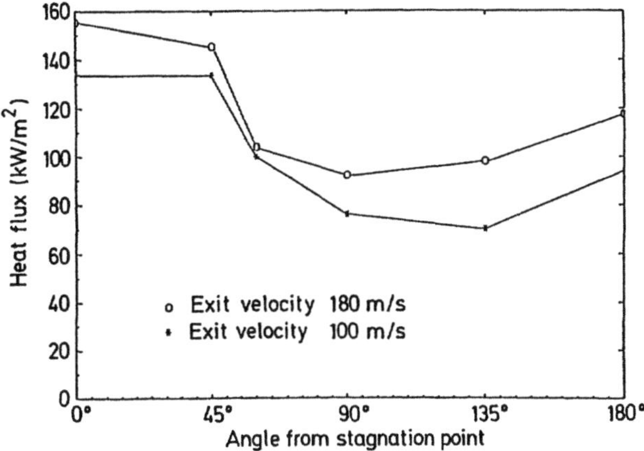

Fig. 13:   Heat flux distribution around the cylinder in a propane flame.

From the total heat flux distribution, an average value can be calculated to be 116 kW/m^2 for the 180 m/s case and 96.5 kW/m^2 for the 100 m/s case. Assuming uniform distribution of the radiation heat flux to the pipe, the average convective heat flux and thus the average Nusselt number for the pipe circumference can be calculated. The results are shown in Figure 14 together with the results of various literature correlations:

$$\overline{Nu} = 0.23 \cdot Re^{0.6} \cdot Pr^{0.35} \cdot Tu^{0.15} \left(\frac{Pr_b}{Pr_p}\right)^{0.25} \tag{9}$$

(Zukausk. and Ziug.)

$$\overline{Nu} = \left(0.42 + 0.48 \cdot Re^{0.5}\right) \cdot Pr^{0.35} \cdot \left(\frac{Pr_b}{Pr_p}\right)^{0.25} \tag{10}$$

(Krall and Eck.)

$$\overline{Nu} = 0.62 \cdot Re^{0.505} \cdot Pr^{0.35} \cdot \left(\frac{Pr_b}{Pr_p}\right)^{0.25} \tag{11}$$

(Sarma and Sukhatme)

$$\overline{Nu} = 0.6 \cdot Re^{0.5} \cdot Pr^{0.33} \cdot \left(\frac{Pr_b}{Pr_p}\right)^{0.12} \tag{12}$$

(Churchill and Brier)

The Prandtl number terms are added by the authors, except for the equation of Churchill and Brier [4]. Our experimental data fall within the variations of the

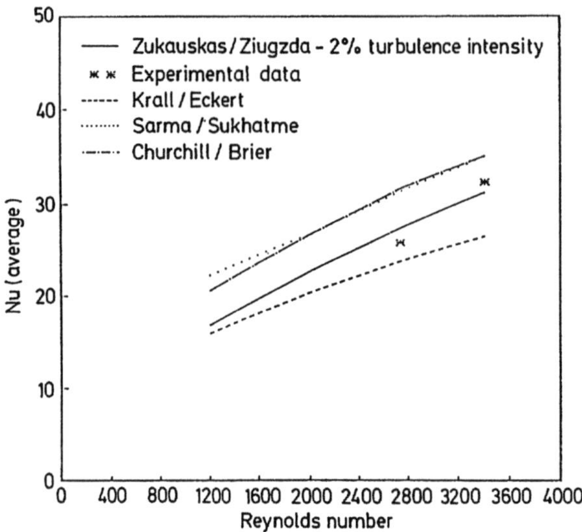

Fig. 14:   Average Nusselt number as a function of Reynolds number.

various model predictions. The best comparison is achieved by using the equation of Zukauskas and Ziugzda [16] including a turbulence intensity of 2%.

SUMMARY AND CONCLUSIONS

Rather extensive measurements of the total heat flux from turbulent jet diffusion flames with methane and propane impinging into a submerged 50 mm diameter pipe have been performed. The total heat flux along the flame centerline is found to be almost symmetrical around the middle of the visible flame height. The maximum heat flux with the probe oriented in the downward stagnation point is found to be at the flame centerline at about the middle of the visible flame height. Further, the maximum total heat flux is very sensitive to the nozzle exit Mach number. For the propane flames, a maximum heat flux of 200 kW/m^2 is observed for a nozzle exit Mach number of 0.8. For the methane flames, heat flux values up to 125 kW/m^2 are measured at exit Mach number of approximately 0.3. Approximate methods for calculating both the convective and the radiative part of the total heat flux for propane flames are proposed. From these calculations, it can be seen that the convective part dominates and represents up to 80% of the total heat flux. This conclusion is in general agreement with results presented by Kilham [6]. The model for calculating the radiative part of the total heat flux must be used carefully and is not valid over the whole flame area. Data on methane flames were limited to Mach numbers less than 0.3 so these data are not separated into a convective and a radiative part. The heat flux distribution around the cylinder shows that the maximum total heat flux is located at the downward stagnation point in the Reynolds number range tested.

The presented experimental results give guidelines for determining the total maximum heat flux from turbulent jet diffusion flames with methane and propane impinging into submerged pipes for various nozzle exit Mach numbers. Data are obtained for flame heights up to 7 m and exit Mach numbers up to 0.8 for propane flames, and flame heights up to 2.5 m and nozzle exit Mach number up to 0.3 for methane flames.

ACKNOWLEDGEMENTS

The authors gratefully acknowledge the important contribution of Stein Storli and Petter Lien, students at the Norwegian Institute of Technology, especially in the experimental work. This work was supported in part by the Royal Norwegian Council for Scientific and Industrial Research.

REFERENCES

1. McAdams W.H.: Heat Transmission. 3 ed., McGraw-Hill Book Company, Inc., New York (1954).

2. Eckert E.R.C., Soehngen E.: Distribution of Heat Transfer Coefficients Around Circular Cylinders in Crossflow at Reynolds Numbers from 20 to 500. Trans. ASME 74 (1952) 343.

3. Giedt W.H.: Trans. Am. Soc. Engrs. 71 (1949) 375.

4. Churchill S.W., Brier J.C.: Chem. Eng. Progr. Symp. Series 17 (1955) 5.

5. Douglas W.J., Churchill S.W.: Chem. Eng. Progr. Symp. Series 18, 52 (1956).

6. Kilham J.K.: Third Symposium on Combustion Flame and Explosion Phenomena. Williams and Wilkins, Baltimore (1949) 733.

7. Milson A., Chigier N.A.: Studies of Methane and Methane-Air Flames Impinging on a Cold Plate. Comb. and Flame 21 (1973) 295-305.

8. Schuller R.B., Nylund J., Sonju O.K., Hustad J.: Effect of Nozzle Geometry on Burning Subsonic Hydrocarbon Jets. Fire Dynamics and Heat Transfer 25 (1983) 33-36.

9. Sonju O.K., Hustad J.: An Experimental Study of Turbulent Jet Diffusion Flames. Dynamics of Flames and React. Systems. AIAA Progress in Astronautics and Aeronautics, New York, 95 (1984) 320-339.

10. Hustad J.E., Sonju O.K.: Radiation and Size Scaling of Large Gas and Gas/Oil Diffusion Flames. Dynamics of Reactive Systems, Part I: Flames and Configurations. AIAA Progress in Astronautics and Aeronautics, New York, 105 (1986) 365-387.

11. Ranz W.E., Marshall W.R. Jr.: Chem. Eng. Progr. 48, 3 (1962) 141-146.

12. Rohsenow W.M.: Handbook of Heat Transfer. McGraw-Hill Inc. (1973).

13. Sislian J.P., Jiang L.Y., Cusworth R.A.: Laser Doppler Velocity Investigation of the Turbulence Structure of Axisymmetric Diffusion Flames. Prog. Energy and Combustion Sci. 14 (1988) 99-146.

14. Schneider P.J.: Temperature Response Charts. John Wiley and Sons Inc., New York (1963).

15. Krall K.M., Eckert E.R.G.: Local Heat Transfer Around a Cylinder at Low Reynolds Number. J. Heat Transfer 95 (1973) 273-275.

16. Zukauskas A., Ziugzda J.: Heat Transfer of Crossflow. Hemisphere, Washington D.C. (1985) 88.

17. Kays W.M., Crawford M.E.: Convective Heat and Mass Transfer. McGraw-Hill Inc. (1980) 141 and 285.

18. Sarma T.S., Sukhatme S.P.: Local Heat Transfer from a Horizontal Cylinder to Air in Cross Flow. Int. J. Heat Mass Transfer, 20 (1977) 51-56.

19. Chun W., Boehm R.F.: Calculation of Forced Flow and Heat Transfer Around a Cylinder in Crossflow. Numerical Heat Transfer, 15 (1989) 101-122.

20. Eckert E.R.G., Drake R.M.: Analysis of Heat and Mass Transfer. McGraw-Hill Kogakusha Ltd., Japan (1972) 407.

ιot separated into a convective and a radiative part. The heat flux distribution around the cylinder shows that the maximum total heat flux is located at the downward stagnation point in the Reynolds number range tested.

The presented experimental results give guidelines for determining the total maximum heat flux from turbulent jet diffusion flames with methane and propane impinging into submerged pipes for various nozzle exit Mach numbers. Data are obtained for flame heights up to 7 m and exit Mach numbers up to 0.8 for propane flames, and flame heights up to 2.5 m and nozzle exit Mach number up to 0.3 for methane flames.

ACKNOWLEDGEMENTS

The authors gratefully acknowledge the important contribution of Stein Storli and Petter Lien, students at the Norwegian Institute of Technology, especially in the experimental work. This work was supported in part by the Royal Norwegian Council for Scientific and Industrial Research.

REFERENCES

1.   McAdams W.H.: Heat Transmission. 3 ed., McGraw-Hill Book Company, Inc., New York (1954).

2.   Eckert E.R.C., Soehngen E.: Distribution of Heat Transfer Coefficients Around Circular Cylinders in Crossflow at Reynolds Numbers from 20 to 500. Trans. ASME 74 (1952) 343.

3.   Giedt W.H.: Trans. Am. Soc. Engrs. 71 (1949) 375.

4.   Churchill S.W., Brier J.C.: Chem. Eng. Progr. Symp. Series 17 (1955) 5.

5.   Douglas W.J., Churchill S.W.: Chem. Eng. Progr. Symp. Series 18, 52 (1956).

6.   Kilham J.K.: Third Symposium on Combustion Flame and Explosion Phenomena. Williams and Wilkins, Baltimore (1949) 733.

7.   Milson A., Chigier N.A.: Studies of Methane and Methane-Air Flames Impinging on a Cold Plate. Comb. and Flame 21 (1973) 295-305.

8.   Schuller R.B., Nylund J., Sonju O.K., Hustad J.: Effect of Nozzle Geometry on Burning Subsonic Hydrocarbon Jets. Fire Dynamics and Heat Transfer 25 (1983) 33-36.

- in the inlet section the temperature of 2800-3000 K, the pressure of 0.15-0.5 MPa, the velocity of 700-1000 m/s,
- in the outlet section the temperature of 2300-2500 K, the pressure of 0.1 MPa, the velocity of 600-800 m/s.

The parameters of plasma cause the occurrence of intensive heat transfer to the walls by convection and radiation, both in the combustion chamber and the channel of the MHD generator.

DETERMINATION OF COMPOSITION OF PLASMA GENERATED IN THE BURNING PROCESS OF CONVENTIONAL FUELS

The elaboration of a method to determine the composition of plasma is the basic problem, when investigating the thermodynamic and electric properties of plasma, obtained in the burning process of burning conventional fuels and composed of molecules, atoms, ions and electrons. All the thermodynamic functions and physical properties of multicomponent plasma depend on physical properties of its particular components and their participation in the composition of plasma.

In the state of chemical equilibrium, the composition of plasma is a complex nonlinear function of temperature. This is brought about by the invertibility of chemical reactions at high temperatures.

The elaboration of a mathematical model of physical and chemical processes, proceeding in plasma, is the basis to define the composition of plasma, generated in the process of conventional fuel burning. This model is reduced to the system of algebraic nonlinear equations, the number of which is defined by the number of assumed components of plasma plus its temperature: The number and kind of components occurring in plasma is established on the basis of the analysis of chemical reactions, proceeding in it. Therefore, equations, defining the course of chemical reactions in plasma, are the fundamental equations of the model. Balance equations of elements composing fuels, the oxidizer and ionizing seed, the balance equation of electric charges (neutrality of exterior plasma), as well as the balance equation of the system energy, are balancing equations. The equation of the sum of partial pressures, determining the total pressure of the state of equilibrium, is an additional equation.

As result of carried out studies and preliminary investigations, the occurrence of the following components in the tested plasma in the gaseous phase has been assumed: K_2CO_3, CH_4, K_2O, KOH, CO_2, H_2O, N_2O, NO_2, SO_2, H_2S, KO, CO, CN, OH, NO, SO, C_2, H_2, O_2, S_2, K_2, C, H, O, N, S, K, Ar^+, NO_2^-, KO^- OH^-, CN^-, C_2^-, O_2^-, O^-, e, and therefore, 38 components generated from 7 elements and electrons. That is why, a model composed of 38 equations should be formulated, in which there are 30 equations describing the course of chemical reactions and 8 balance equations of elements and electric charges. The balance equation of the system energy will be an additional (39) equation completing the system, on satisfying of which the adiabatic temperature of combustion (of generating plasma) will be determined.

Equilibrium constants of chemical reactions have been assumed in the paper as basis of the description of the course of these reactions in plasma. It has been also assumed that all the tested chemical reactions proceed in the gaseous phase and are bidirectional reactions. They can be divided into two basic groups:

a) reactions of thermal dissociation of polyatomic combustion products and ionizing seed,
b) reactions of thermal ionization and electron affinity of some atoms and biatomic molecules.

HEAT TRANSFER IN THE MHD GENERATOR

The extended Newton equation expresses the heat flux, transferred by plasma to the walls of the MHD generator by convection and radiation

$$q = \alpha_t \, \Delta_t = (\alpha_c + \alpha_r) \, \Delta t \tag{1}$$

where:

α_t - total heat transfer coefficient which is the sum of the heat·transfer coefficient by means of convection and of the heat transfer coefficient by means of radiation,

α_c - convection heat transfer coefficient,

α_r - radiation heat transfer coefficient,

Δt - difference in temperatures of media between which heat transfer occurs.

CONVECTION HEAT TRANSFER

The formulation of relations describing convection heat transfer coefficients in the function of physical properties and thermodynamic parameters of the medium (plasma), as well as geometrical dimensions of the system, is the basic problem when solving problems connected with heat transfer. The determination of the dimensionless equation of convection heat exchange was the basis for defining relations describing the convection heat transfer coefficient. This equation was determined on the basis of the three-dimensional model of the developed turbulent flow, taking into account the formation of the flow in the initial part of the channel.

The range of values Re and Pr has been established on the basis of the analysis of thermodynamic parameters variation and of physical properties of plasma in the MHD generators of the open cycle.

On assuming this, the following form of the dimensionless equation was obtained

$$N_u = 0.021 \, Re^{0.8} \, Pr^{0.356} \tag{2}$$

On utilizing the definition of dimensionless numbers Re and Pr, the following relation was obtained for the convection heat-transfer coefficient from plasma to the walls of the MHD generator channel.

$$\alpha_c = 0.021 \left(\frac{v}{\nu}\right)^{0.8} \frac{\eta \, c_p}{D_h^{0.2}} \lambda^{0.644} \tag{3}$$

where:

v   - mean velocity of plasma in the channel,
ν   - kinematic coefficient of plasma viscosity
η   - absolute viscosity of plasma,
C_p - specific heat of plasma,
D_h - hydraulic diameter of the channel,
λ   - thermal conductivity of plasma.

In order to take into account the variations of plasma properties in the cross-section of the channel, the values of plasma transport parameters were determined for the temperature of plasma in the channel axis and for the temperature of plasma close to the walls.

RADIATION HEAT TRANSFER

High temperature of plasma (2400-3200 K) as well as a large number of plasma components cause that the spectrum of emitted radiation is very complex. The following elements can be distinguished in it:

- continuous background in the visible and ultraviolet range, connected with the radiation of ions and free electrons,
- rotational and vibrational - rotational bands in the infrared range, connected with the radiation of biatomic and polyatomic molecules,
- spectral lines of electronic radiation of monoatomic components as well as of biatomic and polyatomic molecules.

It results from the distribution of energy in the spectrum, emitted by the black or grey body at the temperature corresponding to the conditions of the MHD generator operation, that 90-95% of this energy falls to the infrared range.

In connection with the above mentioned, on analysing radiation heat transfer to the walls of the combustion chamber and to the channel of the MHD generator on account of the decisive role of rotational and vibrational - rotational bands, only radiation of molecules entering into the composition of plasma combustion products has been assumed for the considerations.

Heat flux, transmitted by plasma radiation to the walls of the combustion chamber and to the channel of the MHD generator, on assuming that the radiating gas is isothermal and the walls are black, can be described by the relation

$$q_r = \sigma_0 \left[\varepsilon_p T_p^4 - \alpha_{p(w)} T_w^4 \right] \tag{4}$$

where

σ_0 -   radiation constant of Stefan-Boltzman,

ε_p -   emissivity of radiating plasma volume, which is the function of chemical composition, pressure and temperature of plasma as well as of the mean path of radiation,

$\alpha_p(w)$ -   absorption capacity of plasma volume for the radiation emitted by the walls of the combustion chamber or of the channel of the MHD generator at the temperature of T_w,

T_p -   temperature of plasma,

T_w -   temperature of the walls.

Radiation heat transfer coefficient, can be defined as follows

$$\alpha_r = \frac{\sigma_0 \left[\varepsilon_p T_p^4 - \alpha_{p(w)} T_w^4 \right]}{T_p - T_w} \tag{5}$$

The mean value of emissivity of radiating plasma volume for the given temperature can be described by the following relation:

$$\varepsilon_p = \frac{\int_{\lambda=0}^{\infty} \varepsilon_{\lambda,b,p}\left[1 - \exp(-a_\lambda L_d)\right] d\lambda}{\sigma_o T_p^4} \tag{6}$$

where

$e_{\lambda,b,p}$ - monochromatic density of black body radiation at the temperature of plasma

a_λ - monochromatic absorption coefficient of plasma,

L_e - mean path of radiation in plasma.

Emissivity of plasma ε_p as well as absorption capacity of plasma $\alpha_{p(w)}$ have been determined on assuming that H_2O and CO_2 are the only components of plasma emitting and absorbing radiation, and therefore

$$\varepsilon_p = C_{CO_2}\varepsilon_{CO_2} + C_{H_2O}\varepsilon_{H_2O} - \Delta\varepsilon \tag{7}$$

where

$\varepsilon_{CO_2}, \varepsilon_{H_2O}$ - emissivities of CO_2 and H_2O according to their partial pressures, the mean path of radiation (L_e) and to the temperature of plasma,

C_{CO_2}, C_{H_2O} - correction factors which take into account the partial pressure of CO_2 and H_2O in plasma as well as the total pressure of plasma,

$\Delta\varepsilon$ - correction which takes into account partial coincidence of emission bands of CO_2 and H_2O.

Emissivities of CO_2 and H_2O as functions of partial pressures CO_2 and H_2O and the mean path of radiation (L_e) and the plasma temperature are shown in Figures 1 and 2 [1,2]. Correction factors C_{CO_2} and C_{H_2O} as function of the partial pressures CO_2 and H_2O and the mean path of radiation (L_e) and total pressure of plasma are shown in Figures 3 and 4 [1,2]. Correction $\Delta\varepsilon$ as a function of partial pressures CO_2 and H_2O and the mean path of radiation (L_e) is shown in Figure 5 [1,2].

Absorption capacity of plasma layer $\alpha_{p(w)}$ has been determined from relation

$$\alpha_{p(w)} = C_{CO_2}\varepsilon'_{CO_2}\left(\frac{T_p}{T_w}\right)^{0.65} + C_{H_2O}\varepsilon'_{H_2O}\left(\frac{T_p}{T_w}\right)^{0.45} - \Delta\varepsilon' \tag{8}$$

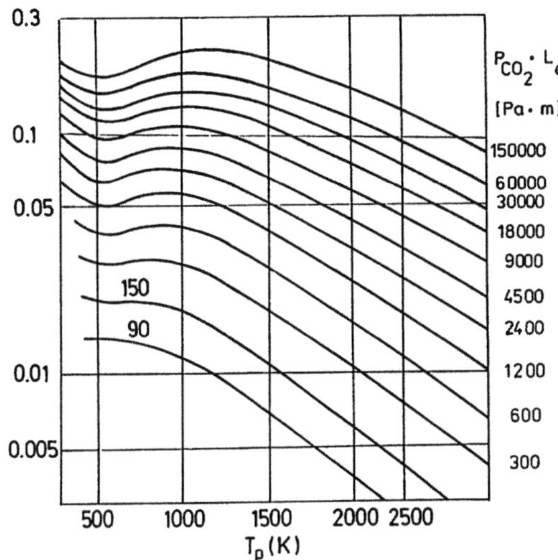

Fig. 1: Emissivity of carbon dioxide (ε_{CO_2}) as a function of its partial pressure (p_{CO_2}) and mean path of radiation (L_e) and plasma temperature (T_p).

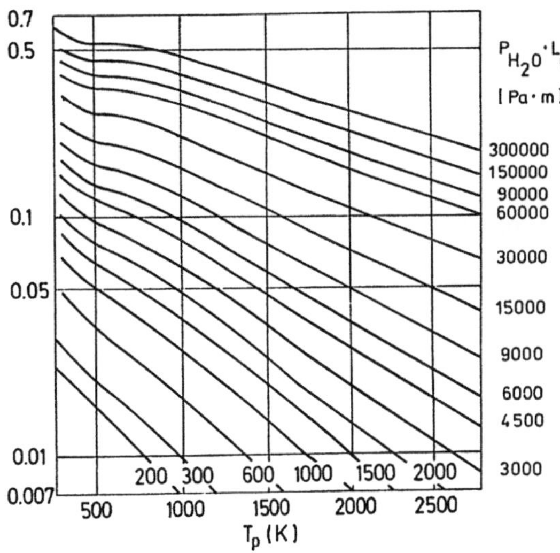

Fig. 2: Emissivity of water vapor as a function of its partial pressure (p_{H_2O}) and mean path of radiation (L_e) and plasma temperature (T_p).

where

$\varepsilon'_{CO_2}, \varepsilon'_{H_2O}$ — emissivity of CO_2 and H_2O for the temperature of walls and mean path of radiation ($L_e \cdot T_w/T_p$),

$\Delta\varepsilon$ — correction factor, which takes into account partial coincidence of absorption bands of CO_2 and H_2O, defined for the mean path of radiation ($L_e \cdot T_w/T_p$).

498

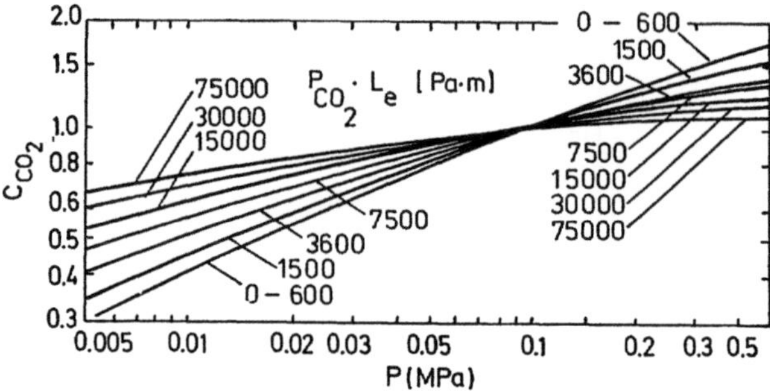

Fig. 3:   Correction factor (C_{CO_2}) as a function of its partial pressure and mean path of radiation (L_e) and total pressure of plasma (p).

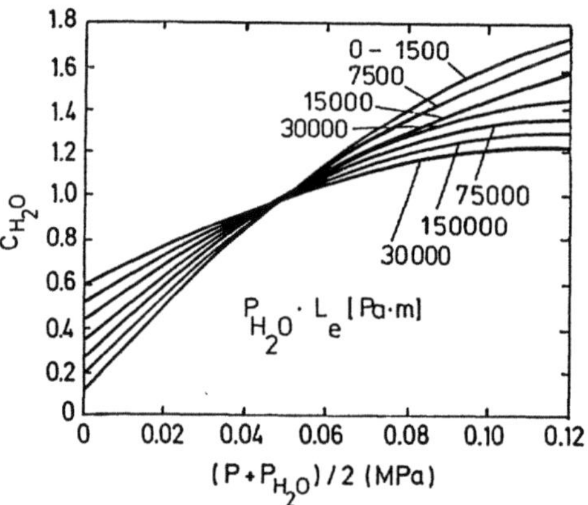

Fig. 4:   Correction factor (C_{H_2O}) as a function of its partial pressure and mean path of radiation (L_e) and total pressure of plasma (p).

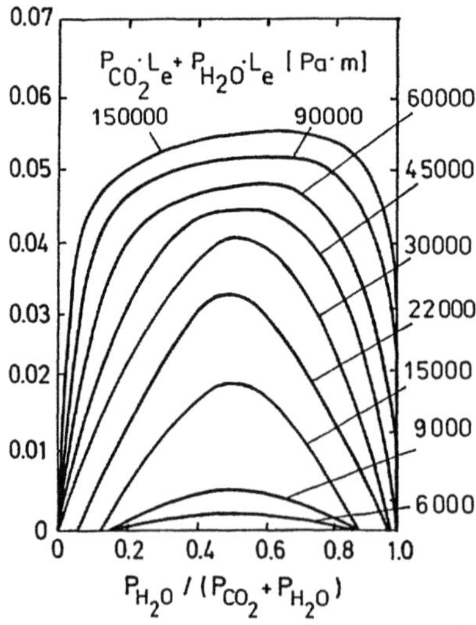

Fig. 5:   Correction ($\Delta\varepsilon$) as a function of the partial pressures CO_2 and H_2O and mean path of radiation (L_e).

INFLUENCE OF MAGNETOHYDRODYNAMIC PHENOMENA ON INTENSITY OF HEAT TRANSFER IN THE MHD GENERATOR

The turbulent flow of plasma in the MHD generator channel in the lateral magnetic field causes

- occurrence of Hartman's effect which increases the intensity of heat transfer to the walls of the channel,
- laminarization of plasma flow which decreases the intensity of heat transfer to the walls of the channel.

Investigation on the influence of the external magnetic field on the intensity of heat transfer in the MHD generator channel was performed experimentally, on account of difficulties in carrying out theoretical investigations in this range.

For experimental investigations, the convection heat-transfer coefficient in the magnetic field was expressed by relation

$$\alpha_{c(M)} = \alpha_c \left(1 - n \frac{H_a^2}{Re^{0.75}} \right) \tag{9}$$

where

$\alpha_c(M)$ - convection heat transfer coefficient in the magnetic field,

α_c - convection heat transfer coefficient without an external magnetic field,

n - coefficient determined experimentally (n=0.204 - experimentally determined for the MHD generator of the Technical University of Poznań),

Ha - Hartman's dimensionless number, which is the function of magnetic field induction and electrical conductivity of plasma,

Re - Reynold's dimensionless number.

The influence of Joule effect on convection heat transfer coefficient was also investigated experimentally. Before starting the investigations, the relation was formulated that took into account the influence of Joule effect on convection heat transfer coefficient utilizing equations of heat transfer in media with internal heat sources. When determining relations, describing the influence of Joule effect on convection heat transfer coefficient, the following assumptions were assumed

- electric energy is generated in the whole volume of the channel, whereas dissipation of the part of generated energy on Joule's heat occurs only in the layers close to the electrode,

- total temperature drop follows in the laminar boundary layer,
- in the laminar boundary layer, heat is transferred by conductance.

On the basis of the above mentioned assumptions, the relation was formulated for the whole convection heat transfer coefficient $\alpha_c(M,J)$, taking into account both the influence of the magnetic field and Joule effect. The determinated relation was expressed by Hartman's (Ha), Prandtl's (Pr) and Eckert (Ec) dimensionless numbers

$$\alpha_{c(M,J)} = \alpha_{c(M)} \left(\frac{1}{1 - Ha^2 Ec\, Pr\, \frac{a}{2}\, (1-K)^2\, \frac{\delta_t}{D_h}} \right) \tag{10}$$

where

a   - coefficient determined experimentally,
K   - load factor of the MHD generator channel,
δ_t - thickness of the laminar boundary layer,
D_h - characteristic flow dimension (hydraulic diameter)

RESULTS OF THEORETICAL AND EXPERIMENTAL INVESTIGATIONS

RESULTS OF CALCULATIONS AND MEASUREMENTS OF HEAT EXCHANGE IN AN EXPERIMENTAL MHD GENERATOR

On the basis of relations (3) and (5), conventional and radiational coefficients of heat transfer in the combustion chamber and channel of an experimental MHD generator of the Technical University of Poznań, of 4 MW thermal power, were determined. For the calculations, the following characteristic geometrical measurements were assumed:

- combustion chamber - D_h=0.14 m, L=0.6 m,
- channel of the MHD generator - (0.03x0.1x0.5) m^3, D_h=0.046 m.

In the calculations, the influence of the magnetic field and Joule's effect on the process of heat exchange in the MHD generator channel, were not taken into account.

The calculations were performed in the function of temperature of the walls of the combustion chamber and the MHD generator channel, on assuming the following parameters of plasma, obtained in the process of kerosene combustion in air enriched with oxygen:

a) for combustion chamber - pressure of plasma 0.294 MPa; temperature of plasma 3040 K and 2290 K; velocity of plasma flow 30 m/s and 90 m/s,

b) channel of the MHD generator - pressure of plasma 0.118 MPa; temperature of plasma 2890 K and 2090 K; velocity of plasma flow 850 m/s and 500 m/s.

Diagrams in Figures 6, 7 and 8 show the results of calculations of α_c and α_r for the combustion chamber and channel of the MHD generator. On the basis of determined heat transfer coefficients α_c, α_r, mean densities of the q_w heat flux, transferred to the walls of the combustion chamber and channel of the MHD generator, were determined (Figures 9 and 10).

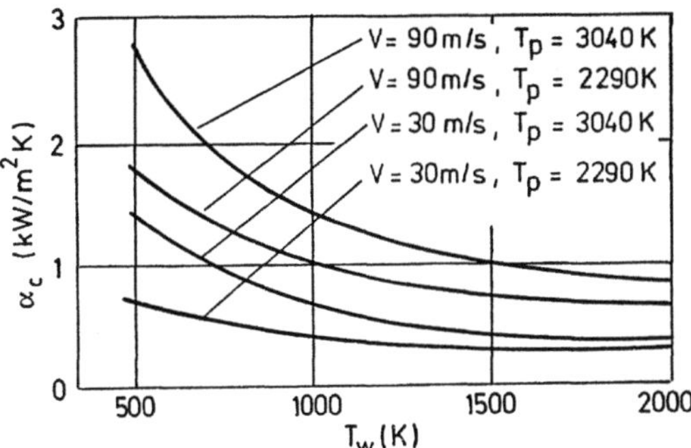

Fig. 6: Convection heat transfer coefficient in the MHD generator combustion chamber (α_c) as a function of plasma temperature (T_p), velocity of plasma (v) and temperature of wall (T_w).

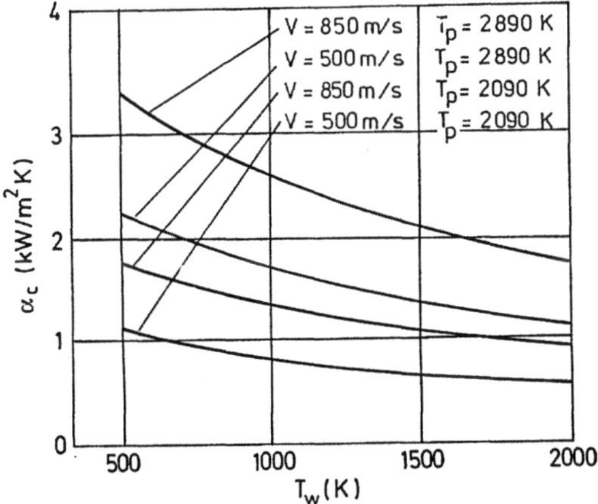

Fig. 7: Convection heat transfer coefficient in the MHD generator channel (α_c) as a function of plasma temperature (T_p), velocity of plasma (v) and temperature of wall (T_w).

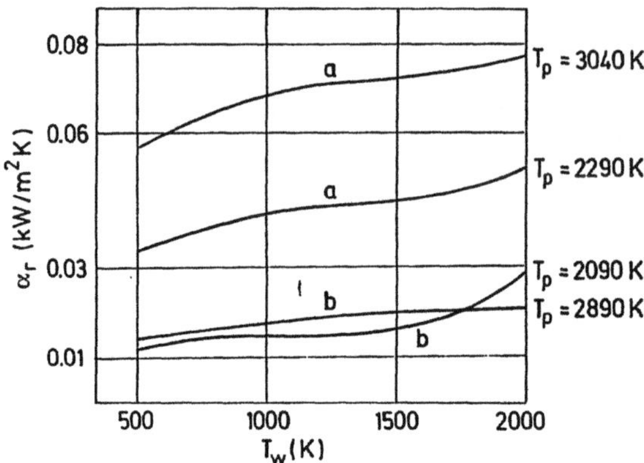

Fig. 8: Radiation heat transfer coefficient (α_r) in the MHD generator combustion chamber (a) and in the channel (b) as a function of plasma temperature (T_p) and temperature of wall (T_w).

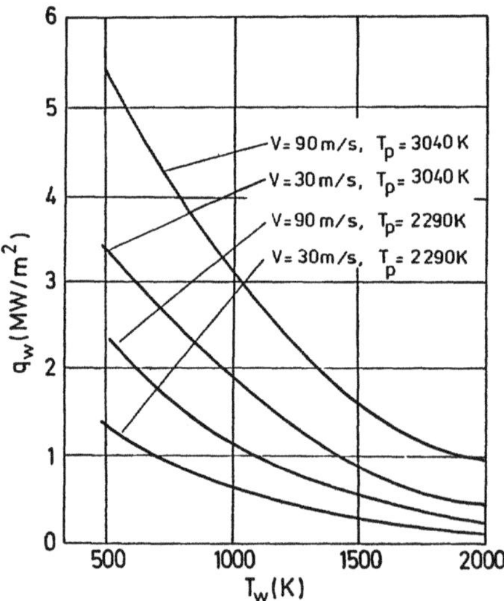

Fig. 9: Heat flux density transferred to walls of MHD generator combustion chamber (q_w) as a function of plasma temperature (T_p), velocity of plasma (v) and temperature of wall (T_w).

Results of experimental investigations of heat fluxes, transferred to the walls of the experimental MHD generator of 4 MW thermal power, are presented in Table 1.

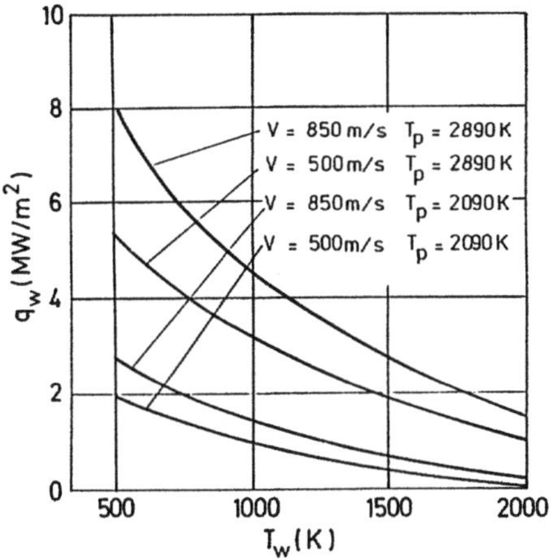

Fig. 10: Heat flux density transferred to walls of MHD generator channel (q_w) as a function of plasma temperature (T_p), velocity of plasma (v) and temperature of wall (T_w).

Table 1. Chosen results of experimental investigations on heat fluxes transferred to the walls of the MHD generator channel

Temperature of pasma	Pressure of plasma in the channel	Mean temperature of electrode walls	Mean heat flux to the electrode walls	Mean temperature of insulating walls	Mean heat flux to the insulating walls	Induction of the magnetic field	Channel load factor
$T_p[K]$	$p[MPa]$	$T_{we}[K]$	$q_e[kW.m{-2}]$	$T_{wi}[K]$	$q_i[kW.m^{-2}]$	$B[T]$	K
2900	0.147	2050	873.19	2300	646.09	0	-
2790	0.145		766.63		557.47	0	-
2900	0.147		806.39		597.41	2.7	1
2780	0.146		733.73		522.11	2.7	1
2890	0.147		951.48		605.14	2.7	0.5
2790	0.148		838.00		522.45	2.7	0.5
2890	0.147		1106.99		604.15	2.7	0
2800	0.148		935.34		528.86	2.7	0

RESULTS OF CALCULATIONS OF RADIATIONAL HEAT EXCHANGE IN A COAL FIRED GENERATOR

On the basis of relations describing radiational heat exchange, calculations of α_r and q_r were performed in the combustion chamber and channel of an experimental and industrial scale MHD generator, on using coal as fuel. These calculations were carried out in the function of wall temperature. Tables 2 and

504

3 show the setting-up of fundamental parameters of plasma, for which the calculations were performed. The results of calculations are presented in Figures from 11 to 16.

Table 2. The calculation parameters of plasma in the experimental MHD generator (4 MW$_{th}$).

Parameter	MHD combustion chamber	MHD channel
Plasma temperature [K]	3050	2600
Plasma pressure [MPa]	0.197	0.099
Partial pressure of CO_2 [MPa]	0.031	0.011
Partial pressure of H_2O [MPa]	0.029	0.097
Mean path of radiation L_e [m]	0.104	0.040

Table 3. The calculation parameters of plasma in the industrial scale MHD generator (400 MW$_{th}$).

Parameter	MHD combustion chamber	MHD channel
Plasma temperature [K]	3250	2750
Plasma pressure [MPa]	0.987	0.493
Partial pressure of CO_2 [MPa]	0.195	0.132
Partial pressure of H_2O [MPa]	0.154	0.093
Mean path of radiation L_e [m]	0.494	0.188

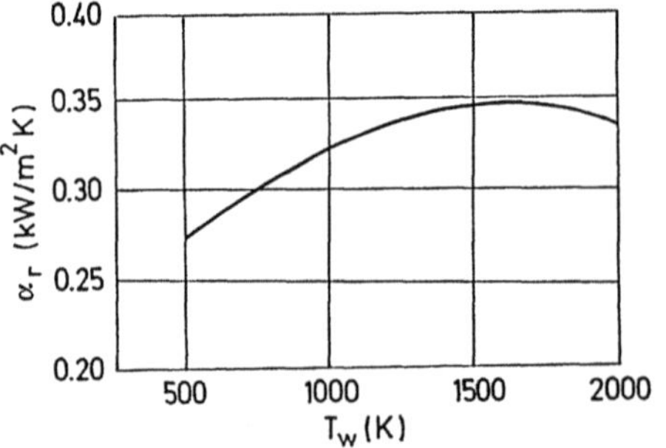

Fig. 11: Radiation heat transfer coefficient (α_r) in combustion chamber of industrial scale MHD generator as a function of wall temperature (T_w).

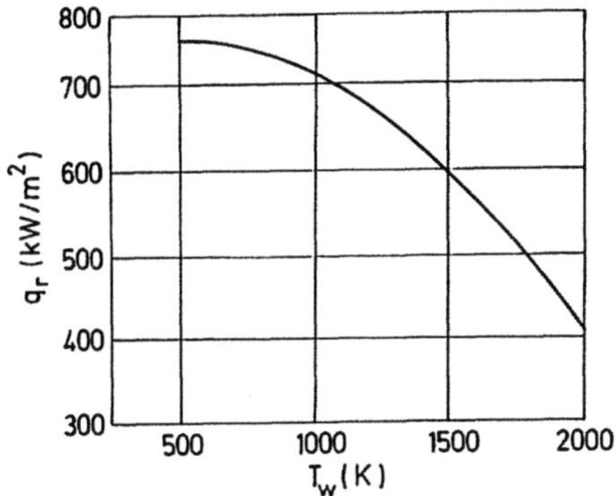

Fig. 12: Radiation heat flux density (q_r) transferred to walls of combustion chamber of industrial scale MHD generator as a function of wall temperature (T_w).

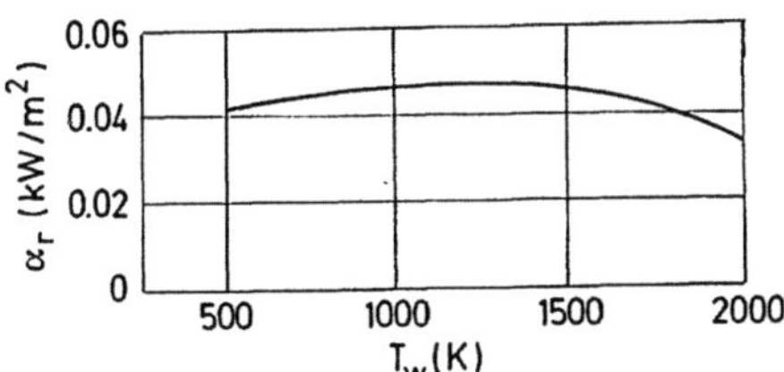

Fig. 13: Radiation heat transfer coefficient (α_r) in combustion chamber of experimental MHD generator as a function of wall temperature (T_w).

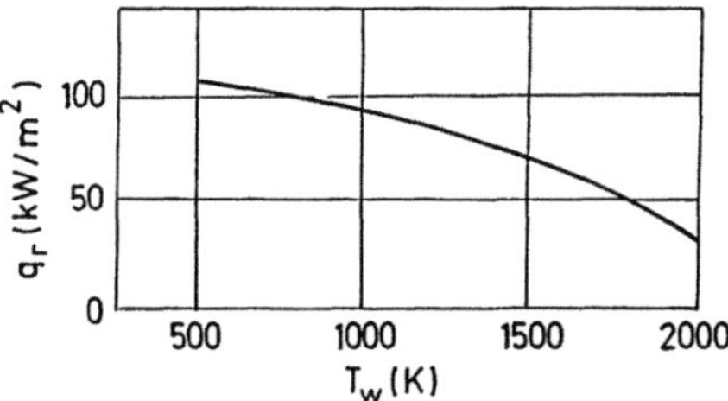

Fig. 14: Radiation heat flux density (q_r) transferred to walls of combustion chamber of experimental MHD generator as a function of wall temperature (T_w).

506

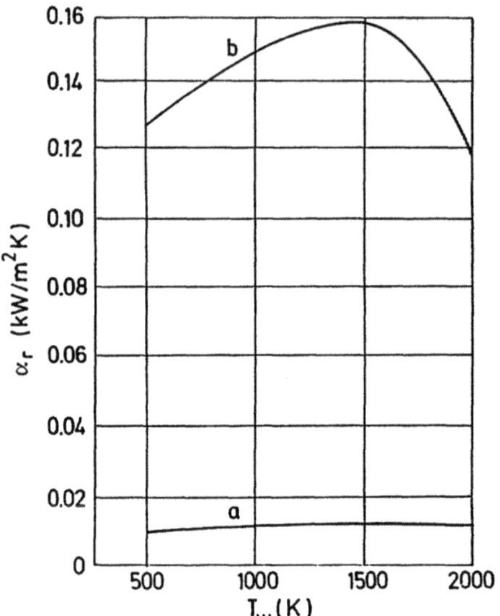

Fig. 15: Radiation heat transfer coefficient (α_r) in the MHD generator channel as a function of wall temperature (T_w).
(a) - Experimental MHD generator,
(b) - Industrial scale MHD generator.

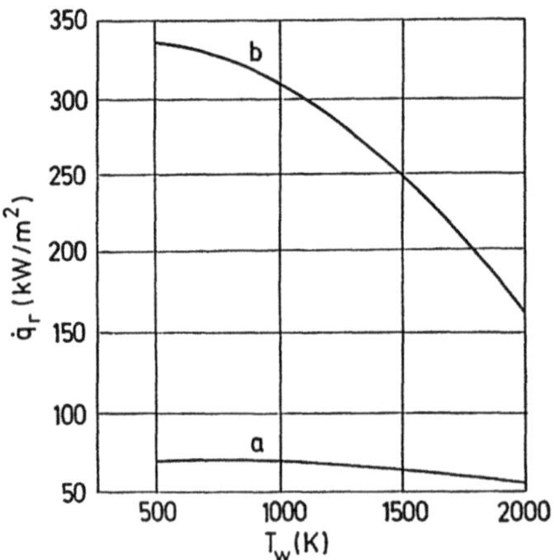

Fig. 16: Radiation heat flux density ($\dot{q}_r$) transferred to walls of MHD generator channel as a function of wall temperature (T_w).
(a) - Experimental MHD generator,
(b) - Industrial scale MHD generator.

CONCLUSIONS

The intensity of radiative heat transfer from plasma to walls of combustion chamber and channel of MHD generator depends on partial pressures of CO_2 and H_2O, temperature of plasma, temperature of walls of combustion chamber and channel and on their geometrical dimensions.

The intensity of radiative heat transfer is greater in the MHD combustion chamber than in MHD channel. It is due to higher pressure of plasma and greater mean path of radiation (L_e) in combustion chamber than in channel.

The intensity of radiative heat transfer, in experimental low scale MHD generator, is much lower (several tens times in channel and more than ten times in combustion chamber) than the intensity of convective heat transfer.

In industrial scale MHD generators the intensity of radiative heat transfer is several times (6-10) higher than in low scale experimental MHD generators.

REFERENCES

1. Hottel H.C., Sarofin A.F.: Radiative Transfer. MacGraw-Hill Book Company. New York (1967).

2. Siegel R., Howell J.R.: Thermal Radiation Heat Transfer. MacGraw-Hill Company, New York (1972).

3. Lin J.C., Greiff R.: Theoretical Determination of Absorption with Emphasis on High Temperatures and a Specific Application to Carbon Monoxide. Transactions of the ASME, Journal of Heat Transfer, 95 (1973) 535-538.

4. Zaporowski B.: Temperature and Physical Properties of Combustion Gases in MHD Generators. Proceedings of a Symposium on Magnetohydrodynamic Electrical Power Generation, Warsaw, IV (1968) 2195-2209.

5. Zaporowski B., Pawlaczyk H., Roszkiewicz J., Switala B.: Investigations of the 4 MW_{th} Experimental MHD Generator. Proceedings of VIth International Conference on Magnetohydrodynamic Electrical Power Generation, Washington, I (1975) 419-434.

AN IMPROVEMENT IN CALCULATING THE THERMAL FIELDS IN THE PISTON-CYLINDER SYSTEM

L. Bignardi, R. Cipollone

Dipartimento di Energetica, University of l'Aquila
Monteluco di Roio, 67040 Roio Poggio, L'Aquila - Italy

ABSTRACT

An improvement on the knowledge of the gas-walls heat transfer coefficient in the Reciprocating Internal Combustion Engines (ICE) requires the development of a procedure able to evaluate accurately the surface temperatures of the components facing gases.

With this aim a method to study the thermal behaviour of the system cylinder-piston is presented, which considers the thermal coupling between these two components, taking into account the motion of the piston.

NOMENCLATURE

c   - gas velocity
d,e - exponents defining h_g
f,g - exponents defining h_g
h   - convective heat transfer coefficient
m   - number of steps along the cylinder axis used to calculate, for each position of the piston, the exchanged energy
n   - number of time steps along the cycle used in the calculation of the exchanged energy
p   - pressure
q   - heat flux
r   - radial position along the cylinder
t   - time
z   - axial position along the cylinder
A   - constant defining h_g
L   - linear dimension
Q   - thermal energy
R   - cylinder radius
S   - surface
T   - temperature
k   - thermal conductivity

$\tau_{1,2}$ - time constants

ϕ - angular position along the cylinder

$C_{1,2}$ - constants shown in the closed temperature model (eq. 7)

Subscripts

c - cylinder

g - gas

h - cylinder head

o - lubricating oil

p - piston

r - radiation

w - wall

ref - reference state. It may be the initial state or the first asymptotic state calculated by eq. (7).

INTRODUCTION

The heat transfer in ICE plays an important role in calculating the thermal and mechanical conditions of the components facing gases as well as in defining the gas thermodynamic state.

Radiation and forced convection, as known, are the two types of energy exchanges between gases and walls: in the specific application they both require further improvements to obtain a wider and universally applicable modeling.

The radiative portion which is the subject of new researches [1-3] is studied mainly from an experimental point of view to improve the knowledge on the emissivity and absorptivity of the gas under the typical ICE pressure and temperature conditions.

Forced convection, on the contrary, requires additional knowledge mainly on the modeling of unsteady heat exchange conditions due to the presence of high time rate of change phenomena (combustion).

On the basis of the high interest on forced convection shown in literature [4-14], this paper presents an insight on the convective gas-wall heat transfer coefficient, $h_g(t)$.

The numerous expressions proposed for the h_g coefficient may be placed in a common general formulation given by

$$h_g(t) = A\, p_g^d(t)\ T_g^c(t)\, c^f(t)\, L^g \tag{1}$$

510

correlating the main thermo-fluodynamic gas properties. The above-mentioned form assumed the validity of a quasi-steady process and the fact that the working fluid has uniform properties $T_g(t)$ and $p_g(t)$.

The heat flux may be, therefore, calculated by means of

$$q_{gw}(r,z,\phi,t) = h_g(t) \, [T_g(t) - T_w(r,z,\phi,t)] + q_r(r,z,\phi,t) \tag{2}$$

Focusing only on the convective term, the integral of expression (2) evaluated, during time, over all the surfaces facing gases represents the exchanged energy.

In a discrete form and according to the symbols in Figure 1, it takes the form

$$Q = 2\pi R \sum_{j=1}^{n} \left[\sum_{i=1}^{m_j} h_{g,j}[T_{g,j} - T_{cpi}]\Delta z_i \right] \Delta t_j + Q_p + Q_h \tag{3}$$

The first term represents the convective heat transferred between gases and the cylinder surfaces, varying during time, considering an axisymmetric distribution of the wall temperature and neglecting its fluctuations over the mean value. The first hypothesis is usually made to simplify the calculations: more detailed analyses may consider the real temperature distribution as in [15]. The temperature oscillations, on the contrary, may be neglected because they assume small values (4-8°C) with respect to the mean values. Q_p and Q_h refer to the surfaces of the piston and of the cylinder head. For both of them, often, it is not possible an explicit form as in the case of the cylinder, due to the absence of an axisymmetric geometry or to the difficulties in considering axisymmetric temperature distributions.

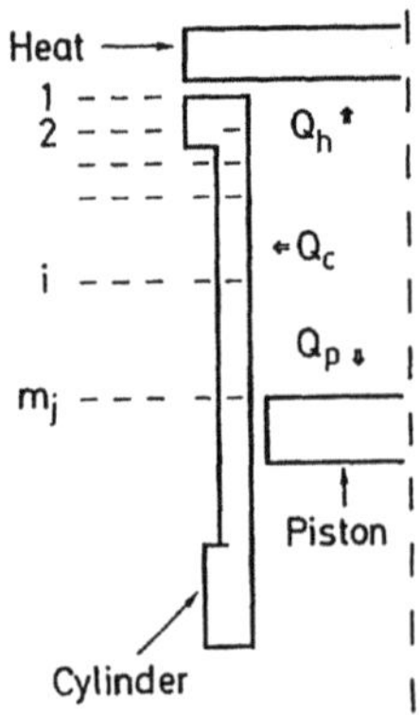

Fig. 1: Diagram of the cylinder piston system

The energy calculated by eq. (3), referred to a given period Δt, is usually compared with that measured, in the same interval, on the refrigerating and lubricating fluids. The gas temperature is usually calculated by a thermodynamic model which uses the indicated pressure diagram. The wall temperatures in eq. (3) are assumed often referring to experimental data. It is important to notice that the correlations for h_g so obtained respect only the energy balance: this condition does not imply that there will be an only one solution for the thermal problem.

A deepening on the meaning of the gas-walls heat transfer coefficient in ICE may be proposed, preserving the form (1), by procedures which respect the temperatures at the gas-wall interface. Thus the problem is solved in terms of conduction on all the components facing gases, adopting, on the remaining parts of the geometrical boundaries, conditions which may be easily assumed or measured. In order to do that an accurate description of the real thermal boundary conditions occurring in ICE components is required considering also the alternate motion of the piston.

With this purpose a procedure which represents more closely the thermal conditions in the cylinder-piston system, interfaced with a finite element (FE) code, is developed in this paper.

Considering the motion of the piston requires, also, a close simulation of the thermal conditions on those surfaces (piston skirt and rings zone) faced to the cylinder.

In calculating the temperatures at the walls and inside the components, it is fundamental to reach a stabilized thermal state on which only periodic temperature oscillations occur. As in the case of other thermal problems with high time rate of change on the boundary conditions, this occurrence requires heavy computer resources and a procedure to ensure the effective reaching of the periodic state.

This paper, improving and updating a method already discussed in [16], offers an insight in the study of the thermal fields for the cylinder-piston system. Usually, in fact, the thermal problem on the ICE components is solved considering them decoupled, using experimental or assumed temperature profiles on the cylinder surfaces or on the piston skirt and ring zone [17-19].

This hypotheses prevents the calculation of the temperatures as consequent to the thermal interaction between the two components under discussion.

THE BOUNDARY CONDITIONS MODEL

When thermal fields are calculated using discrete methods, it is rather difficult to simulate the presence of geometries in motion because for each position (*i.e.* time step) the geometrical meshes must usually be redrawn to respect the congruence between the nodes of the two geometries. This causes very long and useless computational times, which are already heavy considering that the simulation of the engine for significant time intervals requires a number of time steps. This is due to the very high time rate of change of the boundary conditions (gas temperature, convective heat transfer coefficient) which require, during the calculations, short time intervals to avoid losing significant portions of them.

The above outlined difficulties have been overcome by considering fixed meshes of the piston and of the cylinder and by simulating the alternate motion with time and space dependent boundary conditions.

A model for the thermal exchange between the piston in motion (skirt and ring zone) and the surfaces of the cylinder is then required. In these zones the temperatures of the element facing each other are unknown, therefore an iterative procedure must be employed.

Obviously, the final formulation of the problem in terms of boundary conditions becomes more complex, but it can be calculated once for ever outside of the FE code.

The necessity of considering such a complex formulation is due to the fact that the boundary conditions on that coupling zone influence significantly the exchanged energy [20].

The mathematical formulation of all the boundary conditions has, therefore, been applied as follows.

BOUNDARY CONDITIONS ON THE CYLINDER

At the gas and piston side walls the application of the boundary conditions on the cylinder has been divided, for each position of the piston, in three zones: the first exposed to the gases, the second to the lubricating oil, the third facing the piston skirt and the ring zone.

On the surfaces $S_g(t)$ facing gases a convective heat transfer coefficient has been applied to each element i

$$-k_c \frac{\partial T}{\partial r}\bigg|_{w,i} = h_g(t)\left[T_g(t) - T_{w,i}\right] \tag{4}$$

Similarly, to the surfaces $S_0(t)$ facing the lubricating oil a convective cooling may be described as

$$-k_c\frac{\partial T}{\partial r}\bigg]_{w,i} = h_0\big[T_0 - T_{w,i}\big] \tag{5}$$

To the remaining surfaces an equivalent convective heat transfer has been assumed. Considering the contact situation shown in Figure 2, the i-th element of the cylinder is subjected to the following conditions

$$-k_c\frac{\partial T}{\partial r}\bigg]_{w,i} = h'_{i,j-1}\big[T_{j-1} - T_{w,i}\big] \tag{6a}$$

$$-k_c\frac{\partial T}{\partial r}\bigg]_{w,i} = h''_{i,j}\big[T_j - T_{w,i}\big] \tag{6b}$$

$$-k_c\frac{\partial T}{\partial r}\bigg]_{w,i} = h'''_{i,j+1}\big[T_{j+1} - T_{w,i}\big] \tag{6c}$$

The three relations refer to the portions A_1, A_2 and A_3 of the areas facing respectively the j+1, j and j-1 elements of the piston the position of which is known at that very moment. Generally speaking, conditions as in (6a) to (6c) should be applied, according to the geometrical discretization of the piston, to all its elements facing the i-th element of the cylinder.

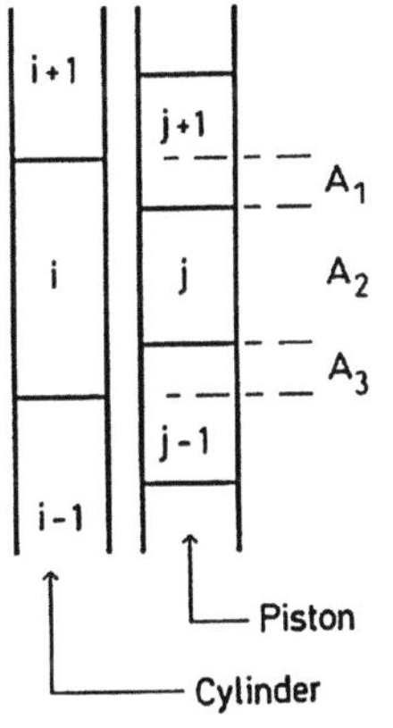

Fig. 2: Diagram showing the contact between the cylinder and piston elements

The temperatures appearing on the right hand side in (6a) to (6c) are, obviously, unknown because they are part of the requested solution. The solution must then be obtained by means of an iterative procedure inside the FE code respecting a given tolerance.

The equivalent convective heat transfer coefficients introduced by (6) have been subdivided in seven zones according to Figure 3. In these zones the coefficients have been assumed space dependent according to each position on the basis of literature data [17], [20-25].

On the surfaces of the cylinder facing the cooling fluid, temperature profiles have been applied according to experimental values [26]. In the cylinder head gasket and in the cooling fluid flange zones adiabatic conditions have been imposed. These hypotheses apply when a low conductive material and a clearance between the liner and the engine block occur.

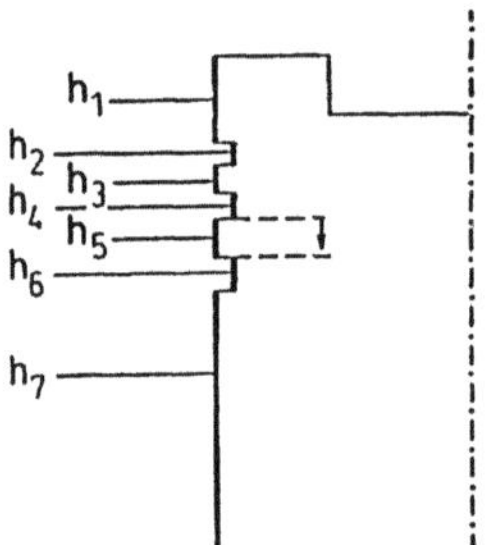

Fig. 3: Diagram of the different zones considered in the piston

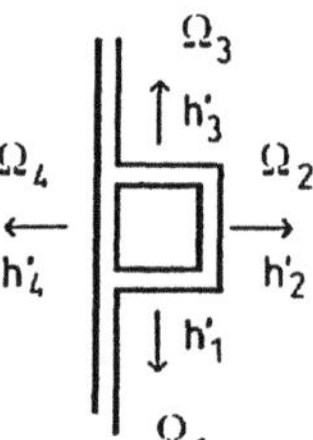

Fig. 4: Diagram of the thermal contact between the ring and the groove.

BOUNDARY CONDITIONS ON THE PISTON

Also for this component the model considers three zones facing the gas, the lubricating oil and the cylinder. For the first two zones boundary conditions similar to those described in (4) and (5) have been applied while for the third one conditions as in (6) have been considered.

The presence of the rings, as known, applies very complex thermal conditions mainly with reference to the piston ring groove surfaces. In these zones a simplified model have been considered. As shown in Figure 4, it considers, for each ring, three equivalent convective heat transfer coefficients applied on Ω_1 Ω_2 and Ω_3 surfaces, taking into account, in the values assumed, the alternate contact of the ring on Ω_1 and on Ω_2. On Ω_3 conditions similar to those described in (6) have been applied. Similarly, the thermal exchange between the ring and the cylinder (simulated by means of the h'_4 coefficient) has been considered.

RESULTS

STABILIZED THERMAL STATE

A stabilized thermal state has been accomplished improving a procedure already developed in [16]. It is based on a close integration between a conduction model applied to each element of the geometry

$$T(t) = T(t_{ref}) + C_1\left[1 - e^{-(t-t_{ref})/\tau_1}\right] + C_2\left[1 - e^{-(t-t_{ref})/\tau_2}\right] \tag{7}$$

and the discrete solution derived from the FE code. The procedure is based on the following steps:

1) given an initial temperature state, a first FE analysis is started and performed simulating few seconds of the working engine applying the above-described boundary conditions;

2) on the basis of FE solutions at time intervals multiple of the engine period, the constants C_1, C_2, τ_1 and τ_2 have been calculated by simple regression techniques. The minimization of an error function based on the values calculated using (7) and the FE analysis values has been imposed as condition;

3) for each element of the geometry the expression (7) gives an asymptotic value which overestimates the final mean stabilized temperature state [16]. This value may be used to define a new fictitious initial state from which a FE analysis may be started again. The solution immediately shows a monotone decreasing trend the values of which may be used to redefine a new asymptotic temperature state as described in point (2).

4) the intersection between the two temperature behaviours according to a double application of eq. (7) gives, for each element, a final value which is close to the stabilized temperature state.

Figure 5 and Figure 6a) present a result of this procedure for two elements of the cylinder-piston system. Figure 5 refers to an element inside the liner, Figure 6a) to an element on the piston crown. These results have been calculated for a research engine AVL 530, S.I. single cylinder running at 2500 RPM and full load using for h_g the Eichelberg correlation [13], starting from an initial state (point 1) equal to 150 °C. The lower curves represent the results of point 2): their asymptotic values coincide with the ones from which the upper curves start (point 3).

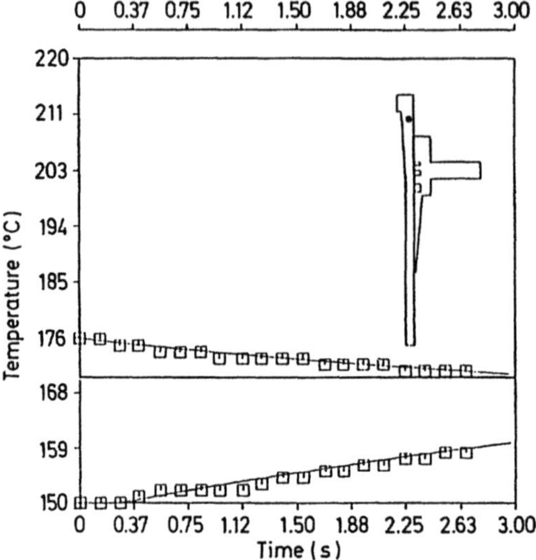

Fig. 5: The reaching of the stabilized state. The point shown is inside the cylinder liner
    □   FE calculations
    —— model eq. (7)

Figure 6b) gives also the temperature oscillations obtained as a result of point 4. These oscillations do not occur in the case illustrated in Figure 5 because there they are completely damped. Nevertheless, a delay in the temperature increase may be observed.

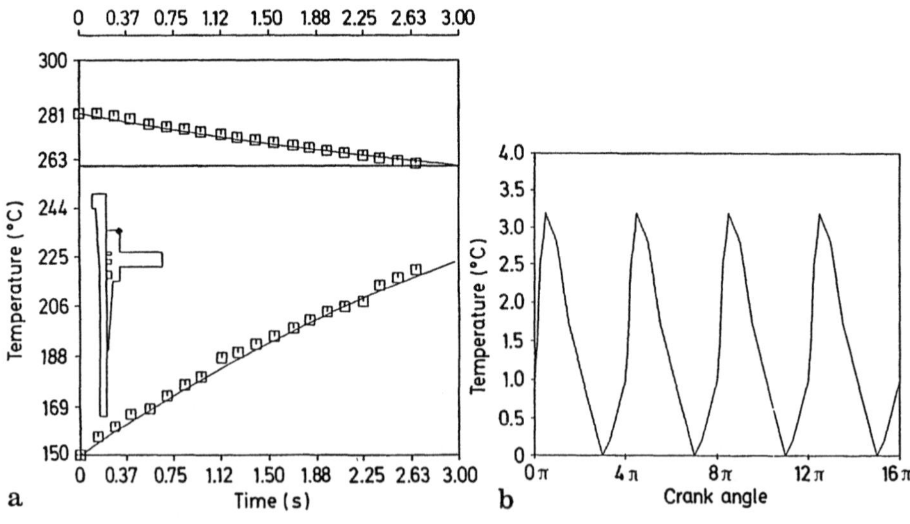

a

Fig. 6a): The reaching of the stabilized state. The point shown is on the piston crown surface.
    □   FE calculations —— model eq. (7)

b

Fig. 6b): Temperature oscillations above the minimum value.

SURFACE TEMPERATURE DISTRIBUTIONS

The proposed model which proved sensitivity and capability of a deeper insight of ICE heat transfer, has been applied in order to evaluate some influences due to the thermal conditions at the lubricating oil side and the ones at the ring zone. The authors are now investigating the effects due to the variations of the other parameters introduced by the proposed model.

With reference to the convective gas walls heat transfer correlation proposed by Eichelberg [13] calculated at 2500 RPM and full load for the mentioned engine, Figure 7 and Figure 8 show the surface temperature profiles on ε_1 and ε_2. Curve 'a' has been calculated considering h_0=600 W/m^2 °C and with the heat transfer coefficients values shown in Figure 9 and reported in Table 1, row a). Curve 'b' differs from the previous case only for the h_0 value which has been doubled. Curve 'a' has been computed using the heat transfer coefficients values given by the row c) in Table 1, keeping the h_0 value equal to curve 'a'.

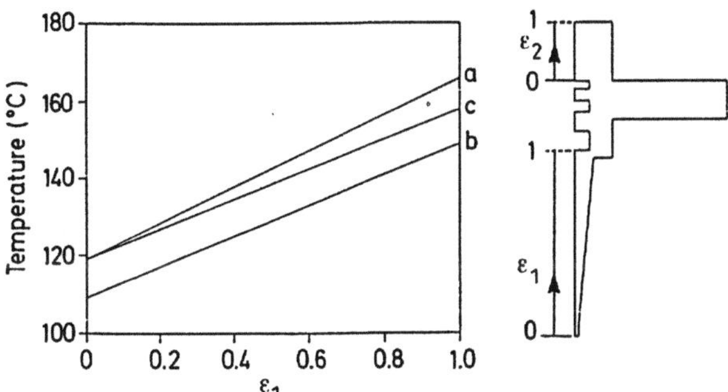

Fig. 7: Surface temperature on the piston skirt.

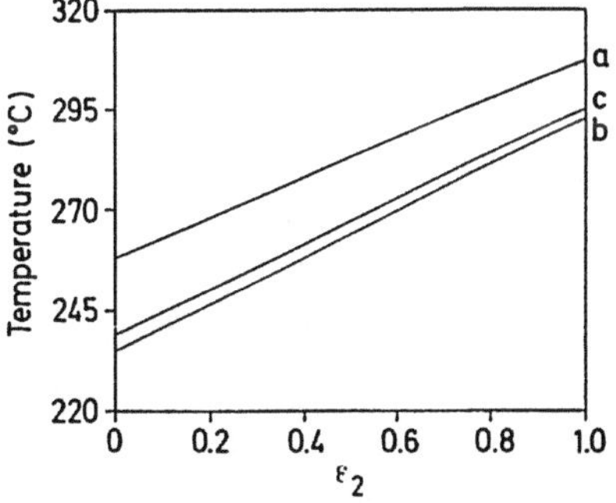

Fig.8: Surface temperature on the piston skirt.

As it may be observed in both zones (ε_1 and ε_2) the values of h_o is important, while the influence of the other heat transfer coefficients seems important only in the lower part of the piston skirt (ε_1).

Table 1:   Values [W/m^2 °C] used for the heat transfer coefficients.

	1st RING				2nd RING			
	h_1	h_2	h_3	h_4	h_1	h_2	h_3	h_4
a	13000	1050	1050	4000	7000	700	700	3000
c	7500	550	550	2000	3500	350	350	7500
	3rd RING				BETWEEN RINGS			
	h_1	h_2	h_3	h_4	h_5		h_6	
a	3500	350	350	2000	1000		800	
c	1750	350	350	1000	500		400	

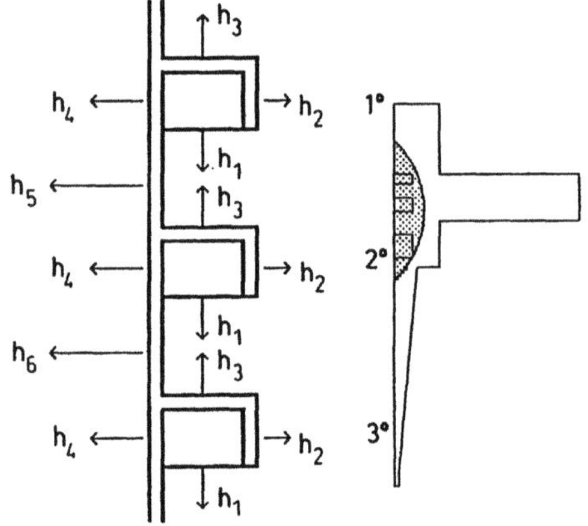

Fig. 9:   Model of the heat transfer for a three rings piston

Figure 10 shows the influences of these three conditions on the surface temperatures along the cylinder at the gas side. Contrary to what occurs in the piston, these profiles remain almost unchanged.

The temperature profiles on ε_1, ε_2 and ε_3 appear significantly influenced by the different gas-walls heat transfer coefficients. As an example, Figure 11 and Figure 12 show the cases of Eichelberg (E) [13] and Woschni (W) [7] correlations at 4500 RPM, full load for the mentioned engine. Even though the tuning of the constants in the original expressions requires a further improvement, it should be noticed the remarkable influence of $h_g(t)$ on both ε_2 and ε_3.

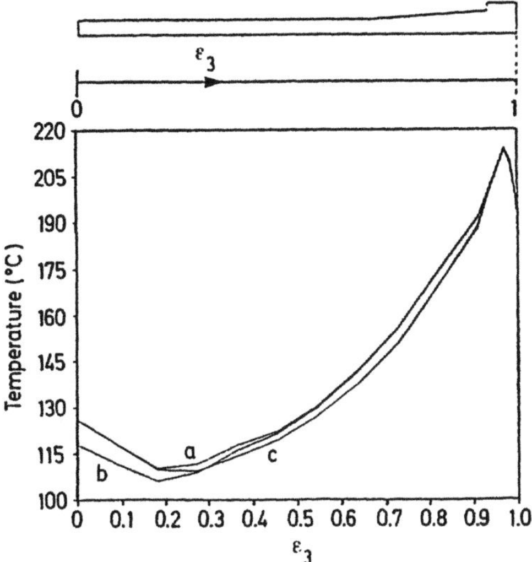

Fig.10: Surface temperature on the cylinder.

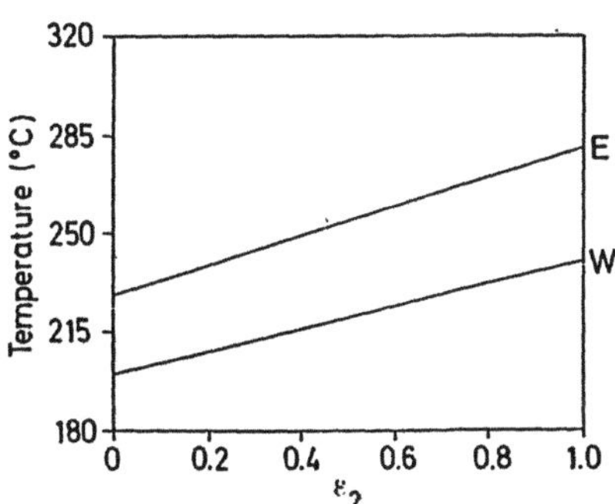

Fig. 11: Surface temperature on the piston skirt

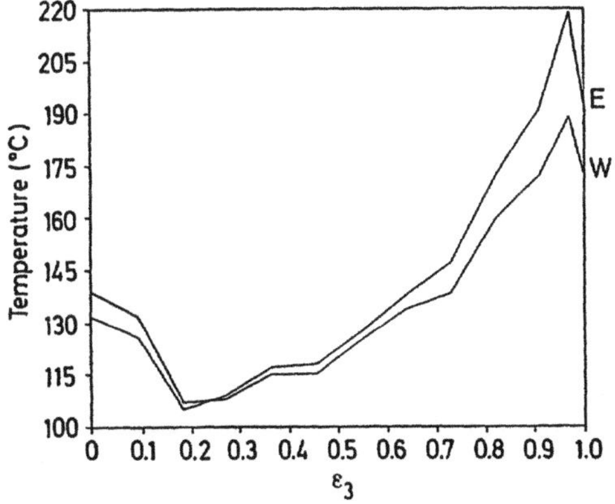

Fig.12: Surface temperature on the cylinder.

CONCLUSIONS

The proposed method proves to be particularly suitable to calculate the thermal fields of the cylinder-piston system with the piston in motion thermally coupled with the cylinder.

Therefore, the convective gas-walls heat transfer coefficients are no longer considered with reference, as in the literature, to a global heat transfer but to local heat exchanges depending on the piston position.

By means of an experimental activity on the surface temperatures, the method may offer an improvement of the gas-walls convective heat transfer coefficients and may help to define new correlations.

ACKNOWLEDGEMENTS

This work was performed under contract n. 880282307 from National Research Council.

REFERENCES

1. Wahiduzzaman S. *et al.*: Experimental and analytical study of Heat Radiation in a Diesel Engine. SAE Paper 870571 (1987).

2. Morel T., Keribar R.: Heat Radiation in D.I. Diesel Engines. SAE Paper 860445 (1986).

3. Bethel S., Anderson C.L.: An Infrared Technique for Measuring Cycle Resolved Transient Combustion Chamber Surface Temperatures in a Fired Engine. SAE Paper 860240 (1986).

4. Kornhauser A.A., Smith J.L.: Application of a Complex Nusselt Number to Heat Transfer During Compression and Expansion. The Winter Annual Meeting of ASME, Chicago, Illinois, FED 65 (1988).

5. Hohenberg G.F.: Advanced Approaches for Heat Transfer Calculations. SAE Paper 790825 (1979).

6. Stradomiskiy M.V. *et al.*: Experimental Studies of Heat Transfer in Cylinders of Spark Ignition and Diesel Engines. Sov. Res. Heat Transfer, 11 (1979) 2.

7. Woschni G.: A Universally Applicable Equation for the Instantaneous Heat Transfer Coefficient in the Internal Combustion Engine. Paper n. 670931, presented at SAE Combined National Meeting, Pittsburgh (1967).

8. Pflaum W.: Warmeubergang in der Verbrennungsakraftmachine. Springer Verlag, Wien (1977).

9. Dent J.C., Soliman S.J.: Convective Heat Transfer in High Swirl Direct Injection Diesel Engine. SAE Paper 770407 (1977).

10. Sitkey G.: Heat Transfer Loading in Internal Combustion Engines. Akademiai Kiodo (1973).

11. Annand W.J.D.: Heat Transfer in the Cylinders of Reciprocating Internal Combustion Engines. Proc. Inst. Mech. Engrs. 177, 36 (1963) 973.

12. Oguri T.: On the Coefficient of Heat Transfer Between Gases and Cylinder Walls of the Spark Ignition Engine. Bull. of Jap. Soc. Mech. Engineers, 13, 11 (1960).

13. Eichelberg G.: Investigations on Combustion Engines Problems. Engineering 148 (1939) 463, 547, 603, 682.

14. Nusselt W.: Acts of VDI Zeitschrift 264 (1923).

15. Cipollone R.: On the Thermal Fields of I.C.E. Cylinder Liner. SAE Paper 900455 (1990). Submitted for publication.

16. Cipollone R., D'Amato E.: On the Evaluation of the Periodic Thermal State in I.C.E. Components.

17. Woschni G., Fieger J.: Determination of Local Heat Transfer Coefficients at the Piston of a High Speed Diesel Engine by Evaluation of Measured Temperature Distribution. SAE Paper 790834 (1979).

18. Singh V.P. *et al.*: Some Heat Transfer Studies on a Diesel Engine Piston. Int. Journal Heat and Mass Transfer 29, 5 (1985).

19. Morel T., Keribar R.: Warmup Characteristics of a Spark Ignition Engine as a Function of Speed and Load. SAE Paper 900683 (1990).

20. Bignardi L., Cipollone R.: Ricerca delle Condizioni di Regime Termico in Organi Fissi e Mobili di Motori a Combustione Interna Alternativi. MARC90 University of Padova (1990).

21. Hug K.: Messung und Berechnung von Kolben Temperaturen Im Diesel Motorem. Diss. Druckerey A.G. Gebr. Leebmann Co. (1936).

22. Sanders J.C., Schramm W.: Analysis of Variation of Piston Temperature with Piston Dimensions. NACA Rep. 895 (1948).

23. Abdelfattah I.: Piston Temperature. A Method of Calculation for Water Cooled Petrol Engine Units. Aut. Engineer (1954) 335.

24. Whitwhouse N.D. *et al.*: Piston Thermal Loading. Proc. Inst. Mech. Engrs. 179 (1965).

25. Pachernegg S.J.: Heat Flow in Engine Piston. SAE Paper 670928 (1967).

26. Bignardi *et al.*: Temperature Measurements in Internal Combustion Engine Components. II International Congress of Automotive Testing and Innovation, Firenze, Italy (1988).

CHAPTER 6
HEAT TRANSFER IN FIRES

HEAT TRANSFER IN BUILDING FIRES

Philip H. Thomas
Division of Building Fire Safety Technology
Lund University
P.O. Box 118, Lund S221
Sweden

ABSTRACT

The paper gives, at the level of current engineering practice, a brief introduction to problems arising from the thermal interaction between a fire and an exposed combustible material or water used in extinguishing. The current development of fire safety engineering requires that many conventional fire tests for ignition, flame spread, heat release, etc. be reformulated or new tests be adopted to provide data for boundary conditions used with computational fluid dynamics (CFD).

1. INTRODUCTION

Computational fluid dynamics (CFD) known in fire research as "field" modelling, is a new productive tool used in research, in fire investigation and in design.

However analysis of the gas phase is only part of the answer to the fire problem. Differential equations require for their solution both initial and boundary conditions. The former in fire studies are usually part of the technology of heating and ventilating and perhaps of meteorology. They determine the conditions prevailing when a fire starts, the latter are conventional for inert rigid surfaces but present problems for fire in a compartment with boundary surfaces that can pyrolyse, evaporate, perhaps even swell and crack.

This coupling between the fuel and the fire even if there is no extinguishing agent present is an essential part on the fire problem.

Without proper statements of these boundary conditions the application of CFD is limited largely to studies of smoke and gas movement. Because its potential is so far reaching across a very broad range of engineering endeavour, considerable effort has been concentrated on its development in areas remote from fire research in field perhaps of greater economic and environmental importance. In comparison the effort devoted to the special problems associated with fire has received little support. The boundary conditions peculiar to fire problems need to be developed within the smaller fire research community and these obviously impinge on the problems associated with testing flammable materials where there is an obvious requirement that fire tests provide information for modelers if CFD is to be exploited fully. The purpose of this paper is therefore to give a very general outline of various aspects of the boundary problem in the context of the development of fire technology as influenced by the growth of fire science.

The importance of heat transfer as a means of spreading fires has long been appreciated. Town planners in past centuries recognized the need for non - combustible facades on buildings and for streets wide enough to act as fire breaks (as well as serving military and ceremonial needs). The fire break reduces the radiation view factor, provides space for the dilution of convected streams and sometimes reduces the passage of flying embers.

Historically, the peacetime fire problem is associated with temperate climates where heating is required in buildings at night and in winter. However travel, industry and the advent of combustible furnishings have brought fire risks to other places.

Fire spreads between adjoining and within large buildings, but is retarded by well designed partitions, walls, ceilings and floors which have to withstand the thermal load imposed by the fire.

2. FIRE RESISTANCE OF FULLY DEVELOPED FIRES

A fire in a partially enclosed space e.g. a room with an open door, has its burning rate (at anyone instant in its development) limited either by the availability of air or by the space area and arrangement of the fuel [1].

The protection offered by the partitions is assessed by a test of the wall, ceiling, etc., in a furnace large enough to accommodate a specimen of order 10 m^2 size.

This test of fire resistance, as it is called, has a long history and is reasonably well standardized internationally [2] - the imperfections and variations being often of more importance in commerce than for fire safety.

This test which does not simulate a particular fire, the properties of which vary with the geometry of the enclosure, the arrangement and extent of openings, etc., can be regarded as representative of the growth but not the decay of many fires. Different standards now exist for hydrocarbon fires where severe fires [3] develop more rapidly from an oil spillage than from dry goods. The standard conditions have been expressed in terms of temperature, not heat flux, although much consideration has been given to this question, and to the measurement of the heat flux.

Amongst research workers and designers the tendency is to regard the test as a means of obtaining estimates of the thermal properties for use in calculating the behaviour of variations of the structure e.g. different thicknesses of insulation on a loaded steel construction.

The thermal exposure shown in Figure 1 is frequently described by

$$T - T_0 = 345 \, \text{Log}_{10}(8t + 1) \tag{1}$$

where T_0 = normal ambient temperature
t = heating time in hours

Other forms e.g. a series of exponentials [4] are used sometimes for algebraic convenience

$$T - T_0 = 1200 - 550 \, e^{-0.6t} + 200 \, e^{-3.0t} - 850 \, e^{-12.0t} \tag{2}$$

Few furnaces test interacting structures i.e. joints and a great deal of effort has gone into attempts to match the test performance to that to be expected in a fire and to defining equivalence.

It is largely the efforts of structural engineers which have led to fire safety design of structures becoming accepted.

They are quantitative and seek to incorporate fire as a load on the building additional to wind, earthquake and other such loads. Despite their sophistication they are, however, still using, in design codes, a very simple fire model i.e. a well

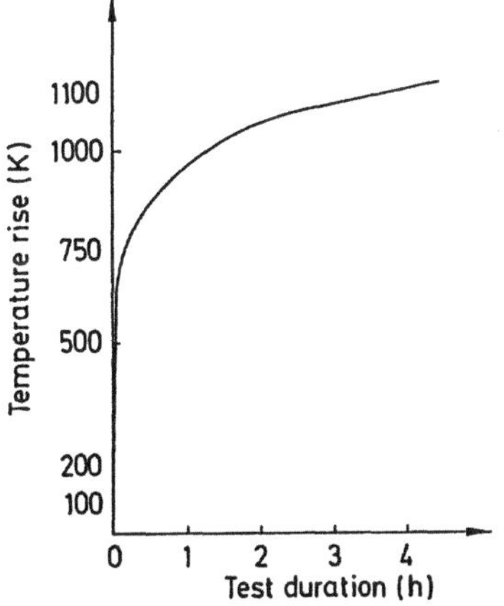

Fig. 1: Standard exposure for elements of construction

stirred reactor producing uniform gas phase temperatures. Experiments and calculations show the existence of "hot spots" [5] and vortices [6] which produce higher than average fluxes. However the errors are thought to be small by comparison with other uncertainties (e.g. the disposition and nature of fuel) but there remains a source of some anxiety.

Heat transfer calculations tend to assume constant or effective mean thermal properties (sometimes for convenience, often because they are not known precisely enough).

Phase change effects of moisture movement in concrete are often allowed for an "ad hoc" basis.

Grey body gas emissivity is usually assumed; it is usually dominated by the soot content and treated empirically. In recent years the emphasis in research has moved from the fully developed fire to the earlier stages of ignition and flame spread largely because most life loss occurs in the early phase of fire growth. It is to these we now turn.

3.   IGNITION

If the thermal degradation of a material is exothermic or it is porous enough for oxidation to take place internally or there is for example, a biological source of heating, internal heating can occur and can cause fires.

There are several industrial and agricultural conditions vulnerable to this hazard [7]. Depending on the levels of internal reactivity, external heating, the surface to volume ratio any one of several situation can obtain.

   (i)   a subcritical equilibrium in which the interior becomes hotter than the ambient conditions permit for an inert material,
   (ii)   a thermal instability beginning internally near the centre, or
   (iii)   if the external heating is strong enough, runaway can begin at or near the surface [8]. (see Figure 2 and 3).

Combustible materials may however degrade or change phase producing gaseous products which react exothermically outside usually in air.

Wood and most materials used in buildings ignite this way [9], [10], [11]. Although the processes may be both physical and chemical the strong dependence of the rate of chemical reaction on temperature rise means that in a fire where temperatures are rising in time it is usually sufficient in practice to discuss

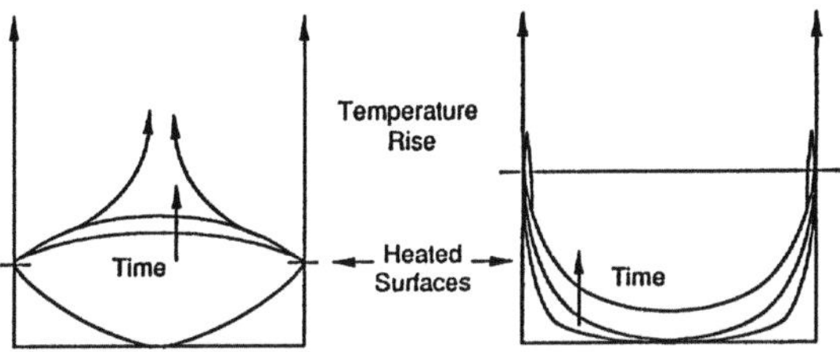

Fig. 2:   Weak exposure   Fig. 3:   Strong exposure

ignition as the result of attaining a particular surface temperature. This is also often done for pyrolysis, especially when these are parts of a more complex process such as flame spread over surfaces.

The problem of ignition is therefore often treated as a problem of conduction into a material which is inert at least up to temperatures very close to ignition.

Strictly it is of course only possible to define a temperature rise that can serve as an "ignition" or "flame front" temperature when the spread is at constant speed in a quasi - steady state, a condition as we see below, which is often relaxed. Because many building materials are not homogeneous, are anisotropic with thermal properties which vary with temperature, one finds calculations tend to be based on "effective" values.

In the case of wood, charring presents special problems and although density changes prior to ignition are commonly neglected the change in surface absorptivity is less easily neglected.

Radiation blocking by emitted particles can also be significant.

4.   SPREAD OF FLAME ON SOLIDS

Horizontal and downward flames spreads on solids have been widely studied (see, for example [12], [13] and [14]).

Near to ignition and extinction chemical kinetics can be dominant as can radiation loss but in a building fire the major term appears to be conduction from the flame through the gases near its base and over a wide range of conditions use is now made of a simple model decoupling combustion from heat transfer (see Figure 4).

Quintiere [15] has exploited the one dimensional heating of a thick homogeneous solid to considerable engineering advantage. The temperature rise is

$$\theta = \frac{2}{\sqrt{\pi}} \frac{\overline{q''}}{K} \sqrt{kt}$$

(3i)

and is identified as the temperature increase whilst the strip of heating passes over a fixed point. $\overline{q}$ " is the mean net flux, k is the diffusivity of the solid and t is the heating time. K is the thermal conductivity.

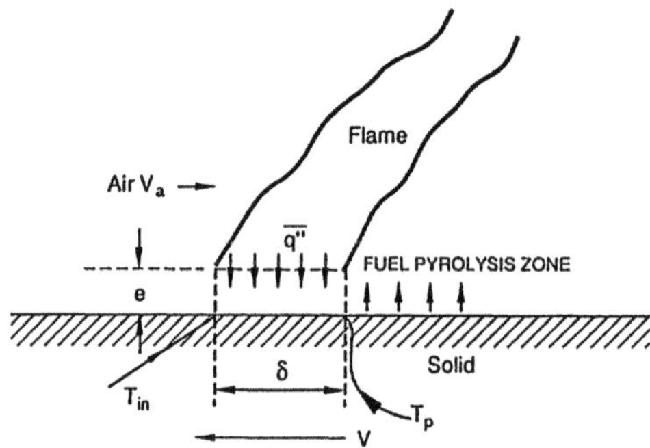

Fig. 4:   Thermal model. Opposed flow flame spread

If V is the velocity of flame spread over the surface

$$t = \frac{\delta}{V}$$

(3ii)

where the distance δ is, in some models

(i)   an independent quality, or in others,
(ii)   a characteristic conduction distance k_g/V_a

where V_a is the velocity of the air flowing over the surface towards the flame (an opposed flow)

Eqs. (3i) and (3ii) give

$$V = \frac{4}{\pi} \frac{\overline{q}''^2 \delta}{K\rho c \left(T_p - T_{in}\right)^2}$$

(4)

where ρ and c are density and specific heat respectively and Tp is the pyrolysis temperature or ignition temperature in the presence of a flame, T_{in} is the initial surface temperature which can be above ambient if external heating is employed.

The flux $\overline{q}$ " is, in principle a dependent variable in a theory incorporating kinetics of flame chemistry.

Parker [16] wrote $\overline{q}''$ as $\dfrac{K_g\,(T_f - T_s)}{\ell}$

and in experiments on thin fuel determined the total energy delivered $q''\delta = K_g$ $(T_f - T_s)\dfrac{\delta}{\ell}$

and measured both "ℓ" and "d".

In a theory in which both δ and ℓ are dependent variables this ratio appears as a non-dimensional coefficient (which de Ris [17] found theoretically to be $\sqrt{2}$) $q''\delta$ can be equated to $\rho.c.\Delta.V$ where Δ is the thickness of thin material.

Except near ignition and extinction a reciprocal law between Δ and V. obtains for many thin materials. de Ris's theory for these materials was of the same form as eq. (4) except δ appeared as kg/Va). He showed that the result did not involve conduction in the solid parallel to the flow and so no loss of energy from the system. He did not include radiation loss from the hot surface.

The neglect of longitudinal conduction because it is small or zero makes the use of eq.(4) in transient conditions more plausible.

Quintiere has examined many data for flame spreading sideways in normal air in which T_{in} was varied by heating the fuel ahead of the flame (this has been a standard test in the U.K. and elsewhere for wall boards before any successful model was made of the process). He found that, over a substantial range of T_{in} and V, $\dfrac{\overline{q}''^2}{K\rho c}\delta$ for many materials could be treated as a constant for many purposes. Many building materials are not homogeneous. Architects and interior decorators often choose to use laminated and coated materials. So one might expect that the effective $K\rho c$ will differ with different degrees of external heating: the above simplification may then prove inadequate as it might in ignition. de Ris did consider the radiation from flames in aiding their spread. One can represent the radiation by a flux acting over a distance of the same order as the flame height.

In general this exceeds k_g/Va and "δ" in eq. (4) becomes comparable to flame height. The flames are however relatively thin and their radiation δ may be relatively small.

However in upward (concurrent) spread radiation may become the dominant term (see Figure 5).

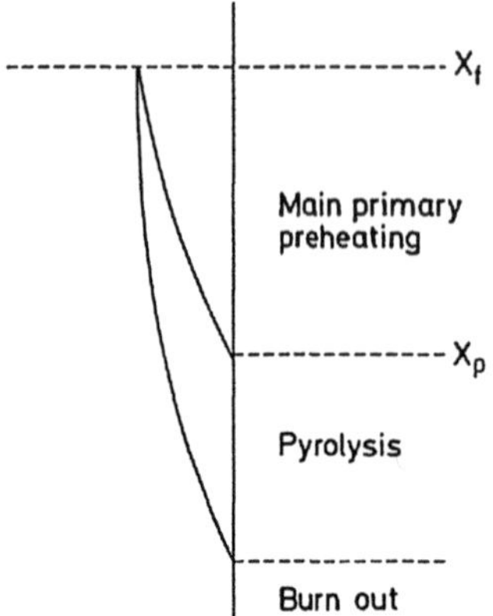

Fig. 5:   Thermal model of upward spread

Saito, Quintiere and Williams [18] have used eq. (4) with

$$\delta = X_f - X_p \tag{5}$$

where X_f is the distance to the flame tip; X_p is the distance to the end of the pyrolysis zone, to describe a two dimensional flame. Both X_f and X_p are measured from a fixed level, burnout of the rear end being disregarded as an influence on the flame pyrolysis.

They wrote

$$V = \frac{dx_p}{dt} \tag{6}$$

and

$$X_f = K \left[Q' + q \int_\delta^{x_p} m'' \, dx \right]^{.n} \tag{7}$$

where K is a constant

Q'(t) an external source making a contribution to the flame, e.g. a burner - per unit width

m" the pyrolysis rate per unit area

q the heat release per unit mass of fuel

[n] is a coefficient which for two dimensional flow is conventionally 2/3 [19].

Saito et al. found n = 1 gave as good a fit to their data and this permitted them to derive a linear integral equation

$$V(E) = \int_{\delta}^{E} F(E - E') . V(E') \, dE' + G(E)$$
(8)

where

$$E = \frac{t}{\tau}$$

$$\tau = \frac{\pi}{4} \frac{K\rho c (T_r - T_o)^2}{q''^2}$$

$$V(E) = \frac{V(t)}{V(0)}$$

$$F = Kq \, m''(t) - 1$$

and

$$G = 1 - Kq \left(m''(o) - m''(t)\right) \left(\frac{X_{po}}{X_{fo} - X_{po}}\right) - K \left(\frac{Q'(o) - Q'(t)}{X_{fo} - X_{po}}\right)$$

where the suffix "o" denotes an initial value. For certain forms of m"(t) and Q' (t) analytic solutions to eq. (8) can be obtained [20] (by means of Laplace transformation) giving three types of solutions:

(i)   flame starts to spread but dies out asymptotically in time.
(ii)   flame dies out at a finite time.
(iii)   flame propagates towards an exponentially increasing rate.

For $Q'(t) = Q'(o)'$
(9)

and $m'' = m''_{max} \, e^{-\gamma t}$

the solutions can be obtained algebraically.

It is a commonplace in fire studies that flame will not spread in normal atmospheres and temperatures up the surface of a thick piece of wood, but restricting radiation loss permits spread.

Simple experiments show spread down a thin vertical cylinder inside a reflecting cylinder proceeds about twice as fast as when the cylinder is black inside [21].

The theory allows one to examine the role of the material properties in particular m"(o), $K\rho c$, γ & Kq. However the importance of the early stages (more important than the asymptotic accelerating spread) requires developments in modelling. For example Saito et al. used $\bar{q}$ = 25 Kw/m^2 as a net flux [22]. The use of the quasi-steady eq. (4) presupposes a particular initial spatial temperature distribution and conditions ahead of the flame have some influence: Hasemi [22] has correlated measured heat fluxes against distance normalized to flame length. Few flames start as two - dimensional.

Delichatsios [23] has described the basis of a model where combustion, fluid dynamic, heat transfer and the thermal properties of the solid are all included.

Before leaving the subject of flame spread over solid surfaces a few comments can be made about the fire on the escalator at the King's Cross underground station. The computational fluid dynamics analysis showed that the trajectory of hot gases resulting from a fire burning within a trench inclined at an angle of 30° to the horizontal, instead of rising vertically, could lie down along and within the trench. The initial CFD study did not however investigate the condition prior to when the fire (confined between two vertical sides) became in effect two dimensional like a wide grass fire on a hillside.

Some discussions have taken place about scaling a small scale experiment and in so far as there is some relationship between upward flame spread on thick solids and spread over an inclined thick surfaces as under a ceiling one can see that simple relationships might obtain for, say, the asymptotic behaviour which might not cover the whole process. There is much activity now in this area, heat transfer rates were higher than the 25 Kw/m^2 referred to above have been measured but they are not constant throughout the whole process. Over 100 KW/m^2 has been measured [25].

There are many enclosed staircases in building with flammable floor coverings. However the absence of any experience of dangers identified with this "trench" behaviour suggests the risks of occurrence as perceived are historically low.

5.   FIRE SPREAD THROUGH FUEL BEDS

Fire spread through a variety of fuel beds, by means of radiation transfer mainly from the hot solids in the combustion zone. This mode of fire spread applies to porous assemblies of wood, and in the absence of wind to conflagrations in cities [26]. Radiation from the flames is then usually secondary, largely because for high fuel beds such as whole buildings the ratio of flame height to opening height is usually of order unity (unless very flammable fuels are present) and flame emissivities may be too low.

The forward flux can be written as

$$q'' = V\rho C\,(T_p - T_o) \tag{10}$$

where V is the velocity, C_p the mean fuel specific heat, T_p is the temperature producing pyrolysis and ρ is the effective density of the material heated to T_p.

Only for a thermally thin material is ρ the actual density.

6. PYROLYSIS

So far we have referred, if only briefly to the thermal flux onto the inert structure, onto the fuel ahead of the pyrolysing fuel but not that onto the fuel which is pyrolysing.

The contribution of non-charring fuels is dealt with by concepts familiar in mass transfer e.g. the Spalding stagnant film hypothesis and to variants using experimentally determined values of the denominator, the heat required to produce unit mass of fuel volatiles. However this relationship becomes less and less relevant as radiation increases. Gases released from charring materials flow through non-uniform pores sometimes endothermically, sometimes exothermically and various empirical relationships exist. De Ris and Delichatsios [27] have discussed a simple diffusion model for a constant surface temperature and no loss of surface mass. The depth of char "d_c" increases as

$$d_c \,\alpha\, \sqrt{kt} \qquad\qquad (t \Rightarrow \infty)$$

However, except for structural timber the material is rarely thick enough to avoid interference between the "forward" and the "reflected" thermal wave, and experimental rates of heat release "$\dot{q}$" and mass loss $\dot{m}$ appear more like that in Figure 6 and a variety of ad hoc simplifications can be justified.

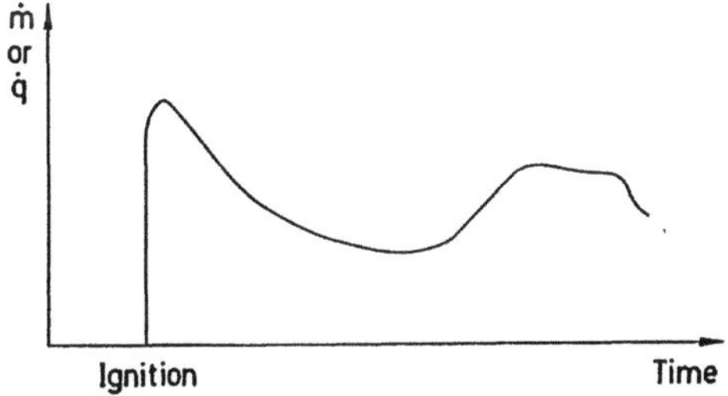

Fig. 6:   Typical consequence of exposure to constant heat flux

7. NON - LINEARITIES AND CRITICALITIES

For building fires there is a general experience that fires are initially benign; people go in search of them sometimes look at them but eventually there is a "flashover" or "flameover" and people nearby have to evade an immediate threat to life. After this sudden "flashover" energy is produced much faster, twice or more times.

There is in general a rapid rise in the temperature of the upper gas layer.

"Flameover" is that type of "flashover" which is the result of the heating up of fuel away from the fire so that it either ignites spontaneously or permits flame to spread very quickly limited by the pyrolysis rate or by the rates of diffusion of air and fuel.

Waterman [28] proposed a general criterion of $20Kw/m^2$ for "flameovers".

The radiation on the floor of a large room is mainly from the descending smoke layer [29], [30].

As the layer thickens its emissivity increases but the influence of the hot ceiling decreases.

In ignition theory the rates of heat generation by chemical reaction (or biological heating) which are "non-linear" with temperature lead to critical conditions and commonly to three equilibria, the middle one being unstable. In the same way radiation transfer, although its non-linearity is weaker than that of chemical reaction, can also produce similar effects. The instability is a form of flashover [31] and the fire "jumps" to the ventilation controlled mode.

A highly simplified model [32] of a uniform room fire (or a uniform stable hot gas - layer) shows features that merit further examination.

The room energy balance in its simplest form is written as

$$\left(\left(m_a+m_f\right)C_p+\overline{h.A_T}\right)\theta = m_f\Delta H \qquad \left(\frac{m_a}{r}>m_f\right) \tag{11 a}$$

$$=\frac{m_a}{r}\Delta H \qquad \left(\frac{m_a}{r}<m_f\right) \tag{11 b}$$

where m_a and m_f are respectively the mass flow of air and fuel. C_p is a mean specific heat, θ a mean gas temperature rise at exit, $\overline{h}$ a mean heat transfer coefficient for the walls, ceiling and ΔH is the area of heated wall, ceiling, etc. ΔH

is the effective calorific value and r is the stoichiometric air/fuel ratio. An approximate linearised mass transfer relationship may be written as

$$m_f = a\,A_f\left[\frac{\Delta H}{r}.\lambda\,\frac{Y_{ox}}{0.23} + b\theta + c\,\theta^p\right] \tag{12}$$

where λ is the fraction of heat release allowing for radiation loss from the flame, c θ^p $(1 < p < 4)$ an approximate representation of radiation loss.

For a well stirred reaction

$$\frac{Y_{ox}}{0.23} = \frac{m_a - r\,m_f}{m_a + m_f} \quad (m_a > m_f) \tag{13 a}$$

$$= 0 \quad (m_a < r\,m_f) \tag{13 b}$$

Eqs. (11a), (12) and (13a) lead algebraically to one or three solutions. Eqs. (11 b), (12) and (13b) lead only to one. Taking the appropriate pairs of solutions one can obtain various relationships between "m_f" and "m_a" of which two are shown diagrammatically in Figure 7. The form of Figure 7a is typical for fuel beds insensitive to radiation feedback e.g. assemblies of elements of low porosity.

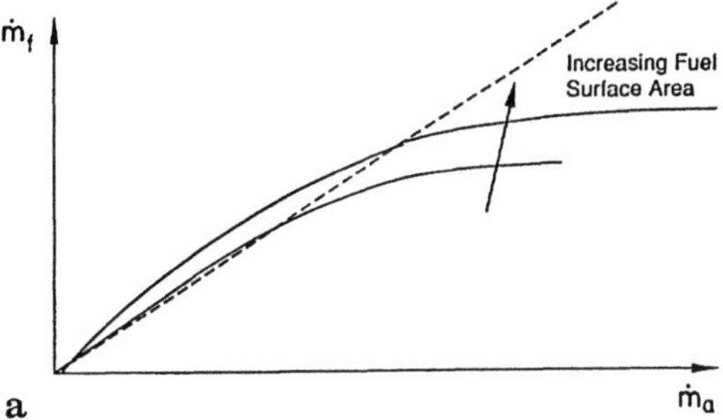

Fig. 7A: Weak coupling between fire and fuel.

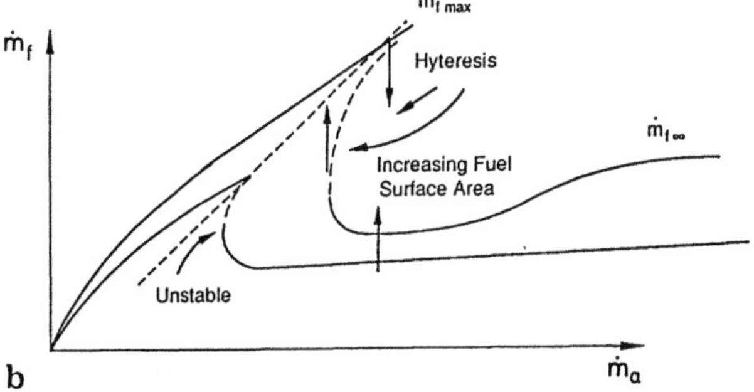

Fig. 7B: Strong coupling between fire and fuel.

Figure 7b is typical of the effect of strong radiation enhancement e.g. a tray of liquid fuel or wall linings. Although the equations are for steady states one can see how an instability can arise as the fire grows. Radiation enhancement has been demonstrated many times [33], [34] and factors of 6 or more have been reported. Hasemi [35] has experimentally shown (see Figure 7b) the existence of hysteresis.

One can identify a narrow range of conditions where there are two criticalities as in Figure 8. The lower one exhibits some of the features of flashback in a vitiated fire, a sudden ignition of the hot gaseous layer.

The model can be developed [35] by the better representation of mass transfer, the inclusion of transient heat conduction, etc.

Zone models [36] incorporating heat loss, internal radiation exchange, layering and a plume have been developed. CFD is being increasingly used but mostly describes the fire by means of an independent heat release input.

8.   CONTROL & SUPPRESSION

In recent years some sprinklers have been made with lower thermal capacity than conventional. The use of too low a temperature of operation of course can lead to accidental water damage.

Describing a sprinkler, especially one with water in the connecting pipes by a single time constant is not always a sufficiently good approximation for design purposes.

Whereas sprinklers operate only when the fire is well established detectors are required to operate sooner.

Estimating the conditions for operation may therefore require a representation of the initial conditions in a room, the effects of heat sources and ordinary ventilation.

Sprinklers control fires by discharging water droplets. There is more than one mode of interaction with the fire and its control [37].

i)   Drops if large enough may penetrate flames and reach the hot evaporating or pyrolysing surface. They will be able to penetrate burning liquids. If the "flashpoint" is below ambient temperature cooling the liquid cannot in principle effect extinction.

ii) The production of water vapour can separate the source of the fuel gases from oxygen.

iii) Drops which do not penetrate the flame may evaporate there both cooling the flame and interfering with the processes that sustain it. Although rapid flame extinction can be effected there are dangers from re-ignition by inadequately cooled solids.

Combinations of these processes occur [38], [39], extracting heat from the gases and from the hot walls, etc. producing steam which expels air from a ventilated room or inerts a large space [38], [39].

In general sprinklers produce a mixture of drop sizes and in many situations a fireman can move a jet to break it up on impact with walls or ceiling, etc. In general the amounts of water used are far greater than estimated [40] from heat transfer theory: The limiting factor is frequently the problem of access, of getting water to a hot surface to prevent reignition. Those surfaces which are shielded and not easy to reach tend to be those where the shielding reduces radiation loss and assists burning.

Two phase CFD [41], [42] is now being applied to the study of the interaction between hot gases and water droplets.

CONCLUSION

The needs of the fire modeler include data about building materials which allow a proper representation of boundary conditions between the condensed phase and the gaseous phase. Fire tests are being developed which are expected to provide some of these data at least at an engineering level but whether they become standard tests legally required by the authorities is another question. If not some connections between the two types of tests are essential. The nature of the problems arising from building materials, the variability of fuel geometry even, sometimes the definition of what the fuel is presents many challenging problems ranging across a wide technical field. Nevertheless the "art" and "craft" of fire safety engineering is being supplemented by and overtaken by technical developments.

REFERENCES

1.   Thomas P.H., Heselden A.J.M., Law M.: Fully Developed Compartment Fires: Two Kinds of Behaviour. Fire Research Technical Paper No. 18, HMSO, London (1967)

2. International Standards Organisation: Fire Resistance Tests Elements of Building Construction, ISO 834, International Organization for Standardization, Geneva (1975).

3. Shipp M.: A Hydrocarbon Fire Standard: An Assessment of Existing Information. Department of Energy, Offshore Energy Technology Board, OT/R/8294 (1983).

4. Lie T.T.: Temperature of Protected Steel in Fire Behaviour of Structural Steel in Fire. Symposium No. 2 HMSO, London (1967) 99.

5. Lockwood F.C., Malalasekera W.M.G.: Fire Computations: The Flashover Phenomena. 22nd Symposium (International) on Combustion. The Combustion Institute, Pittsburgh (1989) 190.

6. Williamson R.B., Mowrer F.W., Fisher F.L.: Combustion Science and Technology 41 (1984) 83-99.

7. Bowes P.C.: Self-Heating: Evaluating and Controlling the Hazards. HMSO, London (1984).

8. Zeldovich Y.B.: Dockl. Academy Nauk SSSR 150(2) (1963) 1283-85.

9. Kanury A.M.: Ignition of Cellulosic Materials: A Review. Fire Research Abstracts and Reviews 14 (1972) 24-52.

10. Rasbash D.J.: Relevance of the Firepoint Theory to the Assessment of Fire Behaviour of Combustible Materials. in International Symposium on Fire Safety of Combustible Materials. Edinburgh University (1975) 169-178.

11. Kashiwagi T., Kashiwagi T.: A Study of the Radiative Ignition Mechanism of a Liquid Fuel Using High Speed Holographic Interferometry. in 19th Symposium (Int.) on Combustion. The Combustion Institute (1982) 1511 - 1521.

12. Williams F.A.: Mechanism of Fire Spread. in 16th Symposium (International) on Combustion. The Combustion Institute, Pittsburgh (1977) 1281-1294.

13. Quintiere J.G.: Surface Flame Spread. The Society of Fire Protection Engineers. in Handbook of Fire Protection Engineering S.F.P.E. - Boston, Mass U.S.A., Section I - Chapter 24 (1988).

14. Di Blasi C., Crescitelli S., Russo G., Fernandez-Pello A.C.: Predictions of the Dependence on the Opposed Flow Characteristics of the Flame Spread Rate on Thick Solid Fuel. in Proceedings of the 2nd International Symposium on a Fire Safety Science. Hemisphere Publishing Corporation (1989) 119.

15. Quintiere J.G.: A Simplified Theory for Generalizing Results from a Radiant Panel Rate of Flame Spread Apparatus. Fire and Materials 5 (1981) 52-60.

16. Parker W.J.: Flame Spread Model for Cellulosic Materials. J. Fire and Flammability 3 (1972) 254-269.

17. De Ris J.N.: Spread of a Laminar Diffusion Flame. 12th Symposium (International) on Combustion. The Combustion Institute, Pittsburgh (1969) 241-252.

18. Saito K., Quintiere J.G., Williams F.A.: Upward Turbulent Flame Spread. in Proceedings of 1st International Symposium on Fire Safety Science. Hemisphere Publishing Corp. (1986) 75.

19. Thomas P.H.: The Size of Flames from Natural Fires. 9th Symposium (International) on Combustion (1963) 844-859.

20. Thomas P.H.; Karlsson B.: To be published.

21. Mygind J.: The Burning of Wood. (Thesis in Danish), J. Jorgensen & Co. Kopenhagen (1951).

22. Hasemi Y.: Thermal Modelling of Upward Flame Spread. in Proceedings of the 1st International Symposium on Fire Safety Science. Hemisphere Publishing Corporation (1986) 87.

23. Delichatsios M.A.: Burning and Upward Flame Spread on Vertical Surfaces: An Outline of a Comprehensive Simulation Model and Simplified Correlations. Factory Mutual Research. Technical Report FMRC, I.O.Q.O.J. BU, August (1988).

24. Fennell D.: Investigation in the King's Cross Underground Fire. HMSO, London (1988).

25. Drysdale D.D.: Personal Communication.

26. Thomas P.H.: Some Physical Aspects of the Spread of Fire. F.O.U. Brand, Fire Research Development News, Swedish Fire Protection Association 1 (1975).

27. Delichatsios M.A., De Ris J.: An Analytic Model for the Pyrolysis of Charring Materials. The Modelling of Pre-Flashover Fires CIB W14 Workshop. CIB Report Publication 81 (1983) 34-41.

28. Waterman T.E.: Room Flashover-Criteria and Synthesis. Fire Technology 4 (1968) 25-31.

29. Orloff L., Modak A.T., Markstein G.H.: Radiation from Smoke Layers. 17th Symposium (International) on Combustion. The Combustion Institute, Pittsburgh (1979) 1029-1038.

30. Tien C.H., Lee K.Y., Stretton A.J.: Radiation Heat Transfer. The Society of Fire Protection Engineers. Handbook of Fire Protection Engineering, SFPE Boston, Massachussets, USA, Section 1, Chapter 5 (1988).

31. Thomas P.H., Bullen M.L., Quintiere J.G., McCaffrey B.J.: Flashover and Instabilities in Fire Behaviour. Combustion and Flame 38 (1980) 159-171.

32. Thomas P.H.: Fire Safety Journal 3 (1980/81) 67-76.

33. Friedman R.: Behaviour of Fires in Compartments. International Symposium on Fire Safety of Combustible Materials. Edinburgh University (1975) 100-113.

34. Bullen M.L., Thomas P.H.: Compartment Fires with Non-Cellulosic Fuels. 17th Symposium (International) on Combustion. The Combustion Institute, Pittsburgh (1979) 1139-1148.

35. Hasemi T.: Theoretical and Experimental Study of Flashover. CSNI Meeting on Interation of Fire and Explosion with Ventilation Systems. Nuclear Facilities, Los Alamos National Laboratory, April (1983) 283-300.

36. Friedman R.: Survey of Computer Models for Fire Smoke. Factory Mutual Research, Norwood, MASS May (1990).

37. Rasbash D.J.: The Extinction of Fire with Plain Water: A Review: Proceedings of 1st International Symposium on Fire Safety Science. Hemisphere Publishing Corporation (1986) 1145.

38. Layman L.: Attacking and Extinguishing Interior Fires. National Fire Protection Association, Boston (1952) 134.

39. Pietrzak L.M., Johanson G.A.: Analysis of Fire Suppression Effectivness Using a Physically Based Computer Simulation. in Proceedings of 1st

International Symposium, Fire Safety Science. Hemisphere Publishing Corporation (1986) 1207.

40. Thomas P.H.: Use of Water in the Extinction of Large Fires. Inst. Fire Eng. 19 Pietreak (35) (1959)130-2.

41. Albert R.L., Delichatsios M.A.: Calculated Interaction of Water Droplet Sprays with Fire Plumes in Compartments. National Bureau of Standards, Centre for Fire Research, NBS - GCR 86-520 (1986).

42. Gardiner A.J.: The Mathematical Modelling of the Interaction Between Sprinkler Sprays and the Thermally Buoyant Layers of Gases from Fires. in Ph.D Thesis, South Bank Polytechnic, London, December (1988).

ON THE FIELD MODELLING OF FIRE USING PARALLEL PROCESSORS

G. Cox[1], P. Cumber[1], F.C. Lockwood[2], C. Papadopoulos[3] and K. Taylor[4]

[1] Fire Research Station, UK
[2] Imperial College, UK
[3] Centre for Renewable Energy Sources, Greece
[4] Topexpress Ltd, UK

ABSTRACT

The paper describes a first step in the production of a "parallel" version of a computational fluid dynamics (CFD) model currently in use for fire simulation purposes. The CINA CFD code has been successfully ported onto a system of nine transputers solving an isothermal flow problem. A 5.5 times speed enhancement over its "serial" performance was achieved.

Further work is required to examine the consequences of incorporating the essential features of buoyancy, combustion and thermal radiation.

INTRODUCTION

Computer fire models of the 'field' type have made substantial progress in recent years. Reasonable accuracy has been demonstrated for a range of smoke movement problems [1] encouraging their use for design evaluation purposes. Current research concentrates on improving treatments for thermal radiation and combustion phenomena [2,3] in order to extend their capabilities to include flame spread. Field models, based on the rapidly growing new technology of Computational Fluid Dynamics (CFD) can offer substantial advantages over alternative strategies because of their general formalism. This versatility can be appreciated by the range of problems that can be tackled by a single model. These include compartments as diverse as, for example, aircraft cabins, air supported domes, atria and tunnels.

An impediment to the more widespread, routine application of these models has been their high computational cost relative to the simpler zonal models. Even though computer hardware speed has increased and costs reduced substantially over recent years, tens of hours of minicomputer time are still required to solve time dependent, three dimensional, smoke movement problems.

It is now widely recognized that serial computers are approaching their limit in terms of further possible speed improvement. This is because conventional sequential machines are limited by the speed with which information can be exchanged between processor and memory. Since this cannot exceed the speed of light in the medium involved, travel distances have to be reduced thus exacerbating power dissipation problems. This is what determines the characteristic cylindrical shape of the CRAY series of supercomputers.

Recognition of these performance limitations imposed by the basic laws of physics has led to a number of developments to overcome the difficulty. One of the most promising is the use of parallel processors. If a number of processors, each with their own local memory, can be configured in such a way as to undertake various computational tasks simultaneously, then the communications bottleneck experienced with the traditional sequential machine can be overcome.

This is the achievement of the INMOS transputer chip. Each transputer is a single chip microprocessor with both store and processor implemented on a one centimetre square of silicon. Each chip has four communication channels for linking a number of processors together in a parallel network.

Being factory-produced in large volume, transputers offer, potentially, not only much greater speed but also low cost. There is, however, a penalty. To exploit the potential of parallelism, the computer software needs to be rewritten to divide the computational task in the most appropriate way.

Ideally, new programs would be written using the special purpose, high level, language OCCAM to derive maximum benefit. However, much investment in terms of experience and validation has already been made in existing FORTRAN programs such as JASMINE and TEACH.

This paper describes work undertaken to examine the potential benefit to be derived from attempting to run one such model on a number of parallel transputer processors.

A number of similar studies have been undertaken in an attempt to improve the performance of algorithms related to computational fluid dynamics for different computer architectures. Ierotheou, Richards and Cross [4,5] vectorised the SIMPLE algorithm [6] and reported a seven-fold maximum speed enhancement for the test problem they considered. Cross, Johnson and Chow [7] implemented an enthalpy based solidification algorithm on a vector processor and a transputer system. For the vector implementation, they report an enhancement of between 3 and 5. The parallel implementation increased this to between 6.7 and 7.5 on 9 transputers, depending on the number of numerical control volumes chosen. Shioji, Dent and Wright [8] adapted the fluid dynamics package KIVA, developed for modelling internal combustion engines, to run on a transputer system. They achieved a maximum speed-up of 2.5 on 4 transputers. This relatively poor enhancement is due to the fact that KIVA models fuel sprays using a particle tracking routine which was run on one transputer preventing good load balancing between processors.

The model chosen to undertake the evaluation in this paper is a new CFD code, CINA [9], specifically written with user-friendliness in mind. It uses the same numerical algorithms as those referred to earlier and thus conclusions from the work will be equally applicable to those and to most other engineering 'field' models of the fire process.

THEORETICAL FOUNDATION OF FIELD MODELS

These models solve, locally, at many thousands or perhaps tens of thousands of nodes throughout the compartment of interest, the time dependent conservation equations for mass, momentum, energy and chemical species.

These each take the general form

$$\frac{\partial}{\partial t}(\rho\phi) + \frac{\partial}{\partial x_i}(\rho u_i \phi) - \frac{\partial}{\partial x_i}\left(\Gamma_\phi \frac{\partial\phi}{\partial x_i}\right) = S_\phi$$

(1)

where ϕ is the generic variable which represents the three (Cartesian) velocity components (u_i), the enthalpy h, the mass fraction of a particular chemical species m_i or the conserved mixture fraction f.

For turbulent flows two further transport equations are solved for k, the turbulence kinetic energy and ϵ, its rate of dissipation.

The term $S\phi$ is a source or sink term describing, for example, the effects of chemical production and radiative heat loss.

These equations are solved subject to appropriate boundary conditions using the finite domain methods of Spalding and his co-workers [10]. The boundary conditions include prescriptions of, for example, thermal properties of compartment walls and ventilation conditions such as natural or powered ventilation and wind effects etc. A fuller description of the equations specifically related to fire phenomena has been given by Cox and Kumar [11]. The numerical solution of these equations can be summarised as follows.

The equation set (1) is solved approximately by superimposing a, usually, rectangular finite difference grid on the domain. Each equation is integrated over each control volume defined by the finite difference mesh, assuming the dependent variables have a certain profile between control volume centres and, for transient problems, over a time interval. The profile assumptions used in CINA correspond to a fully implicit hybrid formulation, where the transient term is described by a backward difference scheme and the convection and diffusion terms by a hybrid difference scheme, (*see* Spalding [10]). Integrating equation (1) in this way for point P gives a system of algebraic equations of the form,

$$(\Sigma A_i + A_P - S_P)\,\phi_P = \Sigma A_i\,\phi_i + A_T\,\phi_T + Su \qquad (2)$$

where the subscript i represents the East (E), West (W), North (N), South (S), High (H) and Low (L) spatial neighbours of P. Figure 1 shows the notation used for a two dimensional grid. In a three dimensional grid, H and L are the high and low values above and below the plane of the paper.

In a three dimensional, transient calculation each control volume equation (2) normally connects values of the dependent variable (ϕ_P) at a grid point with seven others, its neighbours in space (ϕ_i's) and the value of ϕ at P at the previous time ϕ_T. The influence coefficients A_i comprise convective and diffusive links between control volume P and i. A_P and A_T are dependent on the transient term. $S_u + S_P\,\phi_P$ is a linear approximation of the source term integrated over the control volume.

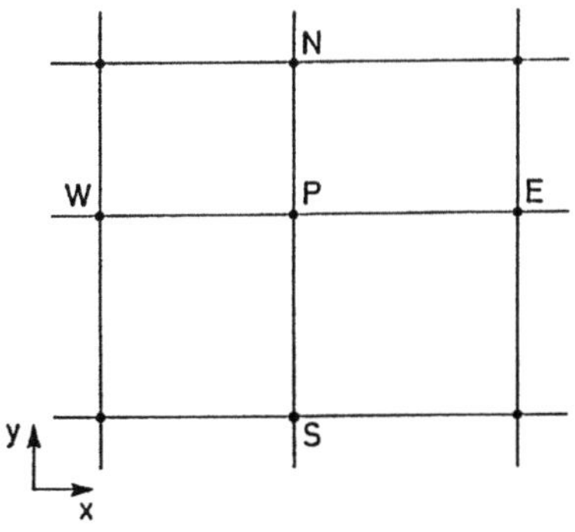

Fig. 1: A portion of a two dimensional grid

548

The solution of the control volume equations approximates the variable ϕ at the control volume centres of the rectangular mesh. The control volume equations (2) are solved using the SIMPLE algorithm (Semi Implicit Method for Pressure Linked Equations) see Patankar and Spalding [6]. SIMPLE is an iterative procedure, where, given an initial guess for the velocity and pressure fields, they are then successively corrected such that local mass balance is approximately conserved. Given the new velocity fields the other field variables are solved for, and the correction cycle is repeated until a convergence criterion is satisfied such that the change in the field variables from one iteration to the next is small or the mass continuity error is small.

At each interaction several linear systems must be solved, one for each field variable, in addition to the pressure correction equation. In CINA the algorithm used to solve these linear systems is a line-by-line procedure based on the tridiagonal matrix algorithm, *see* for example Patankar [12].

In the alternating direction, line relaxation method applied to all the field variables, apart from pressure, each x-y plane of control volumes is considered in turn. All off-plane values are fixed and lines of control volumes in the plane parallel to the co-ordinate directions are solved for in turn. Using the usual notation, in x-y planes we have east-west lines and north-south lines.

Considering north-south lines, the control volume equations (2) can be rewritten in the form

$$A_P \, \phi_P = A_N \, \phi_N + A_S \, \phi_S + [A_E \, \phi_E + A_W \, \phi_W + A_H \, \phi_H + A_L \, \phi_L + b] \tag{3}$$

where the terms in square brackets are assumed to be known. The above system (3) can be solved using the tridiagonal matrix algorithm. This is repeated for all north-south lines. To solve along east-west lines the set of eqs. (2) can be rewritten in the form,

$$A_P \, \phi_P = A_E \, \phi_E + A_W \, \phi_W + [A_N \, \phi_N + A_S \, \phi_S + A_H \, \phi_H + A_L \, \phi_L + b] \tag{4}$$

and again the tridiagonal matrix algorithm is used. During the solution of the set of eqs. (3) and (4) the most recently calculated values in the square brackets are used. Each plane of control volumes is solved for in turn. Once all planes have been calculated, if the solution is not sufficiently converged, the first plane is recalculated and the process continues. To increase the rate of convergence, the pressure correction equation is solved using a whole field solver rather than calculating the solution slab by slab. The whole field solver extends the line-by-line solution to all three co-ordinate directions by using the tridiagonal matrix algorithm to solve for east-west lines, north-south lines and upstream-downstream lines.

IMPLEMENTATION ON TRANSPUTER ARRAY

The finite difference mesh divides the domain of interest into a large number of control volumes, here arranged into a cuboid. The character of the calculations conducted during an iteration of the SIMPLE algorithm determines how the whole task is to be split up over many processors. Here, the computational domain was decomposed into sub-domains which were distributed over each processor. To keep communication between processors and the configuration of transputers as simple as possible it was subdivided by partitioning only one of the co-ordinate directions, in this example the z direction. This guarantees that slices of contiguous array elements are the only data to be communicated to neighbouring processors and that the domain maps neatly onto a line of transputers. Clearly to extract the maximum parallelism from this arrangement, the number of planes of control volumes in the z direction must be large. Configuring the transputer array into a line places a limit on the number of transputers that can be used efficiently to the number of control volumes in the z direction, N_z.

For efficiency, the subdivision of the domain is as even as possible to balance the load on the transputers. As transputers do not share memory, copies of the latest field variable values at the edges of each processor's area-of-responsibility, required to calculate fluxes for example, must be exchanged from time to time. The data involved in this communication are often termed 'halo' data as they form a halo to those calculated by a particular processor.

Cross-sections are added to the beginning and end of the portion of the flow variable arrays stored on each processor to contain the "copied data". Figure 2 shows the domain of the test problem chosen and the domain decomposition onto three processors.

The main computational effort is devoted to solving the linear systems at each iteration, described above. The serial version of the line-by-line method of solving linear systems is solved in all three co-ordinate directions.

The serial code for the lines in the x and y directions is unchanged for the parallel version, since all the information on x-y planes of control volumes is held on one processor. However in the serial version, x-y planes of control

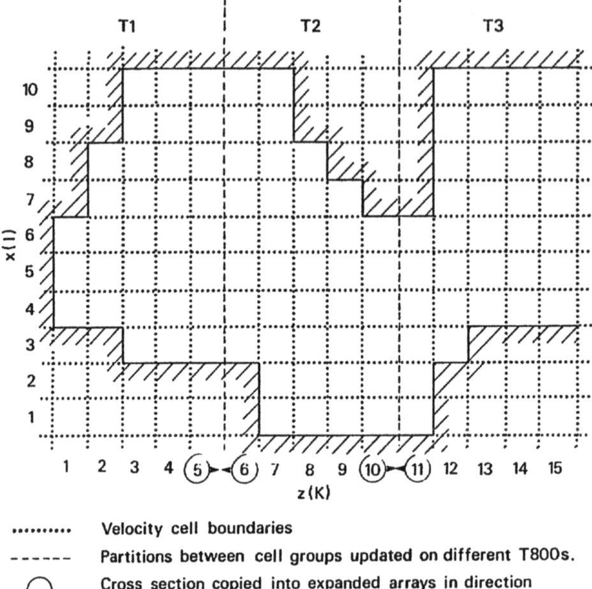

.... Velocity cell boundaries

------ Partitions between cell groups updated on different T800s.

○ Cross section copied into expanded arrays in direction of arrow.

Fig. 2: A cross section of the 10 x 6 x 15 test case distributed over 3 processors.

volumes are dealt with sequentially from left to right, and field variables on both adjacent planes are referred to in order to calculate some of the coefficients in the tridiagonal system. Thus updates are based on current values from the left and those from the previous iteration on the right. For the parallel version, the coupling between sub-domains is reduced so that processors can work more independently, increasing the parallel nature of the calculation. The most left hand plane on a processor uses data from the previous iteration on both the left and right planes (*see* Figure 2). Note the left hand plane is stored as halo data. For all other planes on the processor, the calculation is similar to the serial version. With this modification, it was found for the test case considered in the next section that the parallel version required 87 iterations to achieve convergence compared with 86 in the serial case.

For solution of the pressure correction equation, the tridiagonal solver is used for lines parallel to the z direction, requiring data stored on all processors for all lines. This portion of the code was "parallelised" by taking advantage of the fact that N_x x N_y similar tridiagonal systems are to be solved, setting up two pipelines. In the first, a processor carries out its part in the forward elimination [12] for a particular line in the z direction, hands over the partial result to the processor on its right to carry on, then takes up the elimination for the next z line. The second pipeline is set up to calculate the back substitution, which is similar to the elimination pipeline. Apart from start up and run down time, which becomes less significant as N_x x N_y increases, the pipelines are executing simultaneously on all processors.

RESULTS AND DISCUSSION

The problem considered for initial test purposes was a cold turbulent simulation within a furnace. The geometry and the grid used are given in Figure 2. The relatively coarse grid of 10 x 6 x 15 was used. This simple test problem was run for a fixed number of sweeps on a Quintek Fast 9 board containing nine T800 transputers. The problem was run several times over on different combinations of transputers. The measured processor time per iteration of SIMPLE as a function of the number of transputers is given in Table 1.

Here Tn denotes the time required for one sweep of CINA on n transputers, then speed enhancement has been defined as:

$$S_n = \frac{T_1}{T_n}$$

and efficiency as,

$$E_n = \frac{T_1}{nT_n} \times 100\%$$

The speed-up parameter gives an indication of the increase in performance gained by using a number of transputers compared to using one. The efficiency parameter gives an indication of the benefits of introducing more transputers.

Table 1. The execution times of the test problem on a Fast 9 Board

Nu. of T800s	Tn (sec)	Sn	En (%)
1 'Serial'	8.995		
1 'Parallel'	9.482		
2	4.758	1.9	94
3	3.249	2.7	91
4	2.793	3.2	80
5	2.156	4.1	83
8	1.598	5.6	70
9	1.607	5.5	62

Some explanation is required concerning the two different iteration times given for one T800 in Table 1. The 'serial' time is the time for the original program. The 'parallel' time is calculated by placing the programs for two transputers into the memory of one transputer. Comparing the two timings we see that there is an overhead of 6.6% associated with the communication and extra calculations required for the parallel implementation. The communication overhead could be reduced by transmitting data from one transputer to another while the processing unit is active.

Figure 3 shows these data graphically. The solid line represents an ideal speed enhancement of 100%. Note that the speed-up is good with high efficiencies, considering the small number of x-y planes of data.

The results for three and five transputers are particularly good as these are factors of N_z (fifteen), enabling good load balancing to be achieved. Four and eight transputers are less efficient for this problem because the first processor has one less cross section to update than the others and is comparatively under-utilised. Similarly there is little or no gain using nine transputers instead of eight.

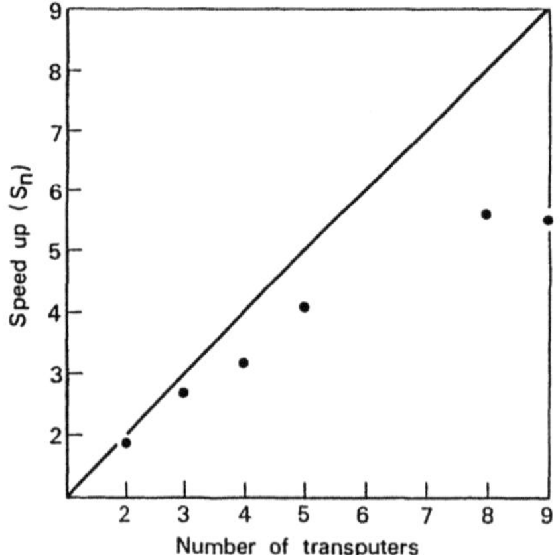

Fig. 3: Speed up parameter for the test problem

CONCLUSIONS

A computational fluid dynamics package, CINA has been successfully ported onto a transputer system and demonstrated significant speed up for a simple test problem. The speedup of 5.5 for 9 transputers is consistent with the results reported by Cross *et al* [7] for a simpler problem but with a similar number of control volumes.

Adding further transputers should increase the speedup further with an efficiency of around 70%. Thus, for example, 18 transputers should reduce execution times by a factor of around 12 and so on.

There are a number of possible ways of improving the performance of the parallel implementation of these models. As already stated the communication overhead could be reduced by transferring data between transputers concurrently with the processor unit being active. One other possibility for improving performance is to configure the transputers into a two dimensional matrix and hence use all four communication channels on the transputers. This would require a more complicated communication harness and would also increase the communication time to the central processing unit, thus degrading efficiency for small problems.

The work described represents only a first step in the parallel implementation of the fire model. It has thus far here only addressed an isothermal problem with none of the complications associated with buoyant flow, chemical reaction or thermal radiation.

If a similar performance can be demonstrated with the full model then the prospects for cheap CFD-based fire modelling are very promising.

REFERENCES

1. Cox G., Kumar S., Markatos N.C.: Some Field Model Validation Studies in Proc. First International Symposium on Fire Safety Science. Hemisphere (1986) 159-171.

2. Lockwood F.C., Malalaskera W.M.G.: Fire Computation; The Flashover Phenomenon in Proc. 22nd Symposium (International) on Combustion, The Combustion Institute (1989) 1319-1328.

3. Kumar S., Cox G.: Radiation and Surface Roughness Effects in the Numerical Modelling of Enclosure Fires in Proc. Second International Symposium on Fire Safety Science, Hemisphere (1989) 851-860.

4. Ierotheou C.S., Richards C.W., Cross M.: Vectorisation of the SIMPLE solution procedure for CFD problems — Part I: A basic assessment. Appl. Math. Modelling 13 (1989) 524-529.

5. Ierotheou C.S., Richards C.W., Cross M.: Vectorisation of the SIMPLE solution procedure for CFD problems — Part II: The Impact of Using a Multigrid Method. Appl. Math. Modelling 13 (1989) 530-536.

6. Patankar S.V., Spalding D.B.: A Computer Model for Three Dimensional Flow in Furnaces in Proc. 14th Symposium (International) on Combustion, The Combustion Institute (1973) 605.

7. Cross M., Johnson S., Chow P.: Mapping Enthalpy Based Solidification algorithms onto Vector and Parallel Architectures. Appl. Math. Modelling 13 (1989) 702-709.

8. Shioji M., Dent J.C., Wright C.: Engine Based Computational Fluid Dynamic Simulation Using KIVA with a Transputer Based Concurrent Computer. Society of Automotive Engineers Paper 89-1986 (1989).

9. Lockwood F.C., Malalaskera W.M.G., Papadopoulos C.: Steel and Glass Furnace Modelling. Applied Energy Research Conference, Institute of Energy, Swansea U.K. (1989) 341-356.

10. Spalding D.B.: A Novel Finite Difference Formulation for Differential Equations Involving First and Second Derivatives. Int. J. Num. Methods Eng. 4 (1972) 551.

11. Cox G., Kumar S.: Field Modelling of Fire in Forced Ventilated Enclosures. Combustion Science and Technology 52 (1987) 7.

12. Patankar S.V.: Numerical Heat Transfer and Fluid Flow. Hemisphere, New York (1980).

RADIATIVE TRANSFER AT THE SURFACE
OF A SMALL SCALE POOL FIRE UNDER THE INFLUENCE OF
EXTERNAL RADIATION

X.L. Zhang and J.P. Vantelon
Laboratoire de Chimie Physique de la Combustion
U.A. 872 au CNRS, Université de Poitiers
Domaine du Deffend, Mignaloux-Beauvoir
86800 Saint Julien L'Ars — France

ABSTRACT

Among the relevant environmental conditions which can influence fires, externally applied radiation appears to be important. In building and compartment fires, for example, these external radiations may be supplied by hot walls, ceiling layers of hot gases, or by other burning objects. In the present work, the influence of external radiation on the radiation flux surface and mass burning rate of a small scale kerosene pool fire is studied. A comparison between flux measurements and calculations from a theoretical approach is made. Despite the experimental difficulties and uncertainties, the agreement appears reasonably good. The results underline that the influence of an external radiative flux on the resulting flux at the surface is relatively weak, due to a blocking effect by fuel vapor and soot. The proposed modeling provides a basic understanding of the surface heat transfer behaviour and a comprehensive account of the role of an external thermal radiation.

INTRODUCTION

The importance of radiation in the heat transfer process from a diffusion flame has been evidenced for a long time. However, all the aspects of this problem have not been studied until now with the desirable attention, more particularly cases involving external radiative fluxes.

This accounts for the small number of past approaches to this problem, mostly qualitative ones [1-6]. These studies have been often concerned essentially with the basic observation of burning rate dependence and the effect of fire size on this influence. Among them, Shih's study [2], which measures the burning rate of an acetone pool fire in the presence of external radiation, appears to be the one most representative of this problem. The results show that the pyrolysis flow rate noticeably increases under the influence of this radiative flux when the fire diameter is small. On the other hand, when the diameter becomes large, the pyrolysis flow rate increase is less significant, as the flame is more and more opaque. It seems therefore that the applied external flux can less penetrate the flame and influence the fuel surface. Rongere [3] developed a numerical analysis using a single dimension integral model to assess the influence of an external radiation induced by the smokes and the walls of a room on the flame and its plume rising above a burning material. His calculations let think that this kind of external radiation does not very much affect these properties. Although this analysis allows the particular behaviour observed in real size tests to be explained, it is however but a first approach. As a matter of fact, the flame model is based on numerous assumptions, which are sometimes too drastic.

The main objective of the present work is to model the radiative transfers from a flame developing above the surface of a fuel, under the influence of variable intensity external radiative fluxes, in order to predict the evolution of the resulting surface thermal flux. The selected configuration is that of a small scale axisymmetrical kerosene fire. Indeed, most of the works covering the properties of buoyant diffusion flames above fuels are performed on small scale fires because they can be studied in laboratory while maintaining the major properties of an actual fire. This kind of configuration, with horizontal axisymmetrical basis, has been widely investigated earlier, as well from an experimental as from a theoretical point of view, but essentially for freely developing fires without external disturbances.

The study of the radiative transfers in question included two stages:

- calculation of the radiative properties of the flame from the preponderant parameter fields: temperature and absorption coefficient due to the soots.

- calculation of the incident radiative fluxes on fuel surface, using a zoning method associating the various contributions of fluxes in the system, and comparison with measurements.

EXPERIMENTAL PROCEDURE

The whole experimental setup is shown in Figure 1. The kerosene is burnt in a circular pan 15 cm in diameter and 5 cm deep. The pan is made of a dual envelope allowing water circulation. This provides an efficient pan cooling, essential to avoid a fuel side heating in the case of experiments carried out with imposed external radiative fluxes. The pan is placed on a stand which keeps 80 cm above ground. The assembly is placed in a room 4 x 3 m wide and 3 m high.

The fire is then isolated from external disturbances. Small inlets in the walls at ground level allow fresh air to flow into this chamber. A fume hood is mounted on the ceiling to direct the smokes out. Smokes are removed by natural heat convection since no mechanical ventilation is used during the trials to avoid any forced heat convection inside the room.

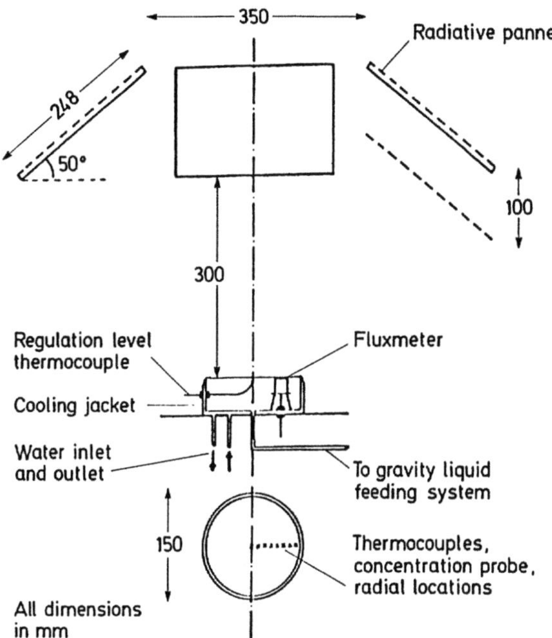

Fig. 1: Pool fire test facility

The study of the influence of an imposed external radiative flux is carried out using radiant panels placed around the flame, with the radiant faces turned towards the pan. The system consists of four panels placed symmetrically with respect to the fire axis. Each panel is made of a set of four transmitters with full emissive face, each one delivering 400 W nominal power. For each radiant panel, the uniformity of emission is obtained at a distance of about 5 cm. In these conditions of use, it can be admitted that there is no distribution curve. The energy is equally distributed over a surface equal to that of the transmitters pack (25 x 25 cm). The power transmitted by the panels can be varied either by modifying their supply intensity, or by changing the distance to the target to be irradiated. Eight intensities of imposed external fluxes were used during the tests. Seven were obtained for a same panel arrangement and the following nominal power values: 40, 50, 60, 70, 80, 90 and 100%. The highest value was obtained with the panel placed 10 cm lower than previously and delivering 100% of the nominal power. This corresponds to the following increasing flux values, measured with fluxmeters, at the fuel surface level: 0.03, 0.05, 0.07, 0.10, 0.12, 0.16, 0.21 and 0.23 $W.cm^{-2}$. These irradiation levels may be regarded as somewhat poor, especially when compared with the external flux levels that can be reached in some compartment fires. They likewise look poor with respect to fluxes due to the flame and received by the surface, as will be seen in the analysis of the experimental results. In fact, these limited values have been imposed by experience. We could observe that too high external fluxes involved such modifications and disturbances of the flame, in connection with its volume as well as with its structure, that it was difficult and even impossible to carry out significant measurements. In particular, the flame tends to project more and more widely from the pan and, owing to vapour condensation on the walls, to develop around the pan in a random way. Let us note besides that in Shih's study, which is the only previous work of this type [2], external flux levels were also limited to approximately the same value: 0.29 $W.cm^{-2}$. The selected operating conditions are the most favourable for a rational study at laboratory scale.

The burning rate is measured with a gravity liquid feeding system based on an electronically controlled On/Off valve. A more detailed description of the device is given in ref. [7]. The fuel level is kept constant by means of a thermocouple whose output signal is compared with the liquid-vapour transition temperature of kerosene and used to energize the valve. The weight of the kerosene fed into the pan is monitored by means of a load cell which follows the consumption of fuel as a function of time. Before each test, the pan is filled to 1 mm below the lip with fresh fuel.

Measurements of the two major parameters, temperature and soot concentration, giving a data base to calculate the radiative interactions, are performed. The procedure of these measurements has been widely described in previous works [8, 9, 10]. Temperature measurements are made with 50 μm diameter chromel-alumel thermocouples. The digitized signals are sampled at a rate of 200 Hz and the measured temperatures are time-averaged with variations in the mean not exceeding 5%. Soot concentration measurements are made by means of a laser light attenuation method using a fiber optic probe system. Since the absorption of light is due primarily to the presence of soot, the monochromatic absorption coefficients measured can be interpreted as soot concentration data. The data are also time-averaged and the uncertainty associated with each value is high (up to 30%), a problem inherent to the experimental technique. No measurements are performed at distances closer than 2 cm to the liquid surface because of the dimension of the laser probe.

Finally, radiative flux measurements at the fuel surface are made by means of water-cooled wide-angle (150°) fluxmeters, fitted with a sapphire window and having a sensing element area of 0.13 cm^2. The detectors are positioned in such a way that their sensing element is at the same level as the liquid. This is possible owing to the constant level system. The measurements are performed at the centre of the pan and at radial distances of 0.75, 2.25 and 4.65 cm for different external radiative fluxes. The operating conditions have already been described in Ref. [11]. Significant values cannot be obtained without ensuring a very efficient cooling of the detector. However, cooling induces condensation of fuel vapours on the protective window and, thus, the formation of a thin layer of liquid which alters measurements. To eliminate this, a continuous suction is created, at a moderate flow rate, by means of a fine probe 1 mm in diameter.

EXPERIMENTAL RESULTS

Combustion rate, temperature, absorption coefficient as well as radiant heat flux measurements were performed in different cases. These measurements were first carried out on a free flame in non disturbed medium. Then, they were repeated on a flame submitted in succession to different external radiative fluxes, imposed according to the protocol previously described.

Figure 2 shows the evolution of the combustion rate in the various experimental conditions described. It should be noted that each reported value is the mean of a great number of tests (ten or more). It can be seen that the increase in the unit mass flow rate of fuel versus the applied radiative flux tends to lessen as soon as the external flux is amplified. This evolution coincides with that it is generally observed: the sensitivity of the pyrolysis rate of a liquid or solid fire, compared with an external flux, is all the less marked as the flux is increased, since the flame is more and more opaque to the applied radiation and causes a true insulating effect.

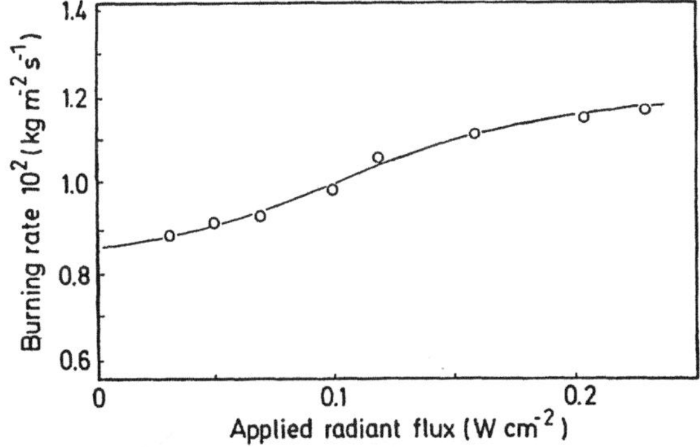

Fig. 2: Burning rate variation with the applied radiant flux

As the flame is statistically axisymmetrical and stationary, the temperature and the monochromatic absorption coefficient were measured in an axial half plane. Nevertheless, with this type of restrictive size flame, it is commonly observed three zones according to height [12]:

- the base of the flame, just above the liquid surface, which is the so-called "persistent" zone, characterized by an annular form. This zone, rich in fuel vapours, is the main chemical reaction and ambient air entrainment zone.
- a transient "intermittent" zone with a fluctuating and very marked turbulent character, where chemical reactions are less important.
- a turbulent thermal flame envelope zone in which chemical reactions are virtually over and the gas velocity and temperature are gradually decreasing with height.

Among the three zones, the "persistent" zone of flame is the most interesting one since it is where the processes which lead to the development and stabilization of the flame on the liquid are generated. The flame structure is characterized by the development of a turbulent envelope which entrains the ambient air. The air is drawn laterally, enters the reactive zone at the bottom and mixes with the vapourized fuel. At the edge of the pan, the inflammation of the combustible gases and air, thus mixed, results in a rapid propagation of a combustion front above the liquid surface. A part of the vapourized fuel is thus consumed. This flame front heats the surface but the production of gaseous products temporarily inhibits the air entrainment until these products enter the envelope, by convection, in large whirling structures. A new volume of fuel vapours can then mix with another volume of air. This new premixed volume is also ignited and develops another combustion front which initiates a new cycle. This cycle: air supply, combustion, propagation, described by Hertzberg [13], causes a series of regular beats of the flame. This entails the existence of ascending annular structures periodically generated at flame bottom. Thus, even in the presence of an initial flux of stable fuel gas, it is in fact a "pulsating" type structure and the low flame zone is only seemingly "persistent". Therefore, fluctuations are initiated close to the surface. Since the flame oscillations are brisk, the temperature and absorption coefficient parameters fluctuate. However, owing to the complex character of these fluctuations and the experimental difficulty which instantaneous measurements exhibit, we have restricted ourselves to the determination of mean values, as was already explained.

Detailed mapping of the temperature and monochromatic absorption coefficient, obtained for undisturbed flame and the four highest radiation levels, has previously been presented in Ref. [14] and is described only briefly here. The temperature field shows a band of temperature maxima that approximately coincides with the luminous flame boundary. The temperature in this band, however, decreases towards the fire axis and is lower than that expected in a kerosene flame. This indicates well that the flame fluctuates around a mean position and that, as a consequence, the measured temperatures are lower averaged values of the fluctuating temperature field. The measured contours of constant monochromatic absorption coefficient, which can be associated with lines of constant soot concentration, show a region of elevated concentration at the axis, near the liquid surface where the pyrolysis of the vapourized fuel takes place and the formation of soot is initiated. Near the pool centreline, the existence of stagnant and recirculating gas would result in the accumulation of fuel vapour, pyrolysis products and soot. The low diffusivity of the soot would reduce its flow. Near the flame lower leading edge, the gas temperature is too high for soot to remain unburned, which explains the low values measured here. It is also interesting to observe a second region of elevated soot concentration at the centreline, above the band of temperature maxima. This indicates the presence of unburned fuel and corresponds to the continuation of the combustion reaction in the intermittent zone of the flame. The results show that the application of external radiant fluxes hardly affects the main properties of the flame. It causes an increase in the flame volume and, consequently, a widening of spatial distribution of the flame structure. However, the overall flame properties, with the three regimes well differentiated, and maximum temperatures and soot concentration levels remain unchanged.

The radiative flux measurements at the surface, corresponding to the four highest radiation levels, lead to the values reported in the following table.

		Radiative flux to fuel surface (W. cm^{-2})		
Radial distance (cm)		0.75	2.25	4.65
External applied flux (W. cm^{-2})	0.00	1.44	1.43	1.34
	0.12	1.58	1.57	1.48
	0.16	1.67	1.67	1.52
	0.20	1.67	1.64	1.55
	0.23	1.78	1.67	1.60

These values show that the measured resulting surface radiative fluxes increase with the external applied fluxes but that this augmentation is not very marked. They show also a diminution of the flux as the radial distance increases. It is an important observation indicating that the rate of gasification is more important in the pool centre region. A more detailed investigation of this kind of evolution is given in Ref. [11]. It shows, by measurements performed using a pan composed of concentric compartments that the burning rate is the largest in the inside region of the pool and confirms thus that the radiative heat transfer is the dominant mode of heat transfer to the surface.

COMPARISON OF CALCULATION WITH EXPERIMENT

The radiative exchanges between the reactive medium, the fuel surface and the environmental elements are dependent on the spectral and directional particularities of the interactive elements. For practical reasons and simplicity of calculation, we have made the following assumptions:

- the radiant panels are opaque, grey and diffuse surfaces.
- the liquid surface is a grey and half-transparent surface.
- the part of radiation due to soots is predominant with respect to that of the gas and the medium can be considered as grey.

Besides, it should be noted that the fuel exhibits a noticeable reflective power.

Then, the evaluation of the fluxes radiated to the fuel surface has been performed using a zoning method, a well adapted method to the interactive system considered. Fundamentals of the zoning method are developed in numerous textbooks such as Refs. [15] and [16]. This method consists in dividing the system which contains a non isothermal gaseous medium in surfaces and volumes that can be individually regarded as being isothermal.

Let us consider a surface zone with area A. In the considered conditions, the radiative flux reaching this surface from a gaseous reactive zone element with volume V, can be calculated by integrating the expression:

$$q_{g\text{-}s} = \frac{\sigma}{\pi A} \int_A \int_V \frac{\cos\beta}{r^2} T^4 k_a \tau dA \, dV$$

$$(1)$$

where:

k_a = absorption coefficient
r = optical distance between volume element V and surface S
T = temperature
β = angle between radius considered and the normal to the surface

$\sigma \qquad$ = Boltzmann's constant

$\tau \qquad$ = transmissivity

A direct gas - surface exchange area is thus defined and, similarly, a direct exchange area between two surfaces can be defined. Then, the following general system representative of the exchanges between the different zones can be expressed as follows:

$$\sum_{j=1}^{N}\left(\frac{1-\varepsilon_k}{A_k}\,\overline{S_jS_k}-\delta_{kj}\right)q_{o,j}=-\frac{1-\varepsilon_k}{A_k}\sum_{y=1}^{\Gamma}\overline{g_YS_k}\,\sigma T_Y^4-\varepsilon_k\sigma T_k^4 \qquad (2)$$

where:

$j \qquad$ = 1,2, ... N represents each surface zone
$k \qquad$ = 1,2, ... represents the considered surface zone
$Y \qquad$ = 1,2, ... Γ represents each volume zones
$\overline{S_jS_k} \qquad$ are the direct surface - surface exchange areas
$\overline{g_YS_k} \qquad$ are the direct surface - volume exchange areas
$q_{o,j} \qquad$ is the radiosity of each surface zone
$\delta_{kj} \qquad$ is Kronecker's symbol
$\varepsilon_k \qquad$ is the emissivity of surface k

The unknowns of this system of equations are the radiosities of the surface zones.

We have divided the studied system into four subsystems: surface of the fuel liquid, radiant panels, flame volume, ambient air. It has to note that the flame cannot be considered as a homogeneous medium since it exhibits variable properties. To use the system of equations (2), the flame has to be divided into Γ homogeneous subvolumes. In system (2), the term:

$$\frac{1}{A_k}\sum_{Y=1}^{\Gamma}\overline{g_YS_k}\,\sigma T_Y^4 \qquad (3)$$

represents the fluxes induced by the flame and incident to the surfaces. Then we observe that the exchange areas $\overline{g_YS_j}$ to be considered are very numerous and make the calculation difficult. Consequently, the fluxes from the flame and reaching any surface zone are not considered according to the general pattern of the zoning method, but by direct integration using expression (3). System (2) can be expressed as follows:

$$\sum_{j=1}^{N}\left(\frac{1-\varepsilon_k}{A_k}\,\overline{S_jS_k}-\delta_{k,j}\right)q_{o,j}=-(1-\varepsilon_k)\,q_{r,i,k}-\varepsilon_k\sigma T_k^4 \qquad (4)$$

564

The ambient air constitutes a particular zone owing to its temperature and its transparent character. This zone has virtually no influence on the other zones. So we disregarded all terms that concern it. After defining the different zones, the system of equations (4) characteristics of the different exchanges may be established. It constitutes a system of linear equations where the unknowns are the radiosities of the surfaces: fuel surface, radiant panel surface.

Before solving these equations, the different exchange areas must be defined. This is made considering appropriate coordinate systems. Appropriate initial and boundary conditions also have to be fixed, and care must be taken in determining the reflectivity of the surfaces. The fuel surface temperature is considered as being uniform and always equal to the temperature assigned to the thermocouple which regulates the fuel level (460K). A previous work [11], on a pan of identical dimensions, showed that the reflectivity of the liquid surface varied from 10 to 6% between the centre and the edge of the pan and we adopted a similar distribution. Concerning the reflectivity of the radiant panels, it has been determined by means of appropriate flux and temperature measurements for the different considered irradiations. As for the emissive power, as previously seen, we considered that the role of the soots was predominant. The soot spectrum is comparatively continuous and the single, classical and empirical equation relating the monochromatic absorption coefficient $k_{a,\lambda}$, as it can be measured, to the volumic fraction f_v and to the wavelength λ, can be used:

$$k_{a,\lambda} = B\, f_v\, \lambda^{-\eta} \tag{5}$$

B being a constant.

The exponent of wavelength varies somewhat with fuel, flame condition and wavelength, but it is usually approximated as unity.

We thus defined an integrated absorption coefficient:

$$k_a = \frac{\int_0^{\infty} \frac{B\, f_v}{\lambda} N_{b,\lambda}\, d\lambda}{\sigma T_k^4} \tag{6}$$

where $N_{b,\lambda}$ is the monochromatic emissivity of the black body.

It is then possible to analytically integrate (6), using appropriate functions, to finally obtain:

$$k_a = 266.4\, B\, f_v\, T$$

The luminance of a path 1 can be then calculated using the expression:

$$i(\iota) = -\frac{\sigma}{\pi} \int_0^1 k_a(\iota') \, T(\iota')^4 \, \tau(\iota',\iota) \, d\iota'$$

The transmissivity is given by the following relation:

$$\tau(\iota'\iota) = \exp\left(-\int_{l'}^1 k_a \frac{T(\iota')}{T(\iota'')} \, d\iota''\right)$$

the ratio T(l')/T(l") being introduced to take the variations of temperature along the considered path l (l' and l" are intermediate integration parameters).

Nevertheless, the above relationships, which include average parameters, can only provide approximate values of the radiative heat flux. In order to consider the fluctuating character of the flame, described above, and to evaluate the contribution of the fluctuating part of the radiative properties or the heat flux, the instantaneous quantities can be decomposed into mean and fluctuating components. However, because of the strong dependence of the radiative flux intensity produced on temperature, it can be admitted, with a good approximation, that the fluctuations of the absorption coefficient are negligible. Then, only for temperature measurements, we have included a probability density function in addition to their averaged values. Assuming a shape for the probability density function P(Ti), we can write:

$$\overline{(\overline{T} + T')^4} = \int_0^\infty P_{(Ti)} \, T_i^4 \, dT$$

where Ti is the instantaneous temperature.

Previous results obtained in similar flames [17, 18] and some of our measurements performed either in the persistent or intermittent and plume zones, showed that the ratio between the mean square of the fluctuation and the mean value, ranges between 0.2 and 0.3 for the persistent zone, and between 0.25 and 0.35 for the intermittent and plume zones. In this work, we considered for $\sqrt{T'^2}/\overline{T}$ a mean value of 0.2 in the persistent zone and 0.3 in the intermittent and plume zones. For the shape of the PDF (Figure 3), we considered as in previous results, rectangles for the persistent region and triangles above. However, the upper part of the flame is characterized by the most marked fluctuations. Their contribution is displayed by the shape of the PDF in this region, which reveals a broadening and a "drag" effect on the high temperature side. Then, it appears that a more refined modeling of this PDF is needed. The solution has been to adjust the shape by the addition of another

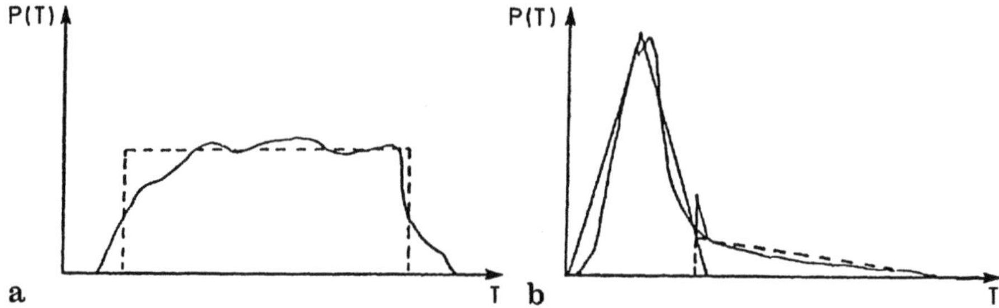

Fig. 3:   Shapes of the PDF a) rectangular form for the base of the flame. b) triangular form
with a "drag" for the intermittent and plume zones.

triangle representative of this "drag", constituting 10% of the overall surface of
the PDF. A sensitivity study shows that to use this approach instead of the exact
shape does not affect seriously the calculated radiative intensity [19].

The resolution of the system of equations (4) yields the radiosity of the surfaces.
The flux incident to the fuel surface can be then determined by the following
relation:

$$\overline{q_{r,i,s}} = \frac{\left(\overline{q_{0,s}} - \sigma\, \varepsilon_s\, T_s^4\right)}{\rho_s}$$

where $\overline{q_{r,i,s}}$, $\overline{q_{0,s}}$, ε_s, T_s and ρ_s are the incident flux, the radiosity, the
emissivity, the temperature and the reflectivity of a surface zone S of fuel,
respectively.

By dividing the liquid surface in concentric zones 1 cm in width, we could
obtain an evolution of the incident flux with respect to the pan radius. In Figure
4, we have plotted the distributions of fluxes to the surface versus the radial
position and we compared them with the experimental value for the various
imposed radiative fluxes. The results relevant to the free flame are also shown
in Figure 4a for comparison. Owing to the difficulties which such
measurements exhibit, it can be observed that they agree satisfactorily.
Nevertheless, the calculated fluxes seem to be slightly underestimated. This is
probably due to the fact that the contribution of soot absorption close to the
surface is overestimated. In fact, as we have seen in the experimental part, no
absorption measurement was performed at a distance less than 2 cm from the
fuel surface, because of the dimension of the fiber optic probe, and the
absorption coefficient curves have been extrapolated to a constant value, not
knowing that the soot concentration generally decreases when nearing the

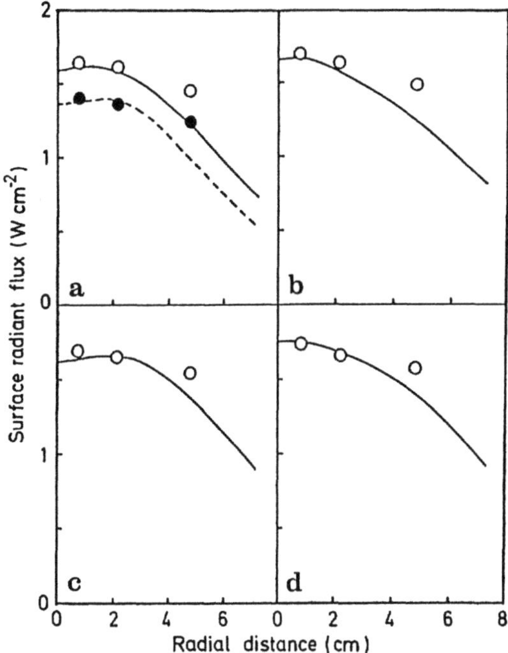

Fig. 4: Radiative flux distribution at the combustible surface as a function of the radial distance - without external radiation (•: measured, ---: calculated) - a, b, c): first position of the radiant panels and respectively 80, 90 and 100% of the nominal power; d) second position of the panels and 100% of the nominal power (o: measured, --: calculated).

surface. The results underline that the influence of an external radiative flux on the resulting flux at the surface is not very marked. Besides, it corresponds in concrete terms to a low rate increase, which proves to be proportional to the increase in incident flux, integrated over the entire surface, as shown in Figure 5. Thus, the flame exercises a true blocking effect on this flux which is all the more marked as its volume increases. Therefore the pyrolysis flow rate of a liquid (or solid) fire is less and less sensitive to this type of influence, inasmuch as that the flame which forms over it has a more marked opaque character (flame noticeably loaded with soot or large fire).

The thermal radiative behaviour revealing a non-uniform feedback flux distribution at the fuel surface, indicates well a radiative energy blockage by the cold fuel vapor and soot mainly in the core.

The proposed radiative modeling provides a comprehensive account of the role of external radiant flux on pool fire compartment. It could be used to investigate quantities of engineering interest such as feedback energy.

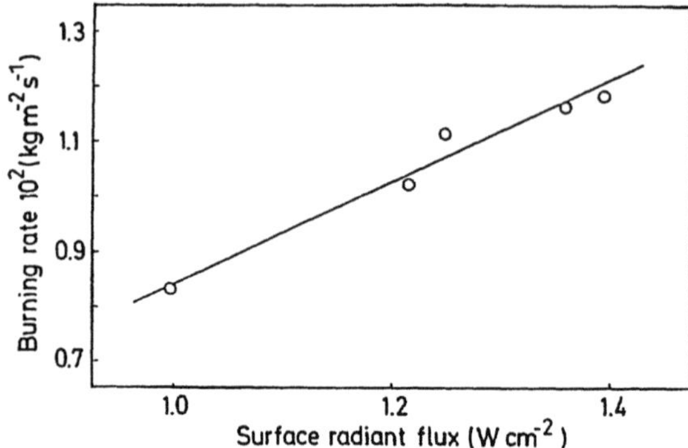

Fig. 5: Burning rate variation with resulting radiant flux at the surface

REFERENCES

1. Tewarson A., Pion R.F.: Flammability of Plastics I. Burning Intensity. Comb. Flame 26 (1976) 85.

2. Shih T.M.: Fire characteristics under the Influence of External Radiation. Paper presented at the Joint ASME/AIChE, 18th National Heat Transfer Conference, San Diego, California (August 6-8, 1979).

3. Rongere F.X.: Influence d'un Rayonnement Externe sur les Caractéristiques d'une Flamme de Diffusion Turbulente, Réunion Société Française des Termiciens. Paris, France (18 mai 1988).

4. Azov A.H.: Etude de la Combustion en Paroi Verticale. Influence du Rayonnement Extérieur, Thèse de Docteur Ingénieur. Université de Poitiers, France (5 octobre 1983).

5. Vantelon J.P., Himdi A., Gaboriaud F.: Bottom Surface Combustion of Polymethylmethacrylate Disks. I. Functional Dependence of the Combustion Rate. Comb. Sci. Technol. 54 (1987) 145.

6. Vantelon J.P., Souil J.M.: Bottom Surface Combustion of Polymethylmethacrylate Disks. II. Diffusion Flame Radiative Properties. Comb. Sci. Technol. 54 (1987) 159.

7. Souil J.M., Vantelon J.P., Joulain P., Grosshandler W.L.: Experimental and Theoretical Study of Thermal Radiation from Freely Burning Kerosene Pool Fires, Progress in Astronautics and Aeronautics, Dynamics and Reactive Systems, Part 1: Flame and Configurations 105 (1986) 388-401.

8. Bouhafid A., Breillat C., Vantelon J.P., Grosshandler W.L.: Predicting Soot Concentration in a Kerosene Pool Fire, Progress in Astronautics and Aeronautics, Dynamics and Reactive Systems, Part 2: Heterogeneous Combustion and Applications 113 (1988) 204-222.

9. Bouhafid A., Souil J.M., Vantelon J.P., Joulain P.: Transferts Radiatifs de Flammes à Base Horizontale. Cas d'une Flamme de Kérosene, Revue Générale de Thermique, mars-avril (1988) 217-227.

10. Bouhafid A., Vantelon J.P., Joulain P., Fernandez-Pello A.C.: On the Flame Structure at the Base of a Pool Fire, Twenty-Second Symposium (International) on Combustion. The Combustion Institute (1989) 1291-1298.

11. Bouhafid A., Vantelon J.P.: Radiative Heat Flux and Energy Balance at the Surface of a Small Scale Kerosene Pool Fire. in 12th International Colloquium on the Dynamics of Explosions and Reactive Systems. University of Michigan July 23-28 (1989).

12. McCaffrey B.J.: Purely Buoyant Diffusion Flames: Some Experimental Results. Centre for Fire Research NBSIR.79-1910 (1979).

13. Hertzberg M., Cashdollar K., Litton C., Burgess D.: The Diffusion Flame in Free Convection. in U.S. Bureau of Mines Report, RF 8263, Washington D.C. (1978).

14. Zhang X.L., Vantelon J.P., Joulain P., Fernandez-Pello A.C.: Influence of an External Radiant Flux on a Moderate Scale Pool Fire (to be published in Comb. Flame).

15. Hottel H.C., Sarofim A.F.: Radiative Transfer. McGraw-Hill Book Company (1967).

16. Siegel R., Howell J.R.: Thermal Radiation Heat Transfer. Hemisphere Publishing Corporation (1981).

17. Crauford N.L., Liew S.K., Moss J.B.: Experimental and Numerical Simulation of a Buoyant Fire. Comb. Flame 61 (1985) 63.

18. Vachon M., Cambray P., Maciaszek T., Bellet J.,C.: Temperature and Velocity Fluctuations Measurements in a Diffusion Flame with Large Buoyancy Effects. Combust. Sci. Technol. 48 (1986) 223.

19. Grosshandler W.L., Hardouin Duparc B.: Thermal Structure of Ethanol Pool Fire. Paper n° 6-11 presented at the Joint Meeting of French and Italian Section of the Combustion Institute, Naples, Italy June 16-19 (1987).

COMBINED CONVECTIVE AND CONDUCTIVE EFFECTS WITHIN AN UPSTREAM FLOW ALONG A VERTICAL FUEL SURFACE

B. Porterie, R. Saurel, J-C. Loraud, M. Larini

Laboratoire des Systèmes Energétiques et Transferts Thermiques.
Université de Provence - Centre de Saint-Jèrôme. Case 321.
Av. Escadrille Normandie-Niemen.
13397 Marseille Cedex 13.

ABSTRACT

A numerical model is developed to study the flame spread processes occuring when a high temperature material is placed in contact with a solid fuel. Gas phase processes including mass, momentum and heat transfer are coupled with solid phase processes, heat conduction and thermal degradation, through conditions at the interface solid-gas. For the gas phase, the unsteady two-dimensional Navier-Stokes equations are written, under Boussinesq approximation, in Stream Function Vorticity formulation. Solid phase processes are described by an energy balance equation. A semi-implicit method, taking coupling conductive-convective effects into account, is used to obtain the numerical solution.

This numerical model is tested assuming the solid product is dry or wet wood. The sensitivity of the flame structure to the moisture content and various parameters is examined.

INTRODUCTION

Studies about upward flame spread over the surface of a combustible material have become more and more active during the last few years due particularly to fire prevention and safety. Among the different modes of flame spread, it is the most rapid and hazardous one. In this case, it is well known that the spread of flame is controlled primarily by the rate of heat transfer from the flame to the unburnt fuel, finite rate chemical kinetic effects having a small influence on this process. Comprehensive review of Fernandez-Pello [1] give an account of the different theoretical studies of the problem.

The aim of the present paper is to establish a heat transfer model that solves the gas and solid phase governing equations coupled together through interface conditions. It concerns the thermal degradation of wood because wood has often been a major fuel in fires. Moreover, a considerable body of theoretical and experimental works exists about wood combustion problems [2-10]. To have a good description of the flame spread processes, mechanisms such as water vaporization, pyrolysis and charring mechanisms are accounted for.

ANALYSIS

PHYSICAL DOMAIN AND ASSUMPTIONS

The two-dimensional problem to be analyzed and the coordinate system used in the analysis are indicated graphically in Figure 1. The ignition of a small region of a thick solid fuel slab placed vertically in an upward gas flow is generated by contact with a high temperature material (HTM). The solid fuel is heated by thermal conduction. So water is vaporized to generate steam from 100 °C and active wood pyrolysis occurs beginning at about 220 °C. Pyrolysis gases mix with the water vapor and flow outward. When the gaseous flux is not strong enough, concurrently oxygen diffuses inward to oxidize the carbon. The hot gases produced in the burning region of the fuel move convectively ahead of the pyrolysis front increasing the transfer of heat to the unburnt material and consequently the spread of flame.

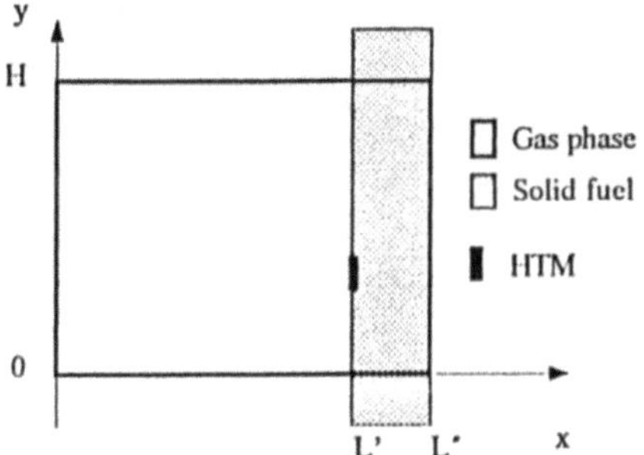

Fig. 1:   Physical model and the coordinate system

To eliminate the complexities of gas flow through a porous media, the gases are assumed to be removed out of the solid instantaneously upon their release and perpendicular to the surface. They do not exchange energy with the wood structure during their migration.

It is assumed that the unconfined fluid is newtonian and that the Boussinesq approximation is valid: the fluid is incompressible with constant properties except for the linear variation of density with temperature in the buoyancy term of the momentum equation.

The simplifying assumption of an exothermic surface reaction is made and the char is treated as pure carbon. A one-step, irreversible, chemical reaction of the form

$$C + \frac{1}{2} O_2 \rightarrow CO$$

is presumed to occur at the surface of the char. The fuel surface is assumed not to change, given the small scale of the process.

Radiation effects are neglected.

BASIC EQUATIONS

The equations governing the motion of the gas phase are those describing the conservation of mass, momentum and energy. Using streamfunction-vorticity formulation the continuity equation is automatically satisfied and the pressure eliminated as a solution variable [11].

The governing equations are

$$-\nabla^2 \Psi + \zeta = 0 \tag{1}$$

$$\frac{\partial \zeta}{\partial t} + u \frac{\partial \zeta}{\partial x} + v \frac{\partial \zeta}{\partial y} = \frac{\mu}{\rho} \nabla^2 \zeta + g\beta \frac{\partial T}{\partial x} \tag{2}$$

$$\rho C_v \left(\frac{\partial T}{\partial t} + u \frac{\partial T}{\partial x} + v \frac{\partial T}{\partial y} \right) = \lambda \nabla^2 T + Q \tag{3}$$

where

$$\zeta = \frac{\partial u}{\partial y} - \frac{\partial v}{\partial x} \text{ and } u = \frac{\partial \Psi}{\partial y}, \ v = -\frac{\partial \Psi}{\partial x}$$

In these equations, Ψ is the stream function, ζ the vorticity, ρ the density, T the temperature, u and v the longitudinal and normal velocity components, λ the thermal conductivity, β the compressibility factor, μ the viscosity.

Within the solid, only energy conservation equation is needed. With solid material properties taken as constant, the energy equation is

$$\frac{\partial T_f}{\partial t} = \frac{\lambda_f}{\rho_f C_{pf}} \nabla^2 T_f + \Gamma \tag{4}$$

where λ_f is the solid thermal conductivity, ρ_f the density, C_{pf} the heat capacity.

At the interface between fuel and gas phase, the following condition is applied:

$$\lambda \frac{\partial T}{\partial x} = \lambda_f \frac{\partial T_f}{\partial x_f} + \Omega \qquad (5)$$

In eqs. (3), (4) and (5), the terms Q, Γ and Ω are source terms resulting of thermal degradation processes.

Boundary conditions and initial conditions are required to complete the specification of the theoretical model. The physical domain contains free boundaries whose conditions must imperatively be compatible together. The interface between solid and gas phase is regarded as a previous wall.

In the solid, boundary conditions are

$$\frac{\partial T_f}{\partial y}(x,0) = 0, \ \frac{\partial T_f}{\partial x}(L'',y) = 0$$

$$\frac{\partial T_f}{\partial y}(x,H) = 0, \ T_f(L',y) = T_w(y)$$

T_w is the temperature of the fuel surface deduced from the interface condition or equal to the temperature of the HTM on the contact surface.

In the gas, the specification of the boundary conditions is of dominant importance because it affects the accuracy of the final solution. But no mathematical rigorous solution is available for our problem. So, drawing mainly upon intuition, literature [11] and computational experimentation, the most successful set of boundary conditions is

$$\text{At } x = L'; \ u = v = 0, \ \Psi = 0, \ \zeta = \frac{\partial^2 \Psi}{\partial x^2}, \ T = T_w$$

but if there is fuel vaporization these last conditions must be replaced by

$$u = u_{eject}, \ v = 0, \ \frac{\partial \Psi}{\partial y} = u, \ \zeta = \frac{\partial^2 \Psi}{\partial x^2}, \ T = T_w$$

where u_{eject} is the velocity of the volatile gases ejected from the fuel surface.

$$\text{At } y = 0: u = 0, \ \frac{\partial \Psi}{\partial y} = 0, v = -\frac{\partial \Psi}{\partial x}, \ \zeta = 0, \ \frac{\partial T}{\partial y} = 0$$

$$\text{At } x = 0: u = 0, \ v = v_0, \ \frac{\partial \Psi}{\partial x} = -v_0, \ \zeta = 0, \ \frac{\partial T}{\partial x} = 0$$

574

where v_0 is the stream velocity parallel to the fuel surface.

$$\text{At } y = H: \frac{\partial u}{\partial y} = 0, \frac{\partial^2 \Psi}{\partial y^2} = 0, v = -\frac{\partial \Psi}{\partial x}, \frac{\partial \zeta}{\partial y} = 0, \frac{\partial T}{\partial y} = 0$$

As for the initial conditions, the flow of air is parallel to the wall ($u=0$ et $v=v_0$) and the two phases are at the ambient temperature.

The source terms of the equations are expressed differently according to the state of thermal degradation. During the first step which corresponds to the water vaporization and pyrolysis period the terms Ω, Γ and Q are expressed as functions of the water vaporization rate m'_{vap} and the pyrolysis rate m'_{pyr} deduced from experimental results [10]

$$\Omega = \frac{2}{\Delta y \, \Delta z \, \Delta x_f} \int_0^{\frac{\Delta x_f}{2}} \left(L_{pyr} m'_{pyr} + L_{vap} m'_{vap} \right) dx$$

$$\Gamma = \frac{L_{pyr}}{C_{pf} m_e} m'_{pyr} + \frac{L_{vap}}{C_{pf} m_e} m'_{vap}$$

$$Q = \frac{1}{\Delta x \, \Delta y \, \Delta z \, e} \int_0^e \left[\Delta H_{pyr} m'_{pyr} + \Delta H_{vap} m'_{vap} \right] dx$$

During the second step relative to the char combustion the terms Ω and Γ are zero while the therm Q is given by

$$Q = \frac{1}{\Delta x \, \Delta y \, \Delta z} \Delta H_{CO} \, m'_{CO}$$

with $\Delta H_{CO} = \Delta H_0 - 410.0 \, (T_w - 298.0)$.

In the case of a carbon-oxygen reaction the mass-production rate of CO, m'_{CO}, can be determined using the first order reaction rate expression of the Arrhenius type given by Srinivas and Amundson [12].

In these source terms, Δx, Δy and Δz are the mesh spacing in the x,y and z directions, Δx_f, Δy_f and Δz_f the fuel ones ($\Delta z = \Delta z_f = 1$), L_{pyr} the heat of pyrolysis, L_{vap} the heat of water vaporization, $m_e = \Delta x_f \Delta y_f \Delta z_f \, \rho_f$ the mass of an elementary fuel cell, ΔH_{pyr} and ΔH_{vap} respectively the heat released by the pyrolysis gases and the water vapor in the gas-phase reaction, ΔH_0 the standard heat of formation of CO and finally e the thickness of the fuel slab.

Assuming as previously mentioned that gaseous products escape through the surface as soon as they are produced, their velocity normal to the fuel surface is defined during the first step by

$$u_{eject} = - \frac{1}{\Delta y_f \, \Delta z_f e} \int_0^e \left(\frac{1}{\rho_{vap}} m'_{vap} + \frac{1}{\rho_{pyr}} m'_{pyr} \right) dx$$

and during the second step by

$$u_{eject} = - \frac{1}{\Delta y_f \, \Delta z_f \, \rho_{CO}} m'_{CO}$$

NUMERICAL PROCEDURE

Referring to Figure 1, the domain of integration consists of two subdomains: a gas one and a fuel one. For each of them the spacing of the rectangular mesh in the y direction is the same, but not the spacings in the x direction. The interface is a mesh line and every grid point of the fuel subdomain is located at the center of the cell.

The elliptic Poisson eq. (1) for the streamfunction is a steady equation solved by using the false transient technique of Frankel [11]. In this method, a time-dependent term $\partial \psi / \partial t$ is added to the left-hand side of eq. (1). The solution of the steady equation by iterations is analogous to solving a time-dependent problem to an asymptotic steady state.

The unsteady solution of eqs. (1)-(4) is computed by means of an ADI technique using forward-time, centered-space differencing and upwind second-order scheme for the convective terms. At each step of the ADI scheme, tridiagonal algebraic equations must be solved by means of the Thomas algorithm. Computations of Ψ and ζ are iterated until convergence is reached. The interface equation is treated explicitly.

RESULTS

The results of numerical simulation of flame spread over dry and wet wood with 50wt% water content under ambient atmospheric conditions with an upstream air flow velocity of 0.1 m/s are discussed. The HTM works during 30 s at a temperature of 3000 K (this value is reached after 5 s) and is 0.02 m long. Available experimental data on the water vaporization rate, pyrolysis rate and thermophysical properties about dry and wet wood are very scarce and practically limited to our knowledge to experiments with oak [10]. Thus, the numerical model is applied to this material. Table 1 shows data used in the simulation. Some of them are those of Ref. 10. A 21x31 mesh for each subdomain (L' = 0.2 m, L" = 0.22 m, H = 0.15 m) and a time step of 0.02 s are found to be satisfactory in terms of both accuracy and computing economy.

Table 1

Properties of dry and wet oak		
	dry	**wet**
ρ_f (kg m^{-3})	666.	1000.
λ_f (J m^{-1} s^{-1} K^{-1})	0.21318	0.326
C_{pf} (J kg^{-1} K^{-1})	1337.	1766.
Moisture content (%)	0.	0.3335
Pyrolysis product content (%)	0.7	0.4665
Char content (%)	0.3	0.2
Other thermophysical data		
Water vaporization temperature (K)	373.	
Pyrolysis temperature (K)	493.	
Carbonisation temperature (K)	1103.	
L_{pyr} (J kg^{-1})	0.	
ρ_{pyr} (kg m^{-3})	0.95	
ΔH_{pyr} (J kg^{-1})	1.10^7	
ΔH_0 (J kg^{-1})	$1.84.10^7$	
ρ_{CO} (kg m^{-3})	1.25	

Time evolution of isothermal lines for dry wood on the whole domain is presented in Figure 2. The four diagrams show the penetration of the thermal wave inside the combustible material. The pyrolysis of the outermost wood layers in contact with the HTM leads to a good development of the flame structure (Figure 2-a and 2-b). The weakness of the ejection velocity of the volatile gases (a few mm/s) from the surface explains why the flame is so thin. When the pyrolysis product supply is diminishing (the inmost wood layers are more and more difficult to pyrolyse) the isotherms of the external flow decrease (Figure 2-c). Moreover, when the HTM is out of order (Figure 2-d) this decrease becomes more pronounced. Even if it is a slow-moving propagation the thermal wave penetration is going on due to the high thermal inertia of the combustible material. From this case it is clear that the energy stemming from wood combustion (pyrolysis and char combustion) is not enough to have a self-sustained combustion after stopping the HTM.

For the wet wood, time evolution of isotherm lines is shown in Figure 3. According to Calvin and Diehl [10] pyrolysis processes are delayed in the wet wood compared to the dry wood ones since a great part of the energy transferred

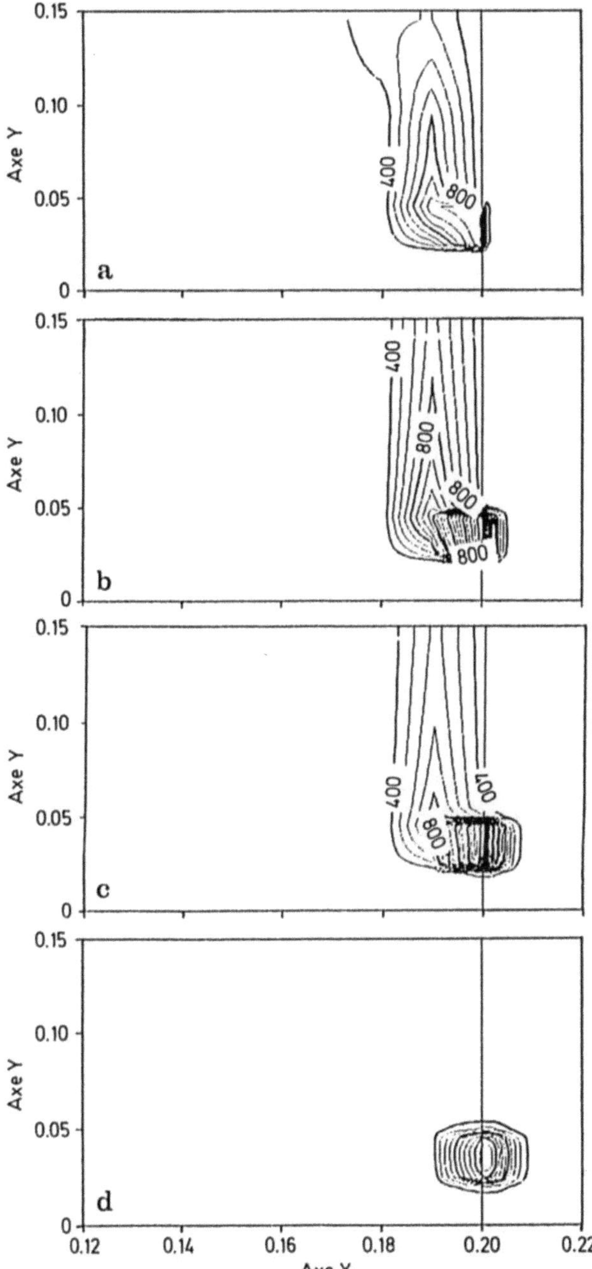

Fig. 2:   Isotherm lines for dry wood at   a) $t = 2s$; b) $t=15s$; c) $t=30s$; d) $t=45s$

to the fuel is used for water vaporization. Thus, the flammability of wood is decreased by the absorbed water in reducing the burning intensity. That explains the reduction of the pyrolysis zones in Figure 3 when compared to

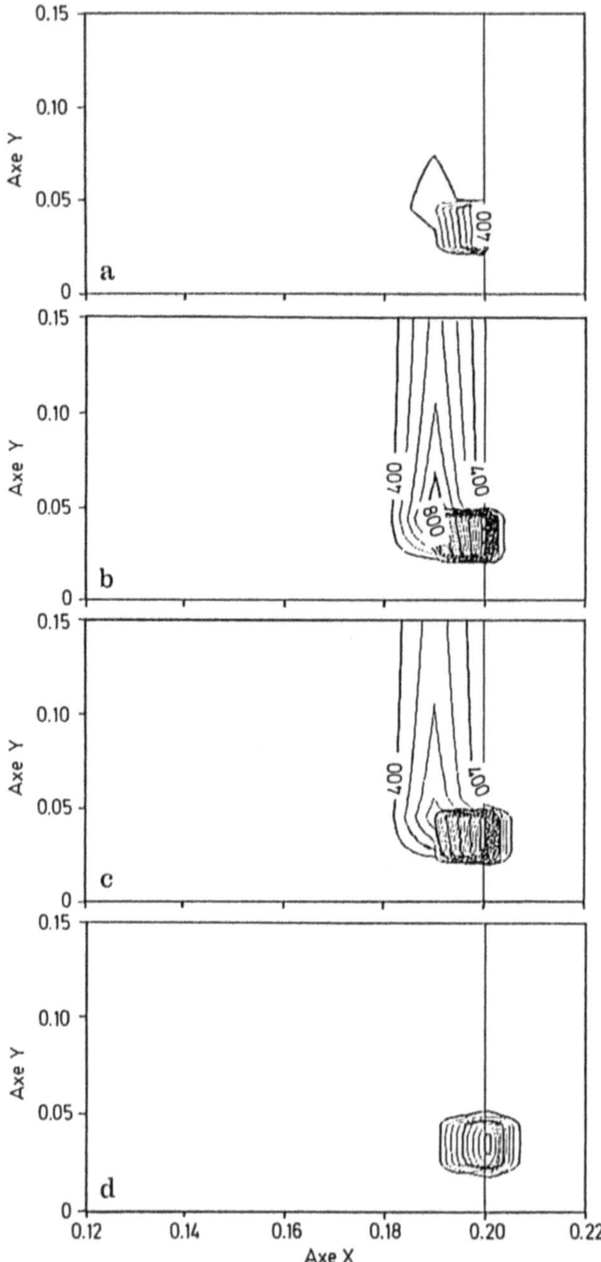

Fig. 3: Isotherm lines for wet wood at a) t = 2s; b) t=15s; c) t=30s; d) t=45s

those of Figure 2 at the very same time. Added to the cooling effect of the water vapor, this reduction which induces a lower volatile gas generation leads the external flow isotherms to a faster decrease in the wet wood case than in the dry wood one.

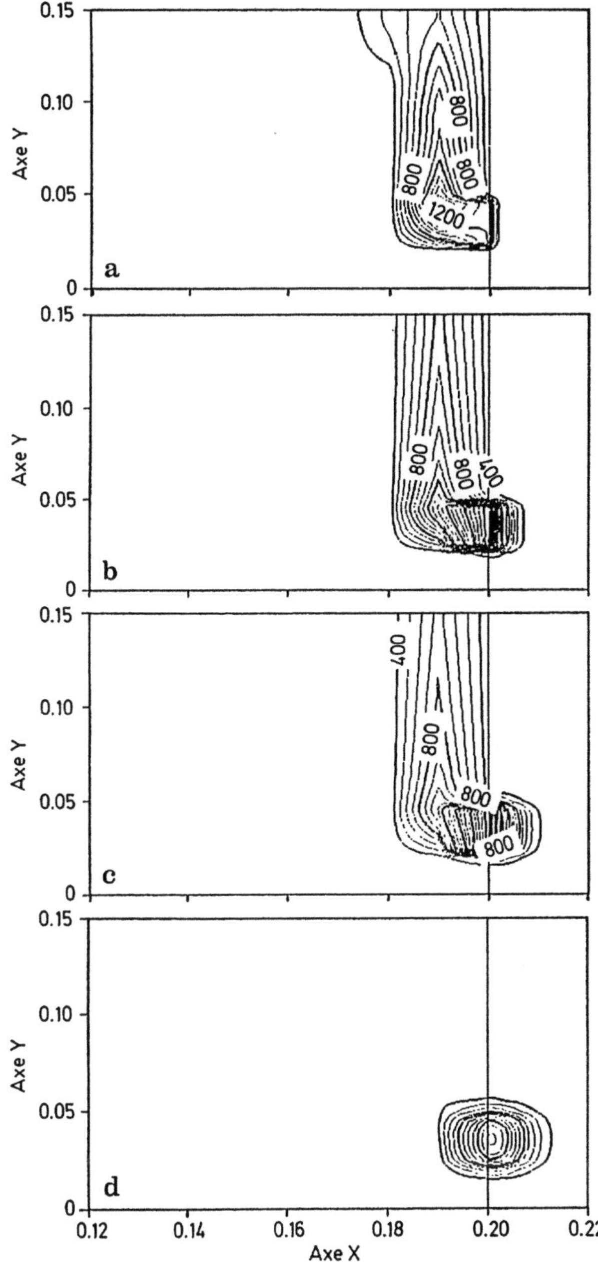

Fig. 4: Isotherm lines for dry wood and $\lambda_f \times 2$ at a) t = 2s; b) t=15s; c) t=30s; d) t=45s

The effects of the thermal conductivity and the heat capacity of the solid fuel on its thermal degradation state and the flame structure are studied. This study applies only to dry wood. On Figure 4, the effects of multiplying by a factor of 2 the thermal conductivity λ_f are shown. It emerges from the comparative

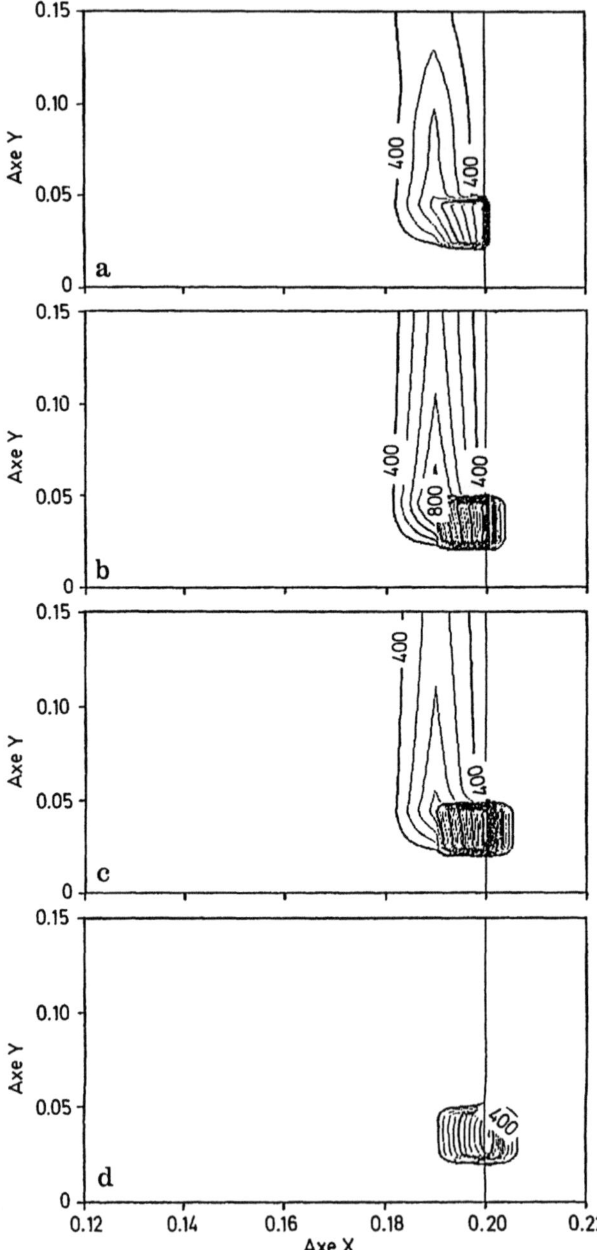

Fig. 5: Isotherm lines for dry wood and $C_{pf} \times 2$ at a) t = 2s; b) t=15s; c) t=30s; d) t=45s

examination of Figure 4 and Figure 5 that the thermal conductivity is a sensitive parameter. A twice value of λ_f leading to higher conductive processes inside the solid fuel increases the heated zone area and consequently the gaseous product generation. The burning intensity is improved.

The same change is made on heat capacity C_{pf} (C_{pf} x 2). The results are indicated in Figure 5 and compared with those of Figure 2. As expected, owing to the worse thermal wave propagation, significant decrease of wood inflammability appears.

CONCLUDING REMARKS

A numerical model for the study of combined convective and conductive effects within an upstream flow along a vertical fuel surface is presented. The model, although ignoring chemical kinetic effects, gives a good understanding of the mechanisms that govern the development of the flame. Further work, however, is needed to achieve better descriptions of thermal degradation processes. An improved treatment of the solid phase analysis including variable properties and, in particular, density variations is necessary. A more refined model should include radiation from the flame to the solid.

REFERENCES

1. Fernandez-Pello A.C.: Flame Spread Modelling. Comb. Sc. Tech., 39 (1984) 119-134.

2. Dusserre P., Otterbein M., Vermande P.: La Combustion du Bois, Contribution à la Modélisation. Proclime Tome 18, 1, janv.-fév. (1987).

3. Murty Kanury A.: Rate of Burning of Wood (A Simple Thermal Mode). Comb. Sc. Tech., 5 (1972).

4. Wai Chun Chan R., Kelbon M., Krieger B.B.: Modelling and Experimental Verification of Physical and Chemical Processes During Pyrolysis of a Large Biomass Particle. Fuel (1985).

5. Miller C.A., Ramohalli K.N.R.: A Theorical Heterogeneous Model of Wood Pyrolysis. Comb. Sc. Techn., 46 (1986).

6. Becker H.A., Phillips A.M.: Burning of Pine in Wind at 357-857 °C and 3-18 m/s: The Wave Propagation Period. Combustion and Flame, 58 (1984) 273-289.

7. Mao C.P., Kodama H., Fernandez-Pello A.C.: Convective Structure of a Diffusion Flame over a Flat Combustible Surface. Combustion and Flame, 57 (1984) 209-236.

8. Vovelle C., Akrich R., Delbourgo R.: Comportement du Bois dans un Incendie Entropie, 87 (1979).

9. Vovelle C., Akrich R., Delbourgo R.: Kinetics of the Thermal Degradation of Cellulose and Wood in Inert and Oxidative Atmosphere. 9th Symp. on Comb. The Comb. Inst. (1982).

10. Calvin K.L., Diehl J.R.: Combustion of Irradiated Dry and Wet Oak. Combustion and Flame, 42 (1981) 123-138.

11. Roache P.: Computational Fluid Dynamics. Hermosa Publishers, Albuberque (1972).

12. Srinivas B., Amundson N.R.: Intraparticle Effects in Char Combustion. Steady State Analysis. The Can. J. of Chem. Eng., 58 (1980).

A "TWO-FLUID" APPROACH TO THE MODELLING OF THREE-DIMENSIONAL TURBULENT FLAMES

K.A. Pericleous[+] and N.C. Markatos[++]

[+]   Thames Polytechnic, London
[++] National Technical University, Athens

KEYWORDS:
Fire Modelling, Turbulent Combustion, Two-Phase Flow

ABSTRACT

The combustion of propane in an enclosure is modelled using a "two-fluid" technique, in which the entrained air is labelled as Fluid 1 whilst the hot gases comprising combustion products, fuel and unreacted oxidant are labelled as Fluid 2. The entrainment process, assumed to be a function of the relative velocity between the fluids, relative density and turbulent eddy size, is modelled explicitly as an interface activity. Conservation equations for mass, momentum, energy and species are solved for each fluid separately, using the shared volume principle.

The two-fluid technique, although on the surface more expensive than the conventional single fluid one (twice as many variables) it is surprisingly economical; the additional degree of freedom provided allows hot fluid parcels to "slip" through the surrounding air promoting faster numerical convergence. It appears to the authors that this approach is physically more realistic, as it accounts for the fragmentariness of turbulent flames and allows counter-gradient diffusion due to density variations to take place.

The CDF code PHOENICS is used to model the fire in a test room, previously modelled by the authors as a single fluid. The results obtained are compared with experimental measurements and the influence of model parameters on the solutions investigated.

INTRODUCTION

It has long been recognised that in a turbulent flame, large fluctuations in temperature and species properties can exist as the oxidant and fuel streams meet and react with one another. This variability can be reduced, or practically eliminated if the reactants are premixed, or if the turbulence of the stream enhanced. In contrast, in non-premixed, buoyancy driven combustion typical in fires, this intermittent behaviour is amplified with, as a result, the flame becoming fragmented and often dominated by a regular large-scale buoyant instability close to the fuel surface Cox & Chitty [1]. This behaviour has been recorded photographically, *e.g.* Lysaght [2], where cold "parcels" of fluid exist within the flame envelope at any time.

Attempts have been made by several investigators to relate this "fragmentariness" to the turbulent behaviour of the reacting gases. Hence, following Magnussen [3] the present investigators and many others, e.g. [4, 5, 6] adopted the so-called eddy-break-up model of combustion to account for the effects of eddy size on reaction rate. This approach proved quite successful, especially in cases where pressure gradients are small or when the flow is not buoyancy driven. When either of these conditions becomes dominant the eddy-break-up model no longer represents the physical situation correctly. Investigators have observed that fire fragments may possess significantly different velocities from surrounding "cold" air; Kuznetsov [7] and Barry and Libby [8] have called the phenomenon "counter-gradient diffusion". Walker and Moss [9] and Shepherd and Moss [10] have provided experimental confirmation of these unequal velocities, and Phillips [11] has published calculations of flame propagation which incorporate this effect.

Spalding [12, 13, 14] as long ago as 1982 proposed that a two-fluid framework usually adopted for two-phase calculations (namely IPSA [15] is used in a generalised turbulence model. This idea was tested on both reacting and inert flows, yielding unequal velocities, temperatures, fuel/air ratios and states of reaction completeness of the gas fragments. Cases studied include a backward facing step [16], ducted flames [17] and combustion in a reciprocating engine chamber [18]. The same method but with different model parameters has also been applied by Malin [19, 29] to simulate turbulence intermittency in jets and buoyant plumes, comparing his findings against conventional statistical models.

The authors first used an early version of the two-fluid method to model flames encountered in fire situations and published a paper on the characteristics of axisymmetric diffusion flames [21], using gas composition as the criterion defining each fluid. The same model, further improved and extended to three dimensions is applied to the room-fire experiment reported in [22] which was also modelled using the conventional single fluid approach [4]. In the current simulation the following distinction is made between the two fluids:

(i)   Fluid 1, comprises the "hot" gas parcels which contain fuel, entrained oxidant and combustion products. Combustion is assumed to occur in this fluid only and for this reason it is assumed to be turbulent.

(ii) Fluid 2, is the oxidant (ambient air) which is assumed laminar and initially at ambient temperature. Mass transfer is permitted from fluid 2 to fluid 1, but not vice-versa; in contrast other investigators, e.g. Malin [19], who used the two-fluid idea to simulate intermittency in turbulence, assume mass transfer to be equal and opposite between the two-fluids, *i.e.* a diffusive process.

The rate of mass transfer has an important influence on the computed combustion behaviour; too little will cause oxygen starvation and extinction, too much will reduce the simulation to a single phase one. The actual rate used is calculated by making physically "realistic" assumptions which relate entrainment of oxidant to velocity difference, fragment size, density and cell volume fraction.

THE PHYSICAL PROCESS CONSIDERED

THE EXPERIMENT

The fire in the test compartment depicted in Figure 1 is simulated. The compartment is 2.4 m high, 2.4 m wide and 3.6 m long with a door opening of 0.8 m x 2.0 m placed at the far end. The dimensions correspond to those of the test compartment used by the Swedish National Testing Institute for calibration experiments [23].

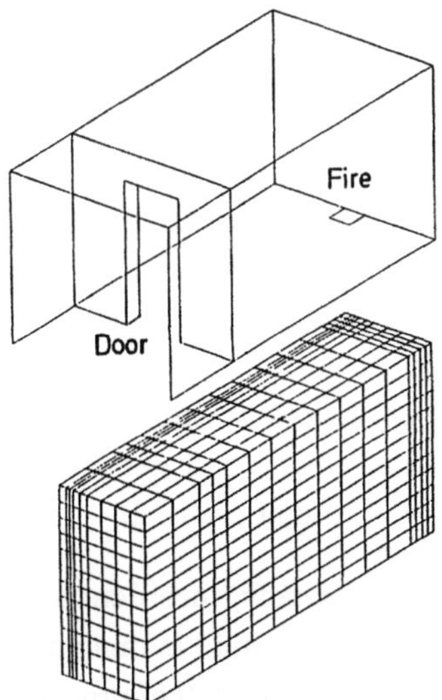

Fig. 1:   Room geometry and solution grid

Fuel (propane) enters the room through a 0.3 m x 0.3 m aperture located centrally on the floor, adjacent to the rear wall. The heat of reaction of propane is taken to be 4.64 x 10⁴ kJ/kg, and a total heat release of 250 kW is simulated. The fuel is ignited at entry and steady combustion is in progress. Temperature and velocity measurements were taken once steady-state conditions were reached, usually about one hour after ignition. The particular experiment chosen here, corresponds to test case "CORNCAL 21".

This case was chosen as it was originally simulated by the authors using a single fluid, eddy-break-up reaction model. Hence, the present work allows comparison of the two methods.

THE SIMULATION

The CDF code PHOENICS (v1.4) [24] was used to perform the computations presented. The following sections describe the mathematical model and its differential equations, bot not the finite-difference procedure embodied in PHOENICS, or the two-phase IPSA algorithm which can be found elsewhere [viz. 15].

(i) *The finite difference domain and coordinate system:* The non-uniform cartesian grid shown in Figure 1 was used in the cases presented. It has 16 cells along the room, 9 cells across and 12 cells in the vertical, a total of 1728 cells. Only half the room was modelled by symmetry. As shown in Figure 1, the domain of calculation extends beyond the door aperture where constant pressure boundary conditions can be reasonably expected to apply. Reference [4] has shown this extension to be necessary, in order to avoid erroneous velocities being computed at the plane of the door.

No grid dependence studies have yet been performed, as the main effort has so far been in the formulation of the model rather than absolute accuracy. Further studies are planned on much finer grids (up to 15000 nodes) to resolve details of the flow, especially in the vicinity of the flame. Even for the modest grid presented the calculations are by no means trivial since the presence of two notional phases and combustion leads to 15 dependent variables, with an equal number of equations to be solved at each cell.

(ii) *Modelling Concepts:* The fire plume is assumed to be steady, turbulent and consisting of two fluids as mentioned earlier. The reacting fluid is imagined to consist of parcels or "balloons" of gas, which enter the room full of propane fuel. The fraction of the volume occupied by these balloons increases continuously as they entrain air. The entrained air subsequently reacts with the fuel within the balloon envelope, releasing products. It should be noted that in general the balloons will continue entraining air even after all the fuel is exhausted, since entrainment is not governed solely by the need of fuel for oxidant but also by the process of turbulent mixing and "engulfment".

Since the same pressure applies to both fluids, hotter, lighter fluid fragments will accelerate faster than the surrounding air in the presence of a pressure gradient and as they do stretch and engulf gas trapped in their wake. The size of the flame fragments will depend on the turbulent characteristics of the fire plume and is directly related to the mixing length.

The question of what turbulence model one uses in a two-fluid simulation such as this one is by no means simple; it is not for example clear what the distinction is between diffusion within phase and inter-phase diffusion. In the present simulation turbulent inter-phase diffusion is ignored in all processes but heat transfer. Within phase, the reacting fluid, being more energetic, is assumed turbulent and it is represented by the two-equation k-ε model [25]. The oxidant stream is by contrast assumed to remain laminar, since velocities within it remain generally quite low.

A single-step diffusion controlled combustion model was used as proposed by Spalding [26], with reaction rates assumed infinite. In effect this model assumes that once reactants are mixed within a grid cell reaction proceeds to completion. For the size of the grid used this was felt to be a reasonable assumption, and one that avoids the need for calculating exponentials as in an Ahrrenius expression.

(iii) *Dependent Variables:*

p$\qquad$the pressure, acting on both fluids

Fluid 1

r1$\qquad$the volume fraction of the first fluid; may be interpreted as a "presence probability"

u1,v1,w1$\qquad$the velocity components in the x, y and z coord. directions

$H1 = CpT1 + m_{fu}H_{fu}$$\qquad$the total enthalpy which includes the heat of reaction H_{fu}; temperature T1 is derived from enthalpy

k,ε	the kinetic energy of turbulence and its dissipation rate
f1	the "mixture" fraction, characterising the ratio of fuel mass entering the domain, to the total mass of gas within fluid 1

Fluid 2

r2	the volume fraction of air (r1 + r2 = 1)
u2,v2,w2	the velocity components in the x,y and z coord. directions
H2=CpT2	the enthalpy of air, whence T2 the temperature
f2	the "virtual" mixture fraction; its value is zero in the oxidant stream, but its solution convenient to allow dilution of f1 through mass transfer

The above variables are the subject of partial differential equations representing convective and diffusive transport and balance of sources as shown in the following chapter. In addition a number of auxiliary variables derived from the above, need to be computed and stored in three dimensions. These are:

T1,T2	the temperature of each fluid
$\rho 1, \rho 2$	the density of each fluid derived from the ideal gas law
m_{fu}	mass fraction of unburned fuel (resides in fluid 1)
m_{pr}	mass fraction of products (resides in fluid 1)
m_{ox}	mass fraction of oxidant (resides in fluid 1)
μ_t	fluid 1 effective viscosity

In addition to the above, auxiliary relations are used to describe interfluid mass transfer, interfluid friction and interfluid heat transfer. The equations used are described in detail in the following section.

THE MATHEMATICAL MODEL

DIFFERENTIAL EQUATIONS

The problem in question is characterised by a set of elliptic partial differential source balance equations that express the conservation of mass, momentum, energy and chemical species in three-dimensional, recirculating, buoyant flows. In time-averaged form these equations can be written in compact form as follows for dependent variable ϕ:

$$\frac{\partial}{\partial t}\left(\rho_i\, r_i\, \phi\right) + \operatorname{div}\left[\rho_i\, r_i\, v_i \phi - \Gamma_\phi\, r_i\, \operatorname{grad}\phi\right] = r_i\, S_\phi \tag{1}$$

where ρ_i, v_i, r_i, Γ_ϕ and S_ϕ are the density, velocity vector, volume fraction, effective exchange coefficient of ϕ and source rate per unit volume respectively for fluid i, as shown in Table 1.

Table 1: Dependent Variable Source Terms

	Variable ϕ	Source, S_ϕ	
u_1	$-r_1\, \dfrac{\partial p}{\partial x}$	$+u_2 \cdot E''' + F_u$	
u_2	$-r_2\, \dfrac{\partial p}{\partial x}$	$-u_2 \cdot E''' - F_u$	
v_1	$-r_1\, \dfrac{\partial p}{\partial y}$	$+v_2 \cdot E''' + F_v$	
v_2	$-r_2\, \dfrac{\partial p}{\partial y}$	$-v_2 \cdot E''' - F_v$	
w_1	$-r_1\, \dfrac{\partial p}{\partial z}$	$+w_2 \cdot E''' + F_w + \rho_1\, g\, r_1$	+ boundary values
w_2	$-r_2\, \dfrac{\partial p}{\partial z}$	$-w_2 \cdot E''' - F_w + \rho_2\, g\, r_2$	
Scalar			
ϕ_1	$\phi_2 \cdot E'''$		
h_1	$h_2 \cdot E''' + C_{12}$		
h_2	$-h_2\, f \cdot E''' - C_{12}$		

INTERFLUID RELATIONS

(a) *Interfluid mass transfer.* In Table 1, E''' is the interfluid mass transfer rate, per unit volume from fluid 2 to fluid 1. It has the form:

$$E''' = k_m . F_{21} . < \rho_2 . r_2 - \rho_1 . r_1, 0 > /\rho \tag{2}$$

where k_m is a dimensionless constant and ρ is the average density. The expression $\langle a,b \rangle$ denotes the maximum of a and b; it ensures mass transfer only occurs from fluid 2 to fluid 1. F_{21} is the interfluid exchange coefficient defined as follows:

$$F_{21} = k_f.\rho_2.r_2.r_1.|V_2-V_1|/ \wedge \tag{3}$$

Equation (3) ensures that F_{21} is finite whenever both fluids are present in a cell and its value proportional to the presence probability and also to the velocity difference, or slip, of the two fluids. The denominator $\wedge$ represents mean fragment size in a cell, its calculation given below. The reciprocal of $\wedge$ represents the surface area per unit volume of an eddy; k_f is a second dimensionless constant given a value of 0.05 (Malin [19]).

Taken together, eqs. (2) and (3) imply that small rapidly slipping fragments entrain oxidant at the greatest rate per unit volume. Eq. (2) becomes identical to the one proposed by Wu and Spalding [17] if the densities of the two fluids are equal and if mass transfer is allowed from both fluids. The introduction of density difference ensures that entrainment is enhanced if the ambient density is higher than that of the plume.

(b) *Interfluid friction.* The source terms F_u, F_v and F_w in Table 1 represent the exchange of momentum between the fluids due to friction. The source expression for x-direction component is:

$$F_u = F_{21} . (u_2 - u_1) \tag{4}$$

and similarly for the other components. Equal and opposite terms appear in the equations for the two fluids. Additional momentum will be imparted to fluid 1 by entrainment (terms $u_2.E'''$ etc., in Table 1).

A consequence of eq(s). (4) is that momentum is imparted by the rising fire plume to the surrounding ambient air.

(c) *Interfluid heat transfer.* Since there is mass transfer from fluid 2 to fluid 1, there is also heat transfer as the fluid carries its energy with it. The enthalpy source as given in Table 1 is $H_2.E'''$. In addition there is convective heat transfer, given by:

$$C_{12} = -k_h.F_{21}.(T_1 - T_2) \tag{5}$$
$$C_{21} = -C_{12}$$

Following Wu [17], the interface heat constant k_h is given a value of 30. The solution appears to be insensitive to variations about this value as the dominant heat transfer mechanism is by bulk transport, *i.e.* entrainment.

(d) *Mixture properties.* The local volume average value of any variable is needed to enable comparison with single fluid computations and measurements. The average density is calculated thus:

$$\rho = \rho_1 r_1 + \rho_2 r_2 \tag{6}$$

Similar expressions are used for all other variables.

(e) *The modelling of turbulence.* As mentioned earlier only fluid 1 is assumed to be turbulent. The two-equation k-ε [25] turbulence model with appropriate modifications to account for buoyancy is used. Noteworthy here is the equation for the fragment dimension $\wedge$, which has been used in the interfluid exchange coefficient F_{21}, above:

$$\wedge = C_\mu^{3/4} \cdot k^{3/2}/\varepsilon \tag{7}$$

(f) The combustion model. The single step "SCRS" reaction model of Spalding is used [25]. Reaction rate is assumed infinite, and combustion takes place solely in fluid 1. A single transport equation for the mixture fraction f, is sufficient to determine the quantity of fuel (burned or unburned) in any cell of the calculation. Linear equations then determine the mass fraction of products, oxidant and fuel. Hence,

$$\text{for } 0 < f \leq f_{st}: \quad m_{fu} = 0$$
$$\& \quad m_{ox} = 1 - f/f_{st} \tag{8}$$

$$\text{for } f_{st} < f \leq 1: \quad m_{ox} = 0$$
$$\& \quad m_{fu} = (f - f_{st})/(1 - f_{st}) \tag{9}$$

where, f_{st} is the stoichiometric fuel to air ratio given the value 0.05482. The mass fraction of products is then given by subtraction, as

$$m_{pr} = 1 - m_{ox} - m_{fu} \tag{10}$$

The infinite rate model used here is not a limitation of the model but simply a means of numerical convenience. More sophisticated models will be used in future extensions of this work.

CLOSING REMARKS ON PROPOSED MODEL

The model involves the following constants:

$$k_f = 0.05, k_m = 1, \text{ and } k_h = 30.$$

The values used have been taken from earlier work, i.e. Malin [19], Markatos, Pericleous and Cox [21] for k_f, Wu [17] for k_h and [21] for k_m (note however slightly different definition of k_m to that given in [21]. The model is most sensitive to the value of k_m and least sensitive to k_h; k_m was varied parametrically between 1 and 10 in this work.

RESULTS

CONVERGENCE AND COMPUTER REQUIREMENTS

Results were obtained for two- and three-dimensional cases. In the two-dimensional case the door opening and the propane inlet grill were assumed to extend to infinity. This case was used to test model assumptions and examine the implications of the various model parameters; since only one x-z plane of cells was present calculations could be performed on a PC type computer. The three-dimensional calculations in contrast were performed on a Norsk-Data ND 950 mini-computer. Time for one hundred iterative sweeps was 1700s for the 3D problem.

Approximately 200 iterative sweeps of the domain were required to achieve a reasonable degree of convergence; convergence was deemed reasonable, when total heat and mass balance at the plane of the door was good to within 0.1% and the sum of absolute residuals of the continuity equation over all cells was below 1% of the inflow through the door. However the criteria chosen did not always ensure that the fields remained unaltered between sweeps. For this reason computations were often run to as many as 1000 sweeps. To procure convergence and prevent a long cycle oscillation in the solution, typical of this type of buoyancy driven flow, the velocity and turbulence fields were under-relaxed using "false-timestep" or inertial relaxation [24]. No relaxation was applied on enthalpy; instead, linear under-relaxation was applied on density to weaken the coupling of the aerodynamic and thermal fields.

With the above measures it was possible to obtain converged solutions in all cases. Convergence was achieved significantly faster than in the single fluid calculations reported in [4]. The reason for this is not clear; a possible explanation is that since fluids can slip, hot fragments in the plume reach the ceiling easier than they would otherwise and they settle in a stable environment within the hot layer. At the same time, ambient air is displaced downwards and since it can slip it does not drag hot fluid with it as it would in a single fluid situation. Fluid 1 temperatures, following our definition will always be at least greater than mixture values. This results to higher buoyancy terms within fluid 1 which again promote good convergence.

A summary of the three-dimensional results obtained is given below.

3D TEST CASE

Figures 2-10 contain selected 3D results in graphical form. Figure 2 first gives an overall view of the fire plume and hot gas layer below the ceiling. What is actually shown is a constant mixture temperature surface, having a value of 600 K. Three distinct zones are clearly depicted; the vertical plume, the horizontal ceiling layer and the region of ambient air. The region above the contour and within the plume has a temperature above 600 K.

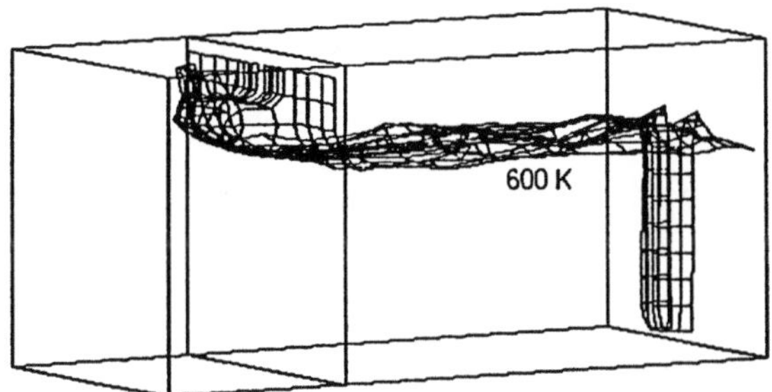

Fig. 2:   Constant temperature surface

Figure 3(a-c) shows velocity vectors on the symmetry plane for fluid 1, fluid 2 and volume average velocity respectively. Highest vertical velocities are in the plume region as expected. The velocity of hot fragments (fluid 1) is always higher than that of ambient air. The difference in velocity diminishes as the ceiling is approached. Within the ceiling layer, ambient air that has not been absorbed tends to flow downwards since it is still "heavy"; in contrast fluid 1 has positive vertical velocities, highest in regions of high volume fraction gradient. In the horizontal direction, highest velocities (3 m/s-5 m/s depending on case) occur at the top of the door aperture as the hot ceiling layer escapes outwards. At levels below about 1 m in height, air enters drawn towards the fire and hence creates a "cold" zone within the room.

594

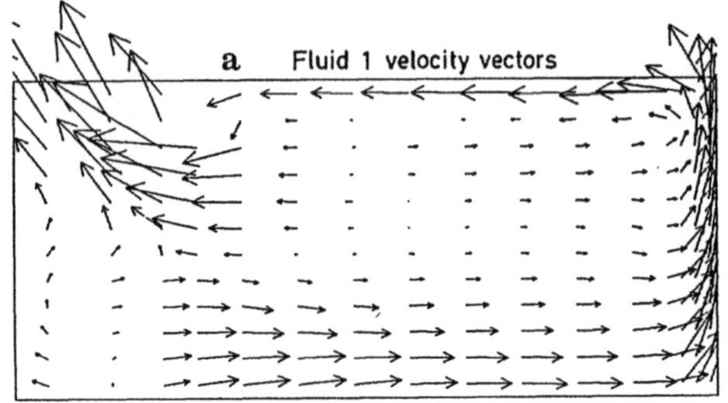

a) Fluid 1 Velocity vectors

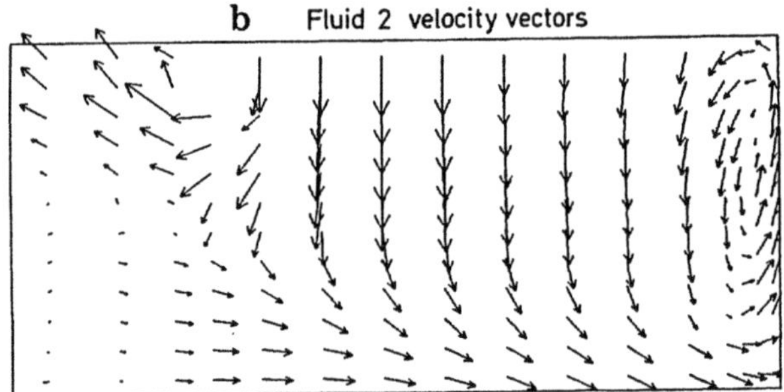

b) Fluid 2 Velocity vectors

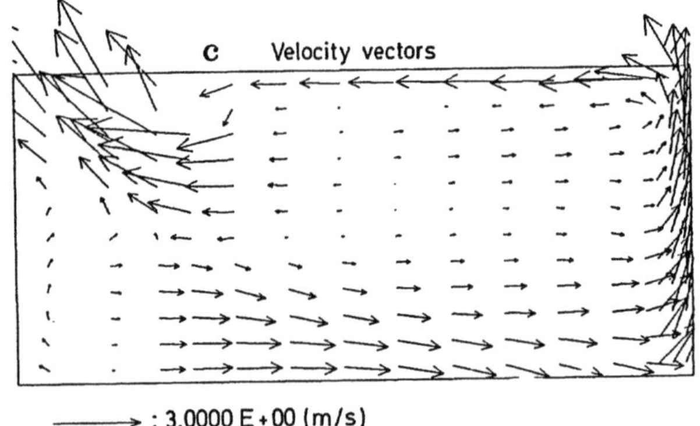

c) Mixture velocity vectors

Fig. 3:   Symmetry plane vectors

For values of k_m less than that 0.5 calculations indicate that because the amount of ambient air entrained is reduced, this incoming flow of air is impeded. The fire plume entrains instead air that has been expelled from the ceiling layer and a large vertical recirculation develops.

Figure 4 shows contours of mixture temperature, at the symmetry plane. The three zones identified above are again apparent. Highest temperatures (up to 1900 K) occur in the plume region. However, as remarked in [21] highest values tend to lie away from the plume axis. The contours are horizontal within the ceiling layer, and the interface between the hot and cold zones is sharper than that predicted by the single fluid approach.

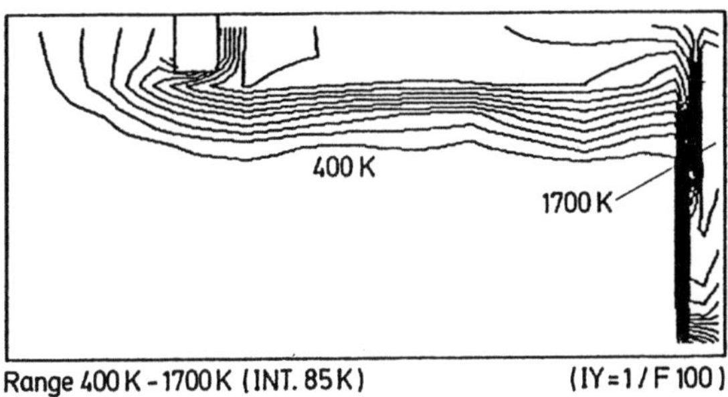

Fig. 4:   Mixture temperature contours

Figure 5 shows contours of product mass fraction. In form they are very similar to temperature contours; however, it should be remembered that products belong to fluid 1 only. Again the distinction between contaminated and ambient air is much clearer in the present calculation than in the single fluid case of [22], where products were found to diffuse throughout the room.

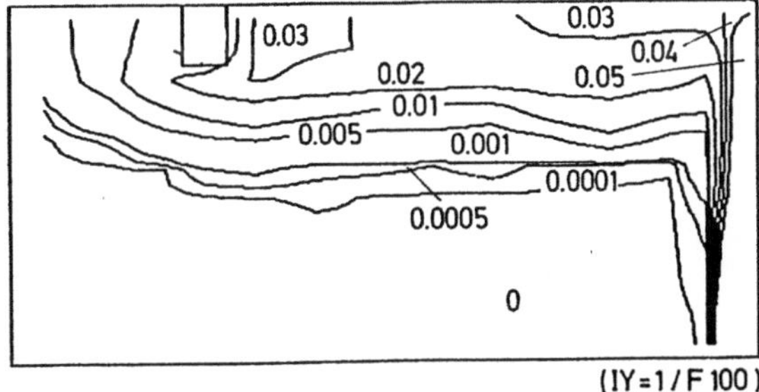

Fig. 5:   Fluid 1: Product mass fraction contours

596

Figure 6 shows contours of volume average mixture fraction f. The flame envelope can be identified as the heavy line corresponding to the stoichiometric value, $f_{st}=0.0548$. A flame height of approximately 2 m is predicted. However this value is sensitive to the combustion model used and the value of k_m. Preliminary calculations using the eddy-break-up combustion model in conjunction with the two-fluid assumption produced excessively long flames with combustion taking place in the ceiling layer.

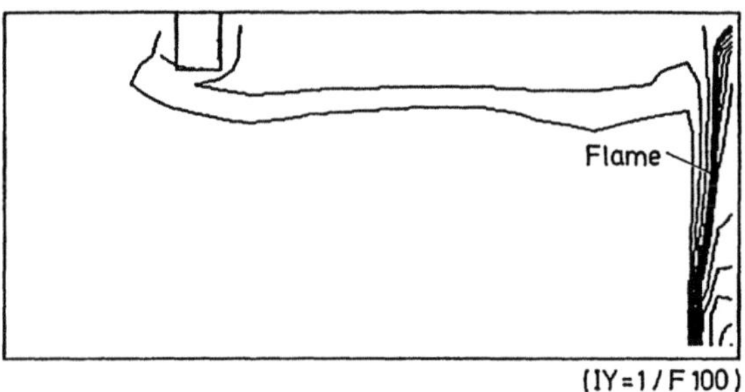

Fig. 6:   Volume average mixture fraction contours

It was however felt that the implicit assumption of "fragmentariness" within the EBU was inappropriate since the combustion scale of fluid 1 is already the eddy size. Similarly flame length increases with decrease in k_m.

Figure 7 shows contours of volume fraction for fluid 1. Fluid 1 is absent in the region close to the floor, marking the entry of ambient air.

Conversely, fluid 1 almost exclusively occupies the ceiling layer. There is a wide transition band between the two regimes starting around 1 m above the floor, and the band widens outside the room as high turbulence promotes mixing of the two fluids. Air also appears to penetrate upwards in the vicinity of the plume, carried along by the fast flowing combustion gases.

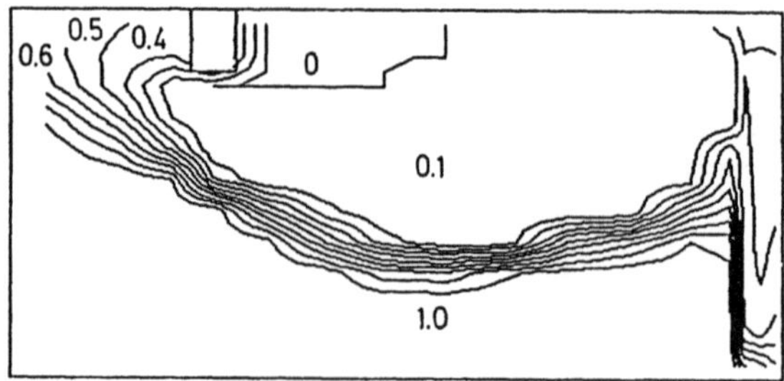

Fig. 7:   Volume fraction contours for fluid 1

Figure 8 shows velocity vectors at a constant x-plane, close to the back wall of the room. The rapidly accelerating column of hot gases spreads along the ceiling on impact, generating a vortex on either side of it, about a longitudinal axis. The limit of ambient conditions is also shown on the same plot by a single temperature contour, at 295 K.

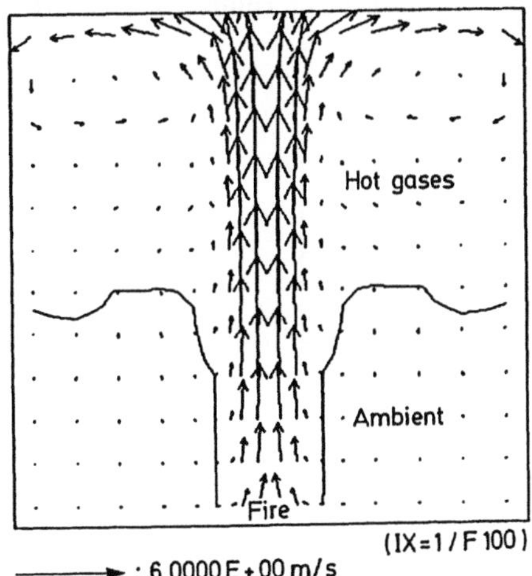

Fig. 8:   Back wall velocity vectors

Figure 9 shows the corresponding temperature distribution. There is a rapid temperature increase as the flame is traversed and a more gradual one in the ceiling layer. The presence of the vortex pair along the ceiling promotes good mixing in the uppermost part of the ceiling layer.

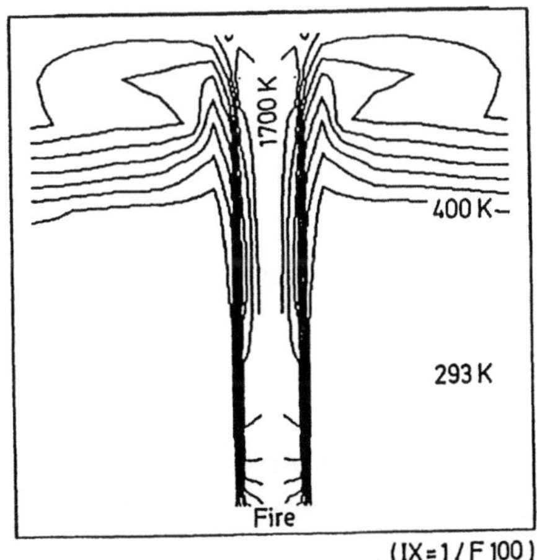

Fig. 9:   Back wall mixture temperature contours

598

Figure 10 shows door velocity profiles at the symmetry plane. Velocities are negative below 1 m (approximately) and positive elsewhere. Hence air is drawn into the room close to the floor at velocities in excess of 1 m/s. At the top end a maximum velocity of 4 m/s is calculated and this represents the rush of hot combustion gases out of the room. The velocities have been volume-averaged to compare with experimental values. Comparison with experiment is a least as good as that obtained with the single fluid simulation of reference [4], with maximum velocities better predicted.

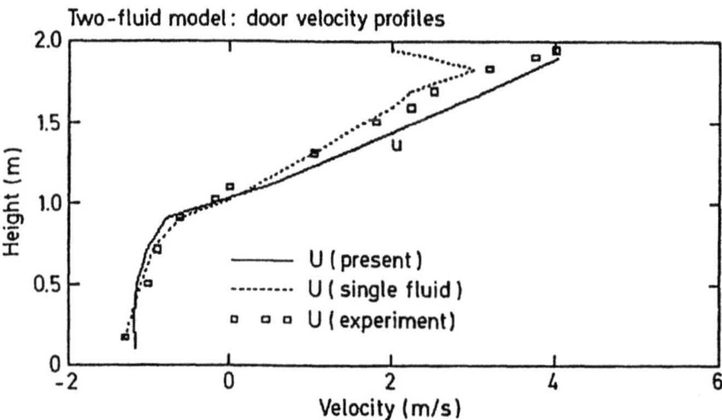

Fig. 10:   Door velocity profiles

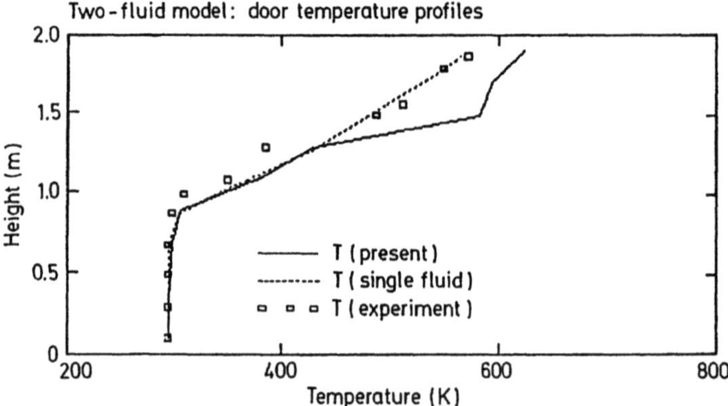

Fig. 11:   Door temperature profiles

Figure 11 shows the corresponding temperature profiles. As expected ambient temperatures (293 K) mark the region of incoming air. However in the hot layer, temperatures in excess to those observed in the experiment are computed. This is possibly due to the absence of radiation and due to residual combustion predicted in the ceiling layer close to the door, a result of the simplified combustion model employed. This behaviour can be prevented by increasing the mass transfer rate between the two fluids in the plume region.

CONCLUSIONS

A novel, two-fluid approach for simulating turbulent fires has been presented. The full set of two-phase transport equations is utilised to describe fire behaviour and in particular intermittency. The resulting model is more complex than the conventional one, but it has certain advantages both in convergence behaviour and in physical realism.

In the two-fluid approach it is recognised that within a non-premixed flame, exist two distinct regimes differing in composition, velocity and temperature. Reaction takes place at the interface of the two regimes and therefore relations describing behaviour at the interface are very important. It is desirable to know the temperature of hot flame fragments in addition to mixture values, because flame fragments are hotter and radiate more intensely (by the fourth-power law). In addition, certain pollutants, namely NOx only become significant above a threshold temperature [27]; average temperatures will underestimate production of NOx.

The results obtained are quantitatively realistic in all respects. The existence of two velocity and two temperature fields adds another dimension to the idealisation of the combustion process and plume dynamics, which aids understanding of the physics involved. Reasonable quantitative agreement has also been demonstrated between measured and computed door velocity and temperature profiles; the fact that no specific attempt has been made to tune the three model constants used to encourage agreement demonstrates that the two-fluid model has now come of age, and can be used with reasonable degree of confidence for fire simulations.

Further work is continuing to introduce radiation and refine the combustion model used and to investigate grid effects on the results.

REFERENCES

1. Cox G., Chitty R.: Some Stochastic Properties of Fire Plumes. Fire and Materials 6 (1982) 127.

2. Lysaght A.S.R.: Visualisation of Mixing in Turbulent Diffusion Flames. Combustion Flame 46 (1982) 105-108.

3. Magnussen B.F., Hjertaker H.B.: 16th Int. Symp. on Combustion, p. 719. The Combustion Institute, Pittsburgh PA (1976).

4. Markatos N.C., Pericleous K.A.: An Investigation of Three-Dimensional Fires in Enclosures. Rev. Gen. Thermique 23 (266) (1984) 67.

5. Huhtanen R.: Numerical Fire Modeling in Turbine Hall. Fire Safety Science - Proc. 2nd Int. Symp., Hemisphere, (1989) 771-779.

6. Kjaldman L.: Numerical Simulation of Dust Explosions. Proc. 2nd Int. PHOENICS Conf. (1987) 117.

7. Kuznetsov V.R.: USSR Fluid Dynam. 14 (1979) 328.

8. Bray K.N.C., Libby P.A.: Countergradient Diffusion in Pre-Mixed Turbulent Flames. AIAA Journal 19 (1981) 205.

9. Moss J.B.: Simultaneous Measurements of Concentration and Velocity in an Open Pre-Mixed Flame. Combust. Sci. Tech. 22 (1980) 115.

10. Shepherd I.G., Moss J.B.: Measurements of Conditional Velocities in a Turbulent Pre-Mixed Flame. AIAA 19th Aerospace Sciences Meeting, St. Louis 81-0181 (1982).

11. Phillips H.: Towards a Two-Fluid Model for Flame Acceleration in Explosions. HSE Buxton. 19th Int. Colloq. on Dynamics of Explosions and Reactive Systems (1983).

12. Spalding D.B.: Chemical Reaction in Turbulent Fluids. Physics Chem. Hydrodynam. 4 (4) (1983) 323.

13. Spalding D.B.: The Two-Fluid Model of Turbulence Applied to Combustion Phenomena. 22nd AIAA Meeting, Reno, Nevada (1984).

14. Spalding D.B.: CFDU Report CFD/82/8, CFDU Imperial College, London (1982).

15. Spalding D.B.: Development in the IPSA Procedure for Numerical Computation of Multi-Phase Flow Phenomena with Interphase Slip and Unequal Temperature. Proceedings 2nd Nat. Symp. Numerical Properties and Methodologies in Heat Transfer. Ed. T.M. Shih, Ch. 6 (1981) 421-436.

16. Spalding D.B., Wu J.: Application of the Two-Fluid Model of Turbulence to Flows over a Backward Facing Step. Proc. 2nd Int. PHOENICS user Conf., London (1987) 728.

17. Wu J.Z.Y.: The Application of the Two-Fluid Model of Turbulence to Ducted Flames. The PHOENICS Journal of Computational Fluid Dynamics and its Applications, 1 (1988) 1.

18. Spalding D.B.: Computer Simulation of Turbulence Combustion in Reciprocating Engines. Proc. 2nd Int. PHOENICS user Conf. (1987) 214.

19. Malin M.R.: Progress in the Development of an Intermittency Model of Turbulence". CFDU report 83/7 Imperial College, London (1983).

20. Malin M.R., Spalding D.B.: A Two-Fluid Model of Turbulence and its Application to Heated Plane Jets and Wakes Physico Chem. Hydrodynamics 5 (1984) 339-362.

21. Markatos, N.C., Pericleous K.A., Cox G.: A Novel Approach to the Field Modelling of Fire. Physico Chem. Hydrodynamics, 7 No. 2/3 (1986) 125-143.

22. Markatos N.C., Pericleous K.A.: An Investigation of Three-Dimensional Fires in Enclosures. ASME, HTD 25 (1984).

23. Sundström B., Wickström U.: Fire: Full Scale Tests — Calibration of Test Room — Part 1. NORD-TEST project 143:78,2. Tech. Rep. SP-RAPP 1981:48, 1981 Nat. Testing Inst. Borås, Sweden.

24. Rosten H.I., Spalding D.B.: The PHOENICS reference manual. CHAM TR/200 (1987) CHAM Ltd. London.

25. Rodi W.: Turbulence Models and their Application in Hydraulics. SFB 80/T/127 (1978) Univ. of Karlsruhe.

26. Spalding D.B.: Combustion and Mass Transfer. Pergamon Press (1979).

27. Pericleous K.A., Clark I.W., Brais N.: The Modelling of Thermal Nox Emissions in Combustion and its Application to Burner Design. Proc. 2nd Int. PHOENICS User Conf. London (1987).

THREE-DIMENSIONAL NATURAL CONVECTION-RADIATION INTERACTIONS IN A DIFFERENTIALLY HEATED CUBE FILLED WITH GAS-SOOT MIXTURES

T. Fusegi
Institute of Computational Fluid Dynamics
1-22-3 Haramachi, Meguro, Tokyo 152, Japan

B. Farouk
Dept. of Mechanical Engineering and Mechanics
Drexel University
Philadelphia, Pennsylvania 19104, USA

and

K. Kuwahara
Institute of Space and Astronautical Science
3-1-1 Yoshinodai, Sagamihara, Kanagawa 229, Japan

ABSTRACT

A high-resolution, three-dimensional finite difference numerical study was performed on interactions of natural convection and surface/gas/soot radiation in a differentially heated cubical enclosure over the Rayleigh number of $10^5 \leq Ra \leq 10^9$. A robust gas/soot radiation model used in the analysis was based on the P_1-differential approximation method and the weighted sum of gray gas model. The three-dimensional characteristics of the thermal and flow fields were examined in detail by the state-of-the-art three-dimensional numerical visualization techniques. The effects of each mode of radiation were described. Overall, radiation was found to enhance the three-dimensionalities of the fields.

NOMENCLATURE

Bo   Boltzmann number, $\rho_o Cp_o U_o / \sigma T_o^3$

Cp   specific heat at constant pressure

Fr   Froude number, $u_o^2 / g_o L_o$

g   gravitational acceleration

J   irradiance, $J = \int_{\Omega=4\pi} I\, d\Omega$, where I is the radiative intensity

k   thermal conductivity

L_0 reference length (enclosure height)

p   pressure

p_0 reference pressure (hydrostatic pressure)

Pr   Prandtl number, $Cp_o\, \mu_o / k_o$

Ra   Rayleigh number, $g_o\, \beta_o\, Cp_o\, \rho_o^2\, L_o^3\, (T_H - T_C)/\mu_o k_o$

Re	Reynolds number, $\rho_0 u_0 L_0/\mu_0$
t	time
T	temperature
T_0	reference temperature $(T_C + T_H)/2$
T_C, T_H	cooled and heated side wall temperatures
u_0	reference velocity, $[g_0\beta_0 L_0(T_H - T_C)]^{1/2}$
u,v,w	velocity components in the x, y and z directions, respectively
W_m	weighting function for the m-th component gray gas
x,y,z	Cartesian coordinates

Greek Symbols

β	thermal expansion coefficient
δ	overheat ratio $(T_H - T_C)/T_0$
ε_w	surface emissivity
κ	absorption coefficient
μ	viscosity
ρ	density
σ	Stefan-Boltzmann constant
τ	optical thickness, κL

Subscripts

m	m-th component gray gas
o	reference quantities (dimensional)

Superscript

*	dimensional quantities

INTRODUCTION

Analysis of natural convection-radiation interactions in an enclosure is a crucial class of problems in thermal engineering, which has a variety of applications; to cite a few examples, cooling of electronic equipment and nuclear reactors, design of solar energy collectors and storage devices, and control of compartment fires. In spite of the technological importance, relatively little research has been reported in the literature due to the complicated nature of the problem: 1) the three models of heat transfer (conduction, convection and radiation) are strongly coupled in the process; 2) the integral-differential nature of gas radiation requires considerable computational efforts; 3) gases of engineering importance, such as carbon dioxide and water vapor, exhibit strong non-grayness for radiation; and, 4) when soot, which is a strong absorber and emitter, is present, its radiation poses another complexity. Hence, the exact treatment of natural convection-gas/soot radiation is extremely tedious and time-consuming. In order to make a compromise between the accuracy of analysis and the affordability of computing resources, radiation models for gases and soot are usually employed for engineering studies of natural convection-radiation interactions [1].

The present paper examines interactions of natural convection and gas/soot radiation in a cubical enclosure heated differentially by its two vertical side-walls. The schematic of the enclosure is depicted in Figure 1. The enclosure is heated to a temperature of T_H at the plane of $x^* = L_o$ and cooled to T_C at $x^*=0$. The remaining walls are thermally insulated. All the inner surfaces are assumed to be black for radiation. Mixtures of carbon dioxide gas and water vapor are considered as the medium. In addition, cases with the gas mixtures and soot are analysed. Soot is assumed to have uniform distribution over the flow field. The geometry and boundary conditions of the present problem are mathematically well-posed and they serve as a basic model for various thermal engineering systems.

Previous works on this class of problems were mostly limited to two-dimensional configurations [2-6]. These numerical studies predict significant changes in characteristics of flow fields and heat transfer rates across enclosures due to radiation when the overheat ratio, δ, becomes large. In a three-dimensional enclosure, the presence of end walls will have large effects on the temperature and flow fields through radiative exchange among surfaces. Therefore, three-dimensional analysis should be considered to study situations of practical importance. In order to keep large efforts for three-dimensional computations to a tractable amount, a robust radiation model, which gives a compromise between the accuracy and efficiency, is necessary.

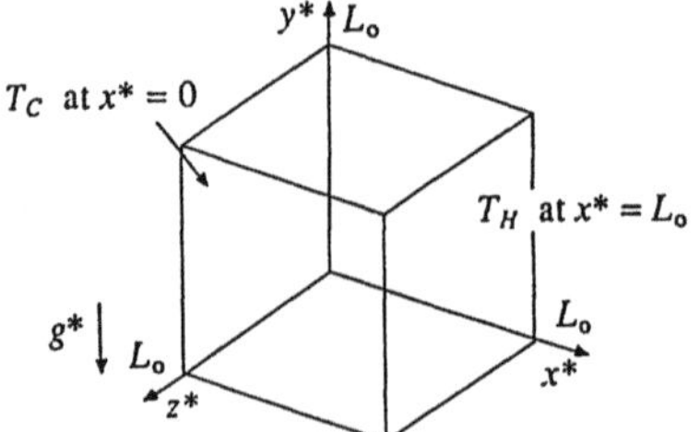

Fig. 1.   The flow geometry and the boundary conditions. The walls are radiatively black. They are also thermally insulated, unless otherwise indicated.

This may be accomplished by a non-gray spherical harmonics P_1 approximation method, which incorporates the weighted sum of gray gas (WSGG) model. This approach reduces the transfer equation for the radiative intensity of a non-gray gas to a set of differential equations for the irradiance of component gray gases. Hence, the resultant equations are compatible in their form with the governing equations for flows. This method has been utilized for a turbulent convection-radiation interaction analysis of flows in a furnace by Song and Viskanta [7]. In their paper, the absorptivity of combustion gases (mixtures of carbon dioxide, water vapor and air) is approximated by that of two component gray gases. The present study adopts the treatment for gas/soot radiation described in the references [8,9]. The same WSGG model was used earlier by the present authors in two-dimensional as well as three-dimensional analysis [5,6].

Computations were performed over a wide range of the Rayleigh number, that extends from 10^5 to 10^9. Predictions of the flow and temperature fields were obtained by directly integrating the time-dependent governing equations. Calculations were carried out by using a higher order upwind scheme for the convection terms of the Navier-Stokes equations and sufficiently small time steps. Results are presented by various numerical visualizations of the flow and temperature fields. Changes in field characteristics due to each mode of radiation (surface, gas and gas-soot radiation) are examined in detail.

MATHEMATICAL MODEL

GOVERNING EQUATIONS

The governing equations consist of the Navier-Stokes equations, the energy equation and the transfer equation for the radiative intensity. The last equation is transformed to a system of differential equations for the irradiance of the component gray gases by using the P_1 approximation method and the WSGG model. The non-dimensionalized governing equations can be expressed in tensor form as:

$$\frac{\partial \rho}{\partial t} + \frac{\partial}{\partial t_j}(\rho u_j) = 0 \tag{1}$$

$$\frac{\partial}{\partial t}(\rho u_i) + \frac{\partial}{\partial x_j}(\rho u_j u_i) = -\delta_i 2\frac{1}{Fr}(\rho - 1) - \frac{\partial \rho}{\partial x_i}$$
$$+ \frac{1}{Re}\frac{\partial}{\partial x_j}\left(\frac{\partial u_i}{\partial u_j} + \frac{\partial u_j}{\partial u_i} - \delta_{ij}\frac{2}{3}\frac{\partial u_k}{\partial x_k}\right) \tag{2}$$

$$\frac{\partial}{\partial t}\left(\rho T\right)+\frac{\partial}{\partial x_j}\left(\rho u_j T\right)=\frac{1}{Re\,Pr}\frac{\partial^2 T}{\partial x_j \partial x_j}+\frac{1}{3\,Bo}\sum_{m=1}^{N}\frac{1}{\tau_m}\frac{\partial^2 J_m}{\partial x_j \partial x_j} \tag{3}$$

$$\frac{\partial^2 J_m}{\partial x_j \partial x_j}+3\,\tau_m^2\left(4W_m T^4 - J_m\right)=0, \quad m=1,2,...,N \tag{4}$$

where δ_{ij} is the Kronecker delta ($\delta_{ij}=1$ if $i=j$, and $\delta_{ij}=0$ otherwise) and N is the number of component gray gases. The fluid properties are assumed to be invariant except for the density. In order to study flows at a large overheat ratio, density variation is accounted for in all the terms of the governing equations. The density is computed from the ideal gas law.

The physical variables are non-dimensionalized in the following manner:

$$(x,y,z) = (x^*,y^*,z^*)/L_0, \quad (u,v,w) = (u^*,v^*,w^*)/u_0,$$

$$t=t^*\,u_0/L_0, \; p=(p^* - p_0)/\rho_0 u_0^2, \; \rho=\rho^*/\rho_0, \; \mu = \mu^*/\mu_0,$$

$$k=k^*/k_0, \; g=g^*/g_0, \; T=T^*/T_0, \; J = J^*/\sigma T_0^4$$

A convective velocity $(g_0\beta_0 L_0(T_H-T_c))^{1/2}$, is chosen as the reference velocity, u_0. The reference temperature, T_0, is set at the film temperature, $(T_C + T_H)/2$.

NON-GRAY GAS RADIATION MODEL

The WSGG model postulates that the absorptivity of non-gray gases (and gas-soot mixtures) can be approximated by the sum of component gray gas absorptivities, weighted with a temperature dependent factor. In the present analysis, three component gray gases are considered to represent the emissivity of the non-gray gas (mixture of carbon dioxide, water vapor and air). Two components are employed for soot by following Smith et al [8,9]. In case of the gas-soot mixtures, eight equations (i.e., N=8) are required for irradiance absorption coefficients of which are determined from the combination of absorption coefficients of the component gray gases. When soot is not considered, N is reduced to three. The absorption coefficient of the gas-soot mixture, κ_{mix}, is computed as $\kappa_{mix}(T^*) = \sum_{m=0}^{3} \sum_{n=1}^{2} (P\,\kappa_{m,gas} + \kappa_{n,soot}).W_{m,gas}.W_{n,soot}$ with $\kappa_{n,soot} = f_v\,C_n$, $W_{m,gas} = \sum_{k=1}^{4} a_k\,T^{*k-1}$ and $W_{n,soot} = \sum_{k=1}^{4} b_k\,T^{*k-1}$ where P is the partial pressure, f_v is the volume fraction of soot and C_n is the soot coefficient. The expressions for the coefficients of WSGG model are taken from Smith et al. [8,9].

BOUNDARY CONDITIONS

In accordance with the problem description, the boundary conditions are:

$$u = v = w = 0 \tag{5}$$

$T=(2-\delta)/2$ at $x = 0$, $T = (2+\delta)/2$ at $x = 1$, and

$$\frac{\partial T}{\partial n} = \frac{Re\,Pr}{Bo}\left(\pm q_s - \sum_{m=1}^{N} \frac{1}{3\tau_m}\frac{\partial J_m}{\partial n}\right)$$

at $y=0$ and $z=0$ (with the positive sign),

and at $y=1$ and $z=1$ (with the negative sign) $\tag{6}$

$$\frac{\partial J_m}{\partial n} = \pm\, 3\tau_m \frac{\varepsilon_w}{2\,(2-\varepsilon_w)}\left(J_m - 4W_m\,T^4\right), \quad m=1,2,\ldots,N$$

at $x=0$, $y=0$ and $z=0$ (with the positive sign),

and at $x=1$, $y=1$ and $z=1$ (with the negative sign) $\tag{7}$

where n denotes the coordinate normal to the surface and q_s is the surface radiative flux, which is computed as $q_s = W_0\,\Sigma_i F_{1-i}(T_w^4 - T_i^4)$, where F_{1-i} is the shape factor for radiation between surface elements 1 and i. The summation is taken for all the surfaces that the element 1 can see. For the gas mixtures, W_0 is defined as $W_0 = 1 - \Sigma_{j=1}^{3} W_j$.

SOLUTION METHOD

The governing equation system (1)-(7) are discretized by a control-volume based finite difference procedure. Numerical solutions are obtained by an iterative method, together with the pressure correction algorithm, SIMPLE [10]. The present study employs the Strongly Implicit Procedure [11] to accelerate convergence characteristics of solutions. SIP is applied to the planes of constant z in order to simultaneously determine dependent variables in the x and y directions on each plane.

The convection terms in the momentum equation (2) are approximated by the QUICK scheme modified for non-uniform grids [12], while those in the energy equation (3) are discretized by a hybrid scheme [11]. The QUICK scheme involves a third-order accurate upwind differencing, which possesses the stability of the first-order upwind formula and is free from substantial numerical diffusion experienced with the usual first-order schemes.

The entire enclosure is considered as the computational domain. The number of grid points for computations is 41x41x41 for lower Rayleigh number ($Ra < 10^7$), and it is increased to 51x51x51 for computations with higher Rayleigh numbers. Variable grid spacing is used to resolve steep gradients of the velocity and the temperature near the walls. Care is taken to distribute several grid points inside the boundary layer formed near the walls. In this manner, the minimum value of grid spacing is reduced to approximately 10^{-4} (normalized by the enclosure height) near the isothermal walls.

A non-dimensional time step of as small as 10^{-3} is used for the computations. At each time level, convergence is assumed when the following criterion is met:

$$\frac{|\phi_n - \phi_{n-1}|}{|\phi_n|_{maximum}} \leq 10^{-4} \text{ for all } \phi \tag{6}$$

where ϕ represents any dependent variable, and n refers to the value of ϕ at the n-th iteration level.

In order to calculate the surface radiative flux, the shape factor for an elemental surface with respect to other elemental surfaces needs to be evaluated. This is done by first obtaining an expression of the shape factor between two differential elements which are taken on the control volume surfaces of the finite difference mesh system. The expression is integrated numerically to compute the desired shape factor between the control volume surfaces. Simpson's formula is applied for multiple integration.

RESULTS AND DISCUSSION

Computations are performed to investigate interactions of natural convection and radiation in a cubical enclosure by employing the mathematical model described in a previous section. The reference pressure is considered to be at 1 atm. Solutions are acquired over the Rayleigh number range of $1.67 \times 10^5 \sim 10^9$. In order to isolate effects of each radiation mode (surface, gas and gas-soot radiation), several computations were performed by selectively neglecting one or more of the radiation modes under the same conditions for the remaining parameters. In the computations, the temperature difference of the isothermal walls is held fixed at 555 K. The same value is chosen as the reference temperature, at which value the fluid properties are evaluated. Consequently, the overheat ratio, δ, is unity and the cooled and heated side wall temperatures are set at T_C=278 K and T_H=833 K, respectively. The reference Prandtl number is equal to 0.68 for all the cases. Under these conditions, the height of the enclosure varies from 2.57×10^{-2} m to 4.66×10^{-1} m and the Boltzmann number (interaction parameter) ranges from approximately 50 to 200. The optical thickness of the mixture is approximately one order of magnitude smaller than the physical path length.

The entire enclosure is treated as the computational domain. In all the computations, the reference pressure of the mixture is 1 atm. For the gas-soot mixture, the partial pressure of the carbon dioxide-water vapor mixture and the volume fraction of soot are, respectively, 0.25 atm and 10^{-7}. They are typical values for combustion of hydrocarbon fuels with moderate soot generation [9]. The partial pressures of carbon dioxide gas and water vapor are equal in the mixture.

The computations were carried out on the HITACHI S820 supercomputer at the Institute of Computational Fluid Dynamics (ICFD) in Tokyo, Japan. The system has a maximum computational speed of 3 GFLOPS. The computer code was vectorized for the supercomputer. The typical CPU time for obtaining converged solutions was approximately 30 minutes *per* run with grid points of 51x51x51 for the natural convection cases. However, the CPU time is increased to approximately 3 hours in the radiation cases. About 200 MB of memory space was used for a complete computation. The graphic display of the results was obtained by a three-dimensional interactive graphic software [13].

Changes in the field characteristics due to radiation are examined by various three-dimensional numerical visualization techniques. First, results for the natural convection mode (without radiation) and the surface/gas radiation mode for carbon dioxide gas are inspected.

TEMPERATURE AND FLOW FIELDS AT Ra=4.5 x 10^6

In this sub-section, results for Ra=4.5 x 10^6 are examined as a representative case. In the obtained results, the fields in the half domain in $0.5 < z < 1$ are the mirror images of those in the remaining flow field $0 < z < 0.5$. Figure 2 presents perspective views of the temperature and the absolute values of the vorticity fields in the cube.

As shown in the figures, if radiation is neglected (the natural convection mode), the temperature and flow fields are almost symmetric with respect to

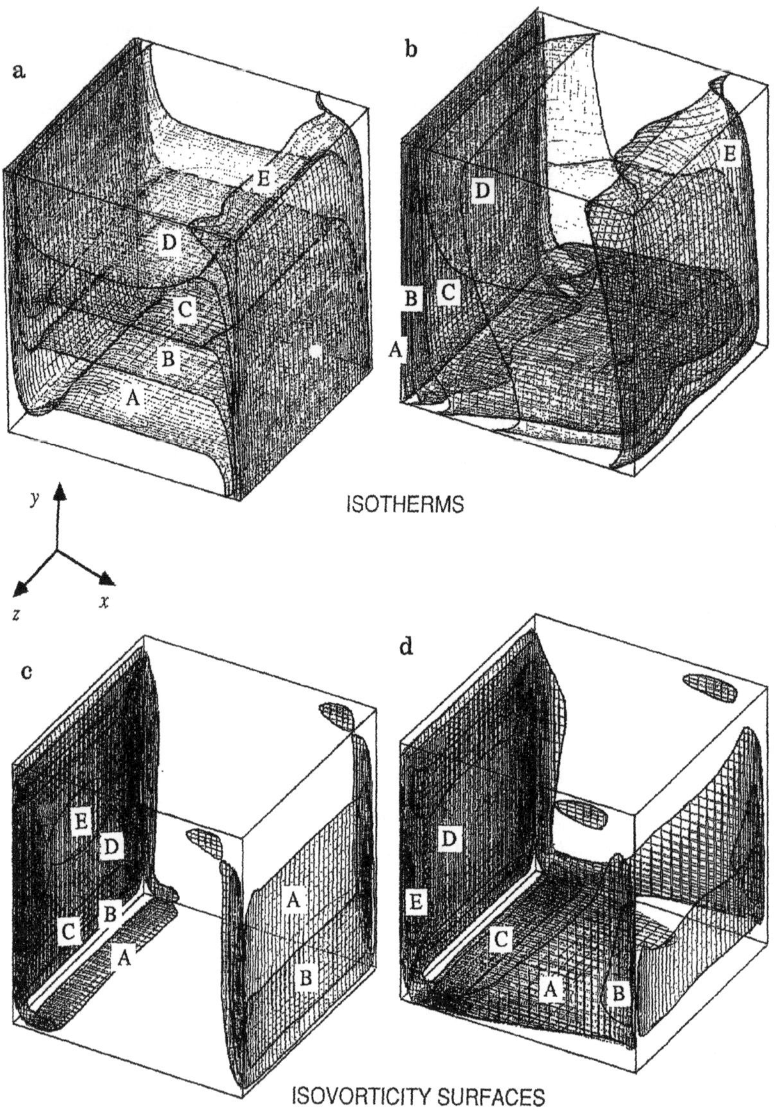

Fig. 2: Perspective views of the surfaces of isotemperatures and isovorticities (absolute values) at $Ra=4.5 \times 10^6$ [contour levels: (a) & (b) 0.6667 (A), 0.8333 (B), 1.0 (C), 1.167 (D), 1.333 (E); (c) 15 (A), 30 (B), 45 (C), 60 (D), 75 (E); (d) 18 (A), 36 (B), 54 (C), 72 (D), 90 (E)].

the plane of (x=0.5, y=0.5). Slight deviations from the exact centro-symmetry are due to the density variation effect. Hydraulic and thermal boundary layers develop along the isothermal walls. The temperature field is stratified in the remaining part of the enclosure. The flow field exhibits a distinct boundary layer-stagnant core structure.

When the surface and gas radiation modes are included, significant changes occur in the fields. The field in this plane is no longer symmetric due to an increase in the temperature of the thermally insulated planes by the surface radiation exchange. As seen in the downward shift of the location of the isotherm which corresponds to the average temperature (T=1.0), the overall fluid temperature is higher than the previous case. The boundary layer-stagnant core structure of the flow field is still retained in the surface/gas radiation mode. However, deviation from the centro-symmetry of the flow field is pronounced. In order to scrutinize the three-dimensionalities of the fields, projected velocity vectors in constant x and y planes are depicted in Figures 3 and 4, respectively. The reference velocity scale is indicated for each case. The results for the natural convection and surface/gas radiation for carbon dioxide are presented side-by-side for comparison, as was attempted in Figure 2.

The combined effects of the vertical (v-) and transverse (w-) velocities are illustrated in Figure 3, which shows the cross-sectional flow field patterns in the planes parallel to the isothermal side-walls (x=0 and 1). All the results exhibit the exact symmetry with respect to the mid y-plane (y=0.5). The fluid in a large portion of the plane of x=0.1 is located within the boundary layer near the cooled wall at x=0, which develops in the negative y-direction. For the natural convection case, the cold-wall boundary layer flow is seen to be discharged into the almost stagnant central core near the bottom plate (y=0). In the mid x-plane (x=0.5), weak secondary vortices are visible near the edges. Mostly anti-symmetric flow patterns, with respect to the line of x=0.5 and y=0.5, of those at x=0.25 and 0.1 can be observed at x=0.75 and 0.9, respectively, in the natural convection mode. The breakdown from the exact anti-symmetry is due to the effect of variable density, as previously remarked. In all the cases, significant z-variations of the velocity vector field are noticeable.

With surface/gas radiation, the anti-symmetry completely disappears; the projected field patterns that are displayed in the figure differ considerably at each x-plane. Overall, the flow intensity increases; compare the velocity vector scales attached to each case. The magnitudes of the fluid motion in the interior core region are appreciable and the flow patterns become uniform. The discharging of the boundary layer flows into the core, as clearly seen in the

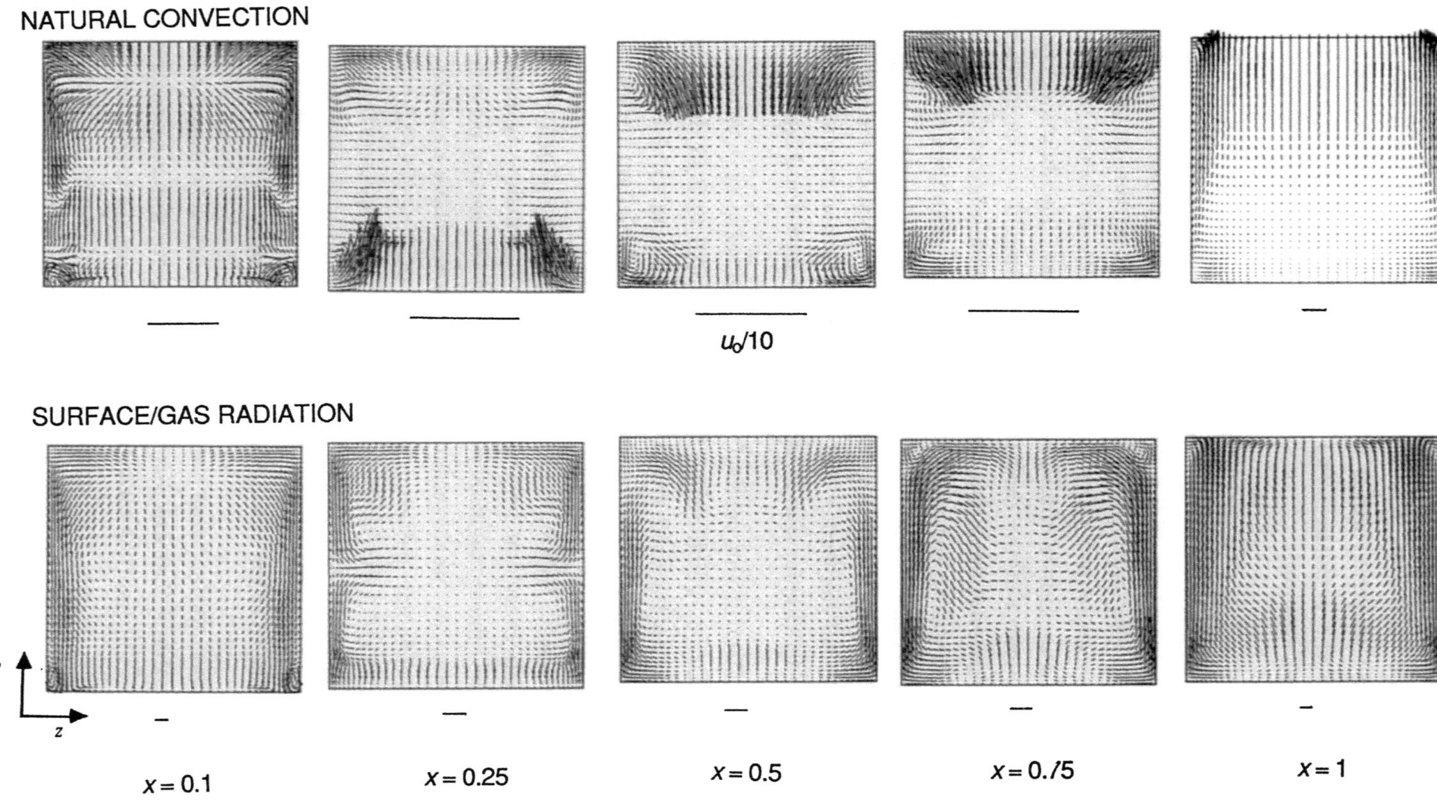

Fig. 3: Projected velocity vectors on planes of x=constant at Ra=4.5x10^6. The scale shows the magnitude of the reference velocity.

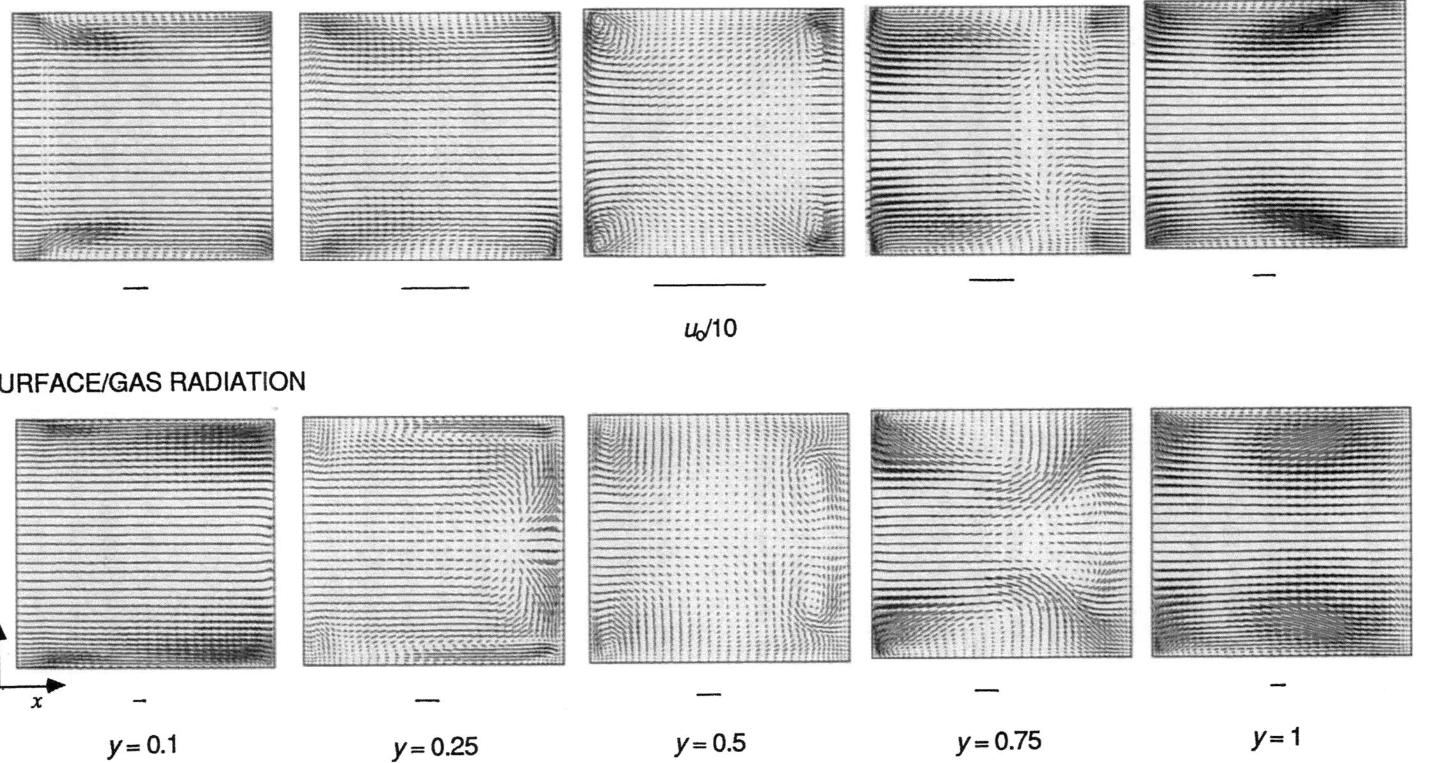

Fig. 4: Projected velocity vectors on planes of y=constant at Ra=4.5x10^6. The scale shows the magnitude of the reference velocity.

natural convection case, is not apparent in the surface/gas radiation mode. Three-dimensionalities of these cross-sectional fields are notable.

The velocity-vector field patterns at different elevations (y=constant) are shown in Figure 4. In both modes of radiative transfer (no radiation, i.e., natural convection, and surface/gas radiation), the results at y=0.1 and 0.9 illustrate the flows directed from the cooled wall (x=0) toward the heated wall (x=1), and *vice versa*. In the natural convection case, the fields are nearly symmetric with respect to the mid y-plane (y=0.5). As a strong contrast, this is not found in the surface/gas radiation mode. An interesting flow pattern is noted in y=0.25 in this mode of radiation. A multi-layer flow structure is observable, which consists of thin layers flowing in the positive x-direction near the end walls (z=0 and 1), and the dominant layer in the center with its flow direction opposite to the secondary layers. The generation of such multi-layer flows is associated with the strong bending of the constant temperature surfaces in the vicinity of the thermally insulated end-walls. In both radiation modes, weak secondary vortices appear near the edges in the y=0.5 plane. The vortices close to the heated plate move toward the symmetry plane (z=0.5). Overall, three-dimensionalities of the flow fields are more pronounced when radiation is included.

Qualitatively similar trends are observed in changes due to radiation for the cases of the mixture (carbon dioxide-water vapor) and the soot-mixture.

TEMPERATURE AND FLOW FIELDS AT Ra = 10^9

Figure 5 illustrates the temperature and the absolute vorticity fields in the cube. Comparing with the lower Rayleigh number case discussed previously, the thickness of the boundary layers is considerably reduced. In the natural convection mode, a distinct boundary layer-stagnant core structure is formed in the enclosure. With radiation, destruction of the symmetric fields is evident as the lower Rayleigh number case.

HEAT TRANSFER CHARACTERISTICS

The heat transfer rate at the isothermal walls is computed from temperature fields. Table 1 presents changes in the average Nusselt number for carbon dioxide gas for cases with and without radiation. The Nusselt number (Nu_{total}) at the heated wall located at x=1 is defined as:

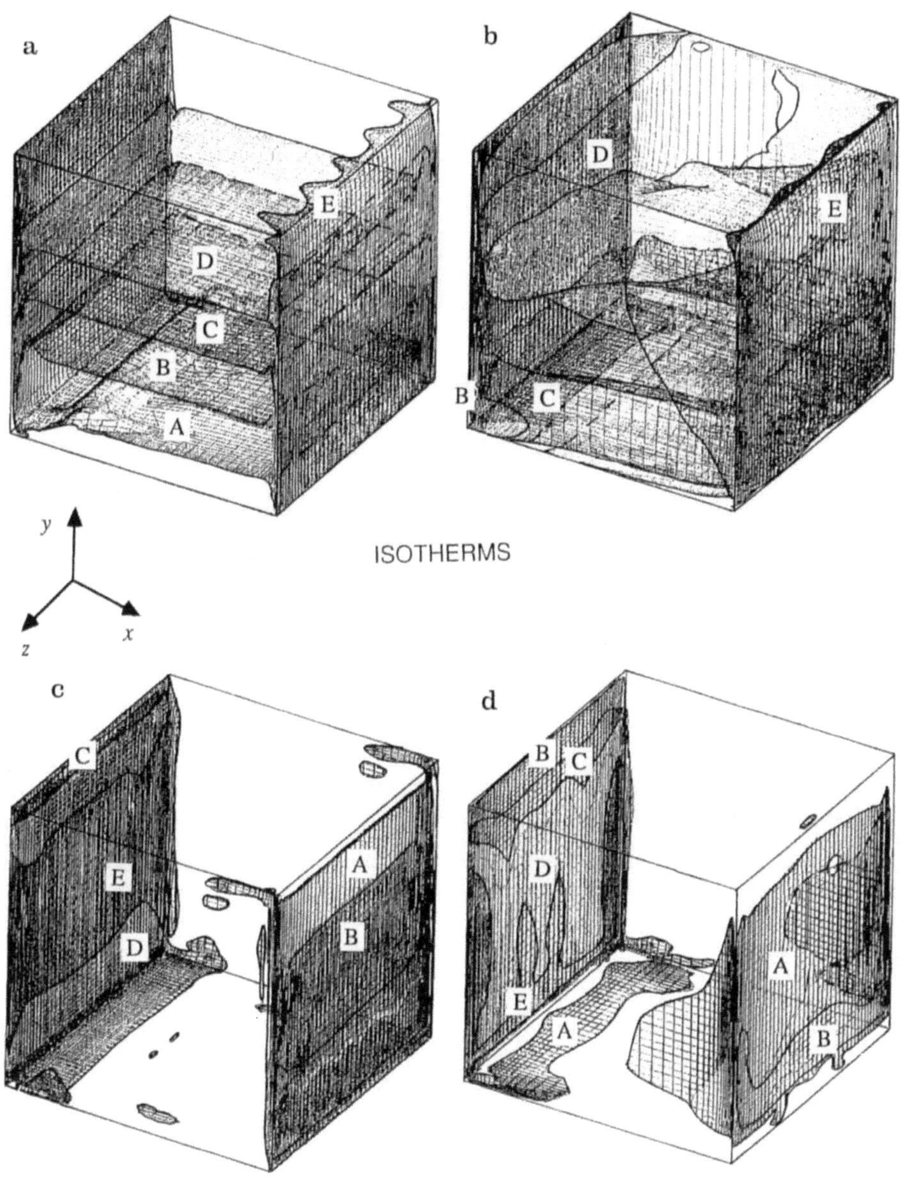

Fig.5: Perspective views of the surfaces of isotemperatures and isovorticities (absolute values) at Ra=10^9 [contour levels: (a) & (b) 0.6667 (A), 0.8333 (B), 1.0 (C), 1.167 (D), 1.333 (E); (c) 10 (A), 50 (B), 100 (C), 200 (D), 300 (E); (d) 50 (A), 100 (B), 200 (C), 300 (D), 400 (E)].

$$\text{Nu}_{\text{total}} = \frac{1}{S}\left[\frac{1}{\delta}\frac{1}{S}\int_s \left.\frac{\partial T}{\partial x}\right|_{x=1} dS + \frac{\text{Re Pr}}{\delta\,\text{Bo}}\left(\int_s \left.q_s\right|_{x=1} dS + \int_s \sum_{m=1}^{N}\frac{1}{3\tau_m}\left.\frac{\partial J_m}{\partial x}\right|_{x=1} dS\right)\right] \quad (7)$$

where S is the area of the isothermal plane. Each term in the right-hand side of the equation represents, from the first to the third terms, the Nusselt number for conduction, surface radiation and gas radiation. With the surface radiation mode, the heat transfer rate increass approximately four times the values of the natural convection cases. The Nusselt number decreases slightly due to absorption of energy by the gas when gas radiation is also accounted for.

Table 1   The average Nusselt number at the isothermal walls (carbon dioxide)

Ra	L_0 [x10^{-1} m]	Bo	Natural Conv.	Radiation Surface	Radiation Surface/Gas
1.67×10^5	0.254	52.6	5.45	20.1	19.3
4.50×10^6	0.762	91.0	14.6	64.6	62.9
10^8	2.17	153	32.4	176	171
10^9	4.66	224	65.0	397	384

For comparison purposes, the Nusselt number predictions of the past investigations are included in Figure 6, which illustrates the Nu_{total} as a function of the Rayleigh number. Generally, the present results attain slightly lower values than the corresponding two-dimensional data due to the end effects.

From the present results, heat transfer correlations are determined over the investigated Rayleigh number range as

$$\text{Nu} = 0.187\,\text{Ra}^{0.282} \qquad \text{(natural convection)} \qquad (8)$$

$$\text{Nu} = 0.323\,\text{Ra}^{0.342} \qquad \text{(surface/gas radiation)} \qquad (9)$$

The above correlations for carbon dioxide gas are valid for $10^5 < \text{Ra} \le 10^9$ with $\delta=1$ and $T_0=555$ K. The length of the enclosure varies from 0.0254 to 0.466 m over the range.

Effects of gas/soot radiation mode are examined at Ra=4.5 x 10^6 by inspecting Table 2. Note that these results are based on the enclosure length, L_0, of 0.11 m, which differs from the previous set (Table 1) for carbon dioxide gas (L_0=0.0762 m at the same Rayleigh number). This is due to the fact that, in the present

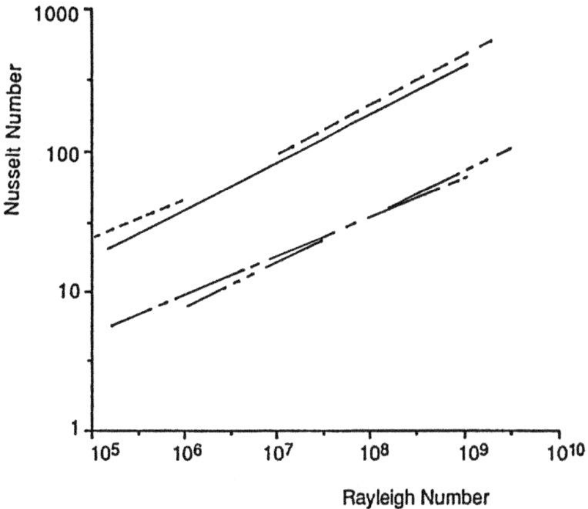

Fig.6: The Nusselt number vs. Reyleigh number. [————— , present 3-D study (surface/gas radiation); — · —, present 3-D study (natural convection); ----------------- 2-D surface/gas radiation [3]: 2-D surface/gas radiation [5]: — ·· — 2-D natural convection [14].

Table 2 The average Nusselt number at the isothermal walls at Ra=4.5x10^6 (gas-soot mixture)

Gas Radiation	Gas/Soot Radiation
72.3	80.5

calculations, the Rayleigh number is held fixed at 4.5 x 10^6 for all the cases considered (carbon dioxide and carbon dioxide-water vapor mixtures); the difference in the fluid properties is adjusted by L_0 to maintain the same Ra value. The Prandtl number of the two fluids remains the same, i.e., 0.68, at the reference temperature, T_0. Inclusion of soot is found to increase the heat transfer rate. It should be realized that in all the previous results for carbon dioxide gas radiation, surface radiation effects through the optical windows of the gases are accounted for; however, by assuming a gray medium for the gas-soot mixture, surface radiation is not taken into account explicitly in the calculations.

CONCLUSIONS

Interactions of natural convection-radiation in a three-dimensional differentially heated cubical enclosure are investigated numerically for gas-soot mixtures. In the thermal boundary condition used in the present study, radiation is found to enhance the three-dimensionalities of the fields. The thermally insulated surfaces (the horizontal and vertical end-walls) reach at high temperatures, which generates large magnitudes of fluid motion in the z-direction near these walls. An increase in the overall temperature of the insulated plates is mainly attributed to surface radiation effects when soot is absent. For gas/soot radiation case, the computed field patterns are similar to those in the surface/gas radiation mode. The gas-soot mixture is assumed to be gray in the model used for the present investigation.

Effects of each radiation modes (surface, gas and gas/soot radiation) are demonstrated in the present computations. When a relatively large overheat ratio is specified, as $\delta=1$ in the present case, surface radiation through non-gray gases is considered to play the dominant role for the heat transfer enhancement. Gas radiation modifies slightly the heat transfer rate due to absorption of energy by the participating gases.

The applicability of the present robust radiation model, based on the P_1-approximation method and the WSGG model, to a three-dimensional differentially heated cubical enclosure filled with gas-soot mixtures is demonstrated by the present study.

Various thermal engineering systems can be modeled adequately by the present geometry. The obtained predictions serve as reference data for the design of such systems.

REFERENCES

1. Yang, K.T.: Numerical Modeling of Natural Convection-Radiation Interactions in Enclosures. in Proc. 8th Int. Heat Transfer Conf., 1 (1986), 131-140.

2. Chang, L.C., Yang, K.T., Lloyd, J.R.: Radiation-Natural Convection Interaction in Two-Dimensional Complex Enclosures. J. Heat Transfer, 105 (1983) 89-95.

3. Zhong, Z.Y., Yang, K.T., Lloyd, J.R.: Variable-Property Natural Convection in a Tilted Enclosure with Thermal Radiation. in Numerical Methods in Heat Transfer, III, John Wiley, New York (1985), 195-214.

4. Fusegi, T., Farouk, B.: Radiation-Convection Interactions of a Non-Gray Gas in a Square Enclosure. Heat Transfer in Fire, ASME-HTD, 73 (1987) 63-68.

5. Fusegi, T., Farouk, B.: Laminar and Turbulent Natural Convection-Radiation Interactions in a Square Enclosure Filled with a Nongray Gas. Num. Heat Transfer, Part A, 15 (1989) 303-322.

6. Fusegi, T., Ishii, K., Farouk, B., Kuwahara, K.: Three-Dimensional Natural Convection-Radiation Interactions in a Cubical Enclosure Filled with a Nongray Gas. Num. Heat Transfer, Part A (1991) (in press).

7. Song, T.H., Viskanta, R.: Interaction of Radiation with Turbulence: Application to a Combustion System. J. Thermophys. Heat Transfer, 1 (1987) 56-62.

8. Smith, T.F., Shen, Z.F., Friedman, J.N.: Evaluation of Coefficients for the Weighted Sum of Gray Gases Model. J. Heat Transfer, 104 (1982) 602-608.

9. Smith, T.F., Al-Turki, A.M., Byun, K.-H, Kim, T.K.: Radiative and Conductive Transfer for a Gas/Soot Mixture between Diffuse Parallel Plates. J. Thermophys. Heat Transfer, 1 (1987) 50-55.

10. Patankar, S.V.: Numerical Heat Transfer and Fluid Flow. Hemisphere, Washington, DC, chapter 6 (1980).

11. Stone, H.L.: Iterative Solution of Implicit Approximation of Multi-Dimensional Partial Differential Equations. J. Numer. Anal., 5 (1968) 530--558.

12. Freitas, C.J., Street, R.L., Findikakis, A.N., Koseff, J.R.: Numerical Simulation of Three-Dimensional Flow in a Cavity. Int. J. Numer. Methods Fluids, 5 (1985) 561-575.

13. Shirayama, S., Kuwahara, K.: Patterns of Three-Dimensional Boundary Layer Separation. 25th Aerospace Sciences Meeting, AIAA Paper 87-0461 (1987).

14. Markatos, N.C., Pericleous, K.A.: Laminar and Turbulent Natural Convection in an Enclosed Cavity. Int. J. Heat Mass Transfer, 27 (1984) 755--772.

If you have any concerns about our products,
you can contact us on
ProductSafety@springernature.com

In case Publisher is established outside the EU,
the EU authorized representative is:
Springer Nature Customer Service Center GmbH
Europaplatz 3, 69115 Heidelberg, Germany

Printed by Libri Plureos GmbH
in Hamburg, Germany